SUPERCRITICAL FLUID CHROMATOGRAPHY

HANDBOOKS IN SEPARATION SCIENCE

Edited by

COLIN F. POOLE
Department of Chemistry, Wayne State University, Detroit, Michigan, USA

Series Editor

COLIN F. POOLE

ELSEVIER elsevier.com

Elsevier
Radarweg 29, PO Box 211, 1000 AE Amsterdam, Netherlands
The Boulevard, Langford Lane, Kidlington, Oxford OX5 1GB, United Kingdom
50 Hampshire Street, 5th Floor, Cambridge, MA 02139, United States

British Library Cataloguing-in-Publication Data
A catalogue record for this book is available from the British Library

Library of Congress Cataloging-in-Publication Data
A catalog record for this book is available from the Library of Congress

ISBN: 978-0-12-809207-1

Publisher: John Fedor
Acquisition Editor: Kathryn Morrissey
Editorial Project Manager: Amy M. Clark
Production Project Manager: Anitha Sivaraj
Designer: Mark Rogers

Typeset by MPS Limited, Chennai, India

SUPERCRITICAL FLUID CHROMATOGRAPHY

Contents

List of Contributors

A.M. Ares University of Valladolid, Valladolid, Spain

M. Ashraf-Khorassani Virginia Tech, Blacksburg, VA, United States

T. Bamba Osaka University, Suita, Osaka, Japan; Kyushu University, Fukuoka, Japan

T.A. Berger SFC Solutions Inc., Englewood, FL, United States

J. Bernal University of Valladolid, Valladolid, Spain

T.L. Chester University of Cincinnati, Cincinnati, OH, United States

M. Combs Celanese Corporation, Narrows, VA, United States

V. Desfontaine School of Pharmaceutical Sciences, University of Geneva, University of Lausanne, Geneva, Switzerland

A. Dispas University of Liège, Liège, Belgium

W.P. Farrell Pfizer Inc., San Diego, CA, United States

E.R. Francotte FrancotteConsulting, Basel, Switzerland

C.M. Galea Vrije Universiteit Brussel (VUB), Brussels, Belgium

M. Ganzera University of Innsbruck, Innsbruck, Austria

D. Guillarme School of Pharmaceutical Sciences, University of Geneva, University of Lausanne, Geneva, Switzerland

P. Hubert University of Liège, Liège, Belgium

P. Lebrun Arlenda SA, Louvain-la-Neuve, Belgium

D. Mangelings Vrije Universiteit Brussel (VUB), Brussels, Belgium

R. McClain Merck Research Laboratories, West Point, PA, United States

A. Murauer University of Innsbruck, Innsbruck, Austria

L. Nováková Charles University, Hradec Králové, Czech Republic

K. Plachká Charles University, Hradec Králové, Czech Republic

D.P. Poe University of Minnesota Duluth, Duluth, MN, United States

K. Taguchi Osaka University, Suita, Osaka, Japan

A. Tarafder Waters Corporation, Milford, MA, United States

D. Thiébaut CNRS, UMR8231 CBI CNRS–ESPCI, PSL Research University, Paris, France

L. Toribio University of Valladolid, Valladolid, Spain

A.A. Tymiak Bristol-Myers Squibb Company, Princeton, NJ, United States

Y. Vander Heyden Vrije Universiteit Brussel (VUB), Brussels, Belgium

J.-L. Veuthey School of Pharmaceutical Sciences, University of Geneva, University of Lausanne, Geneva, Switzerland

C. Wang Bristol-Myers Squibb Company, Princeton, NJ, United States

C. West University of Orleans, Orléans, France

T. Yamada Osaka University, Suita, Osaka, Japan

Y. Zhang Bristol-Myers Squibb Company, Princeton, NJ, United States

Handbooks in Separation Science Series

The goal of the series and volume editors is to develop a new vehicle for collating, interpreting, and disseminating the essential fundamental and practical information of separation science for future generations of separation scientists and to do this by creating the seminal work in the field. Each volume is designed to cover a specific topic and contains relatively succinct chapters with a sharp focus and clear presentation contributed by leading scientists in the field. The target audience for these volumes is professional scientists with responsibility for managing or participating in research projects in either academia or industry. Included in this group are graduate students and professionals in disciplines other than separation science seeking insight into a topic at a level associated with current capabilities. The current volume follows on from the success of earlier volumes with additional volumes in production or planned for the future.

2012	C.F. Poole (Editor), Gas Chromatography
2013	S. Fanali, P.R. Haddad, C.F. Poole, P. Schoenmakers, D. Lloyd (Editors). Liquid Chromatography: Fundamentals and Instrumentation
	S. Fanali, P.R. Haddad, C.F. Poole, P. Schoenmakers, D. Lloyd (Editors). Liquid Chromatography: Applications
2015	C.F. Poole (Editor). Instrumental Thin-Layer Chromatography
	A. Gorak, E. Sorensen (Editors) Distillation: Fundamentals and Principles
	A. Gorak, H. Schoenmakers (Editors). Distillation: Operation and Applications
	A. Gorak, Z. Olujic (Editors). Distillation: Equipment and Processes
2017	C.F. Poole (Editor). Supercritical Fluid Chromatography

CHAPTER

1

Milestones in Supercritical Fluid Chromatography: A Historical View of the Modernization and Development of Supercritical Fluid Chromatography

R. McClain

Merck Research Laboratories, West Point, PA, United States

OUTLINE

Supercritical Fluid Chromatography.
DOI: http://dx.doi.org/10.1016/B978-0-12-809207-1.00001-X

1.1 INTRODUCTION

The chromatographic technique most widely known as supercritical fluid chromatography (SFC) came into existence in 1962. Since its inception, the technique has been touted for its exceptional performance while simultaneously being demonized for deficiencies experienced during its evolution. SFC milestones encountered over the past 55 years, traverse common themes such as instrument capabilities and limitations, scientific understanding and explanation of the performance parameters and resulting data acquired, and realistic comparison of the techniques performance relative to other separation methodologies utilized at the same point in time. This chapter will highlight a selection of pinnacle milestones that have at times driven more widespread adoption and support of the technique and at times suppressed its acceptance by the scientific community. The milestones will be presented chronologically in hopes of educating the reader about the history of SFC and identifying some of the individuals who are responsible for delivering the capabilities we have today. While key references are included to allow further reading, the bibliography is not meant to serve as an all-inclusive list of references as many of these topics will be covered in greater detail in later chapters.

1.2 THE 1960s

In the 1960s gas chromatography (GC) was considered the premier mode of chromatography due to the high efficiency of the technique especially when coupled with the power of flame ionization detection (FID). Analyte volatility as well as thermal stability where two of the limitations of GC that led researchers to pursue other alternatives for separations. Ernst Klesper's article entitled "High Pressure Gas Chromatography Above Critical Pressure" published in 1962 was the first account of what is known today as SFC and combated the analyte stability limitation of GC [1]. Klesper successfully separated a mixture of Ni etioporphyin II and Ni mesoporphyrin IX dimethylester on a polyethylene glycol stationary phase using dicholorodifluoromethane as a mobile phase at a pressure of 131 bar. Porphyrins were previously found to decompose during traditional GC analysis but x-ray powder patterns of the compounds recovered from Klesper's high pressure GC (HPGC) analysis showed the compounds remained structurally intact. This experiment demonstrated that operating pressures above the critical point of the mobile phase enhance mobility on a chromatographic column, thus enabling lower operating temperatures compared to traditional GC, and thus SFC was born.

The Sie and Rijinders research group from the Koninklijke/Shell Laboratorium in Amsterdam was also very active in HPGC at this time but favored the use of carbon dioxide (CO_2) as a mobile phase due to its lower critical pressure (73 atm). Their research focused on evaluating the effects of pressure on partition coefficients (K) for numerous solutes analyzed on squalane and glycerol columns [2]. This groundbreaking work revealed the favorable effect of decreasing retention factors (k) with increasing mobile phase pressure as seen in Fig. 1.1. This finding was a pinnacle as previously a practitioner of gas chromatographer could decrease retention factors only through increasing temperature, which was an issue not only with respect to analyte stability but also for the upper operating temperature limit for GC columns and instruments. The stage was set to evaluate SFC as a possible avenue to extend the upper molecular weight limits of the existing GC technique.

Two years after the group from Amsterdam hypothesized the possibility of handling higher molecular weight compounds via HPGC, Giddings and coworkers at the University of Utah provided experimental data confirming this hypothesis to be true. The Giddings' group performed solubility studies where compounds such as silicone gum rubber possessing a

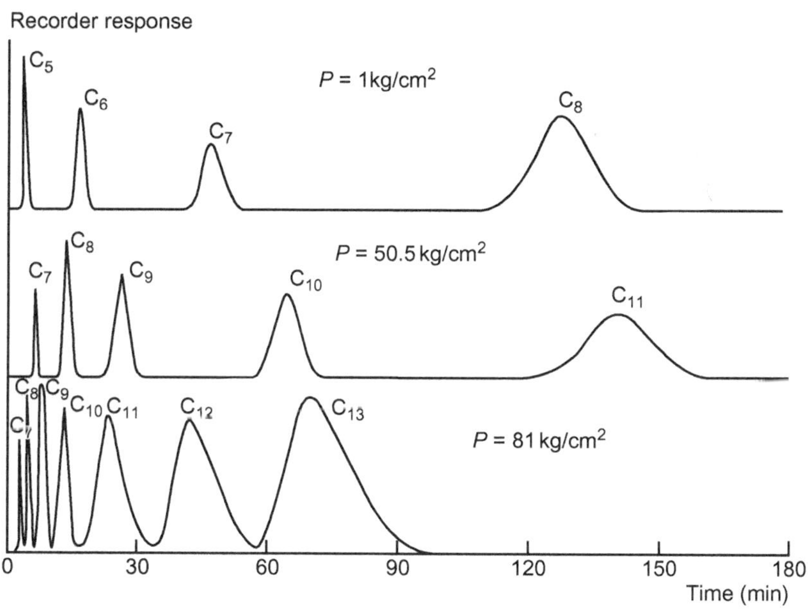

FIGURE 1.1 Separations of *n*-paraffins on a squalane column at 40°C with carbon dioxide as a carrier gas at differrent pressures and comparable linear mobile phase velocities. Column: 1-m length of 3-mm-i.d. tube, filled with 25% w/w squalane on Sil–O–Cel 100/120 mesh support. Sample size: approximately 15 μL. *Source: Figure reproduced from Sie S, Beersum W, Rijnders G. Sep Sci 1966;1:459–90 with permission.*

molecular weight of 400,000 was solvated in carbon dioxide at 1200 atm and 40°C [3]. They also performed separations of high molecular weight biomolecules such as β- and α-carotene with carbon dioxide at 500 atm pressure. These analytes were previously incapable of being analyzed by traditional GC techniques due to inadequate thermal stability. The ability to solubilize material through the use of a supercritical fluid (note: the name of the paper containing dense-gas chromatography) was explained as a result of enhanced intermolecular interactions provided by the compression of the gas to a liquid-like density.

Through a comprehensive study of enhanced migration of over 80 large molecules in compressed gases, Giddings' group was able to prepare a table attempting to correlate the eluotropic strength of compressed gases to the known eluotropic strength of traditional liquids [4]. Ironically, the proposed assignment of carbon dioxide with a liquid-like density to possess eluotropic strength similar to isopropyl alcohol could be the first widely accepted, but major misconception about SFC. Experts later disproved the over-estimated solvent strength of supercritical carbon dioxide and suggested it could have delayed wider adoption of the technique [5,6]. Today it is common knowledge that the solvent strength of supercritical carbon dioxide is closer to that of a small hydrocarbon such as pentane or hexane, not an alcohol such as isopropanol [6,7].

Within its first several years of practice, SFC proved capable of separating thermally unstable compounds not conducive to existing GC conditions as well as extending the operating range to higher molecular weight compounds previously inseparable by traditional GC methods. These two attributes allowed SFC to compete with GC, which at this time was still the dominant, high-performance separation technique. The study of the elution strength of the mobile phase in SFC prompted Giddings' group to postulate that pressure programming during a chromatographic analysis would tune the solvating power of the mobile phase, resulting in elution of the compounds as they become soluble [8]. The ability to perform such a pressure gradient in SFC would allow this emerging technique to compete with liquid chromatography (LC) which commonly utilized a binary mobile phase and composition gradient. This capability would be realized in less than 1 year from Giddings original work, further testament of the attention and effort being dedicated to SFC in its infancy.

1.3 THE 1970s

Jentoft and Gouw at Chevron continued the advancement of SFC thorough incorporation of a pressure programmer into an LC set-up, utilizing high-pressure nitrogen as the pressure source, enabling the execution of an SFC pressure gradient. A 900 average molecular weight polystyrene sample was separated using *n*-octane bonded to Poracil C as the stationary phase with *n*-pentane containing 5% methanol as the mobile phase. The pressure programmer allowed a 6 psi/minute pressure gradient to be performed over 60 minutes, starting with an initial pressure of 650 psi and concluding with a final pressure of 1000 psi [9]. The polystyrene sample was resolved into 32 oligomers over the course of a 60 minute analysis, Fig. 1.2. Several of the individual peaks were captured post UV detection and chromatographed a second time to ensure they were not artifacts, very similar to Klesper's confirmational analysis after the inaugural SFC separation [1]. The ability to deliver pressure ramps to control elution provided a "new" mechanism to increase peak capacity, while operating at linear velocities up to an order of magnitude greater than those in LC and at the time SFC was suddenly being compared and competing with not only GC, but also LC.

The effects of pressure and temperature on chromatographic performance were studied by Novotny at the University of Houston. Various analytes such as durene, naphthalene, biphenyl, and chrysene were separated on packed columns of different particle size using *n*-pentane as a mobile phase. Changes in retention factors and plate heights (HETP) were monitored while changing pressure and temperature. The prediction of chromatographic performance by changing temperature was found to be more difficult than pressure. Novotny's findings, at that time, suggested that an increase in the column pressure drop resulted in a decrease in

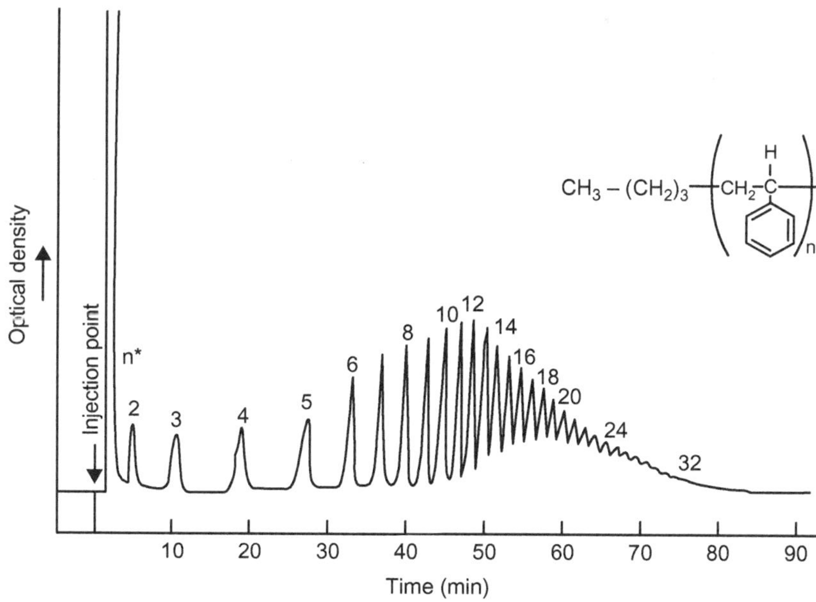

FIGURE 1.2 Chromatogram of a 900 average molecular weight polystyrene sample. *Source: Reproduced from Jentoft R, Gouw T. J Chromatogr Sci 1970;8:138–42 with permission.*

chromatographic performance [10]. This led to numerous inaccurate conclusions regarding the performance of packed column SFC, the effects of pressure drop on column performance, and the separation speed. The chromatography community, as a result of these conclusions transitioned the development of SFC from using the packed column format to embracing open tubular columns. This transition, primarily centered around column type, would be questioned frequently over the next decade.

1.4 THE 1980s

Capillary SFC originated through the work of Milos Novotny and Milton Lee who believed the pressure drop experienced in packed column SFC had "disastrous consequences as far as the column efficiency goes" [11]. These groups sought to design an SFC system that would maintain constant pressure across the entire column length and detector as well as use a stationary phase of minimal thickness to reduce resistance to mass transfer. They used a 58 m long × 0.2 mm i.d. glass capillary with a phenylmethyl polysiloxane coating as the stationary phase to resolve a mixture of polycyclic aromatic hydrocarbons using supercritical *n*-pentane as a mobile phase, Fig. 1.3. The high peak efficiency

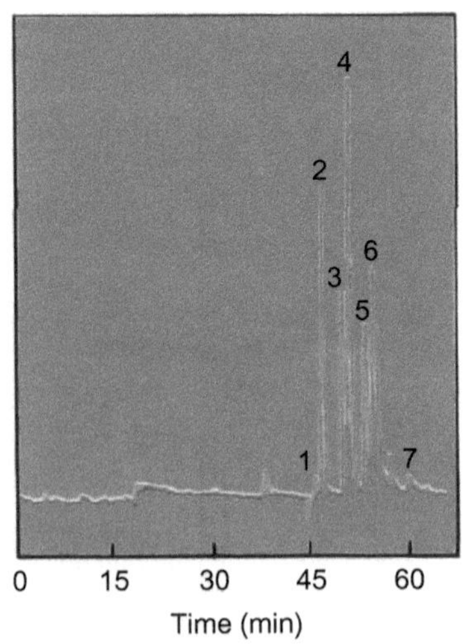

FIGURE 1.3 Isobaric separation of a polycyclic aromatic hydrocarbon (PAH) standards on a 58 m × 0.20 mm i.d. capillary column. Conditions were indicated "as in previous figure" of cited work (mobile phase *n*-pentane at 210°C and 32 atm pressure). The standards were: (1) anthracene, (2) pyrene, (3)benzo[k]fluoranthene, (4) benzo[e]pyrene, (5) dibenz[a, c]anthracene, (6) benzo[ghi]perylene, and (7) coronene. *Source: Reprinted with permission from Novotny M, Springston S, Peaden P, Fjeldsted J, Lee M. Anal Chem 1981;53:407A–414A.*

validated the use of capillary columns in SFC which, at the time were widely embraced by the GC community. Capillary SFC dominated the landscape for the remainder of the decade.

The pioneers of SFC fabricated the instrumentation used in their research as off-the-shelf systems were not yet commercially available. Not surprising, it was common for GC ovens and syringe pumps to be incorporated into these early instrument designs due to the dominance GC possessed in the field of chromatography at that time. A group at Hewlett Packard, later to become Agilent Technologies, modified a 1084B LC to facilitate SFC capabilities by cooling the pump heads via a clamp-on heat exchangers and the addition of a back-pressure regulator [12]. This modified SFC system was used to re-evaluate the effect of particle size and its associated pressure drop across the column in packed column SFC. Polycyclic aromatic hydrocarbons were analyzed on 3, 5, and 10-μm particle size columns and revealed that the efficiency actually increased with decreasing particle size, as seen in Fig. 1.4. In addition van Deemter plots confirmed that SFC and HPLC produce the same minimum HETP for the same particle size, but the optimal linear velocity for SFC was approximately five times greater than for HPLC. These findings kept the concept and implementation of packed column SFC alive during the heyday of capillary SFC.

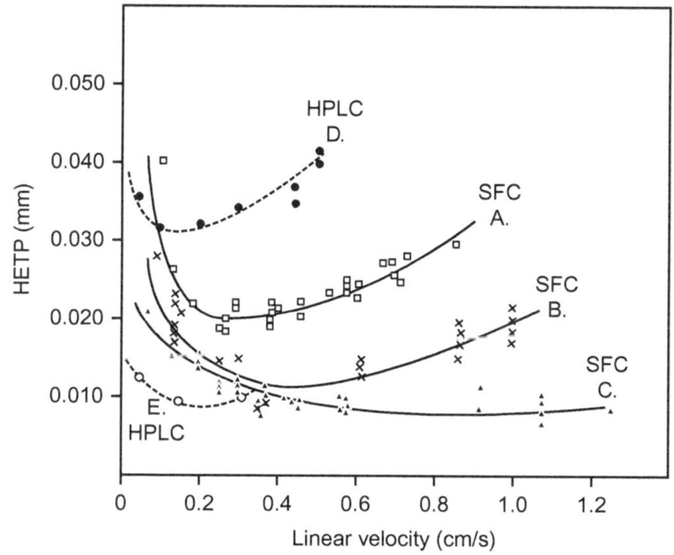

FIGURE 1.4 HETP versus linear velocity: curve A, data from SFC on 10-μm column packing; curve B, SFC on 5-μm column packing; curve C, SFC on 3-μm column packing; curve D, HPLC on 10-μm column packing; curve E, HPLC on 3-um column packing. *Source: Reproduced from Gere D, Board R, McManigill D. Anal Chem 1982;54:736–40 with permission.*

The utilization of SFC for chiral analysis and purification was considered by many to be the most dominant application of SFC, at least in the pharmaceutical industry, during its commercialization. Mourier postulated that supercritical carbon dioxide, possessing the relative polarity of hexane, could be substituted for hexane (or other nonpolar solvents) in a typical chiral chromatographic system [13]. The dramatic increase of the binary diffusion coefficients observed in SFC compared to HPLC reduced proportionately the exceptionally long separation times in HPLC. Mourier's work on the effects of cosolvent density and polarity and their impact on retention and efficiency: notably the decrease in retention with increasing cosolvent polarity as well as the increase in efficiency with the increase in polarity of the cosolvent were important observations for the general development of packed column SFC. The addition of water to the cosolvent was also first evaluated and found to play a role in decreasing retention factors in chiral SFC through masking of active silanol sites. This groundbreaking work, that demonstrated the ability of water to alter retention in SFC, would not be revisited for another 20 years [14]. The series of solvents (methanol, ethanol, and isopropanol) studied in this work are still the most widely used solvents in chiral SFC to this day.

One of the main differences between HPLC and SFC instrumentation is the necessity to control pressure on the detector side of the column in SFC. As demonstrated by journeying through the milestones of this powerful technique, the capability to solubilize, migrate, selectively differentiate, and elute compounds in SFC is dictated by the density, which is related to the pressure of the mobile phase. The devices used to control back pressure in some of the first SFC systems included restrictors and mechanical regulators. The back pressure for restrictor-type devices was dictated by the flow of carbon dioxide supplied by the pump. An increase in flow would result in an increase in pressure downstream. The mechanical regulator operated under the same premise of increased flow resulting in increased back pressure downstream but provided the user the capability of changing the back-pressure setting. Saito's group at Jasco developed a back pressure regulator (BPR) capable of specifying a desired outlet pressure independent of the mass flow rate of the mobile phase [15]. Jasco's BPR utilized a pressure transducer to monitor outlet system pressure and provide feedback to a needle valve and heated seat as shown in Fig. 1.5. The flow switching being performed by the needle and seat would ensure any precipitated material was flushed through the system while delivering the set point pressure. For the first time, it was possible to perform a pressure gradient in SFC at a constant flow rate.

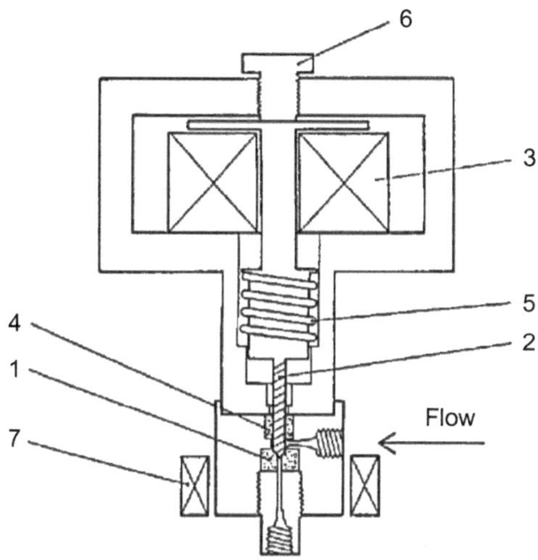

FIGURE 1.5 Cross-section of a flow-switching back-pressure regulator valve. 1, valve seat; 2, valve needle; 3, needle drive solenoid; 4, needle seat; 5, return spring; 6, gap adjustment screw; 7, heater. *Source: Reproduced from Saito M, Yamauchi Y, Kashiwazaki H, Sugawara M. Chromatographia 1988;25:801–05 with permission.*

1.5 THE 1990s

The transition from capillary SFC to packed column SFC experienced in the 1990s reflects the expanding molecular diversity encountered at the time as well as the changing mindset on the general working range of SFC. Capillary SFC was widely successful for the analysis of relatively nonpolar compounds that exceeded the volatility range of traditional GC and consequently was treated more as an extension of GC [16]. Alternatively packed column SFC was found to provide superior performance for polar compounds and was thus treated as a viable alternative to normal phase HPLC [17,18]. An increased focus on the separation of polar compounds via packed column SFC highlighted the need for an additional tool to control peak shape, particularly for ionizable compounds. The use of additives was widely studied in the early 1990s and quickly became the tool enabling further growth and adoption of packed column SFC.

Berger demonstrated that acidic additives were capable of suppressing the ionization of acids provided the additive was more acidic than the solute [19]. He also demonstrated that the additive not only interacted with active sites on the surface of the silica substrate, but also changed the chemistry of the separation system. A comparison of the surface coverage by additives on nonpolar stationary phases, such as

C8, to those of more polar stationary phases, such as diol, revealed that increasing the polarity of the stationary phase resulted in more advantageous complexes being formed between the additive and the stationary phase, improving the peak shape for polar solutes.

Berger continued the additive study by examining the effect of basic additives on the separation of various basic compounds by SFC [20]. Initial work was conducted using 0.1% *t*-butylammonium hydroxide (TBAH) as an additive for the analysis of benzylamines and phenylenediamines. The peak shapes for these compounds was improved through use of the additive yet tailing was still observed. Considering steric effects could impact the chromatographic performance of the additive, the researchers substituted isopropylamine (IPAm) for the TBAH and the peak shape improved significantly. The enhanced performance obtained using IPAm over TBAH was interesting as IPAm is actually the weaker base, confirming the accessibility of the nitrogen as an important factor. Through the combined use of pressure programming, cosolvent gradients, and the use of additives, the working range of SFC has continued to grow as evidenced by Fig. 1.6 [21].

The expansion of the polarity range of SFC through the use of additives further encouraged its use, but the common concern still residing with many potential users at that time was the fear of efficiency loss due to pressure drops caused by "packed" columns with small particles. Terry

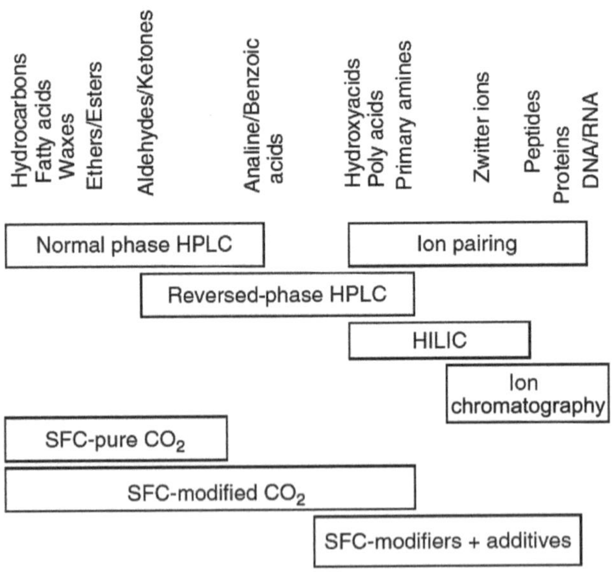

FIGURE 1.6 Schematic representation of the relative application areas of HPLC and SFC in terms of solute polarity. *Source: Reproduced from Berger T, Berger B, Fogleman K. Comprehensive Chirality 2012;8:354–92 with permission.*

Berger, a consistently strong proponent of SFC, designed an experiment in 1993 to finally put the scientific community's fear of large pressure drops at ease. Berger connected ten 20 cm × 4.6 mm i.d. columns packed with 5-μm particles in series to effectively produce a 2-m packed SFC column [22]. Brazilian lemon oil was analyzed on this 2 m column with a cosolvent of 5% methanol, at 60°C, a back pressure of 150 bar, and a flow rate of 2 mL/minute to give the results shown in Fig. 1.7. A plate count exceeding 200,000 was observed with similar extraordinary efficiencies demonstrated for other sample types, including phenylurea herbicides, polycyclic aromatic hydrocarbons, and even a chimney soot extract. While the pressure drop for this specific analysis was not reported, a similar analysis performed using 11 columns in series yielded a pressure drop of approximately160 bar. This groundbreaking work highlighting the power of SFC conclusively ended concerns regarding unfavorable separation performance related to the pressure drop along the column.

Two years after publishing the monumental 220,000 theoretical plate paper, Berger and several colleagues purchased the existing SFC product line from Hewlett Packard and named their new company Berger Instruments [23]. The analytical SFC system available from Berger Instruments delivered the most comprehensive binary pumping system available to SFC users since its inception. The pump heads were cooled by a Peltier chiller to prevent from filling with high pressure, low density gas [24]. Of more importance was the dynamic compressibility

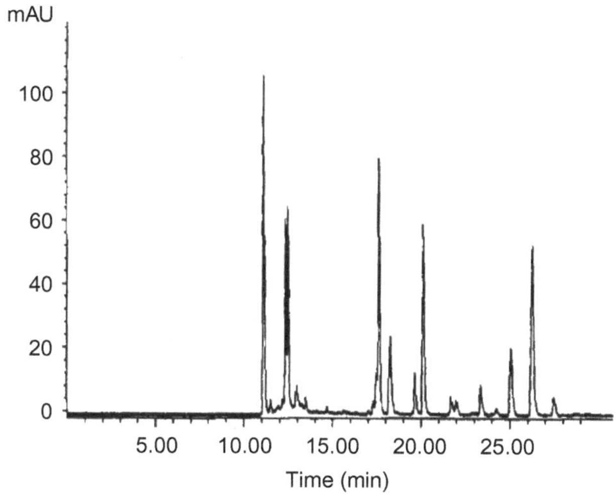

FIGURE 1.7 Packed column SFC chromatogram of Brazilian lemon oil exhibiting >200,000 theoretical plates using 10 columns in series (each 20 cm × 4.6 i.d. packed with 5 μm Hypersil silica), flow rate 2 mL/minute of 5% (v/v) methanol in carbon dioxide at 60°C and 150 bar outlet pressure. *Source: Reproduced with permission from Berger T, Wilson W. Anal Chem 1993;65:1451–55.*

compensation of the CO_2 pump which was calculated by the firmware, based on the temperature and pressure of the carbon dioxide [25]. After filling the pump cavity with compressible carbon dioxide, the speed of the piston is increased to compress the fluid to the proper density as well as make up for the slight decrease in flow experienced during the refilling and re-compression stage. The piston then slowed down to deliver the proper volume of fluid to the chromatographic system. The care exercised with this compression compensation, allowed the Berger Instruments system to deliver accurate (volume:volume) binary mixtures of carbon dioxide and cosolvent. A schematic of the Berger SFC Instrument is shown in Fig. 1.8.

Many pharmaceutical laboratories embraced SFC as a means of performing chiral analysis and purification following Mourier's work in 1985 [26–28]. A group at GSK modified a commercially available analytical SFC system to enable unattended screening of several column and solvent combinations [29]. The preferred column quartet at the time, Chiralpak AD, Chiralpak AS, Chiralcel OD, and Chiralcel OJ was

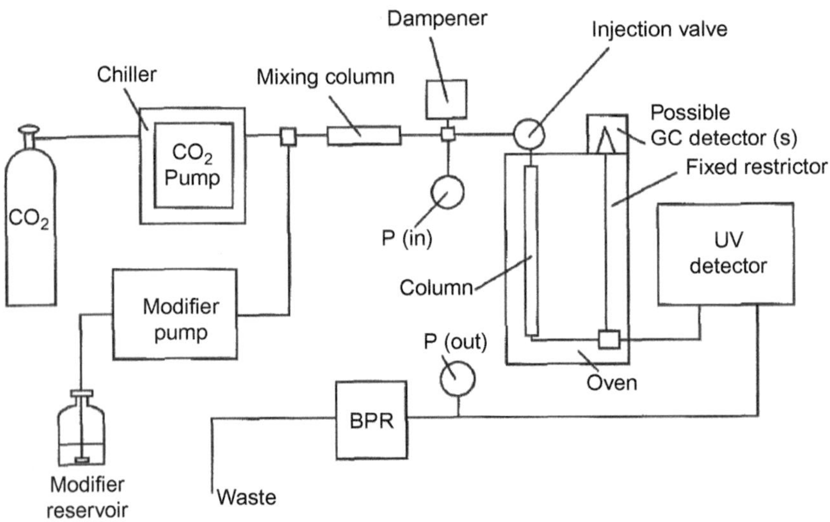

FIGURE 1.8 Schematic diagram of the Berger packed column SFC instrument. One or two pumps are used as flow sources. The pump head used to deliver the compressible fluid is chilled. This pump also has an extended compressibility adjustment range. Binary fluids are mixed and passed through a pressure damper to an injection valve. The column is mounted in an oven with a subambient to elevated temperature range. The columns are the same as used in HPLC. Detection is either through HPLC detectors, using a flow through high-pressure cell, or through a postcolumn split to a fixed restrictor mounted in the base of a GC detector. The outlet pressure is controlled by an electronic back-pressure regulator mounted downstream from the column and detector(s). *Source: Reprinted from Berger T, Greibrokk T. Practical SFC and SFE, In: Caude M, Thiebaut D, editors. p. 107–48 [Chapter 4] with permission.*

plumbed into a six column switching valve located in the column oven. A four port modifier selection valve was also incorporated into the system enabling delivery of four solvents, frequently, methanol, ethanol, or isopropanol, to the cosolvent pump. Both column and cosolvent switching were handled through contact closures. Isocratic methods with a 25-minute run time and a 20-minute re-equilibration could be performed after a change of column or solvent. At the completion of the automated sequence, the resulting array of data was examined to reveal the optimal combination of stationary and mobile phase to yield the best chiral separation or purification conditions. The customization of this instrument, and its rapid commercialization, exemplified the rapid evolution of SFC and the attention researchers were dedicating to improving the technique.

1.6 THE 2000s

The theoretical and practical knowledge shared by early SFC advocates combined with the availability of commercial SFC instruments enabled users to transfer established HPLC separations to SFC with minimal effort by the early 2000s. Purification groups supporting drug discovery laboratories utilizing combinatorial chemistry methods desired the use of SFC for high-throughput application but the prevalence of cyclone technology for separating compounds from cosolvent still presented a problem. The cyclones needed to be cleaned between each sample, a task not amenable to a high-throughput environment. This barrier was overcome when a group led by Berger developed a new semipreparative SFC system using a novel, self-cleaning, heated separator to minimize aerosol formation, and carry-over while delivering recoveries of up to 95% [25]. A pressurized cassette housing four collection vessels, frequently 25 × 150 mm test tubes, was used to capture the liquid portion of the column eluent while allowing controlled release of carbon dioxide through a fixed restrictor on the outlet of each vessel's compartment. The 50 mg samples were separated on a 15 cm × 21 mm i.d. column with a 5 μm cyano column packing and a 50 mL/minute flow rate with a column back pressure of 120 bar. A 5−50% (v/v) gradient over 4.5 minutes with a 50% (v/v) cosolvent hold at the end was used. Each collection vessel was activated to wash the inlet line fully into the vessel effectively maximizing recovery and minimizing carry-over. The initial instrument required the user to exchange the cassette at the completion of each sample but a later version of the system incorporated a robotic Bohdan system for automated exchange [30].

One challenge consistently encountered in high-throughput purification laboratories was minimizing the number of fractions collected during each separation, since the postpurification handling process can become time consuming. Utilization of a selective detector, such as a mass spectrometer as a fraction trigger mechanism has become the proven technique to achieve the desired ratio of one sample in the workflow and one fraction out of the workflow. Wang and Kassel at Dupont Pharmaceutical interfaced a PE-Sciex mass spectrometer to their modified Gilson SFC system to perform mass-directed SFC for the first time in 2001 [31]. Fraction collection was accomplished by simply substituting a higher pressure three-way switching valve on the fraction collector in place of the lower pressure unit originally installed and covering the fraction tubes with foil to capture any aerosolized compound to maximize recovery and to prevent carry-over. A length of 0.0025″ tubing served as a splitter to route a small amount of sample to the mass spectrometer from the column eluent stream. An additional pump delivered 200 μL/minute of a methanol:water mixture containing 0.05% formic acid to aid in analyte ionization. A further additional pump was added to the system for postcolumn addition of 3 mL/minute of methanol to prohibit drying of the solute in the tubing enroute to the fraction collector. The system was able to achieve 77% sample recovery which was considered successful for this first-in-kind instrumentation.

Dupont Pharmaceuticals was not alone in encouraging use of SFC in the drug discovery arena as Pfizer and Eli Lilly also published noteworthy work in the early 2000s declaring adoption of SFC [30,32,33]. The two part series originating from Eli Lilly's United Kingdom facility highlighted the use of both chiral and achiral SFC as a routine support tool in drug discovery [32,33]. Lilly's novel workflow utilized rapid gradients performed at high flow rates to generate independent preparative chromatographic conditions. This was a unique approach as other groups relied on direct transfer of analytical methodology to the preparative scale through geometric scaling calculations. The ability to screen chiral samples on four chiral stationary phases with three cosolvents in 80 minutes led to SFC replacing HPLC as the first-tier screening approach. Scaling up achiral SFC separations was also streamlined through a daily five-point calibration of a 2-minute analytical separation to an 8.5-minute semipreparative method. The ability to predict the preparative elution time from the calibration curve allowed successful isolation of the peak of interest in one of the four fraction vessels triggered by the UV detector signal. The adoption of a novel, highly efficient workflow for both chiral and achiral compounds was a highly regarded accomplishment at this point in time.

The span of compounds separated by SFC continued to grow as the science behind the technique became more widely understood. Zheng in 2006 extended the working range to include peptides of up to 40 amino acid residues. The 40mers were separated on a 2-ethyl pyridine (2-EP) stationary phase with incorporation of trifluoroacetic acid (TFA) as an additive in the methanol cosolvent [34]. Of equal importance was the explanation of the retention mechanisms related to the 2-EP phase. The three mechanisms proposed included hydrogen bonding of the pyridine moiety of the stationary phase with neighboring silanols to decrease the activity of the phase, protonation of the pyridine moiety to give a charge repulsion interaction with the positively charged amino terminus of the analytes, and finally a steric hindrance of active sites on silica by the neutral pyridine moieties [34]. The 2-EP phase was developed in 2001 as the first stationary phase designed specifically for SFC use and was in many industrial and academic arsenals in the years following its release [5]. However, the scientific explanation of possible retention mechanisms led to further adoption as well as highlighting the need for a better understanding of how achiral stationary phases behave in SFC.

Caroline West and Eric Lesellier began one of the most comprehensive studies of SFC stationary phases in 2006 [35–38]. They analyzed over 100 compounds on 24 different stationary phases with the same separation conditions. The experimental result from each stationary phase was evaluated using the solvation parameter model. The model provided information on both the properties of the solutes as well as the resulting interaction between the solute and stationary phase. Several of the solute parameters or attributes examined included hydrogen bond acidity, hydrogen bond basicity, polarizability, and excess molar refraction. The collection of compounds was reduced from 100 to 9 in a follow-up study to allow for characterization of a greater number of stationary phases [39]. By 2008 West and Lesellier realized the need to summarize their stationary phase classification in a format users could employ for selection of a small set of orthogonal stationary phases. They constructed a spider diagram with five vectors representing the stationary phase attributes as shown in Fig. 1.9 [40]. Users could now reference the spider diagram and select phases with different properties to minimize the experimental work required to identify a suitable stationary phase. The knowledge gained through this extensive body of work led several research groups to dedicate efforts to identify, and perhaps even design, a more widely applicable SFC stationary phase that could mimic the almost universal success enjoyed by C18 in the field of reversed-phase HPLC.

1.7 THE 2010s

It was not uncommon for research groups to develop a passion for SFC after their first successful experience with the technique and take on the role of ambassadors to better the field. These improvements

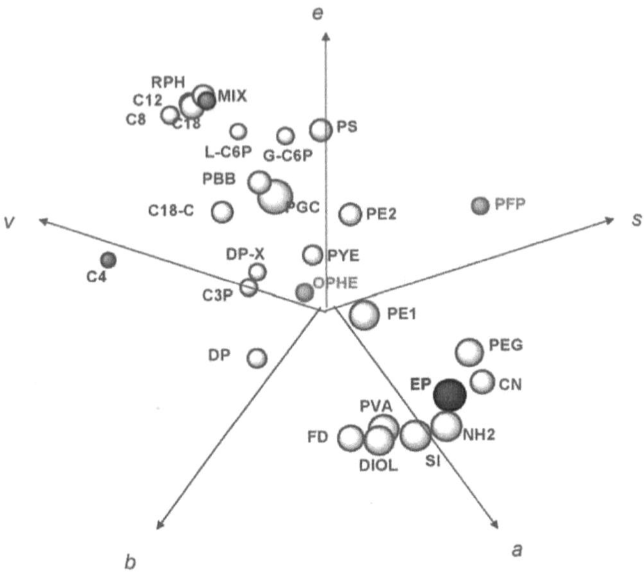

FIGURE 1.9 Spider diagram for a five-dimensional representation of stationary phase selectivity, as evaluated by the solvation parameter model. Bubble size is related to the vector length (u). The stationary phases are identified by abbreviations summarized in [40]. Chromatographic conditions referenced "as in previous figure" (25°C; outlet pressure: 150 bar; mobile phase: CO_2–MeOH, 90:10 (v/v), 3 mL/minute.) *Source: Reproduced from West C, Lesellier E. J Chromatogr A 2008;1203:105–13 with permission.*

many times came in the form of educating other individuals or contributions to the theory and practice of SFC to encourage wider adoption. Ambassadors would make recommendations for instrument supplies or how to better design hardware and/or software for improved SFC performance. Researchers would often optimize experimental conditions to enable SFC utilization in areas previously not adopting the technique. For Pfizer's La Jolla research group, Bill Farrell and Christine Aurigemma did all of the above and added the design of novel SFC stationary phases to the list in hopes of designing an achiral SFC column possessing the selectivity to resolve as many mixtures as a C18 column in reversed-phase HPLC [41]. Realizing the high efficiency and selectivity afforded by stationary phases with some common traits, such as the presence of a basic heterocyclic group, incorporation of nonacidic hydroxyl groups, and/or presence of an aromatic group, the La Jolla group designed numerous stationary phases, evaluated them with complex mixtures, and shared their results with the greater scientific community [41]. In the following years, many of the compounds originating from the Pfizer laboratories were analyzed on these novel phases. The understanding of the retention mechanism derived from this work allowed the scientific community to make informed column selection decisions and enabled successful SFC separations to be performed.

In addition to being active in achiral stationary phase development/evaluation in 2012, the Pfizer group also implemented a comprehensive, open-access, analytical screening system capable of performing SFC-MS and reversed-phase HPLC-MS on the same instrument platform [42]. The incorporation of three valves allowed the user to acquire data from six different columns with a variety of mobile phase conditions through a single sample log-in. A common set of separation conditions encountered on the open-access system included SFC analysis with methanol and carbon dioxide on a pyridine/diol mixed stationary phase, a hydroxyaminopyridine stationary phase, a hydroxyaminodipyridine stationary phase, and a DIOL/MONOL mixed stationary phase, as well as reversed-phase HPLC separations with ammonium acetate at pH 5.5 and trifluoroacetic acid at a pH 1.5 on a C18 column. Virscidian software was utilized to perform peak scoring of the resulting chromatographic data, taking into account peak shape for the compound of interest, resolution of critical peak pairs, and even the number of peaks containing the desired mass to allow for isolation of isobaric compounds. While the end-user submitted this sample and initiated the analysis, a report with the highest scoring separation conditions was sent to the purification group to enable rapid commencement of the purification of the sample with no data interpretation time required. The group referred to this workflow as "FastTrack" as it in fact eliminated the need of manual data interpretation as well as minimized the instrumentation required to acquire the data.

The world financial crisis beginning in 2008 and continuing into this decade required both academia and industries utilizing chromatography to reduce costs. Several ways of reducing costs include incorporating more automation to reduce manpower, performing analyses faster, consuming less solvent, reducing waste disposal costs, and even transition from more expensive solvents such as acetonitrile to less expensive methanol. Eli Lillies' laboratory in Madrid, Spain was one of the first companies to transition a high-throughput purification workflow from reversed-phase HPLC-MS to SFC-MS in order to adapt to these economic pressures. The group revealed 90% of the compounds examined could be successfully purified by SFC with 98% of these being processed on a cross-linked diol stationary phase as the primary column chemistry followed by a 2-ethylpyridine column when necessary [43]. Initially, these researchers utilized a multiple column screen consisting of five columns: diol, 2-ethylpyridine, benzenesulfonamide, diethylaminopropyl, and dinitro. Increased sample throughput experienced by the laboratory led to the necessity for a single-column solution. The analytical method performed on a

15 cm × 4.6 mm i.d. column, 5-μm particle size with a 10% (v/v) isocratic hold for 0.3 minutes, a 10–40% (v/v) linear gradient over 1.8 minutes, a 0.3 minutes 40% (v/v) hold, concluding with a re-equilibration over 0.3 minutes was selected for this task. The total analysis time was less than 3 minutes. The retention time of the desired compound was used to select gradient conditions on a 25 cm × 2 cm i.d. semipreparative column for purifications. The semipreparative column cycle time was less than 8 minutes conclusively proving the success of SFC in HTP laboratories.

Several previous milestones have highlighted the importance of achiral stationary phases in the evolution of SFC over the past decade. Where these milestones have centered on the chemical properties of the stationary phases, there are also two noteworthy works that revolve around the physical properties of the stationary phases, in particular particle size and morphology. The release of the first commercially available ultra-performance liquid chromatography (UPLC) system in 2004 allowed chemists to utilize sub 2-μm particle stationary phases to realize significant gains in separation performance. Berger evaluated the use of these smaller particle size columns in SFC and found the performance to be just as impressive as that in UPLC. Five classes of compounds including steroids, sulfa drugs, profens, xanthenes, and nucleic acids were analyzed on a 1.8-μm silica column under isocratic conditions [44]. Separation times were typically less than 1 minute with adequate resolution of four or five compounds in a designated class with peak efficiencies reaching 22,400 plates per column. Column inlet pressures were found to remain within the working limit of the traditional 400 bar SFC system despite the 10 cm × 3 mm i.d. column being run at the optimum flow rate of 2 mL/minute with 50% (v/v) methanol as the cosolvent and a back pressure of 150 bar. While the reduced viscosity of the carbon dioxide–based mobile phase allows the sub 2 μm particle columns to be run on a conventional SFC instrument, Berger pointed out that the reduced extra-column volume and faster detector response of the UPLC systems was beneficial in preserving column performance.

A year later, Berger performed a systematic comparison of the performance of a 2.6-μm porous shell particle column and a 3-μm totally porous particle column in SFC. A 17-component mixture was separated under identical conditions on both columns to evaluate selectivity, efficiency, and peak symmetry [45]. The selectivity was found to be similar on the two columns while the porous shell column delivered 50% higher efficiency in half the time compared with the totally porous particle column. Plate counts as high as 35,000 were recorded on the porous shell particle column, which is remarkable considering some of the peaks with these high efficiencies were fronted. The cause of the fronting was never revealed but it stands to reason if the cause was found and corrected, even higher efficiencies could be obtained. The findings for the 1.8-μm particle column along with the porous shell particle column results continue to demonstrate that SFC in the second decade of the 21st century is a well-understood science that can be studied systematically just as is the case for HPLC and UPLC.

The first of several generations of analytical SFC instruments exhibited excessive UV detector noise which prevented laboratories operating according to GXP regulations from adopting the technique. One source of the noise was the pressure fluctuations generated by the movement of the BPR [46]. The refractive index of the mobile phase in the UV cell changes with these pressure fluctuations, effectively changing the light intensity in the UV cell, thus producing baseline changes which are observed as noise. Berger developed a new BPR for the Aurora analytical SFC system which minimized pressure fluctuations, particularly at a BPR setting of 200 bar, dramatically reduced detector noise [46]. It was widely known that Aurora's novel pumping mechanism of prepressurizing the carbon dioxide with an external booster pump and then metering out the carbon dioxide by the traditional HPLC reciprocating piston pump, reduced detector noise significantly. The combined noise reduction by the pump and BPR, enabled the peak-to-peak noise level to be reduced to <0.02 mAU, less than the <0.05 mAU required to validate an analytical method according to GXP regulations. An example of an SFC chromatogram with noise minimized to enable a validated method to be performed by SFC can is shown in Fig. 1.10 [47].

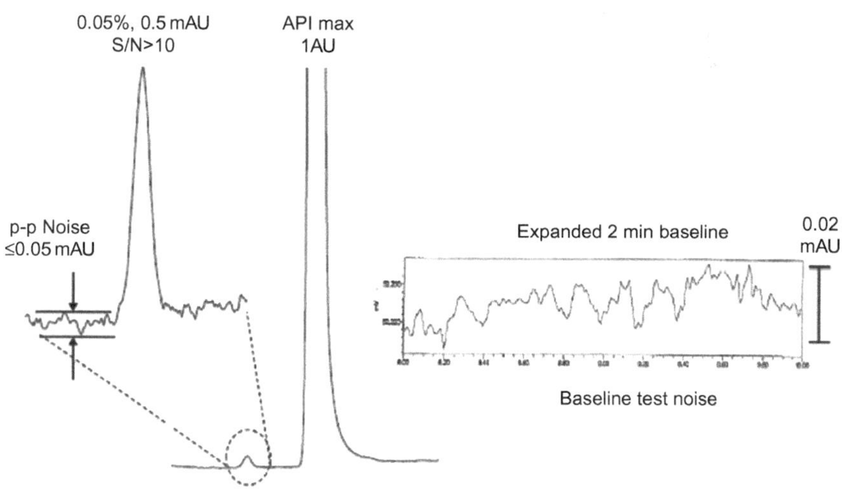

FIGURE 1.10 Separation of enantiomers for a pharmaceutical compound MR-1 by SFC. Low baseline noise and excellent signal to noise allow low-level quantification of minor components. Chiral impurity profiling over 2 minutes with a target API signal of 1AU for the major component MR-1 by the Agilent Baseline Noise Test for the highest and lowest signal across a representative time range from 8.0 to 10.0 minutes. *Source: Reproduced with permission from Hicks M, Regalado E, Tan F, Gong X, Welch C. J Pharma Biomed Anal 2016;117:316–24.*

1.8 CONCLUSION

SFC has matured into a well-understood, widely utilized, and high-efficient technique over the past five decades. The instrumentation used in SFC has achieved the same high quality and robustness as HPLC, a remarkable improvement over the first several generations of SFC systems. The research performed by academic and industrial groups alike throughout its evolution has resulted in a deep understanding of the theory behind practical SFC allowing an increasing number of successful separations to be performed. The development of stationary phases tailored for SFC, such as the ethyl pyridine phase, continue to broaden the scope of SFC and to facilitate novel applications. The combination of all of these innovative progressions has driven SFC to become a premier highly efficient and scientifically validated technique that in many cases surpasses the capabilities and limits of GC and HPLC.

References

[1] Klesper E, Corwin A, Turner D. J Org Chem 1962;27:700–1.
[2] Sie S, Beersum W, Rijnders G. Sep Sci 1966;1:459–90.
[3] McLaren L, Myers M, Giddings J. Science 1968;159:197–9.
[4] Giddings J, Myers M, McLaren L, Keller R. Science 1968;162:67–73.
[5] Berger T, Berger B. LC/GC North America 2010;28:344–57.
[6] Berger T. Chromatography Today 2014;Aug/Sep:26–9.
[7] Dye J, Berger T, Anderson A. Anal Chem 1990;62:615–22.
[8] Giddings J, Myers M, King J. J Chromatogr Sci 1969;7:276–83.
[9] Jentoft R, Gouw T. J Chromatogr Sci 1970;8:138–42.
[10] Novotny M, Bertsch W, Zlatkis A. J Chromatogr 1971;61:17–28.
[11] Novotny M, Springston S, Peaden P, Fjeldsted J, Lee M. Anal Chem 1981;53: 407A–14A.
[12] Gere D, Board R, McManigill D. Anal Chem 1982;54:736–40.
[13] Mourier P, Eliot E, Caude M, Rosset R, Tambute A. Anal Chem 1985;57:2819–23.
[14] Ashraf-Khorassani M, Taylor L, Seest E. J Chromatogr A 2012;1229:237–48.
[15] Saito M, Yamauchi Y, Kashiwazaki H, Sugawara M. Chromatographia 1988;25: 801–5.
[16] Taylor L. LC/GC 2013;31:1–7.
[17] Poole C. J Biochem Biophys Meth 2000;43:3–23.
[18] Berger T. J Chromatogr A 1997;785:3–33.
[19] Berger T, Deye J. J Chromatogr 1991;547:377–92.
[20] Berger T, Deye J. J Chromagr Sci 1991;29:310–17.
[21] Berger T, Berger B, Fogleman K. Comprehensive Chirality 2012;8:354–92.
[22] Berger T, Wilson W. Anal Chem 1993;65:1451–5.
[23] Berger D. LC/GC Europe 2007;20–3.
[24] Berger T, Greibrokk T, Practical SFC and SFE. In: Caude M, Thiebaut D, editors. The Netherlands: Harwood Academic Publishers; 1999. p. 107–148 [Chapter 4]. ISBN 90-5702-409-8.
[25] Berger T, Fogleman K, Staats T, Bente P, Crocket I, Farrell W, et al. J BioChem Biophys Meth 2000;43:87–111.

[26] Terfloth G. J Chromatogr A 2001;96:301–7.
[27] Welch C, Leonard W, DaSilva J. LC/GC North America 2006;23:16–29.
[28] Miller L, Potter M. J Chromatogr B 2008;875:230–6.
[29] Villeneuve M, Anderegg R. J Chromatogr A 1998;826:217–25.
[30] Ventura M, Farrell W, Aurigemma C, Tivel K, Greig M, et al. J Chromatogr A 2004;1036:7–13.
[31] Wang T, Barber M, Hardt I, Kassel D. Rapid Comm Mass Spec 2001;15:2067–75.
[32] White C. J. Chromatogr. A 2005;1074:163–73.
[33] White C, Burnett J. J Chromatogr A 2005;1074:175–85.
[34] Zheng J, Pinkston J, Zoutendam P, Taylor L. Anal Chem 2006;78:1535–45.
[35] West C, Lesellier E. J Chromatogr A 2006;1110:181–90.
[36] West C, Lesellier E. J Chromatogr A 2006;1110:191–9.
[37] West C, Lesellier E. J Chromatogr A 2006;1110:200–13.
[38] West C, Lesellier E. J Chromatogr A 2006;1110:233–45.
[39] West C, Lesellier E. J Chromatogr A 2007;1169:205–19.
[40] West C, Lesellier E. J Chromatogr A 2008;1203:105–13.
[41] Farrell W, Chung L, Cheung M, Ono T, Hirose T, et al, SFC2011 Jul 20, 2011, <www.greenchemistrygroup.org/#!2011-program/wg169>.
[42] Aurigemma C, Farrell W, Simpkins J, Baylisss M, Alimuddin M, Wang W. J Chromatogr A 2012;1229:260–7.
[43] Puente ML, Soto-Yarritu P, Anta C. J Chromatogr A 2012;1250:172–81.
[44] Berger T. Chromatographia 2010;72:597–602.
[45] Berger T. J Chromatogr A 2011;1218:4559–68.
[46] Berger T, Berger B. J Chromatogr A 2011;1218:2320–6.
[47] Hicks M, Regalado E, Tan F, Gong X, Welch C, Pharma J. Biomed Anal 2016;117: 316–24.

CHAPTER

2

Theory of Supercritical Fluid Chromatography

D.P. Poe

University of Minnesota Duluth, Duluth, MN, United States

OUTLINE

Supercritical Fluid Chromatography.
DOI: http://dx.doi.org/10.1016/B978-0-12-809207-1.00002-1

2.1 INTRODUCTION

This chapter provides an overview of retention and kinetic theory of packed column supercritical fluid chromatography (SFC), the physico-chemical properties of supercritical fluids and their mixtures, and connects theory to the contemporary practice of SFC.

A fundamental requirement for success of any chromatographic method is selective retention. A basic understanding of the factors controlling retention is essential to the rational development and analysis of chromatographic processes. A second fundamental requirement is favorable kinetics, which refers to the relative rates of the physical and molecular transport processes involved in the different solute zones in space and time. This latter requirement is generally discussed in terms of column efficiency and overall kinetic performance. Both of these fundamental topics, retention and kinetics, are discussed in this chapter.

While each of the major sections on the theories of retention and dynamics in this chapter begins with a basic introduction, a general background of the principles and terminology of chromatography is assumed. These topics can be found in textbooks of analytical chemistry and instrumental analysis, as well as in several monographs [1,2]. Also a brief tutorial on the modern practice of SFC has appeared recently [3].

2.2 THE MOBILE PHASE IN SFC

2.2.1 Chromatography and Supercritical Fluids

The three major forms of column chromatography differ primarily in the properties of the mobile phase. In GC the mobile phase is a low-pressure gas, and retention is controlled by interactions of the solute with the stationary phase and by temperature. Molecular interactions in the mobile phase are insignificant at the pressures typically used in GC and play no important role in retention. In LC the mobile phase is a (nearly) incompressible liquid. Molecular interactions in the liquid depend strongly on the physico-chemical properties of the liquid, and retention is controlled by the differences in the molecular interaction energies in the two phases. Retention in HPLC is nearly independent of pressure. In SFC the mobile phase is a highly compressible fluid near or (strictly) above its critical temperature and pressure. Molecular interactions in the mobile phase are strongly dependent on the density of the fluid, so that in SFC density plays an important role in retention.

Neat CO_2 is a nonpolar fluid suitable for dissolution and separation of molecules of low polarity, and has found use in open tubular SFC for

TABLE 2.1 Representative Properties of Fluids Used in SFC and HPLC. Fluid Properties from NIST REFPROP [4]

	Viscosity (μPa-s)	Diffusivity[a] ($\times 10^9$ m^2/s)	Density (g/cm^3)
SFC: CO_2 with 5% MeOH, 40°C, 150 bar	76	10.2	0.809
HPLC: MeOH, 25°C	545	1.9	0.786

[a] *For naphthalene: For SFC from reference [5]; for HPLC from Wilke-Chang equation.*

analytes of low to moderate polarity. For more polar analytes a polar organic modifier, such as methanol, is typically added. This not only increases the critical temperature and pressure, but also causes increases in the viscosity and decreases the diffusivity of analytes, both properties that are important for fast, efficient separations. Today most SFC separations utilize modified carbon dioxide mobile phases with packed columns.

The properties of the mobile phase in SFC relative to kinetic performance are intermediate between those used in GC and HPLC. Table 2.1 shows some typical values of some of these properties. The low viscosity and higher diffusivity of CO_2-based mobile phases provide for faster separations in longer columns with a higher plate count relative to HPLC. Although it suffers in this regard relative to GC, the solvating power of supercritical fluids extends some of the advantages of GC to the separation of nonvolatile analytes.

2.2.1.1 *Phase Behavior of CO_2 and Mixtures of CO_2 with Organic Modifiers*

The most common mobile phase used in SFC is carbon dioxide, most often with added modifier such as methanol. The phase diagram for neat carbon dioxide is shown in Fig. 2.1. The critical temperature and pressure are $T_c = 31°C$ and $P_c = 74$ bar. The shaded area shows the conditions commonly used in SFC. The maximum pressure limit is based on the capabilities of modern instrumentation. Although carbon dioxide reaches the supercritical state only at temperatures above 31°C, many SFC separations are conducted at lower temperatures. Note that there is no phase transition when temperature is decreased below the critical temperature when operating above the critical pressure. The method is generally referred to as SFC even when subcritical temperatures are used. The isochores show that for SFC with neat CO_2 the fluid density generally falls between 0.7 and 1.1 g/cm^3.

Addition of organic modifiers such as methanol generally results in a shift of the critical temperature and pressure to higher values, as indicated by the critical points for several mixtures containing up to 38 mol% methanol shown in Fig. 2.1. Conditions at pressures below the

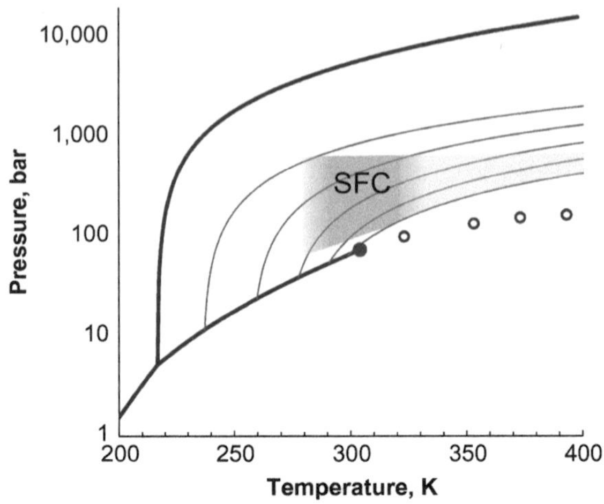

FIGURE 2.1 Phase diagram for neat CO_2 with phase boundaries (heavy blue lines), constant density isochores (light blue lines, density 0.7–1.1 g/cm^3 bottom to top), and critical point for neat CO_2 (solid red circle). Also shown are critical points for CO_2 with 14%, 24%, 32%, and 38% added methanol (open circles, left to right), estimated from Ref. [6]. The blue shaded area shows the P–T region used for SFC, with darker shading indicating commonly used conditions.

critical point may result in separation of the fluid into two phases, one rich in CO_2, the other in added modifier. In SFC with mixed mobile phases this may result in unexpected retention behavior and poor efficiency, and these conditions should generally be avoided [6–8].

2.2.2 Thermophysical Properties of CO_2 and Mixtures of CO_2 and Methanol

In most SFC separations a polar organic modifier is added to decrease the retention of more polar analytes. The added modifier not only increases the solvent strength of the mobile phase, but it also affects the density, viscosity, and phase behavior of the fluid mixture. In this section we examine some of these properties for neat CO_2 and for mixtures of CO_2 and methanol, which is the most commonly used modifier for SFC.

2.2.2.1 Compressibility and Density

An important property of supercritical fluids that distinguishes SFC from GC and LC is their high compressibility. The compressibility of neat CO_2 in the near-critical region is illustrated in Fig. 2.2, which shows how the density of neat CO_2 changes with temperature and pressure in and near the supercritical region. At constant temperature there is always an increase in density with pressure. Because solubility and

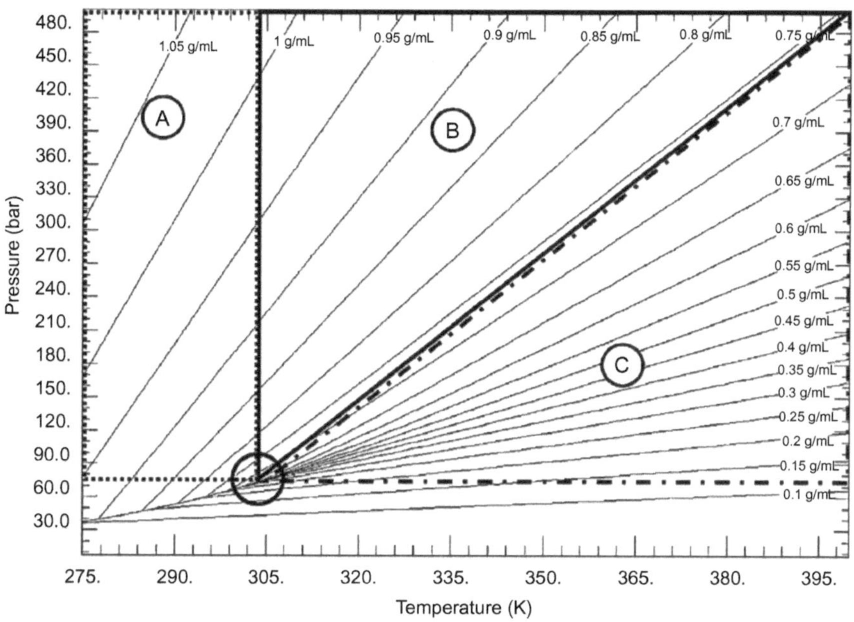

FIGURE 2.2 Supercritical region of neat carbon dioxide with isopycnic lines and different operating zones (see text). The large circle identifies the location of the critical point, and the vertical and horizontal lines show the critical isotherm and critical isobar, respectively. *Source: Reprinted from reference Tarafder A, Guiochon G. Use of isopycnic plots in designing operations of supercritical fluid chromatography: II. The isopycnic plots and the selection of the operating pressure-temperature zone in supercritical fluid chromatography. J Chromatogr A 2011;1218(28):4576–85 [9].*

retention are strongly dependent on the mobile phase density, pressure and temperature are both important operational parameters for controlling retention in SFC. The effects of mobile phase compressibility on retention and efficiency require special approaches to describing retention and efficiency that are not generally necessary in GC and HPLC. These topics are dealt with in the following sections.

Also shown in Fig. 2.2 are three distinct regions or zones of operation. Zone A to the left of the critical isotherm represents subcritical conditions. Zones A and B represent conditions that are suitable for successful SFC separations whereas zone C, where the density is below about 0.7 g/cm^3, generally results in poor separation performance. The loss in performance in zone C is related to the extremely high compressibility of the fluid in this region.

In any SFC column, the pressure drop along the column is accompanied by changes in the temperature and density of the mobile phase. While the approximate changes in density with pressure for isothermal conditions

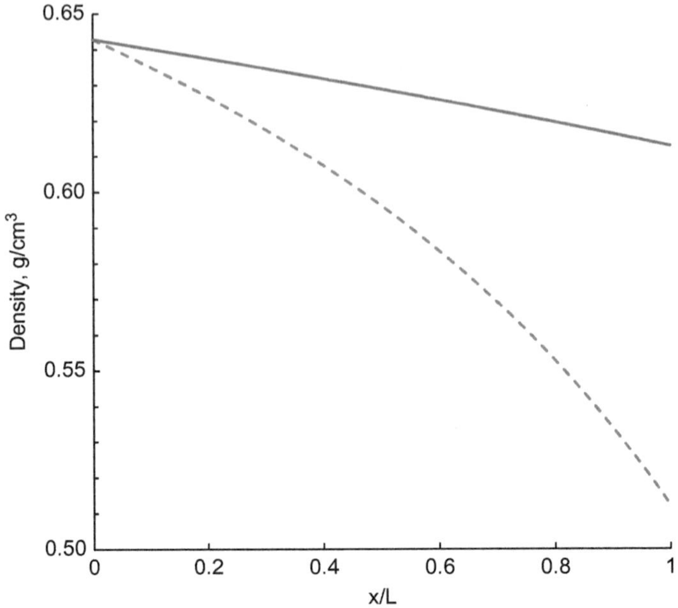

FIGURE 2.3 Density profiles along an SFC column with a 16.3 bar pressure drop for isothermal (dashed line) and adiabatic (solid line) operation. Inlet temperature 50°C, inlet pressure 109.3 bar. Outlet temperature = 46.1°C for adiabatic mode. Mobile phase 0.945 CO_2/0.055 methanol (mol/mol). The profiles were computed assuming a linear pressure drop along the column.

can be gleaned from the isopycnic plots shown in Fig. 2.2, the actual density profiles along an SFC column depend on the thermal environment in which it operates. The density profiles for a column operated under isothermal and adiabatic conditions are shown in Fig. 2.3. A column operated in a thermostatted environment (e.g., water jacket or forced-air oven) can be considered near isothermal, and the density drop along the column is greatest. A well-insulated column can be considered as near adiabatic; the temperature decreases along the column, and as a result the density drop is relatively small. For most SFC operations the thermal environment will lie somewhere between these two conditions.

The fluid compressibility and resulting changes in density with pressure have both negative and positive impacts on operations in SFC. The negative impacts include the following:

- Many important chromatographic parameters are density-dependent, including retention, mobile phase velocity, viscosity, and diffusivity. Compared to HPLC, it is more difficult to characterize these properties in an SFC column, to apply rational interpretations to observed retention factors, and even to construct and interpret van

Deemter plots. Also repeatability of observed retention times requires precise control and monitoring of the outlet pressure and the pressure drop. This is perhaps the most important negative impact of compressibility.

- Large density changes in the high-compressibility region (zone C in Fig. 2.2) are accompanied by significant enthalpic cooling, resulting in axial and radial temperature gradients. While the axial gradients in density and temperature may have some small impact on efficiency, the associated radial temperature gradients are the direct cause of the large efficiency losses that have been observed when operating in this region.

These negative impacts on repeatability and efficiency could be largely avoided if SFC columns and instruments could be designed to operate in purely isothermal or adiabatic modes and with precise and reliable control and monitoring of the pressure and temperature at the column inlet and outlet.

There is also at least one clear positive benefit due to compressibility. The solvating power of a supercritical fluid is strongly dependent on its density. This provides the operator with a powerful tool for controlling retention, and can simplify method development.

2.2.2.2 *Viscosity and Pressure Drop*

The pressure drop along a column is an important limiting factor in ultra-high-performance chromatography. The pressure drop per unit column length $\Delta P/L$ is expressed by the Darcy equation as

$$\frac{\Delta P}{L} = -\frac{\eta u}{K_0} \tag{2.1}$$

where η is the viscosity, u is the superficial velocity, and K_0 is the permeability which is proportional to the square of the particle size. Tarafder et al. [10] pointed out that for SFC Eq. (2.1) should be written in differential form due to the local variations in viscosity and velocity. Substituting $\eta_k = \eta/\rho$ and $F_m = u\rho S$, where η_k is the kinematic viscosity, F_m is the mass flow rate, and S is the cross-sectional area, Eq. (2.1) becomes

$$\frac{dP}{dz} = -\eta_k \frac{F_m}{K_0 S} \tag{2.2}$$

This shows clearly that for a column operated at a constant mass flow rate the pressure drop along the column varies with the local kinematic viscosity. The dependence of η_k on the local P and T for neat CO_2 is shown in Fig. 2.4. The figure shows that the lowest kinematic viscosity, and thus the lowest pressure drop, occurs near the critical isopycnic

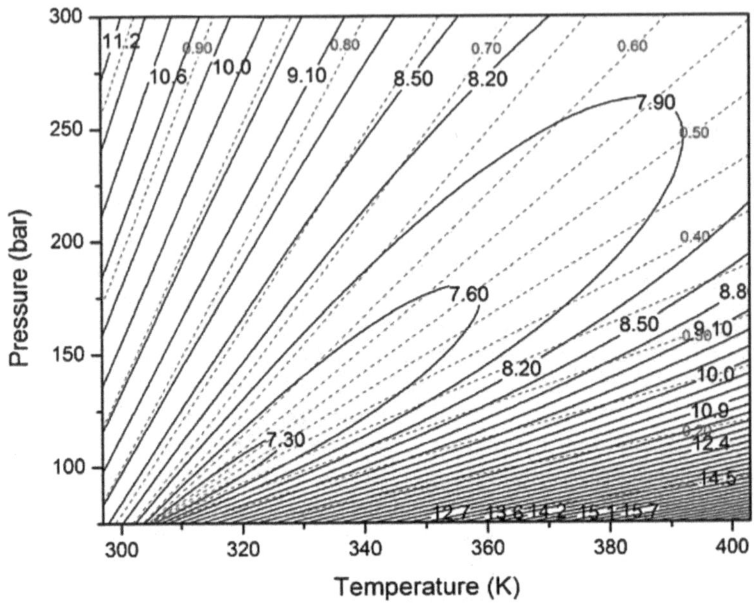

FIGURE 2.4 Contour plots of the kinematic viscosity (solid lines, in Stokes, or cm^2/s, $\times 10^4$) and density (dashed lines, in g/cm^3) of CO_2 on the $P-T$ plane. *Source: Reprinted from reference Tarafder A, Kaczmarski K, Ranger M, Poe DP, Guiochon G. Use of the isopycnic plots in designing operations of supercritical fluid chromatography: IV. Pressure and density drops along columns. J Chromatogr A 2012;1238(0):132–45 [10].*

line, density = 0.468 g/cm^3. Such diagrams can provide guidance to chromatographers in how the column pressure drop will change in response to changes in operating conditions.

2.2.2.3 *Fluid Expansion and Temperature*

In addition to changes in the density, velocity and viscosity of the mobile phase along a column, expansion of the mobile phase can be accompanied by heating or cooling. The incremental temperature change of the mobile phase along a column operated in adiabatic conditions depends on the local pressure gradient and temperature, and is predicted by

$$\partial T = \mu_{JT}\partial P = -\left(\frac{1}{\rho C_p} - \frac{\alpha T}{\rho C_p}\right)\partial P \tag{2.3}$$

where μ_{JT} is the Joule–Thomson coefficient, ρ is density, C_p is the isobaric heat capacity, and α is the thermal expansion coefficient. The amount of heating or cooling depends on the relative contributions of the two terms in parentheses on the right-hand side of Eq. (2.3). The

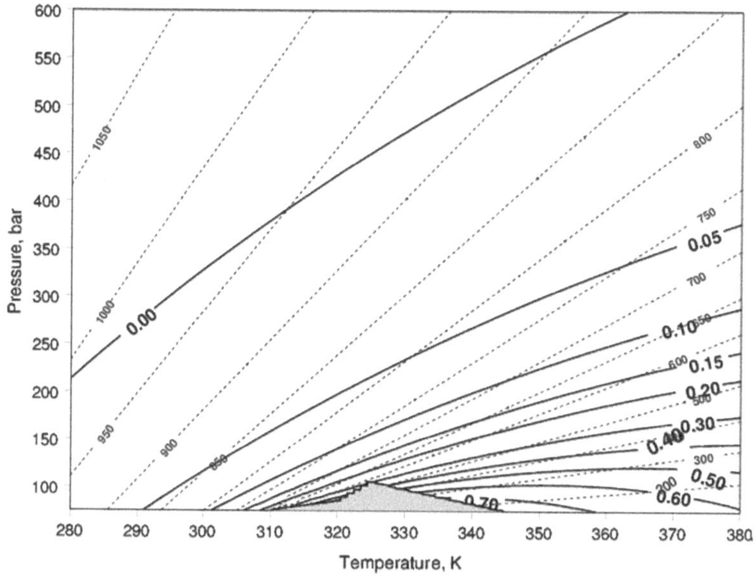

FIGURE 2.5 Joule–Thomson coefficients in K/bar for 95% CO_2/5% methanol (mol/mol) mobile phase (solid contour lines). Also shown (dashed lines) are isopycnic lines for CO_2 with 5% methanol (kg/m^3). The gray area is a two-phase region for the 5% methanol fluid. *Source: Reprinted from reference Poe DP, Veit D, Ranger M, Kaczmarski K, Tarafder A, Guiochon G. Pressure, temperature and density drops along supercritical fluid chromatography columns in different thermal environments. III. Mixtures of carbon dioxide and methanol as the mobile phase. J Chromatogr A 2014;1323(0):143–56 [11].*

first term is related to viscous heating, and the second term to cooling associated with enthalpic expansion. Plots of μ_{JT} versus temperature and pressure for CO_2 with 5% added methanol are shown in Fig. 2.5. A positive value of μ_{JT} corresponds to a decrease in temperature with decreasing pressure. For negative values heating occurs. No heating or cooling occurs at $P-T$ conditions along the zero μ_{JT} isopleth. For SFC separations at temperatures above 310 K and pressures below 400 bar, where most SFC operations are conducted, these plots indicate that net cooling occurs.

Isenthalpic plots show how the temperature of an expanding fluid changes with pressure at constant enthalpy. Fig. 2.6 shows the isenthalpic plots for 95% CO_2/5% methanol. If a column is operated under adiabatic conditions, isenthalpic expansion occurs, and the plots in Fig. 2.6 can be used to estimate the extent of any temperature change along the column. Note again the contour line for $\mu_{JT} = 0$. For this condition no net heating or cooling occurs. Operating at pressures and temperatures near the zero $J-T$ isopleth, where the isenthalpic plots are nearly vertical, results in little heating or cooling, essentially avoiding issues related

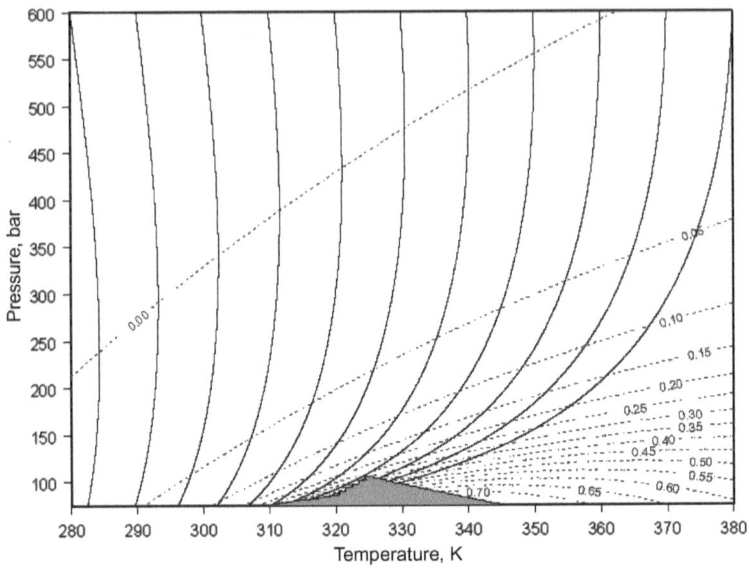

FIGURE 2.6 Isenthalpic expansion curves (solid lines) and $J-T$ coefficients (dashed lines) for 95% CO_2/5% methanol (mol/mol). The gray area is a two-phase region. *Source: Reprinted from reference Poe DP, Veit D, Ranger M, Kaczmarski K, Tarafder A, Guiochon G. Pressure, temperature and density drops along supercritical fluid chromatography columns in different thermal environments. III. Mixtures of carbon dioxide and methanol as the mobile phase. J Chromatogr A 2014;1323(0):143–56 [11].*

to temperature gradients. At higher temperatures and low pressures these isenthalpic plots indicate large temperature drops, especially for $\mu_{JT} > 0.1$ K/bar. If the column wall is thermostatted (e.g., placed in a forced-air oven), the axial temperature gradients can be accompanied by radial temperature gradients, leading to loss of efficiency. Note that the regions of large temperature drop indicated in Figs. 2.5 and 2.6 correspond closely to the danger zone C in Fig. 2.2. This topic is addressed in greater detail in Section 2.5.2.

2.3 FLUID COMPRESSIBILITY AND AVERAGE PARAMETERS IN NONUNIFORM COLUMNS

An SFC column is a classic example of a nonuniform column, in which certain local parameters that impact on the solute band velocity and dispersion, such as density and temperature, vary along the column. Parameters observed at the column outlet such as the retention factor and plate height should therefore be regarded as apparent parameters. Martire [12] developed a general approach to relate apparent parameters to their local values. In this section we examine these approaches to expressing the observed parameters in a meaningful and consistent fashion.

Of the several mobile phase parameters in SFC, the mass flow rate is one that does not vary along the column. The mass flow rate and the mobile phase velocity are related by the expression

$$u_0 = \frac{F_m}{\varepsilon_t \rho S} \tag{2.4}$$

where $u_0 = L/t_0$ is the linear velocity of the mobile phase, F_m is the mass flow rate, ε_t is the total porosity (fraction of total column volume occupied by the mobile phase), and S is the cross-sectional area of the column. The velocity increases along the column as the density decreases. Other important column variables such as the retention factor, mobile phase viscosity, and diffusion coefficients can also be expressed as simple functions of density and temperature. The approach taken by Martire was to express all variables in terms of the mass flow rate, temperature, and density. Spatial and temporal average quantities are related to operational parameters such as temperature and pressure through appropriate distribution functions. Some of the key relationships summarized in a second paper by Martire et al. [13] include:

The spatial average mobile phase density $\langle\rho\rangle_z$

$$\langle\rho\rangle_z = \frac{\int_{\rho_o}^{\rho_i} \rho D_z(\rho)d\rho}{\int_{\rho_o}^{\rho_i} D_z(\rho)d\rho}$$

The temporal average mobile phase density $\langle\rho\rangle_t$ (2.5)

$$\langle\rho\rangle_t = \frac{\int_{\rho_o}^{\rho_i} \rho D_t(\rho)d\rho}{\int_{\rho_o}^{\rho_i} D_t(\rho)d\rho}$$

where ρ_i and ρ_o are the inlet and outlet densities, and $D_z(\rho)$ and $D_t(\rho)$ are the relevant distribution functions [12]. In general, for any parameter Q that can be expressed as a function of density, the spatial or temporal average of Q is

$$\langle Q\rangle = \frac{\int_{\rho_o}^{\rho_i} QD(\rho)d\rho}{\int_{\rho_o}^{\rho_i} D(\rho)d\rho} \tag{2.6}$$

where $D(\rho)$ is the spatial or temporal distribution function. For example, it was shown that the apparent retention factor is the temporal average value of the local retention factor k',

$$k = \frac{t_r - t_0}{t_0} = \langle k' \rangle_t = \frac{\int_{\rho_0}^{\rho_i} k' D_t(\rho) d\rho}{\int_{\rho_0}^{\rho_i} D_t(\rho) d\rho} \quad (2.7)$$

More about the retention factor and a general expression for k' in terms of temperature and density appear in Section 2.4.

Another commonly measured parameter is the average mobile phase velocity. The local velocity, which varies with the local density, is given in Eq. (2.4). The average velocity is

$$\langle u_0 \rangle = \frac{L}{t_0} = \frac{F_m}{\varepsilon_t \langle \rho \rangle_z S} \quad (2.8)$$

With the availability of modern software such as the NIST REFPROP program [4] to evaluate various thermophysical properties of fluids, the functions and integrals described earlier can be easily evaluated using an Excel spreadsheet set up for numerical integration. This would provide for reporting SFC data in a rigorous and consistent fashion, and might help to further advance SFC as a mature analytical method.

2.4 RETENTION IN SFC

2.4.1 Distribution Equilibria and the Retention Factor

For the reader who may not be familiar with the basic concepts and terminology of retention in chromatography, the following brief introduction may suffice as a background to concepts treated later in this section. More detailed background is available in any of several books on separation science and chromatography [1,2]. Reviews of the various factors influencing retention in SFC are available [8,14].

Retention in chromatography is generally expressed in terms of the retention factor k or retention time t_r. Both are based on the equilibrium distribution of the analyte solute between the stationary phase and mobile phase, represented by the process

$$c_m \rightleftharpoons c_s \quad (2.9)$$

where c_m and c_s are the concentrations of the analyte in the mobile phase and stationary phase, respectively. For ideal behavior the equilibrium distribution constant, or Henry constant, is

$$K_d = \frac{c_s}{c_m} \quad (2.10)$$

The distribution constant is a thermodyamic measure of chromatographic retention. Large values of K_d result in strong retention.

A related quantity that is more commonly used in chromatography to express retention is the retention factor, defined by IUPAC [15] as

$$k = \frac{t_r - t_0}{t_0} \tag{2.11}$$

as noted in Section 2.3, in SFC we need to distinguish between local and apparent retention factors. The local retention factor k' is defined in terms of the distribution constant and the phase ratio V_s/V_m

$$k' = \frac{n_s}{n_m} = \frac{c_s V_s}{c_m V_m} = K_d \frac{V_s}{V_m} \tag{2.12}$$

where n_s, n_m are moles of solute in the stationary and mobile phases in a narrow zone along the column, and c and V are the solute concentrations in the corresponding volumes of the two phases. It is also known as the capacity factor, a term originally developed for treating distribution equilibria in solvent extraction processes. For nonuniform columns, which are pretty much the standard case in SFC, the local retention factor k' varies along the column, and is related to the apparent retention factor k by Eq. (2.7).

2.4.2 Dependence of Retention on Temperature and Pressure (Density) in SFC

In supercritical fluid systems, the retention factor varies with temperature and pressure (or density), providing experimentalists with a powerful and convenient means to control retention in SFC. Numerous studies have shown that retention is a smooth function of density rather than pressure. The treatment by Martire and Boehm [16,17], which employs a statistical–mechanical treatment based on a lattice-fluid model, expresses the local retention factor in terms of the reduced temperature and reduced density of the mobile phase. Their treatment resulted in the following expression for the retention factor in SFC with immobilized liquid stationary phases:

$$\ln k' = \ln k^0 = c_2 \rho_R + c_3 \frac{\rho_R}{T_R} + c_4 \frac{\rho_R^2}{T_R} \tag{2.13}$$

where $\rho_R = \frac{\rho}{\rho_c}$ and $T_R = \frac{T}{T_c}$ are the reduced density and temperature, ρ_c and T_c are the critical density and temperature, respectively, and $\ln k^0 = c_0 + c_1/T_R$ is the value of the retention factor for ideal GC conditions. At constant temperature, the general expression may be simplified to

$$\ln k' = \ln k^0 - a\rho_R + b\rho_R^2 \tag{2.14}$$

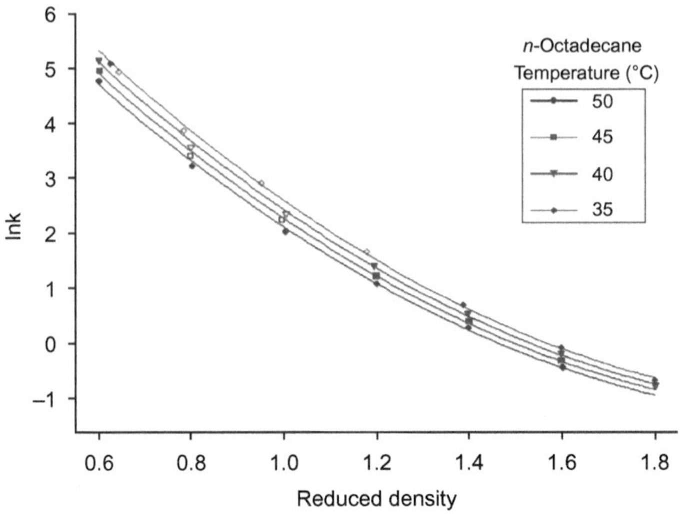

FIGURE 2.7 Retention isotherms of *n*-octadecane on 5-micron Waters Spherisor-C8 particles. Symbols show experimental data. Solid lines fit to Eq. (2.13). *Source: Reprinted from reference Kaczmarski K, Poe DP, Guiochon G. Numerical modeling of the elution peak profiles of retained solutes in supercritical fluid chromatography. J Chromatogr A 2011;1218 (37):6531–9 [20].*

These relationships are the result of a general theory for absorption chromatography that applies to GC, LC, and SFC. Although the theoretical model leading to Eq. (2.13) was developed specifically for absorption chromatography with a single-component mobile phase in contact with an immobilized liquid stationary phase, these equations also serve as a useful model to describe the retention isotherms for SFC using modern silica-based bonded phase packings. As an example, isotherms at different temperatures for elution of a test solute on a bonded phase packing using neat carbon dioxide as the mobile phase are shown in Fig. 2.7. Related expressions were developed for SFC with adsorption on solid surfaces. Reported behavior for SFC using modified mobile phases with bare silica stationary phases [18] indicate that Eq. (2.13) is suitable for modeling these systems as well.

Values for the local retention factor shown in Fig. 2.7 were determined experimentally by careful measurements at low-pressure drop, less than 2 bar, where the observed retention factor provides a good estimate of the local value at the average column density and temperature. This can be a long and tedious process. Alternatively numerical simulations can be applied to experimental data using larger pressure drops to obtain the isotherm coefficients [19].

As noted in the preceding section, the observed (or apparent) retention factor in SFC is equal to its temporal average value. If the local retention factor is expressed in terms of temperature and density as given by Eq. (2.13), the apparent retention factor can be computed by numerical integration using a spreadsheet as described in Section 2.3.

2.4.3 Effect of Modifiers on Retention

SFC with packed columns is almost always performed with polar organic modifiers added to the CO_2 mobile phase in order to elute analytes of moderate to high polarity [7]. Even for analytes of low polarity that could be eluted with neat CO_2, addition of several percent modifier, such as methanol, is useful to minimize injection solvent effects.

There are several mechanisms by which added modifiers decrease retention. On silica-based stationary phases, the polar modifier is strongly adsorbed on the surface, competing with adsorption by analyte molecules. Also the modifier makes the mobile phase more polar, increasing the solubility of the polar analytes. Further discussion of the effects of polar modifiers can be found elsewhere [7,8,14,18].

2.5 KINETIC THEORY

2.5.1 Basic Concepts

2.5.1.1 Peak Variance, Plate Number, and Plate Height

The basic processes of band spreading in a chromatographic column are well-known and have been treated thoroughly in many books and reviews [1,2,21–23]. For those unfamiliar with some of the basic concepts and terminology, and to clarify terminology, we begin with a brief introduction.

The efficiency of a chromatographic column depends on a variety of column parameters and operating conditions, each of which may contribute to dispersion of analyte molecules away from the center of the band zone. This dispersion is measured in terms of the band variance given by the general expression

$$\sigma_z^2 = 2Dt \tag{2.15}$$

where D is an apparent dispersion coefficient and t is the time spent in the column. While band spreading is a dynamic process, the total dispersion is commonly expressed in terms of the plate number, in which one plate corresponds to the axial distance in the column, or the plate height (H), over which a single equilibrium distribution process occurs, each making its own contribution to the band variance. The variances in each of the individual zones i are additive, and the plate height is

$$H = \frac{d\sigma_z^2}{dz} = \frac{\sum_i^N (\sigma_z^2)_i}{L} \tag{2.16}$$

where L is the column length and N is the plate number. Related expressions are

$$H = \frac{L}{N} = L\frac{\sigma_t^2}{t_r^2} \tag{2.17}$$

where σ_t^2 is the variance in time units. For symmetrical peaks the plate number is

$$N = 5.54\left(\frac{t_r}{w_h}\right)^2 \tag{2.18}$$

where w_h is the peak width at half-height. This relation tends to overestimate the plate number for asymmetrical peaks. For accurate work the second statistical moment, which is a measure of the peak variance, is preferred [24].

In a typical packed column the physical contributions to the total variance include longitudinal diffusion (σ_L^2), flow and diffusion in the mobile phase (σ_M^2), and resistance to mass transfer in the stationary phase (σ_S^2). The total variance is simply the sum of these independent processes:

$$\sigma_{total}^2 = \sigma_L^2 + \sigma_M^2 + \sigma_S^2 \tag{2.19}$$

The total plate height is then

$$H = \frac{d\sigma_{total}^2}{dz} = H_L + H_M + H_S \tag{2.20}$$

The H_M term represents contributions from two separate physical processes in the mobile phase. One is eddy diffusion which originates from convective flow processes (H_f), the other from molecular diffusion between stream paths in the mobile phase (H_D). These two processes are coupled as given by

$$H_M = \frac{1}{\frac{1}{H_f} + \frac{1}{H_D}} \tag{2.21}$$

The nature of these dispersion processes are described in greater detail by Giddings [23].

2.5.1.2 Plate Height and the van Deemter Equation

The van Deemter equation expresses the local plate height at any point along the column as a function of the local linear velocity u_0. In its simplest form it is written

$$H = \frac{B}{u_0} + A + Cu_0 \tag{2.22}$$

where $u_0 = L/t_0$ and the coefficients A, B, and C represent column constants and other parameters, including the solute diffusion coefficient and retention factor. While this equation is widely used for fitting the plate height data for van Deemter plots, the fitting constants A and C lack sufficient detail for complete physical meaning. The more accurate coupled form proposed by Giddings is

$$H = \frac{B}{u_0} + \frac{1}{\frac{1}{A} + \frac{1}{C_m u_0}} + C_s u_0 \tag{2.23}$$

where the B, A, C_m, and C_s terms correspond to to H_L, H_f, H_D, and H_S, respectively. Unfortunately this form complicates the fitting of the data for van Deemter plots, and is not commonly used. A simpler empirical formula that represents the mobile phase terms in a single velocity-dependent term is given by the well-known Knox equation for liquid chromatography

$$h = \frac{B'}{\nu} + A'\nu^{1/3} + C_s^{'}\nu \tag{2.24}$$

where h and ν are the reduced plate height and reduced velocity, respectively, given by

$$h = \frac{H}{d_p}; \quad \nu = \frac{ud_p}{D_m} \tag{2.25}$$

where d_p is the particle size, u is the superficial velocity of the mobile phase, and D_m is the solute diffusion coefficient in the mobile phase. These reduced quantities are used in GC, SFC, and HPLC to scale the wide range of plate heights and velocities to a common basis. For well-packed columns typical values at optimum flow rates are $h_{min} \approx 2$ and $\nu_{opt} \geq 2$.

Any of these forms of the van Deemter equation apply in SFC. The expressions for the individual van Deemter coefficients differ depending on the physical geometry of the packing particles. In GC and HPLC these coefficients can generally be regarded as fixed constants for a given solute. In SFC the local values of the van Deemter coefficients B and C vary along the column. The impact of these nonuniform conditions are treated in Section 2.5.2.

2.5.2 Sources of Band Spreading in SFC

In a chromatographic column, we can identify three main sources of band spreading: (1) classical van Deemter dispersion processes described in Section 2.5.1; (2) axial gradients in temperature, density, and related parameters; and (3) radial temperature gradients. The latter two fall into the general category of nonuniform columns, in which axial and/or radial variations in the solute velocity occur due to nonuniform conditions of temperature and/or density. We address each of these contributions in the following sections.

2.5.2.1 van Deemter-Type Band Spreading and Local Plate Height

The simple van Deemter Eq. (2.22) can be used to fit experimental data quite well in many cases, but it greatly oversimplifies the details of the processes controlling band spreading in packed columns. Although the fundamental theories were developed decades ago, significant progress continues to be made to improve the understanding of these physical processes. These gains have been focused primarily on liquid chromatography [22,25–28]. Although these recent findings should also apply to SFC, there is little direct evidence to support this assumption. This is probably due to the swings in the general interest in SFC since its introduction more than 50 years ago, and also to the experimental and theoretical difficulties that arise from the nonuniform properties of SFC columns. The following discussion focuses on the van Deemter *C* term and incorporates recent developments in application of the general rate model to band spreading in HPLC. The resulting explicit expressions may provide additional insight into processes that control efficiency in SFC, especially in the high-velocity regime.

The general rate model is based on separate mass balance equations for the fraction of eluent that is moving in the interstitial volume between the particles and for that which is stagnant inside the pores of the particles. It is the most sophisticated of the various theoretical models of chromatography developed over the past 70 years. However, like any of the various models, it does suffer from several weaknesses. One weakness is that it lumps the physical phenomena described by the A and B terms of the classical van Deemter equation into a single axial dispersion coefficient D_L. It also fails to address the band spreading caused by the heating that can occur at high flow rates in UHPLC, or by the gradients in density and temperature that generally occur in SFC columns. The latter phenomena are addressed later in Section 2.5.2.2. The model does, however, provide the most complete description of the mass transfer processes for exchange of analytes between the stagnant fluid inside the porous particles and the interparticle space.

These are the processes responsible for the C term in the classical van Deemter equation.

2.5.2.1.1 The General Rate Model and Local Plate Height for Packed Column SFC

The introduction and widespread adoption of superficially porous particles in HPLC and SFC has required updated expressions for the plate height of columns packed with these particles. The following expression, which is based on the general rate model and was developed for HPLC [26], should also be valid for SFC. For the separation of small molecules in which the rates of adsorption and desorption are fast, the local plate height can be written as

$$H = \frac{2D_L}{u_e} + \frac{2u_e r_p}{3\phi_e}\left(\frac{k''}{1+k''}\right)^2\left[\frac{1}{k_e} + \frac{\phi_r r_p}{5D_p}\right] \tag{2.26}$$

where D_L is the axial dispersion coefficient, $u_e = u/\varepsilon_e$ is the velocity in the interparticle space where ε_e is the external porosity (fractional volume of the interparticle space), k_e is the external mass transfer coefficient, D_p is the diffusivity in the pore space, $\phi_e = (1 - \varepsilon_e)/\varepsilon_e$, and k'' is the zone capacity factor. The parameter ϕ_r is

$$\phi_r = (r_e^4 + 2r_e^3 r_i + 3r_e^2 r_i^2 - r_e r_i^3 - 5r_i^4)/(r_e^2 + r_e r_i + r_i^2)^2 \tag{2.27}$$

where $r_e = r_p$ is the overall external radius of the particle including the porous surface layer and r_i is the radius of the internal solid core. For totally porous particles, $\phi_r = 1$.

Due to the porous nature of modern particles for HPLC, Eq. (2.26) recognizes that the mobile phase occupies two distinct zones; the moving zone that occupies the interstitial space between the particles, and a stationary zone containing stagnant fluid within the intraparticle space and the stationary phase. This leads to the introduction of the excluded velocity, which refers to the velocity of the mobile phase in the mobile zone, as well as the zone capacity factor (or zone retention factor), which is defined as the ratio of moles of solute in the stationary zone over moles in the mobile zone. The relationship between the two retention factors is $(1 + k'') = (1 + k')(\varepsilon_t/\varepsilon_e)$, where ε_t is the total porosity. Further details about the treatments for columns filled with porous particles and other packings can be found in references [22,25,27,28].

The two terms in the square brackets on the right in Eq. (2.26) represent the relative contributions of resistance to mass transfer: the first term for mass transfer at the external surface of the particle (in contact with the moving zone); the second term for mass transfer in the stationary zone. These terms can be transformed to obtain an expression written in terms of the reduced velocity. The Wilson-Geankoplis model yields the following expression for the external mass transfer coefficient [27,29]

$$k_e = \frac{1.09}{\varepsilon_e} \frac{D_m}{d_p} \nu_e^{1/3} \tag{2.28}$$

where $\nu_e = \nu(\varepsilon_t/\varepsilon_e)$ is the excluded reduced velocity. Dividing both sides of equation Eq. (2.26) by $2r_p$ and introducing the expression for reduced velocity we obtain

$$h = \frac{2(D_L/D_m)}{\nu_e} + \frac{2\nu_e}{3\phi_e}\left(\frac{k''}{1+k''}\right)^2 \left[\frac{\varepsilon_e}{2.18}\nu_e^{-1/3} + \frac{\phi_r D_m}{20 D_p}\right] \tag{2.29}$$

The relation between D_m and D_p is

$$D_p = \gamma_p F(\lambda_m) D_m + (1 - \varepsilon_p) K_d D_s \tag{2.30}$$

where γ_p is the internal obstruction factor to diffusion within the mesopores and $F(\lambda_m)$ is a hindrance factor to diffusion within narrow pores. For small molecules in liquid solvents, and presumably in supercritical fluids, $\gamma_p = 0.6$ and $F(\lambda_m) = 0.8$. In the second term on the right ε_p is the particle porosity, K_d the equilibrium distribution constant, and D_s the diffusion coefficient in the stationary phase. In the limiting case where diffusion in the stationary phase is negligible, $D_p \cong 0.5 D_m$, and Eq. (2.29) becomes

$$h = \frac{2(D_L/D_m)}{\nu_e} + \frac{2}{3\phi_e}\left(\frac{k''}{1+k''}\right)^2 \left[\frac{\varepsilon_e}{2.18}\nu_e^{2/3} + \frac{\phi_r}{10}\nu_e\right] \tag{2.31}$$

The first term on the right in Eq. (2.31), which represents dispersion processes in the moving mobile phase represented by the van Deemter A and B terms, provides no detail about the nature of those processes. Insight into these processes must rely on other approaches, such as Giddings' coupled model represented by Eq. (2.23), which is discussed in detail elsewhere [23,27].

The second term on the right in Eq. (2.31) (the van Deemter C_s term in Eq. (2.23)) shows that the band dispersion due to resistance to external film transfer depends on $\nu_e^{2/3}$, and that mass transfer in the stagnant mesopore space depends on ν_e. Also, variations in the local plate height along the column are controlled by k'' and ν_e. For large k'' these variations would be governed primarily by ν_e. Variations in the latter are relatively small as shown in the following section. Except for weakly retained solutes, we should expect only modest variations in the local plate height along an SFC column.

2.5.2.1.2 Reduced Velocity, Flow Rate, and van Deemter Plots in SFC

With few exceptions, van Deemter studies for SFC separations present the data as plots of plate height versus flow rate at the pump, or as plate height versus the average velocity $u_0 = L/t_0$. While these plots provide important practical information for most studies, some of the meaning of the van Deemter plot is lost. We have seen that van Deemter plots using reduced variables can provide useful information about the general performance characteristics of a column. How is the flow rate at the pump related to the reduced velocity, and how does the reduced velocity vary along the column?

The variation of the reduced velocity along a column can be better understood from the following transformation obtained by multiplying the numerator and denominator by the local density.

$$\nu = \frac{u d_p}{D_m} \frac{\rho}{\rho} = \frac{F_m d_p}{S(D_m \rho)} \tag{2.32}$$

where F_m is the mass flow rate and S is the cross-sectional area of the column. All of the terms on the right side are constant except for D_m and ρ. The product $D_m\rho$ in the denominator is sometimes assumed to be approximately constant for supercritical fluids [30], much like the product D_gP in dilute gasses. Data for substituted benzenes in neat CO_2 at 40°C show that when the density changes from 0.75 to 0.85 g/cm^3, D_m changes by 19%, while $D_m\rho$ changes by about 13% [5]. The variation in the reduced velocity would then change by roughly the same amount along a column for the similar conditions.

The mass flow rate is $F_m = F_p\rho_p$, where F_p and ρ_p are the volumetric flow rate and the fluid density at the pump, respectively. If CO_2 is pumped at 5°C the density of the fluid changes less than 10% as the inlet pressure increases from 100 to 300 bar with increasing flow rate. The mass flow rate therefore increases in a nearly linear fashion with the volumetric flow rate at the pump. Plots of plate height versus F_p are therefore roughly equivalent to plots of plate height versus reduced velocity. By employing appropriate reference conditions [31], it is possible to design an SFC pump that delivers a mass flow rate that is directly proportional to the volumetric flow rate.

Finally, an analysis of Eq. (2.32) also explains why the optimum flow rate for a column of fixed dimensions and particle size remains nearly constant as the operating temperature and pressure are changed. If we replace F_m with $F_p\rho_p$, for a fixed pumping rate the reduced velocity depends only on the ratio $\rho_p/D_m\rho$. This ratio remains nearly constant as the column operating temperature and pressure change.

2.5.2.2 Band Spreading in Nonuniform Columns

The plate height Eq. (2.26 and 2.29) in the preceding discussion apply to uniform columns where the mobile phase velocity and the retention factors are essentially constant along the column. They have their primary application in liquid chromatography. In essentially all SFC separations these parameters vary along the column, and the variance of an eluting peak provides an apparent plate height. In Section 2.3 we applied the concepts of spatial and temporal average quantities in SFC to relate local column variables such as mobile phase density and the retention factor to the corresponding apparent values that are observed at the column outlet. In this section we first extend these approaches to describe the effects of axial gradients on the apparent plate height. We then examine the effects of radial temperature gradients on column efficiency.

2.5.2.2.1 Axial Gradients and Apparent Plate Height

The theory for plate height in nonuniform columns was originally developed by Giddings [23] to describe the influence of pressure drop on the apparent plate height in gas chromatography. He showed that the apparent plate height $\hat{H}$, which is the observed value based on the variance of an eluting peak, is always greater than the local plate height according to the relation

$$\hat{H} = \frac{\left\langle H(1/u_s)^2 \right\rangle_z}{\langle 1/u_s \rangle_z^2} \tag{2.33}$$

where H is the local plate height and u_s is the local solute velocity, and the brackets indicate the spatial averages of the enclosed quantities. All SFC separations involve a decrease in density along the column, as well as a temperature change associated with Joule–Thomson expansion. The resulting efficiency is strongly dependent on the pressure and temperature conditions in the column. Poe and Martire [32] showed that for SFC Eq. (2.33) should be written

$$\hat{H} = \frac{\left\langle H(1+k')^2 \rho^2 \right\rangle_z}{\langle (1+k')\rho \rangle_z^2} \tag{2.34}$$

where k' can be replaced with k'' with the same result. All terms within the brackets, including the velocity and diffusivity terms in the expression for local plate height, are written in terms of the mass flow rate, temperature, and density. The approach described in Section 2.3 can then be used to compute the apparent plate height.

If the local plate height does not vary greatly along the column, as we showed in Section 2.5.2.1, Eq. (2.34) shows that efficiency loss due to axial gradients in temperature and density is due primarily to changes in the local retention factor, which increases exponentially as the density decreases. It is noteworthy that an isothermal environment yields larger density gradients (see Fig. 2.3) and accordingly greater changes in the local retention factor. Based on Eq. (2.34) the increase in plate height due directly to axial temperature and density gradients should be smallest for columns operated in an adiabatic environment.

The *P–T* conditions that lead to large changes in density are associated with large temperature drops (see Figs. 2.2 and 2.6). Experimental evidence has shown that radial temperature gradients associated with these temperature drops can lead to efficiency losses that are much greater than those due to the axial gradients in temperature and density [33]. The origin and effects of radial temperature gradients are treated in the following section.

2.5.2.2.2 Radial Temperature Gradients

Radial temperature gradients in UHPLC and SFC can lead to significant efficiency loss. In UHPLC, where columns packed with sub-2-μm particles are operated at very large pressure drops to achieve fast separations, viscous heating generates radial temperature gradients where the center of the column is warmer than the wall. Gritti and Guiochon added a term to the van Deemter equation for UHPLC [34]. The use of small diameter columns to provide efficient heat transfer and a still air environment can help to minimize the magnitude of these radial gradients and avoid degradation of column efficiency.

In SFC the tendency to form radial temperature gradients and corresponding loss of efficiency is strongly dependent on the operating conditions. These gradients are caused by the heating or cooling associated with expansion of the mobile phase. For a fluid undergoing decompression in a packed column operated in an adiabatic environment, the change in temperature with pressure is given by Eq. (2.3), and the associated temperature changes for carbon dioxide with 5% added methanol are shown in Fig. 2.6. The impact of the resulting radial temperature gradients has been treated in a series of recent papers [11,20,33,35–39].

The isenthalpic plots for carbon dioxide with 5% added methanol in Fig. 2.6 show that large temperature drops can occur in the *P–T* region near the critical point. For columns operated in a forced-air environment, the associated radial temperature gradients can result in catastrophic losses in efficiency. Fig. 2.8 shows the separation of an

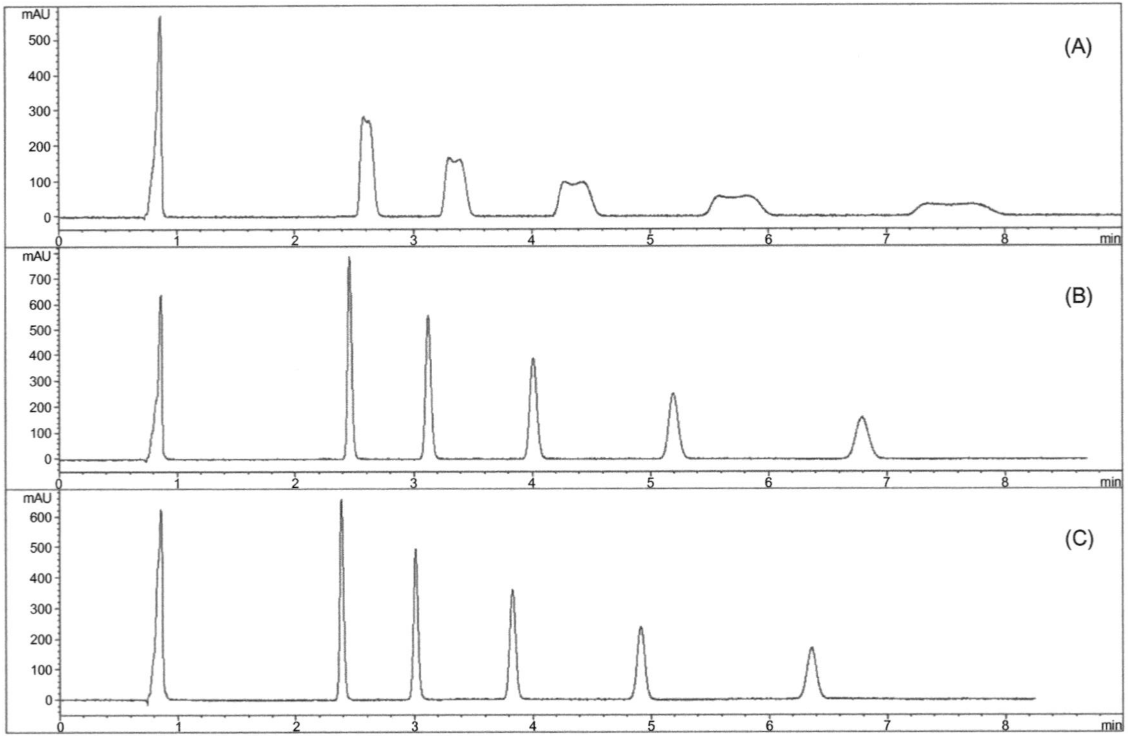

FIGURE 2.8 SFC separation of alkylbenzenes on a 250 × 4.6-mm column with 5-micron Luna C18 particles using three different thermal environments: (A) convective air; (B) foam insulation; (C) still air. Mobile phase: 95% CO_2/5% methanol. Oven temperature: 50°C. Outlet pressure 100 bar, flow rate 3.00 mL/min at pump (271 K). The first peak is the injection solvent, followed by peaks for the five alkylbenzenes with, in order of elution, 10, 12, 14, 16 and 18-carbon chains. Inlet and outlet pressure (bar): (A) 123.0, 100.0; (B) 122.9, 100.2; (C) 124.2, 100.8. *Source: Reproduced from reference Zauner J, Lusk R, Koski S, Poe DP. Effect of the thermal environment on the efficiency of packed columns in supercritical fluid chromatography. J Chromatogr A 2012;1266:149 [40].*

alkylbenzene mixture at 323 K and 100 bar outlet pressure for three different thermal environments. The top trace A, obtained in the convective air mode, shows extremely distorted peaks and poor efficiency. Traces B and C were obtained with the same column either wrapped with foam insulation or placed in still air at 323 K. The latter conditions provide a near-adiabatic environment so that the temperature of the decompressing mobile phase will decrease along an isenthalpic plot beginning at the inlet temperature and pressure, with the *J*–*T* coefficient increasing from about 0.15 to 0.21 K/bar, or an average of about 0.18 K/bar. With a 23-bar pressure drop, the temperature drop would be $\Delta T \approx \mu_{JT} \times \Delta P = 4.1$ K, yielding an outlet temperature of 319 K. In the convective air oven the temperature of the column wall is maintained near the oven set temperature, leading to large radial temperature gradients and the resulting distorted peaks shown in trace A.

The catastrophic efficiency loss described above can be avoided in several ways. For operations in a forced-air environment, selecting conditions with small μ_{JT} values is a simple solution. Recent work in the author's laboratory has shown that for a 250 × 4.6-mm i.d. column packed with 5-μm porous particles, efficiency loss in a forced-air oven is negligible for outlet pressures corresponding to $\mu_{JT} < 0.10$ K/bar at the oven temperature. The *J*–*T* coefficient then serves as a useful guide to safe operating conditions (see Fig. 2.5). For columns operated in still air or in other near-adiabatic environments, the safe operating conditions can be extended to larger values of the *J*–*T* coefficient. These conditions correspond to lower pressure and density where retention is greater, offering improved conditions for separation of weakly retained analytes. Finally the use of superficially porous particles with high thermal conductivity can also help to minimize the magnitude of the radial temperature gradients.

2.5.2.3 Extra-Column Variance

In column chromatography the band spreading can originate from processes within the column itself, and from extra-column sources such as the injector, the connecting tubes, and the detector. The observed peak variance and peak shape will reflect contributions from all these sources. In general, the various contributions to the overall peak variance are additive, and can be expressed as

$$\sigma^2_{total} = \sigma^2_{col} + \sigma^2_{ec} \tag{2.35}$$

where the subscripts *col* and *ec* refer to column and extra-column contributions, respectively. The best measure of the variance of a chromatographic peak is the second statistical moment [21,24]. The

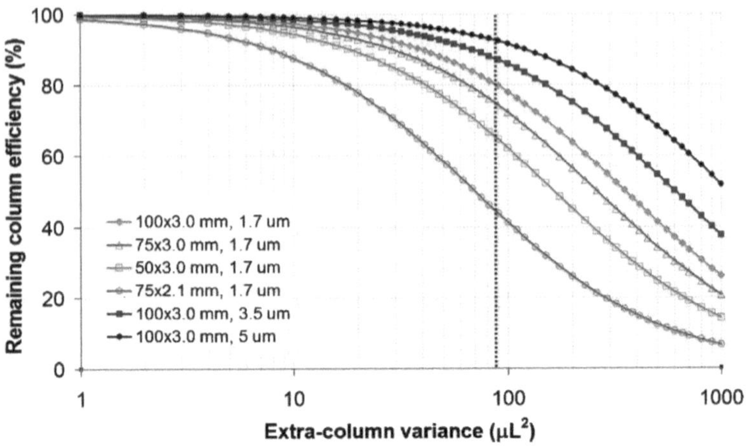

FIGURE 2.9 Loss of efficiency for different columns due to extra-column variance for $k = 5$. Vertical line represents extra-column variance of a typical modern SFC instrument with $\sigma_z^2 = 85\ \mu L^2$. *Source: Reproduced from reference Grand-Guillaume Perrenoud A, Hamman C, Goel M, Veuthey J-L, Guillarme D, Fekete S. Maximizing kinetic performance in supercritical fluid chromatography using state-of-the-art instruments. J Chromatogr A 2013;1314:288–97 [43].*

contributions from the various sources of extra-column variance are additive and can be estimated from known relationships [41,42].

The contribution to the total peak variance from a column with a plate number N depends on the retention factor, and is given in volume units by

$$\sigma^2_{\nu,col} = \frac{V_0^2}{N}(1+k)^2 \tag{2.36}$$

where V_0 is the column hold-up volume. If the extra-column variance is known, Eq. (2.36) can be used to determine the appropriate column geometry and packing to keep the relative extra-column contribution to the total variance at or below the desired level.

Although modern SFC instruments are designed to minimize extra-column contributions, these can amount to 10% or more of the total peak variance when working with small diameter columns packed with very fine particles, as shown in Fig. 2.9. With these modern high-performance packings, instrument manufacturers are challenged to minimize these sources of extra-column variance while also minimizing excessive back pressure due to small diameter connectors and maintaining good detector sensitivity.

Several recent studies provide useful insights for understanding and managing this source of band spreading in SFC [44,45]. One study concludes that decreasing connector tubing from 0.007″ to 0.005″ is a necessary improvement to make modern SFC instruments compatible with sub-2-μm particles [46].

2.5.2.4 *Other Sources of Band Spreading*

2.5.2.4.1 Sample Introduction

If the properties of the injection solvent differ greatly from those of the mobile phase, unfavorable mixing processes after sample injection can lead to additional band spreading and poor peak shapes. Normally the strength of the injection solvent should closely match that of the mobile phase. For separations using CO_2 as the mobile phase, a sample solvent such as heptane meets this requirement. In most SFC applications, in which organic modifiers are added to enhance the solvent strength for elution of more polar compounds, it may be impractical to dissolve the sample in the desired injection solvent due to solubility issues [47].

In addition to solvent strength, a large mismatch in the solvent viscosity should also be avoided. Efficiency losses due to a viscosity mismatch may arise due to wall effects and a phenomenon known as viscous fingering. These phenomena, which have been studied in liquid chromatography, may be even more significant in SFC, especially for the high performance systems that are now available [48].

2.5.2.4.2 Multiple Retention Mechanisms

Multiple retention mechanisms involving slow sorption/desorption processes can lead to severe peak distortion. This is especially common for basic mixtures where basic molecules are readily protonated in the acidic CO_2−methanol environment. For the SFC separation of basic drugs with MS detection, additional modifiers such as ammonium formate have been shown to substantially improve peak shapes [49].

2.6 KINETIC THEORY AND CONTEMPORARY PRACTICE

2.6.1 Performance of Columns Packed with Very Fine Particles

The past decade has seen historic advances in the practice of liquid chromatography and SFC. An important trend is the move toward columns packed with smaller-diameter particles to further improve chromatographic performance. A few selected papers provide a review of these advances. Perrenod et al. discuss the use of columns packed with sub-2-μm fully porous particles and superficially porous particles in SFC [46,50]. Fekete et al. reviewed recent approaches to fast and high resolution separations with HPLC and SFC [51]. Novakova et al. have provided an excellent tutorial on modern SFC with sub-2-μm particles [3].

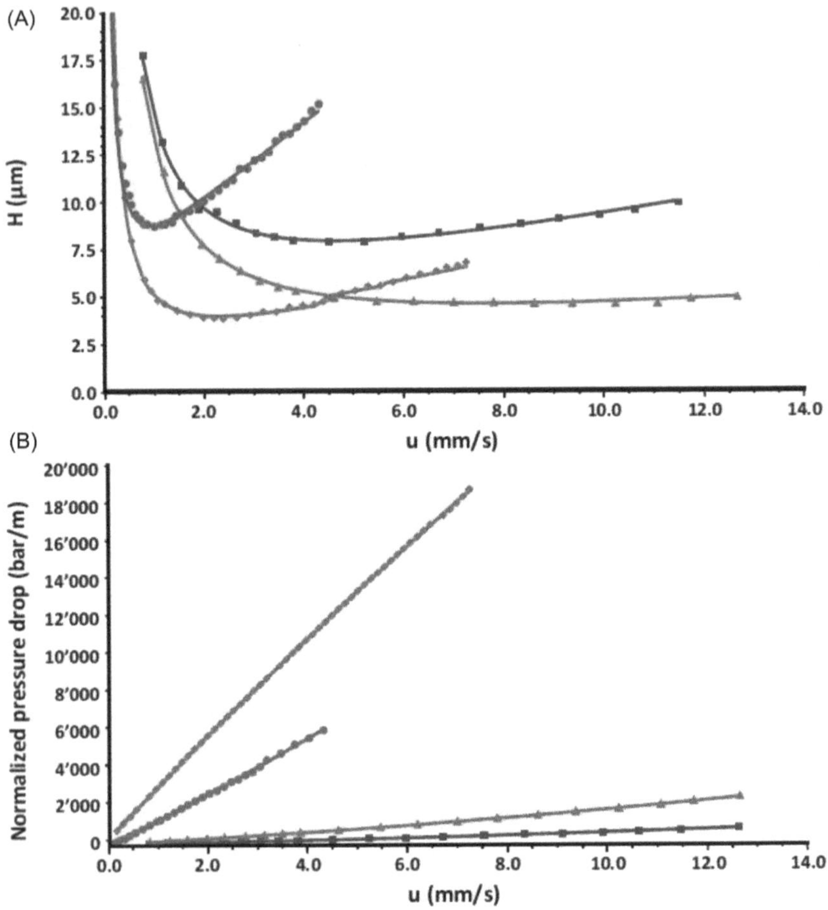

FIGURE 2.10 Kinetic performance and normalized generated pressure drop as a function of linear velocity for 1.7 and 3.5 μm totally porous particles in UHPLC and UHPSFC. (A) van Deemter curves for butylparaben on two systems equipped with 1.7 or 3.5 μm particles columns. HPLC mode: H_2O/ACN (60/40, v/v), 30°C; RP-C18 columns: 50 × 4.6 mm × 3.5 μm (dots); 50 × 2.1 mm × 1.7 μm (diamonds). SFC mode: CO_2/MeOH (94/4, v/v), 40°C, 150 bar back pressure; Acquity UPC2 BEH 2-EP columns: 100 × 3.0 mm × 3.5 μm (squares); 100 × 3.0 mm × 1.7 μm (triangles). (B) Corresponding generated column pressure drop normalized to 1 m of column, to avoid influence of column geometry variations. *Source: Reproduced from reference Grand-Guillaume Perrenoud A, Veuthey J-L, Guillarme D. Comparison of ultra-high performance supercritical fluid chromatography and ultra-high performance liquid chromatography for the analysis of pharmaceutical compounds. J Chromatogr A. 2012;1266:158–67 [52]. (Readers are referred to the web-based version for color image)*

Fig. 2.10A shows van Deemter plots for columns packed with 1.7 and 3.5-micron particles operated on a UHPLC system and on a UHPSFC system. For either system the 1.7-μm particles yield higher optimum velocity and lower minimum plate height compared to 3.5-μm

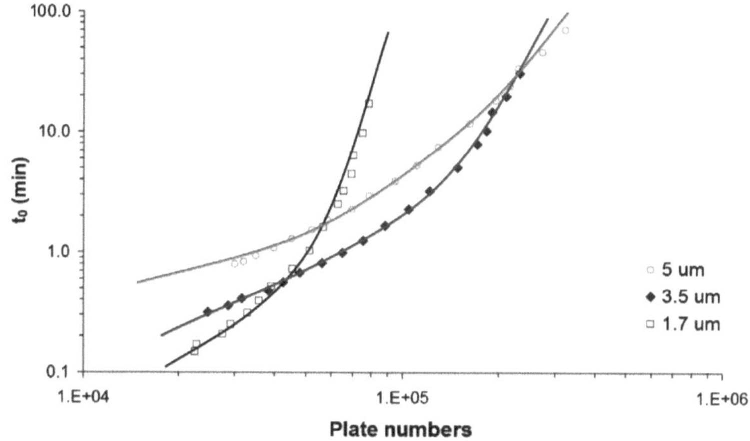

FIGURE 2.11 t_0 versus N-type kinetic plots. Columns: Waters BEH 100 mm × 3.0 mm columns packed with 1.7, 3.5, and 5-μm particles. Mobile phase: 2% MeOH in CO_2, temperature: 40°C. Test solute: butylparaben. Inlet pressure 400 bar with 150 bar outlet pressure. *Source: Reprinted from reference Grand-Guillaume Perrenoud A, Hamman C, Goel M, Veuthey J-L, Guillarme D, Fekete S. Maximizing kinetic performance in supercritical fluid chromatography using state-of-the-art instruments. J Chromatogr A 2013;1314:288–97 [43].*

particles. The lower pressure drop afforded by the low viscosity of the supercritical fluid mobile phase illustrates an important advantage of SFC. Also, the SFC conditions at the column outlet correspond to low compressibility and a maximum $J-T$ coefficient of approximately 0.06 K/bar (see Fig. 2.5). These conditions should be sufficient to avoid any excess efficiency loss arising from the axial density gradients and radial temperature gradients that occur in nonuniform columns.

2.6.2 Limits on Speed and Resolution in SFC

The endless pursuit of faster and more efficient separations has also led to enhanced tools known as kinetic plots that go well beyond the basic van Deemter plots shown in Fig. 2.10. One approach involves plots of t_0 versus plate number shown in Fig. 2.11 [43]. These curves show the elution time for an unretained component for a required plate number when a column is operated at the maximum system pressure drop. For small plate numbers shorter columns are thus operated at high linear velocities. The column length increases and the flow rate decreases as the plate number is increased. The smallest particles (1.7 μm) provide the shortest elution times for $N < 40{,}000$, whereas the 3.5 μm particles and longer columns are preferred for higher plate numbers up to $N = 200{,}000$. Plate numbers $> 80{,}000$ are not accessible with the 1.7 μm particles due to the maximum pressure limitation. For very high resolution ($N > 200{,}00$) 5 μm particles are an option at the expense of long columns and long separation times.

2.7 CONCLUSIONS

The fundamental principles of SFC are essentially the same as those for other modes of column chromatography. There are, however, significant differences that require different approaches to experimental work and to data analysis and interpretation. These differences are due primarily to the compressibility of the fluid and to the effect of pressure and pressure drop on the retention factor and plate height. The fluid density, temperature, velocity, and viscosity all vary as the fluid expands along the column. The local retention factor is a function of local temperature and density, and the local plate height can be written in terms of the mass flow rate, temperature, and density. The apparent values of the retention factor and plate height at the column outlet can then be expressed as the appropriate temporal and/or spatial averages of the local values. An explicit expression for the local plate height in terms of the reduced velocity is given, and it is shown that the reduced velocity is nearly constant along the column. The inherent performance of columns packed with very fine particles is difficult to achieve with currently available instrumentation due to significant extra-column contributions to the total peak variance. The kinetic performance of SFC is superior to that of HPLC when columns are operated near the optimum velocity, but as in HPLC the overall kinetic performance is limited by the maximum operating pressure and flow rates provided by available instrumentation.

Acknowledgments

The author thanks Krzysztof Kaczmarski (Rzeszow University) and Shawn Helmueller (Waters Corporation) for helpful comments.

References

[1] Karger BL, Snyder LR, Horvath C. An introduction to separation science. New York: John Wiley and Sons, Inc; 1973.
[2] Snyder LR, Kirkland JJ, Dolan JW. Introduction to modern liquid chromatography. 3rd ed. Wiley; 2010.
[3] Nováková L, Grand-Guillaume Perrenoud A, Francois I, West C, Lesellier E, Guillarme D. Modern analytical supercritical fluid chromatography using columns packed with sub-2 μm particles: a tutorial. Anal Chim Acta 2014;824:18–35.

[4] Lemmon E., Huber M., McLinden M. NIST Standard Reference Database 23: Reference Fluid Thermodynamic and Transport Properties-REFPROP, Version 9.1. Gaithersburg, MD, USA: National Institute of Standards and Technology, Standard Reference Data Program; 2013.
[5] Sassiat PR, Mourier P, Caude MH, Rosset RH. Measurement of diffusion coefficients in supercritical carbon dioxide and correlation with the equation of Wilke and Chang. Anal Chem 1987;59(8):1164–70. Washington, DC, US.
[6] Berger TA, Deye JF. Efficiency in packed column supercritical fluid chromatography using a modified mobile phase. Chromatographia 1991;31(11–12):529–34.
[7] Berger TA, Deye JF. Composition and density effects using methanol/carbon dioxide in packed column supercritical fluid chromatography. Anal Chem (Washington, DC, US) 1990;62(11):1181–5.
[8] Berger TA. In: Smith RM, editor. Packed column SFC. Cambridge: The Royal Society of Chemistry; 1995. p. 251.
[9] Tarafder A, Guiochon G. Use of isopycnic plots in designing operations of supercritical fluid chromatography: II. The isopycnic plots and the selection of the operating pressure-temperature zone in supercritical fluid chromatography. J Chromatogr A 2011;1218(28):4576–85.
[10] Tarafder A, Kaczmarski K, Ranger M, Poe DP, Guiochon G. Use of the isopycnic plots in designing operations of supercritical fluid chromatography: IV. Pressure and density drops along columns. J Chromatogr A 2012;1238(0):132–45.
[11] Poe DP, Veit D, Ranger M, Kaczmarski K, Tarafder A, Guiochon G. Pressure, temperature and density drops along supercritical fluid chromatography columns in different thermal environments. III. Mixtures of carbon dioxide and methanol as the mobile phase. J Chromatogr A 2014;1323(0):143–56.
[12] Martire DE. Generalized treatment of spatial and temporal column parameters, applicable to gas, liquid and supercritical fluid chromatography. I Theory J Chromatogr 1989;461:165–76.
[13] Martire DE, Riester RL, Bruno TJ, Hussam A, Poe DP. Generalized treatment of spatial and temporal column parameters, applicable to gas, liquid and supercritical fluid chromatography. II. Application to supercritical carbon dioxide. J Chromatogr 1991;545(1):135–47.
[14] Guiochon G, Tarafder A. Fundamental challenges and opportunities for preparative supercritical fluid chromatography. J Chromatogr A 2011;1218(8):1037–114.
[15] International Union of Pure and Applied Chemistry. Compendium of Analytical Nomeclature. Definitive Rules 1997, <http://www.iupac.org/publications/analytical_compendium/>; [accessed 06.07.04].
[16] Martire DE. Unified theory of absorption chromatography: gas, liquid and supercritical fluid mobile phases. J Liq Chromatogr 1987;10(8–9):1569–88.
[17] Martire DE, Boehm RE. Unified molecular theory of chromatography and its application to supercritical fluid mobile phases. 1. Fluid-liquid (absorption) chromatography. J Phys Chem 1987;91:2433.
[18] Berger TA. The effect of adsorbed mobile phase components on the retention mechanism, efficiency, and peak distortion in supercritical fluid chromatography. Chromatographia 1993;37(11–12):645–52.
[19] Leśko M, Poe DP, Kaczmarski K. Modelling of retention in analytical supercritical fluid chromatography for CO_2—methanol mobile phase. J Chromatogr A 2013;1305: 285–92.
[20] Kaczmarski K, Poe DP, Guiochon G. Numerical modeling of the elution peak profiles of retained solutes in supercritical fluid chromatography. J Chromatogr A 2011;1218 (37):6531–9.
[21] Giddings JC. Unified separation science. Wiley-Interscience; 1991. p. 320.

[22] Felinger A, Cavazzini A. Kinetic theories of liquid chromatography. In: Fanali S, Haddad PR, Poole CF, schoenmakers PJ, Lloyd D, editors. Liquid chromatography: fundamentals and instrumentation. Amsterdam: Elsevier; 2013.

[23] Giddings JC. Dynamics of chromatography, part I: principles and theory (chromatographic science series), vol. 1. New York: Marcel Dekker; 1965. p. 336.

[24] Gritti F, Guiochon G. Accurate measurements of peak variances: importance of this accuracy in the determination of the true corrected plate heights of chromatographic columns. J Chromatogr A 2011;1218(28):4452–61.

[25] Guiochon G, Felinger A, Shirazi SG, Katti AM. Fundamentals of preparative and nonlinear chromatography. 2nd ed. Boston: Academic Press; 2006. p. 990.

[26] Kaczmarski K, Guiochon G. Modeling of the mass-transfer kinetics in chromatographic columns packed with shell and pellicular particles. Anal Chem 2007;79:4648–56.

[27] Gritti F, Guiochon G. Mass transfer kinetics, band broadening and column efficiency. J Chromatogr A 2012;1221(0):2–40.

[28] Felinger A, Guiochon G. Comparison of the kinetic models of linear chromatography. Chromatographia 2004;60(1):S175–80.

[29] Wilson EJ, Geankoplis CJ. Liquid mass transfer at very low Reynolds numbers in packed beds. Ind Eng Chem Fundam 1966;5(1):9–14.

[30] Reid RC, Prausnitz JM, Poling BR. The properties of gases and liquids. 4th ed. New York: McGraw-Hill; 1987.

[31] De Pauw R, Shoykhet K, Desmet G, Broeckhoven K. Effect of reference conditions on flow rate, modifier fraction and retention in supercritical fluid chromatography. J Chromatogr A 2016;1459:129–35.

[32] Poe DP, Martire DE. Plate height theory for compressible mobile phase fluids and its application to gas, liquid and supercritical fluid chromatography. J Chromatogr 1990;517:3–29.

[33] Poe DP, Schroden JJ. Effects of pressure drop, particle size and thermal conditions on retention and efficiency in supercritical fluid chromatography. J Chromatogr A 2009;1216(45):7915–26.

[34] Gritti F, Martin M, Guiochon G. Influence of viscous friction heating on the efficiency of columns operated under very high pressures. Anal Chem 2009;81(9):3365–84. Washington, DC, US.

[35] Kaczmarski K, Poe DP, Guiochon G. Numerical modeling of elution peak profiles in supercritical fluid chromatography. Part I–Elution of an unretained tracer. J Chromatogr A 2010;1217(42):6578–87.

[36] Kaczmarski K, Poe DP, Tarafder A, Guiochon G. Pressure, temperature and density drops along supercritical fluid chromatography columns. II. Theoretical simulation for neat carbon dioxide and columns packed with 3-μm particles. J Chromatogr A. Selected papers of 35th International Symposium on high performance liquid phase separations and related techniques, International Symposium on preparative and industrial chromatography and allied techniques, supercritical fluid extraction and chromatography. 2012;1250:115–23.

[37] Poe DP, Veit D, Ranger M, Kaczmarski K, Tarafder A, Guiochon G. Pressure, temperature and density drops along supercritical fluid chromatography columns. I. Experimental results for neat carbon dioxide and columns packed with 3- and 5-micron particles. J Chromatogr A 2012;1250:105–14.

[38] Kaczmarski K, Poe DP, Tarafder A, Guiochon G. Efficiency of supercritical fluid chromatography columns in different thermal environments. J Chromatogr A 2013; 1291(0):155–73.

[39] Tarafder A, Iraneta P, Guiochon G, Kaczmarski K, Poe DP. Estimations of temperature deviations in chromatographic columns using isenthalpic plots. I. Theory for isocratic systems. J Chromatogr A 2014;1366(0):126–35.
[40] Zauner J, Lusk R, Koski S, Poe DP. Effect of the thermal environment on the efficiency of packed columns in supercritical fluid chromatography. J Chromatogr A 2012;1266:149.
[41] Gritti F, Guiochon G. On the minimization of the band-broadening contributions of a modern, very high pressure liquid chromatograph. J Chromatogr A 2011;1218(29): 4632–48.
[42] Sternberg JC. Extra column contributions to chromatographic band broadening. Adv Chromatogr 1966;2:205–70. New York, NY, United States.
[43] Grand-Guillaume Perrenoud A, Hamman C, Goel M, Veuthey J-L, Guillarme D, Fekete S. Maximizing kinetic performance in supercritical fluid chromatography using state-of-the-art instruments. J Chromatogr, A 2013;1314:288–97.
[44] De Pauw R, Shoykhet K, Desmet G, Broeckhoven K. Understanding and diminishing the extra-column band broadening effects in supercritical fluid chromatography. J Chromatogr A 2015;1403:132–7.
[45] Berger TA. Instrument modifications that produced reduced plate heights <2 with sub-2 μm particles and 95% of theoretical efficiency at k = 2 in supercritical fluid chromatography. J Chromatogr A 2016;1444:129–44.
[46] Grand-Guillaume Perrenoud A, Veuthey J-L, Guillarme D. The use of columns packed with sub-2 μm particles in supercritical fluid chromatography. Trends Anal Chem 2014;63:44–54.
[47] Enmark M, Åsberg D, Shalliker A, Samuelsson J, Fornstedt T. A closer study of peak distortions in supercritical fluid chromatography as generated by the injection. J Chromatogr A 2015;1400:131–9.
[48] Shalliker RA, Samuelsson J, Fornstedt T. Sample introduction for high performance separations. Trends Anal Chem 2016;81:34–41.
[49] Grand-Guillaume Perrenoud A, Boccard J, Veuthey J-L, Guillarme D. Analysis of basic compounds by supercritical fluid chromatography: attempts to improve peak shape and maintain mass spectrometry compatibility. J Chromatogr A 2012;1262: 205–13.
[50] Grand-Guillaume Perrenoud A, Farrell WP, Aurigemma CM, Aurigemma NC, Fekete S, Guillarme D. Evaluation of stationary phases packed with superficially porous particles for the analysis of pharmaceutical compounds using supercritical fluid chromatography. J Chromatogr A 2014;1360:275–87.
[51] Fekete S, Veuthey J-L, Guillarme D. Comparison of the most recent chromatographic approaches applied for fast and high resolution separations: theory and practice. J Chromatogr A 2015;1408:1–14.
[52] Grand-Guillaume Perrenoud A, Veuthey J-L, Guillarme D. Comparison of ultra-high performance supercritical fluid chromatography and ultra-high performance liquid chromatography for the analysis of pharmaceutical compounds. J Chromatogr A 2012;1266:158–67.

CHAPTER

3

Practical Approaches to Column Selection for Supercritical Fluid Chromatography

W.P. Farrell

Pfizer Inc., San Diego, CA, United States

OUTLINE

Supercritical Fluid Chromatography.
DOI: http://dx.doi.org/10.1016/B978-0-12-809207-1.00003-3

3.1 INTRODUCTION

Supercritical fluid chromatography (SFC) began competing with high-performance liquid chromatography (HPLC) as a mainstream analytical technique nearly 40 years ago, when it emerged as the technique of choice over gas chromatography for thermally labile and nonpolar compounds with higher molecular weights [1]. Over the next 30 years, SFC has been the focus of numerous research studies to better understand not only the fundamental and theoretical aspects of the technique but to highlight its advantages relative to HPLC for the separation of a diverse range of pharmaceutically relevant compounds. SFC uptake as a chromatography impact tool in academic and industrial laboratories has been a slow but gradual process facilitated by incremental technological advancements such as instrument enhancements and an influx of different column chemistries. Momentum for SFC began to change when it started gaining dominance as the preferred technique for chiral separations, replacing normal phase chromatographic methods and toxic hydrocarbons (e.g., hexanes, heptanes, etc.) with a greener solvent system consisting of CO_2 and an organic alcohol cosolvent (e.g., methanol, ethanol, etc.). In addition, SFC gained wide acceptance in the pharmaceutical industry as complementary to HPLC in numerous cases for high-throughput analyses and purification applications.

Although the physical components of an SFC system closely resemble that of HPLC, the parameters responsible for achieving separations are somewhat different. In HPLC, the main driver of the separation is the partitioning of the solute in and out of the stationary and mobile phases. Other parameters such as solubility in the mobile phase, the stationary phase, temperature, and mass transfer all play specific roles in obtaining a separation. In SFC, however, these same parameters are reported to have contrary and unexpected effects on the separations. The main influencers of retention and separation in SFC include the density variations (caused by pressure or temperature changes), the modifier composition, and the stationary phase chemistry. To expand the utility of SFC to more polar molecules, an understanding of the advantages and limitations of these parameters are crucial to developing a successful separation. The addition of acidic or basic additives in the cosolvent, which attach to active sites on the stationary phase, serves to increase the polarity of the stationary phase and results in enhanced peak shape for polar solutes. While the use of additives did expand the polarity range of molecules for SFC, lower separation efficiencies relative to HPLC did not help instill general confidence in the technique. To overcome these deficiencies and to assist with fine-tuning the separation, the development of different stationary phase chemistries, particularly those with more polar moieties, provided another dimension to the SFC method development process.

Ostensibly speaking, modern SFC is only practiced using carbon dioxide as a solvent, either in the liquid (subcritical) or fluid (supercritical) state. The terms of the defined state of the actual fluid vary depending on the usage of the fluid for separation purposes. The terms subcritical and supercritical are irrelevant as there is no substantive point at which these two regions occur (or coexist) [2]. For the remainder of the text and the sake of simplicity, we will simply refer to the solvent system as SFC. To better understand the effects of the mobile phase on the separation and column phase, one needs at least a basic understanding of the properties of the fluids. Several extensive reviews have been published which cover the vast number of experiments to define the fluid properties and their impact on the separation, efficiency of the peaks, and overall runtime [3–5].

In practical use today, carbon dioxide (CO_2) is the primary solvent used for SFC as it is readily available with sufficient purity for chromatographic separations, is relatively nontoxic and inert and has modest critical temperature and pressure ($T_c = 31°C$; $P_c = 73bar$). CO_2 is also miscible with a broad range of solvents and has both a low UV cutoff and lower viscosity than other chromatographic liquids. These attributes translate to an achievable working range within the constraints of modern chromatographic instrumentation. Typical operating conditions to maintain a stable mobile phase vary from 100 to 200 bar (outlet pressure) and 0–80°C column temperature. This working range is large enough to vary the density of CO_2 from ~0.2 to 1.1 g/mL which can lead to drastic changes in retention, efficiency, and selectivity depending on the chosen conditions. For instance, at low temperature (<20°C) and high pressure >200 bar, CO_2 acts very much like a liquid. Conversely, at high temperature (>80°C) and low pressure (<100 bar), CO_2 is more gas-like. In use as a pure fluid system, e.g., no modifier added, the variation of density is the primary driving force of retention. Once a modifier is added, the density characteristic of the CO_2 changessuch that it becomes a secondary retention mechanism depending on the specific cosolvent used [6]. The reason for this is the relatively poor solubility of all but nonpolar compounds in CO_2. For a more extensive review of the physical properties of supercritical CO_2, see Reference [2].

While SFC has since become an established technique in many laboratories, additional advancements are still needed before it could be considered a replacement for applications long dominated by reversed-phase HPLC. As with the technique itself, column technology designed not only to improve and enhance existing SFC applications but also to broaden the SFC application space have developed slowly. The drive to design and implement column chemistries includes specific goals such as to demonstrate increased or comparable efficiencies to reversed-phase HPLC, as well as to enable SFC use for water-soluble and ionic compounds. The pressure limitations of early SFC instrumentation were a hurdle which restricted adoption of sub-2-μm particles thus leading researchers to reevaluate the potential use of smaller particles. Simultaneously, improvements were made to increase SFC system functionality by enabling higher flow rates and higher pressure applications, thus increasing the suitability of smaller particle columns for use in SFC. Over the last few years, a rapid growth was seen in the number of existing and novel column chemistries for achiral SFC. Different selectivity was provided by superficially porous, sub-3-μm and even sub-2-μm particles, hydroxylated and aromatic phases to name a few. In this chapter, we will focus on these and other column technological advancements as well as some practical considerations for obtaining a successful separation. In depth discussions relating to preparative, chiral, or capillary SFC separations are beyond the scope for this chapter but can be found in several recent reviews [4,7–9].

3.2 KEY PARAMETERS GOVERNING SFC SEPARATIONS

3.2.1 Properties of the Analytes

The key to success using SFC is matching the properties of the compounds to that of the technique and the concept of "like dissolves like" has long been a useful guideline in the application of SFC to various separations. A good understanding of the physicochemical properties of the compound(s) of interest should provide some direction for choosing an appropriate column. Compounds that have at least some modest solubility in organic solvents are likely to result in a successful separation, provided that the right conditions can be found. Typically, these compounds contain some degree of hetero-aromaticity, but that is by no means a mandate for achieving a separation. While moderately polar to hydrophobic compounds are the most amenable to the technique, SFC can also be applied to hydrophilic compounds, e.g., nucleotides, peptides, phosphonates, highly viscous polymers, etc. Despite identifying optimal conditions to provide a separation for these compounds, the nontraditional approaches for separating these compounds make SFC use impractical, relative to other techniques such as HPLC, Capillary Electrophoresis (CE), Ion Chromatography (IC), etc. Hydrophobic and hydrophilic compounds tend to behave differently depending on the nature of the SFC conditions or parameters employed. Water-soluble compounds usually require an aqueous component to the modifier which, depending on the concentration used, may be unsuitable for SFC using CO_2 [10]. However, this aspect could change with future developments in the instrumentation that expand the working pressure range. The distribution analyses of a sampling of approximately 80000 drug-like achiral compounds successfully purified at the 80–120 μmole scale by SFC to >85% purity are presented in Fig. 3.1. From this data, which is similar to those found in other studies [11,12], several key properties of molecules are shown to be amenable to SFC: (i) weak to moderate polarity (cLogP > 1, Topological Polar Surface Area (TPSA) < 180 Å), (ii) MW below 500 Da, and (iii) moderate acidity and basicity.

However, these properties only define suitability with SFC as a technique and may not be useful as a general guide for a specific separations. Factors such as the complexity of the sample and matrix, the structural similarities of the analytes, isobaric isomers (regio- or geometric), and potential intramolecular hydrogen bonds must also be considered. In addition, the most important consideration is most likely the solubility of the analytes in the chosen solvent system [1,13]. As the composition of mobile phase plays such a crucial role in dictating the separation, it is important to provide a basic understanding of its impact before discussing the properties of the stationary phase. Many of these factors and their impacts on SFC separations will be discussed in subsequent sections.

3.2.2 Working with Compressible Fluids

3.2.2.1 Density

Modern SFC systems can operate to inlet pressures up to a maximum of 400 bar before system safeguards are activated. The current trends in instrument development with respect to the pumps, flow cells, valves, etc. will likely drive improvements to enable higher pressure operation over 1000 bar. Pressure changes across the column present some interesting dynamics that can adversely affect peak shapes, especially for more retentive compounds. Regardless of the modifier composition, the properties of the mobile phase are highly influenced by CO_2 density, and any perturbation of pressure or temperature (ΔT) will result in a substantive change in the mobile phase diffusivity and flow rate. The adiabatic processes at work in SFC are contradictory. Adiabatic heating from the compression of the gas requires cooling of the resulting fluid to ensure stable delivery to the column while the adiabatic expansion of the fluid as it decompresses requires heating to prevent dry ice

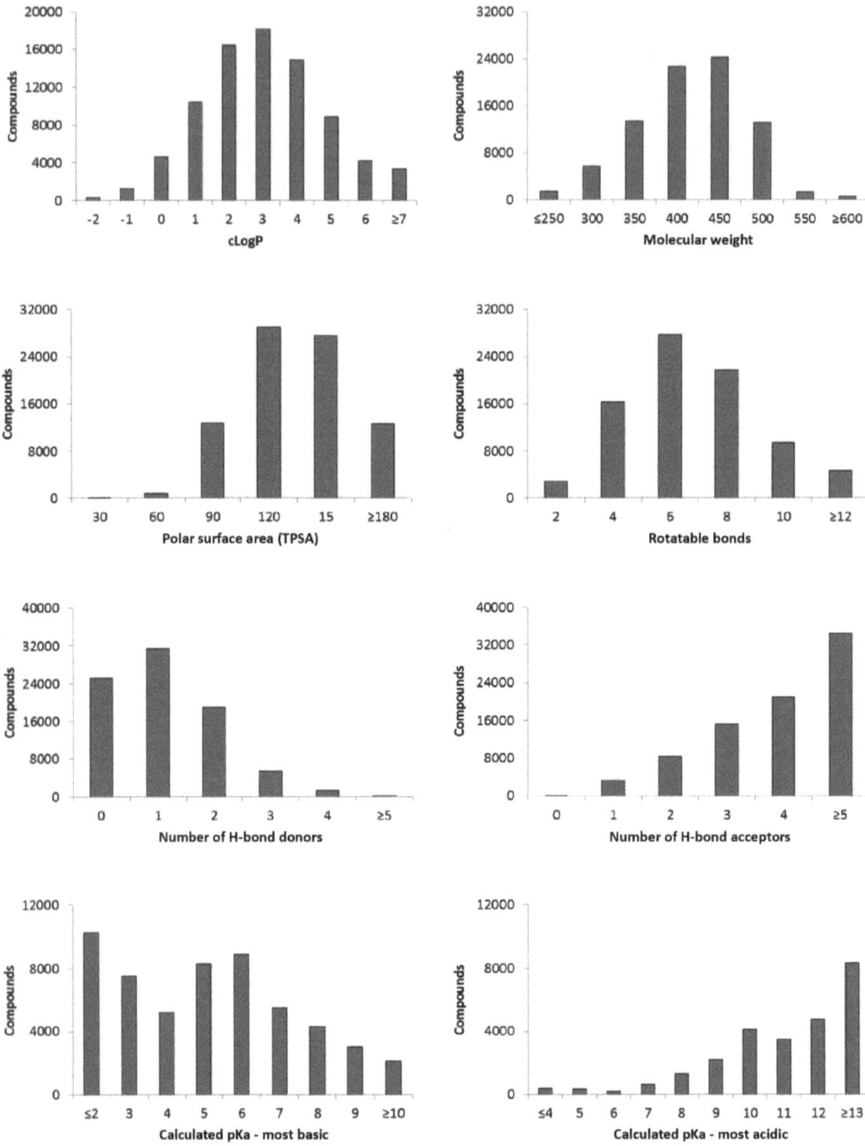

FIGURE 3.1 Physicochemical properties distributions (molecular weight, cLog P, H-bond acceptors, H-bond donors, polar surface area, and rotatable bonds) of approx. A total of 80,000 drug-like compounds analyzed by SFC and meeting post-purification purity requirements. Calculations for cLog P, polar surface area, and p*K*a were determined using ACD/Labs software, version 12.5.

formation. The range of compressibility, which is the measure of the change in relative volume (ΔV) of a fluid in response to a pressure change (ΔP), is higher when temperature and pressure are close to the critical value (31°C, 74 bar) of supercritical carbon dioxide. This density can fluctuate inside the column depending on the ΔP from the column inlet to the outlet, leading to variations in elution strength and fluid viscosity of the mobile phase, axial and radial temperature heterogeneity within the column, and changes to the diffusion coefficient (D_m) of the analyte [14–18]. Flow rate, column dimensions (length and internal diameter), particle size, and porosity (fully porous or superficially porous) can all affect the fluid density resulting in differences in analyte retention, column selectivity, and peak efficiency [19]. Some suggest that maintaining the density of the fluid along the column above 0.75 g/mL will minimize these impacts [2,20].

It has been reported that the axial ΔT across the column will decrease as a function of pressure and these effects will be exacerbated as ΔP increases [21]. For the SFC practitioner, this pressure drop must be considered because the consequence of operating in the isocratic mode is that late eluting solutes tend to broaden, leading to poorly defined peak shapes and a significant loss of efficiency. To counter this effect, it may be necessary to introduce a mobile phase gradient or change the column temperature so that a desired peak shape can be obtained. To illustrate this point, Fig. 3.2 shows stacked chromatograms of fenoprofen and flurbiprofen using SFC conditions in both isocratic and gradient conditions at various flow rates. Although the peak efficiency of fenoprofen and flurbiprofen in both isocratic and gradient modes generally decrease with an increase in flow rate, the extent of the decline in the gradient mode affords a greater peak capacity (peaks/unit time) due to a reduction in peak width (see Table 3.1 for details).

Density is influenced by a combination of flow rate changes, modifier viscosity, column dimensions, and particle size which increase ΔP. However, caution should be exercised in not overinterpreting the pressure drop reading displayed on the pump system readout, as this does not account for the myriad of small I.D. tubing plumbed throughout the system, which has been shown to contribute upwards of 50–100 bar [22].

3.2.2.2 Flow Rate

The flow rate contribution to density should be considered bimodal because the delivery rate of the pump and mass flow through the column are easily determined (the fluid that enters the column at a specific rate, must exit at the same rate); however, the flow rate through the column will depend on the change in density along the column, the viscosity of the mobile phase, the column dimensions and particle size. The

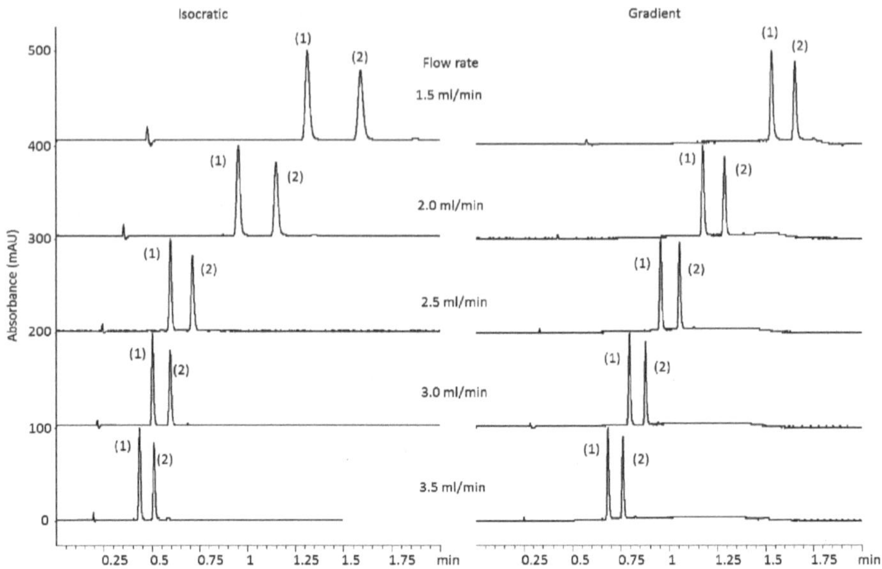

FIGURE 3.2 Consequences of changes in flow rate on peak shape and elution in isocratic and gradient elution modes. Injections of fenoprofen (1) and flurbiprofen (2) in Isocratic mode (left) with 10% modifier; and Gradient mode (right) with 5–25% modifier gradient in 2.0 min. SFC conditions: Sun Shell Diol, 3.0×150 mm^2, 2.6-μm column maintained at 25°C with 20-mM ammonium formate in methanol in CO_2 at a various flow rates from 1.5 to 3.5 mL/min and 160-bar outlet pressure.

actual volumetric flow will be affected by the magnitude of the resulting ΔP across the column as well as the system operating temperature, neither of which necessarily occurs in a linear fashion. This means that the adsorption equilibrium constants, retention factors, and the diffusivity of solutes will vary at different points along the column. For instance, an decrease in the concentration of organic modifier while at constant back pressure (130 bar) and column temperature (40°C) was shown to increase the difference between the set point and average volumetric flow rates because the density of the mobile phase was affected by the addition of a less compressible liquid [23]. For instance, a mobile phase consisting of approximately 15% methanol in CO_2 at a low flow rate was shown to deviate from a consistent solute retention on the order of ~6%, which then jumps to over 10% at a higher flow rate. This occurs because the mass flow of the CO_2 is greatly affected as the pressure of the column increases resulting from say a change in flow rate or column length, while delivery of methanol through the column is relatively static from the pump [24]. The result is a variable change in the local mole fraction of methanol along the column depending on the conditions, which can be further exacerbated when operating near

TABLE 3.1 Peak details for flurbiprofen in isocratic and gradient elution modes, see Fig. 3.1 for conditions

		Gradient mode		
Flow (mL/min)	**k_2**	**pw**	**N′**	**α**
1.5	1.89	0.015	67186	1.13
2	1.98	0.014	46997	1.16
2.5	2.1	0.011	47622	1.15
3	2.14	0.010	39388	1.16
3.5	2.05	0.010	32388	1.18
		Isocratic mode		
Flow (mL/min)	**k_2**	**pw**	**N**	**α**
1.5	2.31	0.028	17278	1.33
2	2.28	0.022	15400	1.33
2.5	2.23	0.018	13506	1.33
3	1.97	0.015	13053	1.32
3.5	1.84	0.013	12112	1.31
4	1.69	0.011	11778	1.31

k_2 = retention factor of peak 2
pw = peak width @ 50% height (s)
N = Plates/column[a]
N′ = apparent plates/column[a]
$\alpha = k_2/k_1$
[a]*As measured using Agilent CS software.*

the critical point of CO_2. To achieve optimal flow rates or increase throughput on instruments that can only reach a maximum of 5 mL/minute requires the use of a smaller diameter column to achieve a higher linear velocity through the column [25].

As modern SFC instruments have a broad range of temperature (~0–100°C) and pressure control, there should be little advantage operating near the critical point of CO_2. Moving further from the critical point should lead to a more robust and reproducible separation. Avoiding operational temperatures above 80°C without a subsequent rise in outlet pressure (e.g., >175 bar) will minimize the potential effects that lead to loss of peak efficiency.

3.2.2.3 *Influence of Column Dimensions and Phase Support*

Recent developments in HPLC column technology have gradually spilled over into the area of SFC. This has generally been the case as manufacturers and academia work toward increasing peak efficiency to separate ever more complicated samples. In SFC, the dimensions of a column have a complex effect on separations, especially when working with pure CO_2. As supercritical fluid density is a contributing factor to retention and efficiency especially when using highly compressible mobile phases (pure CO_2 or $<5\%$ modifier), care must be observed when selecting an appropriate column dimension. Several factors appear to either enhance or diminish the quality of a separation including the induced pressure difference across the column, the operating temperature of the column oven, and the overall flow rate. Therefore, simply decreasing or increasing the column internal diameter or length can adversely affect a given separation. Similarly, reducing the particle size within a column of similar dimensions doesn't necessarily lead to an improved separation if the system conditions drastically change the density of the CO_2.

As in HPLC, the standard column dimension format in SFC has migrated toward smaller internal diameters and shorter lengths. The first uses of SFC actually occurred using capillary columns, either as open tubular or packed particles and found applications for compounds primarily soluble in pure CO_2 [26,27]. Research showed that fine tuning of the CO_2 solvent through variation of density lead to an enhancement (or degradation) of the separation and peak efficiency, as these columns were susceptible to external temperature fluctuations [28]. The results suggest that working with pure CO_2, the most efficient and reproducible results will be obtained at higher density and with absolute control of pressure and temperature. However, the scope of analytes that were compatible with these types of columns limited their adoption.

Monolithic columns, another popular column format used in HPLC, have also not found applications in SFC. The monoliths are prepared as a single rod of silica or from porous polymers encapsulated in a PEEK tube. These columns have a bimodal pore size distribution consisting of macropores that allow for high mobile phase velocity with low pressure drops and smaller mesopores responsible for retention [29–31]. As the column pressure drop is substantially lower, monolithic columns do not suffer the same detrimental effects from density changes in SFC. However, due to the hold-up time, they must be run at significantly lower flow rates or with longer column lengths (e.g., column coupling) [32]. Although these porous phases tend to favor better mass transfer properties to spherical particles, poor eddy diffusion characteristics directly related to the rod heterogeneity as well as poor distribution at the column inlet leads to plate heights that are similar to particle-packed columns containing large particles [29]. Subsequently, the enthusiasm for these highly permeable columns faded upon introduction of sub-2-μm fully porous particles (FPP), which generated separations with a higher resolution per unit time.

Similar to HPLC, the reduction in particle size from the traditional 5 μm (or 3 μm) toward a sub-2-μm particles results in an increase in total plate counts for a similarly sized column [33]. With the introduction of sub-2-μm particles providing a much greater surface area per meter, better efficiency is favored with shorter columns and internal diameters that allow for the dissipation of frictional heating [34–36]. As demonstrated in Fig. 3.3, the separation of a 17-component mixture on a 2-ethylpyridine column containing 1.7-μm FPP can be achieved in less than 2 minutes. In SFC, the migration has been paradoxical in many respects, with improvement in efficiency within the column, but adversely countered by the more pronounced extra column volume contributions of the SFC system [25]. Although the need to reduce the inner diameter of the system components and tubing improves the peak shapes, they also reduce the operating pressure range of an already limited system (recall that at least 73 bar of back pressure is needed to utilize the system as mentioned in Section 3.1). Therefore, unlike in HPLC, there appears to be no benefit in SFC to using 2.1 mm ID columns over 3-mm ID columns to avoid thermal gradients [22].

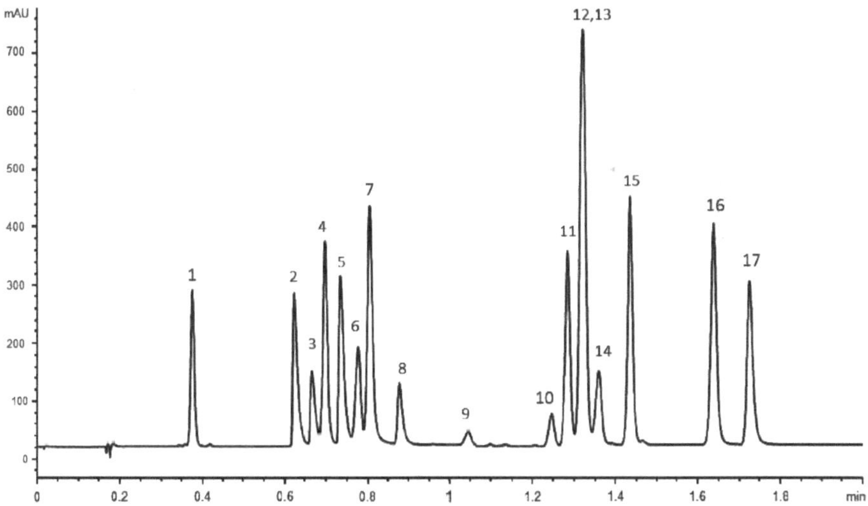

FIGURE 3.3 Chromatogram of the separation for the 17-component mixture using the Waters Acquity Viridis 2-EP, 3.0×100 mm^2, 1.7-μm column. A methanol gradient of <2 min was used on the Agilent 1260 Infinity SFC system. SFC conditions: Flow rate 3.5 mL/min; outlet pressure 160 bar; and column temperature 70°C. Gradient program: 5.0–12.5% in 1.0 min, 12.5% for 0.25 min, then 12.5–20% in 0.75 min. Analytes: caffeine (1), flurbiprofen (2), naproxen (3), thymine (4), ketoprofen (5), etophylline (6), uracil (7), warfarin (8), cortisone (9), hydrocortisone (10), acetamidophenol (11), sulfamethoxazole (12), sulfamethazine (13), prednisolone (14), sulfadimethoxine (15), sulfaquinoxaline (16), and sulfamethizole (17).

As previously mentioned, the mobile phase undergoes changes in kinetic properties corresponding to a change in density; therefore, any contribution that alters the density will be greater in SFC than with liquids. Although the mobile phase may be less viscous, a reduction in column diameter or an increase in length sets up the potential to increase the pressure drop across the column, therefore increasing the radial temperature gradients leading to a reduction in efficiency [15,20,37,38]. That said, the lower overall viscosity of the mobile phase enables the use of longer columns, even the coupling of columns [39]. Coupled columns can provide an opportunity to enhance selectivity more than simply using a single column or phase. For instance, while using six 10-cm core–shell columns, Lesellier was able to achieve upwards of 120,000 plates for a homologous series of alkylbenzenes [40]. This can lead to a simplification of the number of different phase types needed to provide adequate separation. It also enables the possibility of using several phase types in series to achieve unique separations. For instance, Delahaye showed that the concepts of stationary-phase optimized selectivity liquid chromatography (SOS-LC) could be reasonably translated to SFC provided some attention to the pressure and temperature (i.e., density) was observed. The relative order of the columns was found to be important, but with the use of a cartridge-based system, selectivity can be successfully manipulated. Although this approach was only applied to a known set of compounds, with a limited set of phases and required a labor intensive process for general use, the potential for mixed mode separations is promising. The major requirement to refine the model of measuring several iterations of retention factors for the individual sample components on each individual stationary phase, make this approach more useful for separating well characterized samples with known targets/impurities [41].

The next innovation in column technology stemmed from a generation of solid core shell or superficially porous particles (SPP), originally intended for use in HPLC [42]. These particles have a solid, nonporous silica particle center (core), formed with an outer layer (shell) of porous silica available for use as a stand-alone stationary phase or for further bonding with an appropriate silane to create a suitable stationary phase. The size of the core and thickness of the shell can be controlled depending on the process used in formation of the particle and demonstrated superior loading and lower pressure constraints when compared to fully porous sub-2-μm particles. It was shown that 2.7-μm SPP exhibited similar kinetic characteristics to the smaller totally porous particles without the need for special instrumentation to operate at pressures above 400 bar [42,43]. As SFC requires at least 100 bar back pressure to operate, limiting the usable pressure range of the instruments, any technology that reduces pressure drop is of interest and these core–shell particles were eventually applied to SFC separations. Berger first reported that 2.6-μm Kinetex particles significantly outperformed porous Luna 3-μm silica particles, observing that uncorrected plate heights as low as 1.62 μm could be obtained. McCalley reported reduced plate heights near 1.5 with a hydrophilic interaction liquid chromatography (HILIC) core–shell column using a mobile phase consisting of 90% acetonitrile and 10% aqueous formic acid mixtures [44]. Fig. 3.4 demonstrates the same sample mixture shown in Fig. 3.3 separated on a core–shell 2-ethylpyridine column. Particularly interesting to note are the differences in column length, flow rate, and column temperature to achieve comparable separations. These changes were necessary for the totally porous particle phase due to the maximum pressure limitation of the system (<600 bar). As the lower pressure drop of these SPP columns translates to lower resistive heating, the requirement for smaller inner diameter columns used with other sub-2 μm totally porous particles could also be relaxed.

Therefore, the selection of the column dimensions and particle size, or even the type of particle, can affect the separation regardless of the stationary phase chemistry applied.

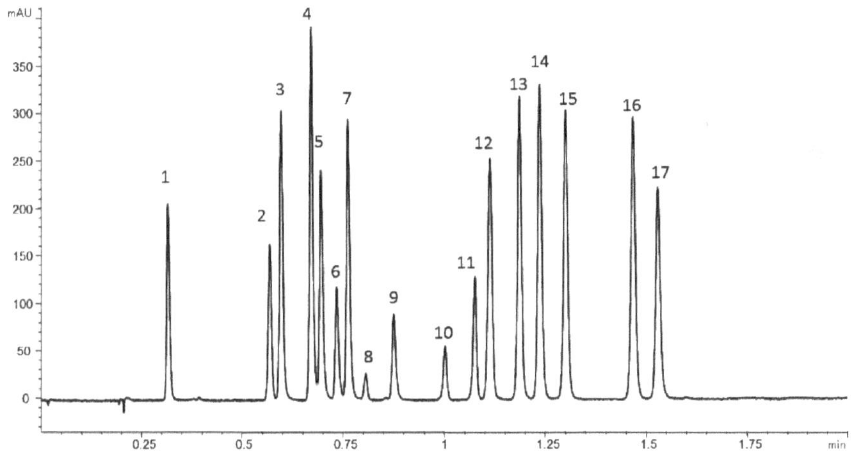

FIGURE 3.4 Chromatogram of the separation for a 17-component mix of drug-like compounds using the core–shell Sun Shell 2-ethylpyridine, $3.0 \times 150\ mm^2$, 2.6-μm column. A methanol gradient of <2 min was used on the Agilent 1260 Infinity SFC system. SFC conditions: flow rate: 4.0 mL/min; outlet pressure 160 bar; column temperature 55°C. Gradient program: 5.0–7.5% in 0.20 min, then 7.5–20% in 1.3 min and held at 20% for 0.2 min. Analytes: caffeine (1), etophylline (2), thymine (3), uracil (4), flurbiprofen (5), naproxen (6), ketoprofen (7), cortisone (8), warfarin (9), hydrocortisone (10), prednisolone (11), acetamidophenol (12), sulfamethazine (13), sulfamethoxazole (14), sulfadimethoxine (15), sulfaquinoxaline (16), and sulfamethizole (17).

3.2.2.4 *Solvent Considerations*

SFC is considered to be faster than HPLC because of the high diffusivity and low viscosity of the CO_2-containing mobile phase. One of the distinctions of reversed-phase HPLC (RP-HPLC) is the role the organic solvent plays in modifying the separation. In RP-HPLC, both the phase and silica support, while playing an important role, are secondary to the solvent effects and especially the pH have on the distribution mechanism. In SFC, solvents (called modifiers) and additives have a substantial effect on the chromatography through: (1) modification to the overall polarity and solvating capability of the mobile phase; (2) elevation of the density of the mobile phase; (3) diminution of the contribution of active sites on the stationary phase; (4) modification of the stationary phase through adsorption; (5) expansion of the net volume of the stationary phase, thus altering the phase ratio; and (6) selectively solvating solutes in the mobile phase by forming clusters with different distribution properties [4,45,46]. Changes to the polarity and solvating capability of the mobile phase are likely the most important of these factors [47].

Carbon dioxide alone has low solvating power and is too weak to elute polar compounds. Therefore, it is commonly paired with a protic solvent such as methanol, ethanol, or isopropanol. The addition of the modifier serves to increase the eluotropic strength which, with gradient programing, promotes the elution of more polar analytes. One of the most interesting findings over the last few years has been the combination of different cosolvents and its impact on separations and their role in enhancing sample detectability. For pharmaceutical applications, experience has shown that small amounts of methanol (<5%) in acetonitrile is also a good solvent combination, as it has just enough polarity to retain hydrophobic solutes while maintaining sufficient solvent for accurate delivery of modifier. Using this combination of methanol/acetonitrile is useful for enhancing the detectability of solutes by MS ionization. Small amounts of water in an alcohol (e.g., <2–5% (v/v) in methanol) also provide enhancements to analyte solubility, which will be discussed in Section 3.2.2.7. The addition of a polar modifier will generally lower the retention of most analytes, except in rare cases where the analytes are very hydrophobic and retention actually increases with the addition of polar modifiers [48]. Carbon dioxide is miscible with many organic solvents except for rather polar solvents such as water. The diversity of solvents that range from protic to aprotic allow for exquisite interactions between the stationary phase and the solvent system. In addition, analyte solubility can be improved by judicious choice of the solvents used. However, there are several solvents not recommended as modifiers: dimethyl sulfoxide, dimethylformamide, dimethylacetamide, and *N*-methylpyrrolidone because they result in significant UV and Ms detector interferences. Solvents such as dichloromethane, tetrahydrofuran, acetone, ethyl acetate, or hydrocarbons (e.g., hexanes, heptanes) can be used; however, due to environmental, safety, and reactivity concerns, these are usually avoided. Solvents that have been shown to work well are alcohols (methanol, isopropanol, ethanol, butanol, etc.) and acetonitrile [48].

3.2.2.5 Solvent Adsorption onto the Phase

The addition of as little as 0.1% (v/v) of a cosolvent such as methanol to the CO_2 changes the surface chemistry of the stationary phase through formation of a layer of entrained methanol around the phase particles. It has been demonstrated that methanol forms monolayers with a thickness of $\sim$4 Å on a Diol column at methanol mole fractions of less than 0.28, at which point the retention factor (log k) vs methanol fraction behave in a nonlinear fashion sensitive to changes in composition. Above a 0.28 mole fraction, the eluent behaves in a more or less linear fashion [49]. Depending on the specific composition of the mobile phase, very different retention profiles for compounds are observed. For methanol mole fractions above 0.50 mole fraction, the adsorption isotherm potentially indicates the formation of multilayer behavior. This phenomena can also be phase-dependent since comparable methanol adsorption exhibits multilayer formation on silica columns with a monolayer thickness of 10.2 Å [45]. Deviation of the actual flow rate and fraction of cosolvent as determined by the instrument set points (e.g., the values set in the control software) are larger at low cosolvent percentages. As such, drawing conclusions concerning solute equilibrium between stationary and mobile phases compositions in the 20–30% (v/v) range may not be valid at low fractions of modifier (c. $<$ 5%). Combined with elution behavior associated with layer formation, a density gradient exists along the column which also affects the equilibrium constants of analytes between stationary and mobile phases depending on the local pressure [23]. Therefore, any attempt to compare the attributes of a specific stationary phase must be met with a full understanding of the conditions used to make that determination.

In addition to adsorption variability, solvent interactions with the phase can result from modification of the residual silanols present in some stationary phases when using silica as a support. Fairchild demonstrated that a time-dependent equilibrium exists between free silanols and silyl esters formed from alcohols that change the particle surface chemistry of two stationary phases which produced significant variations in retention and selectivity [50]. It was further reported that regeneration of these columns to the free silanol form can be achieved by flushing the column with water to hydrolyze the esters, followed by a recommendation of storing column in pure CO_2, or a water-free, alcohol-free solvent. Since the scope of this study was limited to two columns, it has not been demonstrated for a broad selection of stationary phases. Another study using a selection of 24 phases did observe a significant shift in retention over prolonged use, which may have been due to pervasive ionic interactions with free silanols or potentially the same to silyl ester formation [11]. This hypothesis is to some degree confirmed by the observation that small amounts of water added to the modifier were also shown to stabilize the retention drift and will be discussed in more detail in Section 3.2.2.7.

These points are mentioned to serve as a warning that solvents used in SFC may have a lingering effect that can be easily misleading, especially if moving quickly through a series of solvent changes during method development. In addition, as suggested by Fairchild, brand new columns are stored in a variety of different solvents, which can differ from the one on the system, and require substantial flushing before stable performance is observed [50]. Operating below ~3% (v/v) modifier should be avoided, especially if using gradient elution, as the changes to the phase may result in separation conditions that are difficult to transfer or reproduce. The time needed to completely remove the cosolvent from the phase can be significant and may impact separations using pure CO_2 until the phase has reached full equilibrium [49]. In addition, mobile phase mixing at modifier concentrations lower than 5% tends to be inaccurate since mixing is based upon the volumetric flow rates of two individual pumps [23]. If a separation requires modifier below about 5%, switching to an alternative stationary phase that provides more retention is a plausible solution [23]. Compounds that are too polar or are poorly soluble in the mobile phase may have difficulty eluting in mobile phase compositions below ~50–60% modifier. Therefore, an alternate phase or even possibly a different technique (e.g., reversed-phase HPLC) may be required. Several studies have shown that peptide and phosphorylated compounds can be accomplished using SFC, but the conditions require additives and even small amounts water to effect elution [51–53].

3.2.2.6 *Additives*

Additives are constituents added to the modifier to either enhance or decrease the interactions of achiral and chiral analytes with the stationary phase. The possible roles of additives include (1) enhancement of the mobile phase solvating power, (2) ion suppression of the analyte, (3) ion pairing with the analyte, and (4) modification of the stationary phase [54]. These interactions can affect not only the overall elution, but also the selectivity of the phase to a particular analyte. In SFC, additives are typically organic acids, bases, or salts which are usually incorporated into the modifier at low concentrations (<100 mM) before precipitation becomes an issue. Inorganic acids and bases are poorly soluble in CO_2-based mobile phases and are rarely used in SFC. For basic analytes, which tend to form strong hydrogen bond interactions with the free silanols, basic additives tend to improve peak shape for these compounds and may slightly change the selectivity. Triethylamine (TEA), ammonia (NH_3), isopropylamine (IpAm), ammonium hydroxide (NH_4OH), diethylamine (DEA), and ethanolamine (EtOHAm) are examples of basic additives that are often added to the mobile phase to compete with the solutes for the free silanols [55–57]. However, the ammonium salts of common acids help to reduce the peak width considerably when compared with acidic additives especially for compounds such as nucleobases and amines [58]. Acidic additives (trifluoroacetic acid, formic acid, and acetic acid) can improve peak shape for acidic compounds, but may also change the selectivity [58].

Fig. 3.5 shows a separation of basic analytes on a diol column and demonstrates the dramatic improvement in peak shape using ammonium formate as an additive. Although the utility of additives cannot be questioned, there are several issues that need to be considered upon using them. While one of the advantages of using SFC is the relatively short equilibration time when switching between different sets of modifier conditions, there is a consequence to proceeding too quickly. For

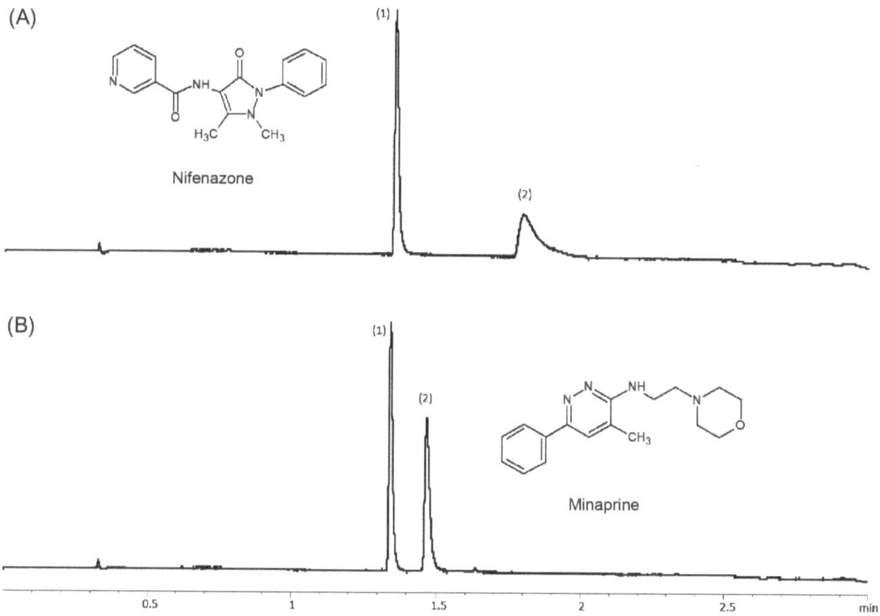

FIGURE 3.5 Positive effect of additive on a diol phase with basic compounds. Injections of nifenazone (1) and minaprine (2) in: (A) methanol, (B) 20-mM ammonium formate in methanol. Conditions: Sun Shell Diol, 3.0×150 mm^2, 2.6-μm column maintained at 25°C with a 5–50% modifier in CO_2 composition; flow rate of 2.5 mL/min and 160-bar outlet pressure.

instance, equilibration time may be longer as sufficient time is required for the additive to permeate the phase and reach a steady concentration in the absorbed layer. In practice, a high percentage (circa 50% composition) of modifier containing the additive represent a facile way to efficiently switch over the solvent system, before allowing the SFC system to equilibrate at the desired composition. This practice enables both solvent and additive to thoroughly saturate the phase and remove residual solvents from the previous conditions faster, by completely disrupting the equilibrium of the system. This is especially important when moving from acidic to basic additives (and vice versa) since the interactions with the phase or base silica can be very strong and exhibit a lingering "memory" effect of the additive. To illustrate this point, Fig. 3.6 demonstrates an inadequate equilibration time after switching solvents despite both the pump pressure and UV baseline appearing to be stabilized after 5 minutes (t_0) of equilibration. It becomes clear after successive injections following this equilibration period that this stabilization was in fact deceptive. The left figure shows that the additive required nearly

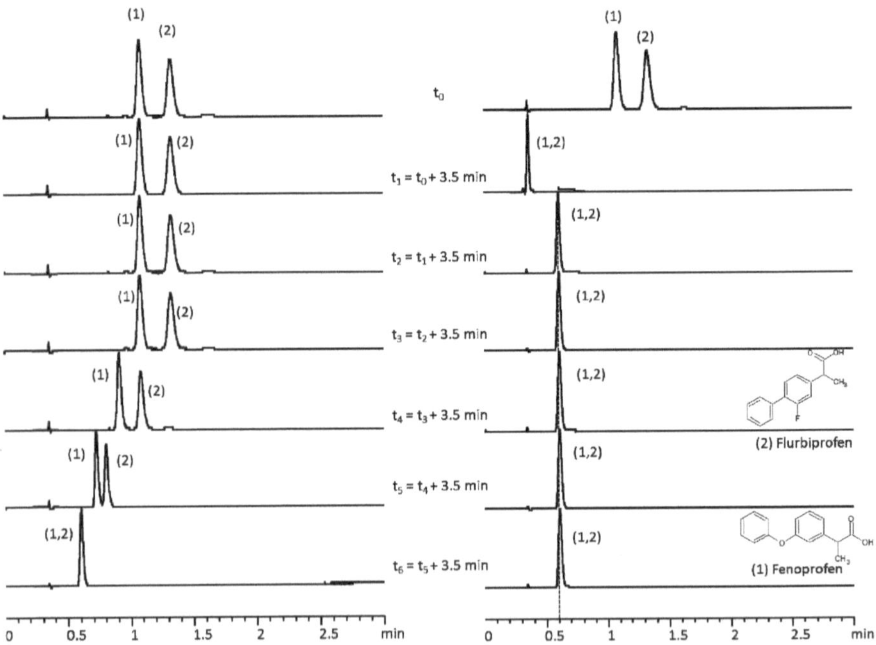

FIGURE 3.6 Illustration of the apparent additive "memory" effect. Injections of fenoprofen (1) and flurbiprofen (2) in (Left) methanol transition to 20-mM ammonium formate in methanol at 3.5-min intervals; (Right) removal of ammonium formate additive with a 3.5-min wash containing 50% methanol (t_1). SFC conditions: Zymor Diol-Monol, 3.0×150 mm^2, 5-μm column maintained at 25°C with a 10% modifier in CO_2; flow rate of 2.5 mL/min and 160-bar outlet pressure.

15 additional minutes to be fully purged from the column, whereas purging an additive from the column using > 50% modifier does, in fact, accelerate the equilibration process (Fig. 3.6—Right).

Another factor one must bear in mind is that for HPLC methods, a 10-mM solution of an additive (i.e., buffer) will constitute a large portion of the mobile phase, especially at the beginning of a gradient separation. This concentration diminishes when the organic solvent is increased though this can depend on whether the solubility of the buffer allows addition into the organic solvent. In contrast, with SFC, the amount of additive/modifier is usually lower at the beginning of a gradient, with CO_2 making up the bulk of the mobile phase. Given that the additive is not added to the CO_2, a concentration gradient will always exist as well. Therefore, the concentration of a 10-mM additive in modifier 5–50% gradient will vary from 0.5 to 2.5 mM. In practice for many applications, stable retention is achieved when using solutions with at least 20-mM additive, when translated to a 5–50% gradient, equates to 1–5 mM. Higher concentrations can also be employed provided the additive remains soluble in the specific mobile phase being used.

3.2.2.7 *Water as an Additive*

While acids or bases are typically chosen as additives in SFC solvent mixtures, the use of water as an additive has been growing in popularity as a solution for addressing separation challenges associated with water-soluble compounds. The CO_2/alcohol/water ternary mixture is considered a type of enhanced fluidity liquid, since the solvent is no longer supercritical [59]. This added dimension introduces a HILIC-like partitioning effect, where the more polar water (relative to the organic cosolvent) is adsorbed deeper on the stationary phase and the analytes pass between the water and the cosolvent layer [52,60]. As a result, water tends to enhance the solvating power of the mobile phase for hydrophilic compounds by increasing solubility of the analytes, thus enabling elution of these compounds while improving peak shapes. For instance, aqueous-based SFC has been used to successfully separate highly polar nucleobases [54,61], peptides [51], and for the separation of flavonoid glycosides [61]. Fig. 3.7 shows example separations of several small polar molecules using ammonium formate, methanol and water as a modifier in SFC. Even though water can be utilized, the amount suitable or stable in CO_2 for SFC use is limited and reportedly less than ~9/1 (v/v) overall [52,59,62].

In summary, the SFC conditions, specifically solvents and additives, can have dramatic impact on a separation, regardless of the stationary phase selected. The best tool is to be as informed and thoughtful not only about the compound properties but the impact that different system parameters may have on these compounds. One of the unintended consequences of smaller particles has been the observed ΔP axially

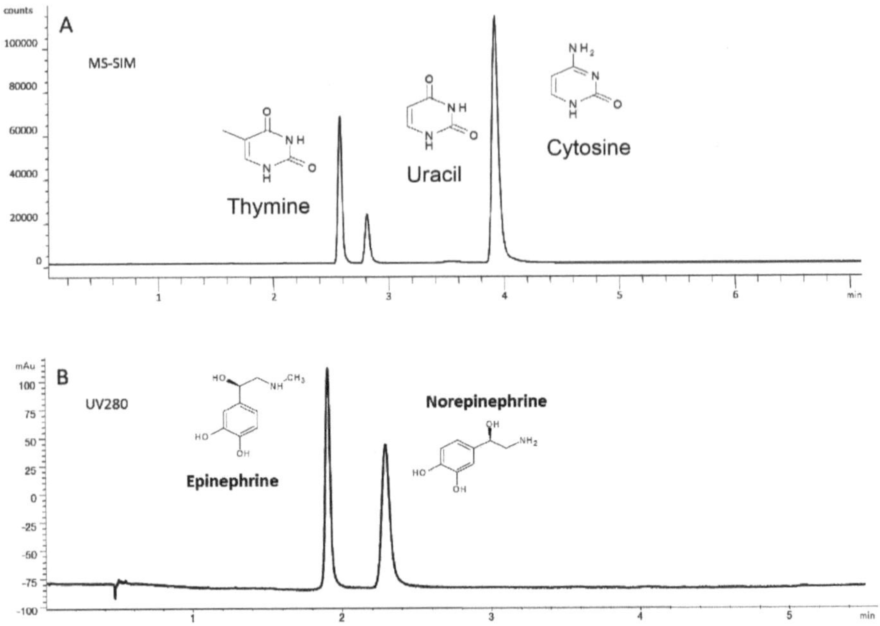

FIGURE 3.7 SFC separation of polar compounds using water as an additive. Conditions: Zymor HA-Morpholine, $4.6 \times 150\ mm^2$, 5-μm column at ambient temperature with 20-mM ammonium formate in methanol and water (95:5 v/v) modifier in CO_2, and 160-bar outlet pressure. (A) Nucleobases (thymine, uracil and cytosine) eluted at 3.5-mL/min flow rate using 5–50% gradient in 5 min with APCI (+) mass spectrometric detection-single ion monitoring (Ms-SIM); (B) Neurotransmitters (epinephrine & norepinephrine) eluted at 3.0-mL/min flow rate using a 20–50% modifier in 2.5 min with an additional hold at 50% for 2.5 min.

across the column, which can dramatically change the dynamics of the fluid, causing early peaks to resolve efficiently but later peaks to diffuse. To counter these effects, gradient separations have been employed successfully, especially when there is a mixture of analytes with competing or distinctive physical properties. The choice of stationary phase, column dimension, and particle size may represent a compromise depending on the task at hand.

3.3 STATIONARY PHASES

3.3.1 Column Classification by Modeling

The linear solvation energy relationship (LSER) parameter model provides a widely used estimation of the contribution of intermolecular interactions and retention of neutral molecules in two-phase separation systems [63–65]. The main hypothesis of this model is that solute transfer between condensed phases in the appropriate form of the solvation parameter model is:

$$\log k = c + eE + sS + aA + bB + vV \qquad (3.1)$$

where k is the retention factor, the lower case letters are system constants, and the capital letters denotes solute descriptors. The solute descriptors represent solute properties such as excess molar refraction—E, dipolarity/polarizability—S, hydrogen-bond acidity—A, hydrogen-bond basicity—B, and volume—V (McGowan's characteristic volume scaled by a factor of 10^{-2}). The system constants are the complementary properties of the separation system described as the polarizability associated with loosely held electron pairs (e), the capacity for dipole-type interactions (s), the hydrogen-bond acidity (a), the hydrogen-bond basicity (b), and the difference in cavity formation and residual dispersion interactions that are not canceled by transfer between phases (v). The equation constant, or c term, contains a number of system and modeling properties, but is usually dominated by the phase ratio for the separation system when the dependent variable is the retention factor. Solute descriptors are available for over 4000 compounds and are usually determined by calculation (V and E) and by experiment (S, A, and B), as explained elsewhere [66,67]. The system constants in Eq. (3.1) represent the parameters used to characterize the contribution of the defined intermolecular interactions to the retention of neutral compounds. They are calculated by multiple linear regression analysis from the experimental retention factors determined for a varied group of compounds with known descriptor values that meet a set of chemical and statistical requirements for modeling [64,67].

To account for ionic interactions with anionic and cationic species in HILIC, Eq. (3.1) has been extended by including two additional terms as follows:

$$\log k = c + eE + sS + aA + bB + vV + d^{-}D^{-} + d^{+}D^{+} \qquad (3.2)$$

where D^{-} represents the negative charge carried by anionic and zwitterionic species, and D^{+} represents the positive charge carried by cationic and zwitterionic species, according to the following equations: [68–70].

$$D^{-} = 10^{(pH^{*}-pK^{*})}/1 + 10^{(pH^{*}-pK^{*})} \qquad (3.3)$$

$$D^{+} = 10^{(pK^{*}-pH^{*})}/1 + 10^{(pK^{*}-pH^{*})} \qquad (3.4)$$

For neutral species, D^{-} and D+ are zero so that Eq. (3.3) reverts to Eq. (3.1). pH* is the effective pH in the mobile phase, also known as

SSpH, according to IUPAC notation. The acidity of carbon dioxide–alcohol mobile phases is unknown. However, there are several indications that it should be somewhat acidic, possibly close to pH 5 [71–73]. LSER approaches have been applied to characterize stationary phases in SFC [67,74]. West et al. [75,76] have constructed models based upon the retention of a standard set of analytes with set of conditions to ascertain differences in the general classification of phases. Although limited in scope by the solutes evaluated, these results present the general trend that alkyl nonpolar, aromatic, polar, and alkyl-polar phases behave differently, even though there are a number of columns that fit into multiple categories. In some cases, these classifications are quite useful for guiding the type of column to pursue for a separation. For instance, aliphatic compounds with low polarity will likely be used with a nonpolar phase, while compounds containing basic groups will likely be used with a polar phase [77]. Unfortunately, given the limited scope of the columns and compounds used for these studies, specific separations cannot be predicted using these simple classifications. In addition, mixtures of polar and nonpolar compounds within the same sample may not be successful on any one particular column. The more structurally similar a compound of interest is to a discrete compound in the training set used, then the suggested column classification and mobile phase combination has a greater likelihood of resulting in a valid predication of retention. However, given the challenges inherent in capturing a broad structural diversity in model compound collections, the vast majority of compounds will likely require a degree of experimentation to determine the optimal separation conditions [74]. Columns that retain the common substrates based upon these models will, to a large degree be similar. Therefore, the classification can group "like" phases together. Although this can provide general guidance of broad implications, it can have limited utility in determining specificity to any particular phase for a given separation challenge. These classifications of phases describe in principle the retention of solutes under a specific set of conditions, with no regard for separation time, peak efficiency, or in many cases, extreme tailing. However, the rudimentary aspects of this system can be useful in defining the type of column needed.

3.3.2 Empirical Column Classification

Although the classification model approach described in Section 3.3.1, Column Classification by Modeling, can be useful for some separations, the limited scope of compounds and conditions evaluated are not as broadly applicable as the diversity of compounds deviate from the studied data sets and do not easily transfer to gradient elution, the primary elution mode used in modern chromatographic techniques. In addition, due to the variety of stationary phases available within each classification of the LSER model, it can be difficult to decide which column to choose for a specific separation.

SFC stationary phases can be categorized into one of three groups based upon the predominate type of interactions that occur to drive separations: hydrocarbon (hydrophobic), amine-containing or hydroxyl-containing (both primarily hydrogen bonding/ionic). The hydrocarbon-type phases interact from a combination of partitioning between the solute in the mobile and stationary phases based upon the hydrophobicity of the analyte relative to the modifier used. But not all interactions of this column classification are the same, with differences primarily arising from the degree of aromaticity (e.g., phenyl) relative to the aliphatic-type (C-18, C-8, cyano, etc.) phases. Similarly, the amine-type (e.g., amino, pyridine, etc.) and hydroxyl-type (e.g., diol) phases introduce stronger hydrogen bonding and ionic mechanisms in addition to their weaker partitioning interactions but also demonstrate specific selectivity differences for a variety of analytes. The type and source of the base silica used in the column also contributes secondary interactions depending on the level of accessible free silanols available to interact with the solute. Combining the attributes related to the density of CO_2, the effects of solvent and additive, and the variety of interactions on the stationary phase, it's no wonder there is so much confusion on which phase to use for SFC.

Any recommendation of which stationary phase to use for a given separation is only as good as the compounds used to evaluate that phase, utilizing a representative set of acidic, neutral, and basic solutes under a variety of SFC conditions appears to still be the most practical approach available at this time. This approach relies on choosing compounds that reflect the region of chemical space useful for the workflow to which they are applied. For instance, use of nonpolar, hydrophobic compounds to evaluate columns for retention and selectivity will be meaningless for polar, hydrophilic compounds because the retention profiles are drastically different. In addition, a consequence of using basic compounds (which are positively charged in the CO_2-based mobile phases) result in strong ionic interactions with the stationary phase in SFC, thus requiring the use of additives to ensure elution and minimize retention [76]. In that context, even the specific compounds used to evaluate columns will likely evolve over time unless working within a limited scope of chemical diversity. The discussion in the remainder of this section will attempt to highlight some fundamental differences between types of stationary phases and solutes to illustrate how to narrow the selection using an empirical approach.

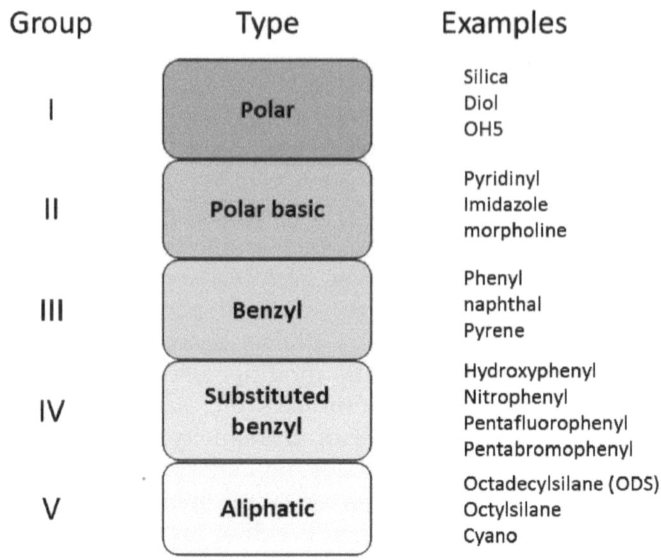

FIGURE 3.8 Column grouping according to their primary functional group.

The simplest approach to differentiating between and characterizing columns is to apply the retention profile of a representative test set of compounds that are stable in solution and can be applied to each new column using a standard set of SFC conditions with the use of additives. The data set should include retention time, peak shape, and elution order of these compounds which provide reasonable indicators of the relative utility of each column.

In an attempt to illustrate this approach, stationary phases from each group will be discussed which are based on the polarity of functional groups of the phase relative to the properties of the analytes using the grouping shown in Fig. 3.8 and summarized in Table 3.2 with a small set of representative compounds.

For simplicity, the most convenient practice of screening compounds by SFC is to run a generic methanol gradient from a low to a high modifier percentage (e.g., from 5 to 50%), though these limits are somewhat arbitrary, but are chosen as an attempt of retaining any hydrophobic analytes, while eluting the polar components. Since the parameters that either enhance or diminish selectivity in SFC can be rather complex, screening each column with several additives or solvents should also be considered as a strategy to optimize a particular separation.

Beginning with hydrophobic compounds, a series of columns from each group under gradient conditions are presented in Fig. 3.9. These specific compounds were chosen due to their structural similarity to minimize any bias and highlight the effects of the modifier or stationary

TABLE 3.2 Phase Selection Guidelines

Family	Examples	Aromatic	Hetero aromatic	Amine	Amide	Acid	Alcohol	Halogen	Ester/ Ether	Suggested Rank	Suggested Solvent	Additive
Aromatic hydrocarbons	Napthalene	Y	—	—	—	—	—	—	—	IV, III, V, II, I	aprotic [a]	n/a
Aliphatic hydrocarbons	Lipids	N	—	—	—	—	Y	—	Y	V, III, IV, II, I	aprotic [a]	n/a
Aromatic with alcohols, amides, esters, ethers	Benzamides	Y	—	—	Y	—	Y	—	Y	II, I, III, IV, V	protic [b]	n/a
Aromatic with amines	Minaprine	Y	—	Y	—	—	—	—	—	II, I, III, IV, V	protic [b]	Y
Aromatic with acids	Ketoprofen	Y	—	—	—	Y	—	—	—	II, I, III, IV, V	protic [c]	Y
Aromatic sulfamides	Sulfanilamide	Y	Y	Y	Y	Y	Y	—	—	II, I, III, IV, V	protic [d]	n/a
Halogenated aromatic	Chloroquinoline	Y	Y	—	—	—	—	Y	—	III, IV, II, V, I	aprotic [a]	n/a
N-heteroaromatic	Nifenzaone	Y	Y	—	—	—	—	—	—	II, I, III, IV, V	protic [b]	Y
N-heteroaromatic with amine	Nucleobases	N	Y	Y	—	—	—	—	—	II, I, III, IV, V	protic [b]	Y
N-heteroaromatic with acid	Nucleic acids	N	Y	Y	—	Y	—	—	—	II, I, III, IV, V	protic [c]	Y
Regio-iosomers I	Diacylglycerols	Y	—	—	Y	—	Y	Y	Y	III, I, IV, II, V	protic [b]	n/a
Regio-iosomers II	Amino acids	Y	—	Y	—	Y	—	—	—	II, III, IV, I, V	protic [b,c]	Y

Comments:
[a] *Or low % protic solvent.*
[b] *Buffer or basic additive; low % water.*
[c] *Buffer or acidic additive; low % water.*
[d] *High % protic solvent.*

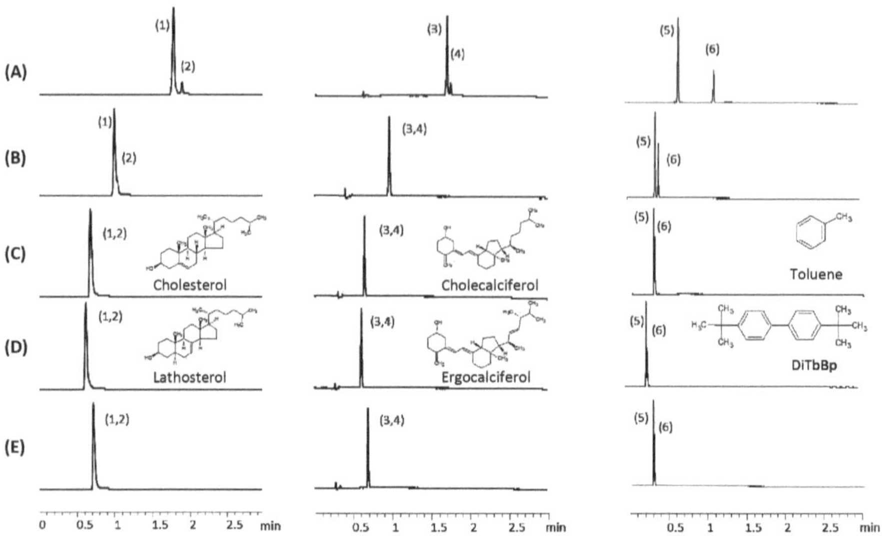

FIGURE 3.9 Example of hydrophobic contributions to separations. Left: Cholesterol (1), lathosterol (2), Center: calciferol (3) ergocalciferol (4), and Right: Toluene (5), 4,4′-Di-tert-butylbiphenyl (DiTbBp) (6) on columns ranging in hydrophobicity. Columns: (A) Kinetex Biphenyl, (B) Cosmosil 3-hydroxyphenyl, (C) Sun Shell Diol, (D) PCI CS-HADP, (E) Sun Shell 2-ethylpyridine. Conditions: $3.0 \times 150\ mm^2$, 2.6–2.7-μm columns maintained at 25°C with a 5–50% methanol in CO_2 composition gradient from 0 to 2.0 min; flow rate of 2.5 mL/min; 160-bar outlet pressure; APCI (+) Ms detection monitored at $m/z = 369$ [$M - H_2O$] and UV detection at 260 nm.

phase. For several of the more polar columns (Fig. 3.9, C–E), the compounds are not retained sufficiently using methanol ($k \ll 1$) as a modifier, while baseline separation is achieved on the biphenyl phase (A) with some retention and separation on the 3-hydroxyphenyl phase (B). For improvements to the separation, focusing on columns from Groups III–V will yield an increase in retention and resolution. As shown in Table 3.2, aliphatic and aromatic hydrocarbon compounds can be retained using low % of methanol and based on the compounds specific functionality, solubility, and size, the retention order can also be predicted reasonably well by the LSER model [40]. Separations on nonpolar aromatic phases (Group III—phenyl, naphthyl, pentafluorophenyl (PFP), pyrene, etc.) tend to improve upon the aforementioned octadecylsiloxane-bonded phases, especially for those compounds with closely related structural features. These phases can be applied for hydrophobic compounds such as hydrocarbons [40,78], fused aromatic systems [48], fats and oils [77]. For more subtle tuning of retention on these phases, use of alcohols, such as 2-propanol, *n*-propanol and ethanol, or an aprotic solvent like acetonitrile, can be effective replacements for methanol.

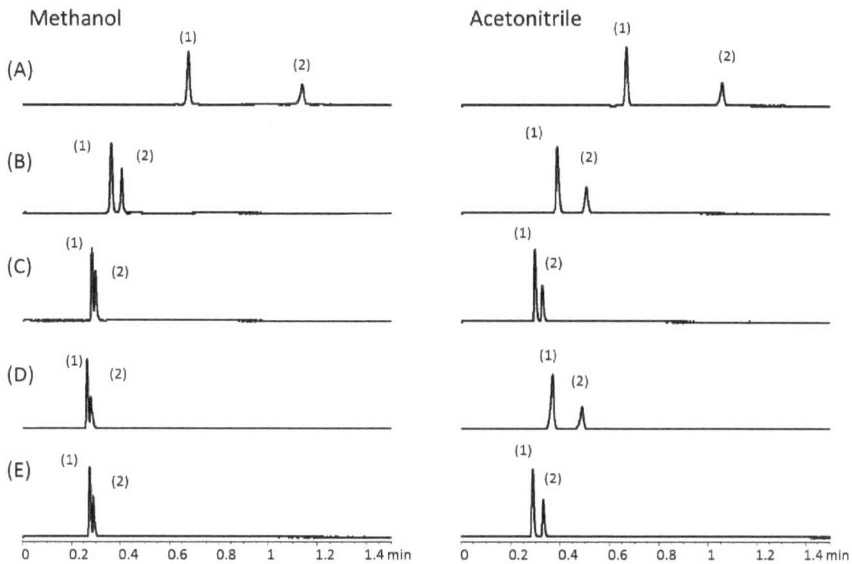

FIGURE 3.10 Separation of hydrophobic compounds using methanol and acetonitrile conditions: Analytes: Toluene (1), DiTbBp (2) on columns ranging in hydrophobicity. Columns: (A) Kinetex Biphenyl, (B) Cosmosil 3-hydroxyphenyl, (C) Sun Shell Diol, (D) PCI CS-HADP, (E) Sun Shell 2-ethylpyridine. Conditions: 3.0×150 mm^2, 2.6–2.7-μm columns maintained at 25°C with a 5–50% methanol in CO_2 composition gradient from 0 to 2.0 min; flow rate 2.5 mL/min; 160-bar outlet pressure with UV 260-nm detection.

As illustrated in Fig. 3.10, the change in modifier can have a dramatic effect on compound elution and separation, even on polar stationary phases, which highlights the limited scope of the model-based predictions mentioned in Section 3.3.1, Column Classification by Modeling. Although the strategy demonstrates separations for a defined pair of neutral analytes, its effect on basic solutes is much less applicable.

As the diversity of analytes increase with the incorporation of polar functional groups the first choice in column changes. In other words, the more polar the compound, the more polar the mobile and stationary phases will need to be to minimize strong interactions and retention [1]. Moving down the rows in Table 3.2, the suggested order of columns to utilize as a first choice migrates toward primarily Groups II and III. Neutral analyte methods are typically the easiest to develop because the retention mechanism is based upon the combination of hydrophobic interactions (with a minor contribution from hydrogen bonds), either with the phase directly, and/or with the underlying base silica. These compounds tend to have good solubility in organic solvents and will generally elute with good peak symmetry as represented by the separation of cortisone and hydrocortisone in Fig. 3.11 (left).

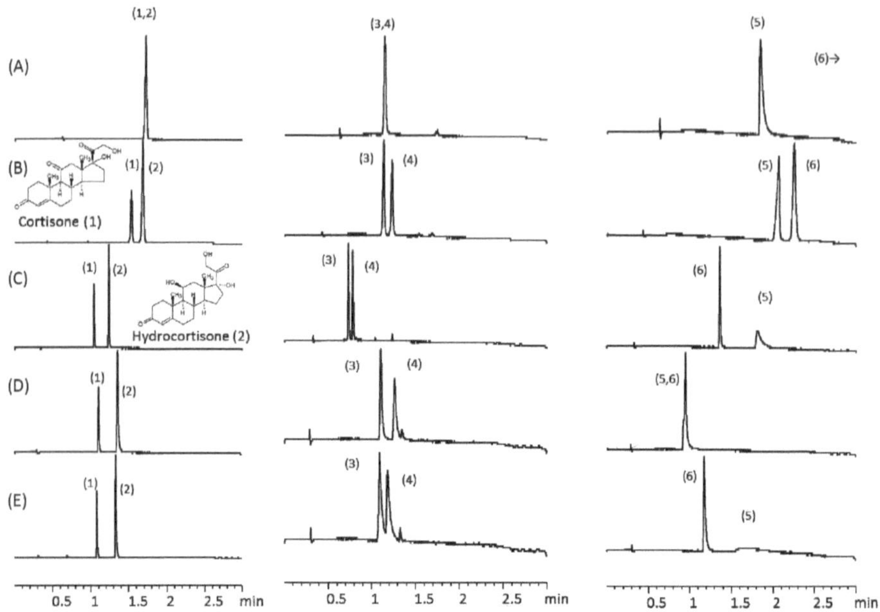

FIGURE 3.11 Example of polar separations. Left: cortisone (1), hydrocortisone (2), Center: fenoprofen (3) flurbiprofen (4), and Right: minaprine (5), nifenazone (6) on columns from Groups I–IV. Columns: (A) Kinetex biphenyl, (B) Cosmosil 3-hydroxyphenyl, (C) Sun Shell Diol, (D) PCI CS-HADP, (E) Sun Shell 2-ethylpyridine. Conditions: 3.0×150 mm^2, 2.6–2.7-μm columns maintained at 25°C with a 5–50% methanol in CO_2 composition gradient from 0 to 2.0 min; flow rate 2.5 mL/min; 160-bar outlet pressure; with UV detection and APCI (+) Ms confirmation.

Upon incorporation of an ionic functional group to a solute, the use of polar columns as a first choice will require less optimization of the conditions to improve the separation. However, the use of an additive may be necessary to minimize excessive retention and improve peaks shape. As demonstrated in Fig. 3.11 (right), the basic compounds minaprine and nifnenzaone display considerable peak distortion with only methanol as a modifier with each type of column and the use of an additive must be considered to reduce these interactions.

For instance, nifenazone did not elute even after repeated injections under these gradient conditions on the biphenyl column without the use of an additive as shown in Fig. 3.11(A). Only upon addition of ammonium formate to the modifier was it possible to elute this analyte (Fig. 3.12). It should also be noted that the retention order of minaprine (5) and nifenazone (6) is different on the 3-hydroxyphenyl phases (b), which further supports the need to screen compounds using a variety of columns to determine an optimal separation.

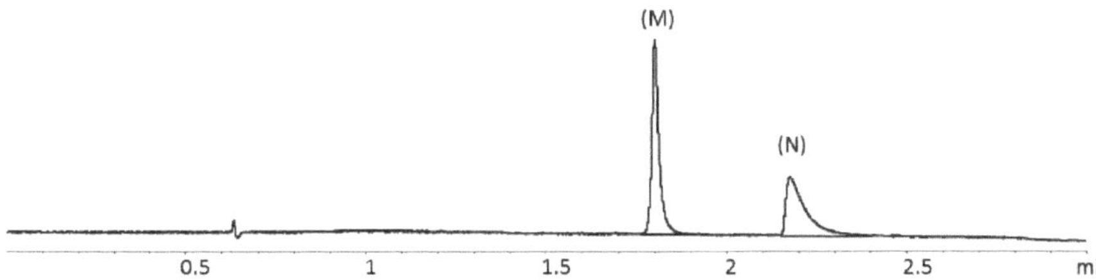

FIGURE 3.12 Example of modifier additives for basic compounds. minaprine (M), nifenazone (N) on Kinetex Biphenyl using 20-mm ammonium formate added to methanol with the same conditions listed in Fig. 3.11.

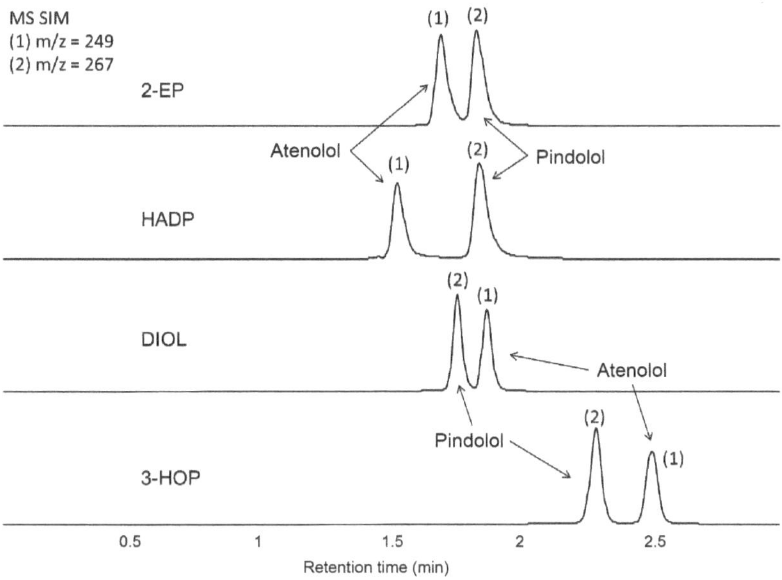

FIGURE 3.13 Example of hydrogen bonding differences between pyridinyl and hydroxyl-type columns. Atenolol (1) and pindolol (2). Columns used: Sun Shell 2-ethyl-pyridine (2-EP), Princeton core shell prototype HADP, Sun Shell Diol (DIOL), and Cosmosil 3-hydroxyphenyl columns (3-HOP). SFC conditions: $3.0 \times 150\ mm^2$, 2.5–2.7-μm columns maintained at 25°C with a 15% methanol with 20-mM ammonium formate (w/v) in CO_2 at a flow rate of 2.5 mL/min and 160-bar outlet pressure with APCI (+) detection monitored at m/z = 249 and 267 [M + H].

Although these interactions are quite difficult to predict a priori, it is precisely the ability to take advantage of these subtle property differences between stationary phases that make SFC such a versatile separation technique. For instance, in the example shown in Fig. 3.13, beta blockers atenolol and pindolol also exhibit a reversal of elution order depending on the type of phase used.

The elution profile on several core–shell phases with hydroxyl or pyridinyl moieties is shown in Fig. 3.13. The overall decrease in k for atenolol and pindolol on the 2-EP/HADP is likely due to a combination of the increased bulk of the phase (~6% more carbon load), the addition of π–π interactions due to the aromatic groups, and the presence of a secondary polarizable group on the HADP phase. Owing to the greater basicity of these phases, the more basic atenolol elutes first. On the contrary, atenolol has a greater capacity for hydrogen bonding than pindolol, which is a primary driver for more retention on the diol-based column. The acidity of the benzylic hydroxyl group of the 3-HOP phase, combined with π–π interactions, enhances the retention of atenolol thus increasing the separation from pindolol.

For the acidic analytes fenoprofen and flurbiprofen shown in Fig. 3.11 (center), the retention mechanism is less clearly attributable to one specific type of interaction. The use of SPP phases in this example affords separation of these compounds without the use of additive (recall the separation interplay depending on the additive used on the totally porous particle diol-monol (DM) column from Fig. 3.6), which can be attributed to the higher peak capacity relative to the 5-μm particles, as well as the increased efficiency relative to the totally porous particles [79]. The increase in retention on the 3-hydroxyphenyl column (Fig. 3.11B) suggests an increase in π–π interactions, while the ionic interactions between the phase and the solute suggest a strong p*Ka* dependence when using polar phases as indicated on the basic columns (Fig. 3.13D and E). Further evidence of the ionic interactions can be seen in Fig. 3.14, where the retention and degree of separation are shown to be influenced by the basicity of the stationary phase relative to the DM (acidic) column. The degree of separation and increased retention for both analytes follows the approximate increase in phase basicity using the same conditions. It should be noted that all four columns in this case study utilized the same particle type and source to minimize any unintended interactions deriving from the silica support.

Not surprisingly, both De la Puente et al. [80] and Aurigemma et al. [81] describe a similar selection of phases for their screening columns, with each group reporting very high success rates for SFC methods using phases primarily from Groups I & II, many of which were subsequently scalable for isolation of analytes. In the case of Aurigemma, the selection of phases stems from experience acquired from roughly 80% of the compounds represented by the properties in Fig. 3.1, which were isolated as pure using either 2-ethylpyridine or diol as the stationary phase. Fig. 3.15 shows the elution profile for approximately 69,000 unique samples on the distribution of elution, which served as a basis for developing phases with mixed hydroxyl and pyridinyl moieties (see Section 5.1). Ebbinger and Weller suggest that a reasonable set of

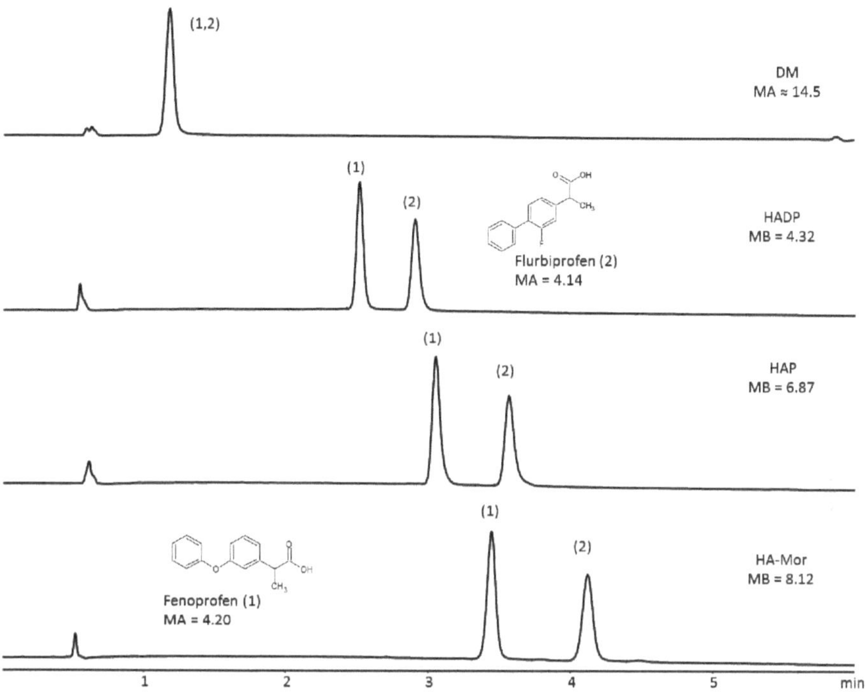

FIGURE 3.14 Example of how ionic interaction can contribution to separations. Fenoprofen (1) and flurbiprofen (2) on columns with different p*Ka*. Phases: 1,2-propandiol-ethanol (DM), hydroxyaminodipyridinyl (HADP), hydroxyaminopyridinyl (HAP), hydroxyamino morpholine (HA-Mor). Conditions: 4.6×150 mm^2, 5-μm columns maintained at 25°C with a 5–50% methanol in CO_2 composition gradient in 5 min at a flow rate of 4.5 mL/min and 140-bar outlet pressure with UV detection monitored at 260 nm. p*Ka* calculations determined using ACD/Labs p*Ka* DB software, version 12.5. (MA = most acidic; MB = most basic)

screening columns for a wide diversity of compounds would contain a cross-linked diol and a high-purity acidic silica (both Group I), a sterically-shielded pyridine (Group II), planar pyrene (Group III), and some polar-embedded stationary phases (Groups IV and V) [11]. Perrenoud found that columns from Groups I and II, the Poroshell HILIC, SunShell Diol and SunShell 2-EP phases, represent the most orthogonal superficially porous (SPP) columns with the highest peak capacities for a small set of pharmaceutical and basic small molecule compounds [82]. De la Puente et al. [83] later refined their screen to use of a single hydroxyl-type (Group I) and pyridinyl-type (Group II) columns. In each case, addition of a ammonium acetate or formate-based additive to the methanol was not detrimental but did limit the number of screening iterations needed to optimize separations. Lemasson found for a small set of unreported structural features that a C-18 (Group V)

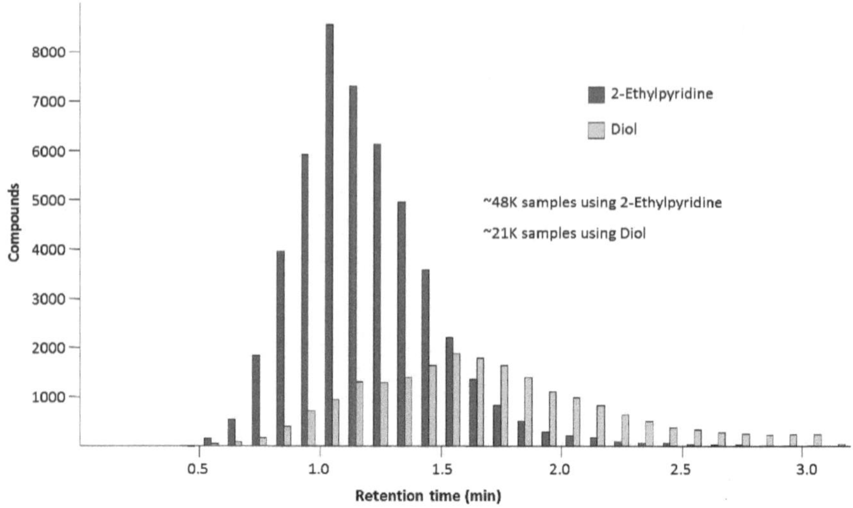

FIGURE 3.15 Retention times for approximately 69-K compounds using 2-ethylpyridine or diol phases. Conditions: Zymor 2-ethypyridine or Diol, $4.6 \times 150\ mm^2$, 5-μm columns at ambient temperature with a 5–50% modifier in CO_2 composition gradient at a rate of 18%/minute and a flow rate of 5.6 mL/min with 140-bar outlet pressure.

and a HILIC (Group I) offered the best results for the majority of their data set [74]. Similar groupings were reported in a review by Galea et al. [84] although these may have been biased by the LSER model.

Common to these studies is the diol phase, of which the versatility can be attributed to some degree by its aliphatic characteristics as well as its high capability for hydrogen bond interactions, though in the latter case, these can be detrimental to the elution of very basic compounds, thus requiring the use of basic additives. Although an additive is necessary for the improvement of peak shape, selectivity is typically unchanged and the high interaction with these basic compounds is likely due to a combination of these factors: the acidity of residual silanols, strong interactions with the hydroxyl groups on the phase, as well as the slight acidity of the mobile phase carbon dioxide/modifier combination. Nonetheless, this column, with judicious selection of solvent, can be utilized for all but the most hydrophobic compounds. Further studies have shown that diol or diol-containing phases have good applicability for moderately polar achiral SFC applications [80,85–90]. Though polar compounds may require high composition of a protic modifier, an acidic or basic additive, or even a small percentage of water to improve peak shape and reduce retention on these columns. At the other end of the polarity spectrum, very hydrophobic compounds tend not to retain on the diol phase even under pure CO_2 or low modifier conditions.

As a concluding suggestion, compiling a database of solute and column behavior for each known compound used, then applying some structure-matching algorithm (e.g., Tanimoto), will further aid in refining a reasonable class of stationary phases to apply firsthand [91]. For instance, if the SFC conditions for this type of evaluation are adequately controlled, similar to those used for the ~80-K compounds in Figs. 3.1 and 3.15, this approach can be quite effective.

3.4 NEW STATIONARY PHASES AND THEIR POTENTIAL IMPACT ON SFC

3.4.1 Novel Stationary Phases

Developing novel stationary phases can be a laborious, trial, and error process without a strategy for quickly evaluating each new phase against a broad set of analytes. Rapidly screening different moieties can also present some synthesis challenges to form a stable stationary phase. Toward this effort, Dunkle et al. [92] demonstrated the utility of using click chemistry to scout potential phases for improved selectivity. Although the resulting triazole moiety formed from the click reaction was also found to contribute to the separation selectivity, further work has demonstrated that traditionally linked phases could be developed which can demonstrate desired selectivity [93]. Fig. 3.16 shows the separation of acidic, basic, and neutral compounds on morpholine-type phases that have been derived from the click chemistry screen to commercial viability. Despite their general availability, so far there are limited publications of these phases.

Along the column development continuum in search of new SFC-specific phases, several column chemistries may not work as broadly as anticipated but nonetheless have demonstrated utility. A mixed mode phase derived from a combination of bonded diol and cyano on the same silica support in a proprietary ratio. Since the vast majority of analyses in Fig. 3.15 utilized either a pyridine or diol, the aptly named pyridine-diol (P/D) mixed mode phase was created, and it was determined that the combination of functionalities offers a unique selectivity relative to the individual phases that is best suited for the separation of moderately polar compounds. In addition, a quinoline phase, originally designed to minimize residual silanol activity through additional aromatic structure (bulk) while retaining the elution characteristics of 2-ethylpyridine, yet failed to gain acceptance for routine use. Although less known, the quinolone phase can be classified with other benzylic (Group III) phases such as the pentafluorophenyl, naphthyl, and pyrene phases.

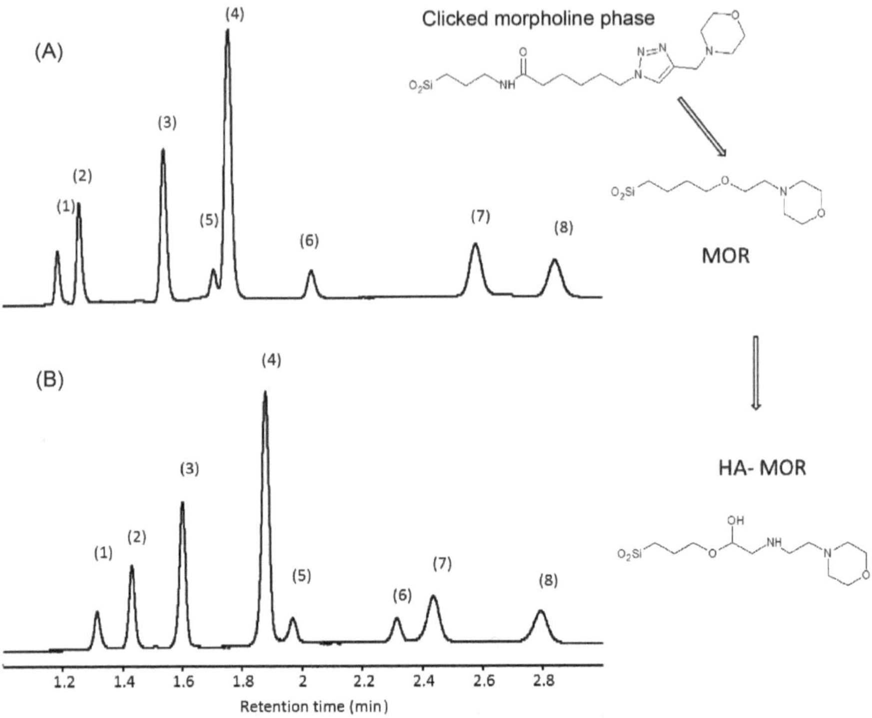

FIGURE 3.16 Progression of morpholine phase development originating from click chemistry screen [92]. Nifenazone (1), minaprine (2), cortisone (3), hydrocortisone (4), atenolol (5), pindolol (6), fenoprofen (7) and flurbiprofen (8) on (A) morpholine (MOR) and (B) hydroxyamino morpholine (HA-MOR) columns. Conditions: $4.6 \times 150\ mm^2$, 5-μm columns maintained at 25°C with a 5–50% methanol in CO_2 composition gradient in 2.0 min at a flow rate of 3.5 mL/min and 160-bar outlet pressure with UV detection monitored at 235 nm.

Given the breath of interactions possible, through combinations of additive, phase and modifier, the potential exists for new phase chemistries positively impacting the field of separations in SFC.

3.4.2 Use of Chiral Phases for Achiral Separations

Although the benefits of chiral columns for enantioselectivity are out of scope for this chapter and will be covered elsewhere in this book, the value of these columns for achiral separations should not be understated. Several groups have reported the separation of achiral samples with chiral columns, especially in the separation of regio-isomers or other structurally similar compounds. Using caffeine and several related compounds as an illustration of this point, Fig. 3.17 shows the

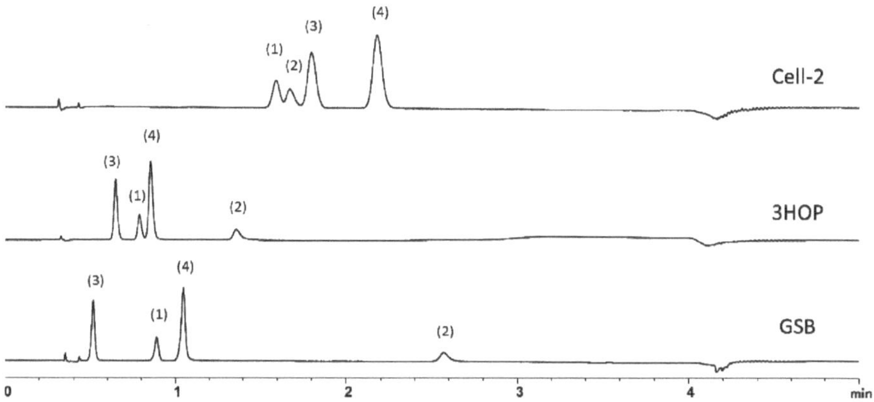

FIGURE 3.17 Chromatograms illustreing the use of a chiral phase for an achiral separation. Etophylline (1), hypoxanthine (2), caffeine (3), theobromine (4), on Phenomenex Lux Celluose-2 (Cell-2), Cosmosil 3-hydroxyphenyl columns (3-HOP) and ES industries GreenSep Basic (GSB) columns. Conditions: $3.0 \times 150\ mm^2$, 5-μm columns maintained at 25°C with a 20–50% methanol in CO_2 composition gradient in 2.5 min, with a 1.0-min hold at 50%, at a flow rate of 2.5 mL/min and 160-bar outlet pressure with UV detection monitored at 272 nm.

separations on both chiral and achiral stationary phases. Interestingly, the variety of separations allows for nearly every combination of elution order using the same chromatographic conditions.

In many cases, the multifunctional diversity of the chiral phase combined with solvents that have low dielectric constants, unlike aqueous solvents used in reversed-phase liquid chromatography, are suitable for enhancing intramolecular hydrogen bonding interactions with analytes [94–96].

3.4.3 Potential Changes in Column Configurations in SFC

Parallel segmented flow (PSF) is an interesting column technology on the horizon that has the potential to advance SFC, especially those with atmospheric pressure detectors like ELSD or Ms [97]. The end fitting of the column has two output flow streams, one centered radially at the head of the column and the other at the periphery, with each flow separated by a series of porous and nonporous frits (Fig. 3.18). In HPLC, this belongs to the broader suite of column technologies referred to as active flow technology (AFT) [98,99]. The commercialization of the PSF column, and for that matter, any of the AFT columns has not yet been realized. As such, there are no examples of their application in SFC and in this work prototype columns have been employed.

A unique feature of the PSF column is that the flow stream at the point of exit from the column is separated into two flow regions,

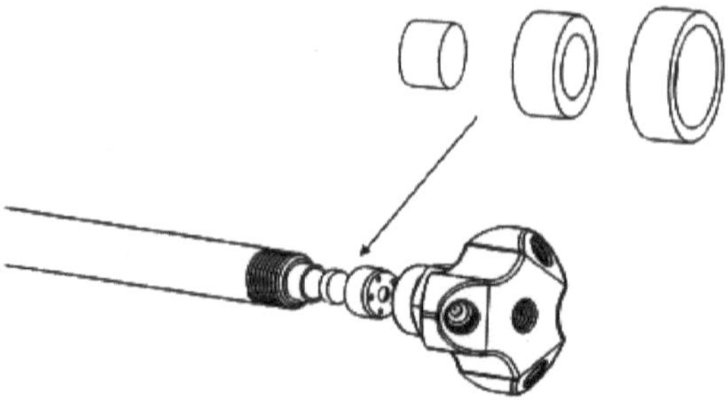

FIGURE 3.18 Illustration of the column outlet showing the three peripheral exit ports and the single central exit port. The annular frit is expanded and inserts into a cap located within the outlet cap. *Source: Reprinted from Camenzuli M, et al., Enhanced separation performance using a new column technology: parallel segmented outlet flow. J Chromatogr A 2012;1232: 47–51 with permission [101].*

courtesy of a radial flow splitting end fitting with a three component frit. The frit comprises two porous zones, an inner radial porous zone and an outer or peripheral porous zone, that is separated from the inner zone by an impermeable barrier. This frit serves as the first step in the process of producing two separated flow regions. The second step is achieved by using a multiport end fitting. A radial central exit port is aligned with the radial central porous region of the frit, and a peripheral port or ports are aligned with the peripheral porous zone of the frit. These peripheral ports direct the flow from the wall region of the column. Thus, the radial central flow and the flow near the column wall are isolated in what is essentially a radial flow stream splitting process. The basic design is illustrated in Fig. 3.18.

One significant advantage of this radial flow stream splitting process is that it overcomes the heterogeneity of the flow through the column. It is well known that bands migrate through HPLC columns with parabolic-like profiles, resembling partially filled bowls (see Reference [99] and references cited therein). The flow in the central zone travels at a higher velocity near the wall because the bed density is lowest in the radial central region of the column and denser nearing the wall, as discussed in Reference [100]. By separating the peripheral flow from the radial central flow, the flow in the radial central region of the column resembles more the ideal cylindrical plug, and in essences a wall-less, virtual column is created [98], which has higher efficiency. Segmentation of the flow accordingly yields an improvement in peak shape, and enhanced Ms detection, including speed in the analysis since typically 80% of the flow is not transported to the detector [97]. An improvement in peak shapes can be achieved by this segmentation.

In SFC, these flow phenomena are not surprisingly reversed from HPLC, since the center of the column is cooler than at the walls as a result of a radial temperature gradient. This radial temperature gradient is thus critical in the establishment of heterogeneous flow. The subsequent band may assume a parabolic-like profile, but with the periphery moving faster than the center, as shown in Fig. 3.19. As the flow is heterogeneous, it seems reasonable that PSF and AFT columns in general may offer advantage in separation efficiency. Hence their application was tested in SFC, but with some minor modification in the general concept. That is, the PSF column was operated in the reverse direction, much like a curtain flow column (another column within the AFT suite) described in Reference [102], but without the PSF end fitting attached. In this operating mode, the flow from the injector is directed to the radial central inlet port, while mobile phase without sample, i.e., bypassing the injector is directed to the peripheral ports. The relative portion of these flows can be varied to produce maximum signal response or separation efficiency. Sample is thus loaded into the central region of the column in a zone dictated by the splitting or segmentation ratio between the radial central flow and the peripheral flow. A curtain of mobile phase flow prevents or limits the migration of the solute from the radial central region of the column to the wall region. Effectively, the sample is concentrated in the radial central zone of the column, and migration is less affected by the heterogeneity generated by the thermal gradient across the column. At the outlet of the column, the sample exits the column as it would do in a conventional column, i.e., there is no flow stream splitting. In this mode of operation, improved separation efficiency and sample detectability was observed. It is worth noting, however, that we consider the observed performance gain to result from two effects, (1) the sample did not penetrate to the wall and the

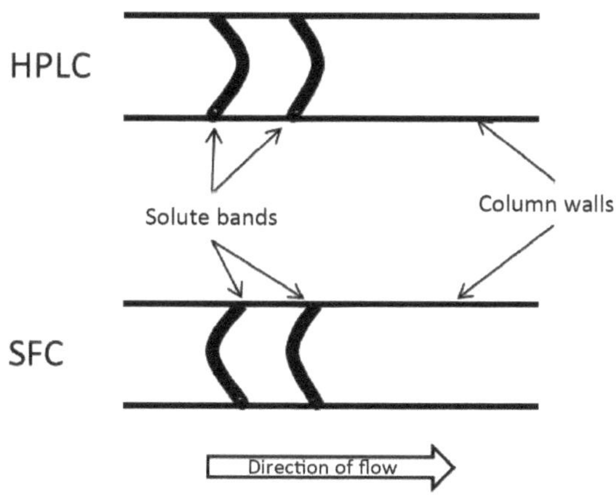

FIGURE 3.19 Illustration of parabolic band profiles in HPLC and SFC caused by radial temperature gradients.

flow profile was more like a cylindrical plug rather than a parabolic-bowl-like profile, and (2) the sample elution process from the column was improved because the sample did not have to flow transverse to the flow direction from the extremities of the tube, thus speeding the process of sample exit from the column.

As demonstrated in Fig. 3.20, the peak shape can be manipulated by changing the flow distribution at the column inlet by varying the volume of tubing from a common point just past the pump (tee). The central port is connected to the outlet of the injector while the periphery port is connected to the tee. Since measuring the actual flow rate in SFC is difficult, estimated segmented flow in this study was determined solely based on the ratio of the tubing volume between the central port and the peripheral port in each leg from the tee. In addition, the conditions depicted are far enough away from the critical point to avoid any

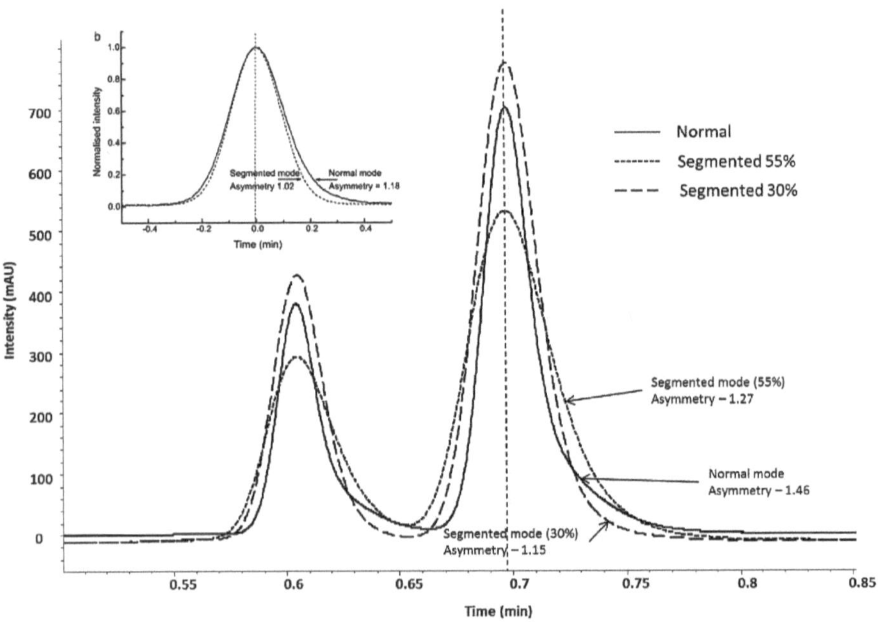

FIGURE 3.20 Band profiles of cortisone and hydrocortisone eluting from the column operating in normal mode (solid), parallel segmented mode (45:55 central:peripheral) (dotted) and segmented mode (70:30) (dashed). Conditions: Hypersil Gold CN, $4.6 \times 100\ mm^2$, 5-μm column maintained at 25°C with 15% modifier in CO_2 composition at a flow rate of 3.5 mL/min and 200-bar outlet pressure. Inset: Normalized band profiles of butyl benzene eluting from the column operating in normal mode (solid) and parallel segmented mode (45:55 central:peripheral) (dotted). Mobile phase: 30/70 water/methanol, flow rate 1.0 mL/min, injection volume 2 μL, detection 250 nm. *Source: Reprinted from Camenzuli M, et al., Enhanced separation performance using a new column technology: parallel segmented outlet flow. J Chromatogr A 2012;1232: 47–51 with permission from Elsevier [101].*

anomalous behavior which could occur closer to the critical point. At the segmentation ratio of 55%, that is 45% of the flow through the radial central inlet port, the peak was observed to broaden and decrease in peak height, although there was less tailing compared to the conventional column. As a result, the separation between the two components in Fig. 5.3 was better than was apparent on the conventional column in which substantial tailing was apparent. When 30% of the flow was directed to the peripheral region of the column and 70% to the radial central inlet port, the peak height and the separation efficiency both increased, with tailing being substantially reduced. It is likely that this gain in performance was due to two factors, (1) the flow velocities in the peripheral and radial central regions of the column were better matched when 70% of the flow was directed to the radial central inlet ports than when 45% was directed there, hence less band expansion was apparent, and (2) having the sample loaded into a smaller cross section region of the column, coupled with less band broadening resulted in a greater concentration of the sample and hence an increase in signal response. Under these conditions, the separation of the two components in Fig. 3.20 was clearly the best. We suspect that further improvements could be obtained with fine tuning of the segmentation ratio, but we caution the effect may also be dependent on the physical state of the fluid.

While these columns, now called active flow technology (AFT), are limited thus far to prototypes for reversed-phase phase applications (e.g., only available in C-18, CN), the application in SFC could bring about a shift in instrument design.

3.5 CONCLUSION

The versatility of SFC as a separation technology can be attributed to the wide range of different stationary phase chemistries available and the influential role the modifier and additives play in optimizing separations. In many cases, the same separation can be achieved using different retention mechanisms, depending on the presence of secondary or tertiary interactions, on vastly different stationary phases. It is precisely the ability to take advantage of these subtle property differences between stationary phases and solutes that have led to the resurgence of SFC a mainstream separation technique. Utilizing a systematic approach for the selection of phases, either using the LSER or empirical approaches, will limit the number of iterations needed to optimize a specific separation. Furthermore, enhanced the knowledge of the properties of the analytes will provide a better opportunity to exploit the subtle differences and optimize a specific separation.

On the other hand, the complexity of the fluid dynamics, those relating both to the instrument operation and their effects on density, can lead to poor chromatographic performance and become a an impediment to adoption. Working with understanding of the impacts changes to the column temperature or pressure drop will have on performance of the system is critical regardless of the stationary phase applied.

Finally, HPLC technology advancements will continue to trickle over into SFC and will result in an exciting future and assure the technique remains viable for years to come.

Acknowledgments

The author wishes to thank Drs. R. Andrew Shalliker and Danijela Kocic, University of Western Sidney, and Tony Edge, Thermo Fisher Scientific, for the contributions on the AFT technology section and the kind use of AFT columns. In addition, sincere appreciation is granted to Dr. Paul Richardson and Ms. Christine Aurigemma for their review of this chapter.

References

[1] Berger TA. Supercritical Fluid Chromatography: Overview. Reference Module in Chemistry, Molecular Sciences and Chemical Engineering. Elsevier, Amsterdam; 2013.

[2] Tarafder A, Guiochon G. Use of isopycnic plots in designing operations of supercritical fluid chromatography: II. The isopycnic plots and the selection of the operating pressure–temperature zone in supercritical fluid chromatography. J Chromatogr A 2011;1218(28):4576–85.

[3] Galea C, Mangelings D, Vander Heyden Y. Characterization and classification of stationary phases in HPLC and SFC – a review. Anal Chim Acta 2015;886:1–15.

[4] Lesellier E, West C. The many faces of packed column supercritical fluid chromatography – a critical review. J Chromatogr A 2015;1382:2–46.

[5] Poole CF. Stationary phases for packed-column supercritical fluid chromatography. J Chromatogr A 2012;1250:157–71.

[6] Goldfarb DL, Fernández DP, Corti HR. Dielectric and volumetric properties of supercritical carbon dioxide(1) + methanol(2) mixtures at 323.15 K. Fluid Phase Equilibria 1999;158–160:1011–19.

[7] Guiochon G, Tarafder A. Fundamental challenges and opportunities for preparative supercritical fluid chromatography. J Chromatogr A 2011;1218(8):1037–114.

[8] Saito M. History of supercritical fluid chromatography: instrumental development. J Biosci Bioeng 2013;115(6):590–9.

[9] West C. Enantioselective Separations with Supercritical Fluids-Review. Curr Anal Chem 2014;10(1):99–120.

[10] Berger TA. Separation of polar solutes by packed column supercritical fluid chromatography. J Chromatogr A 1997;785(1–2):3–33.

[11] Ebinger K, Weller HN. Comparative assessment of achiral stationary phases for high throughput analysis in supercritical fluid chromatography. J Chromatogr A 2014; 1332:73–81.

[12] Lemasson E, et al. Development of an achiral supercritical fluid chromatography method with ultraviolet absorbance and mass spectrometric detection for impurity profiling of drug candidates. Part I: Optimization of mobile phase composition. J Chromatogr A 2015;1408:217–26.
[13] Akin A, et al. An orthogonal approach to chiral method development screening. Curr Pharm Anal 2007;3(1):53–70.
[14] Lou X, et al. Pressure drop effects on selectivity and resolution in packed column supercritical fluid chromatography. J High Resolution Chromatogr 1996;19(8):449–56.
[15] Kaczmarski K, et al. Pressure, temperature and density drops along supercritical fluid chromatography columns. II. Theoretical simulation for neat carbon dioxide and columns packed with 3-μm particles. J Chromatogr A 2012;1250:115–23.
[16] Baker LR, et al. Density gradients in packed columns: II. Effects of density gradients on efficiency in supercritical fluid separations. J Chromatogr A 2009;1216(29):5594–9.
[17] Baker LR, et al. Density gradients in packed columns: I. Effects of density gradients on retention and separation speed. J Chromatogr A 2009;1216(29):5588–93.
[18] Rajendran A, Gilkison TS, Mazzotti M. Effect of pressure drop on solute retention and column efficiency in supercritical fluid chromatography Part 2: Modified carbon dioxide as mobile phase. J Sep Sci 2008;31(8):1279–89.
[19] Berger TA, Blumberg LM. Efficiency losses caused by gradients in solute velocity in packed column supercritical fluid chromatography. Chromatographia 1994;38(1):5–11.
[20] Tarafder A, Guiochon G. Use of isopycnic plots in designing operations of supercritical fluid chromatography. III: Reason for the low column efficiency in the critical region. J Chromatogr A 2011;1218(40):7189–95.
[21] Poe DP, et al. Pressure, temperature and density drops along supercritical fluid chromatography columns. I. Experimental results for neat carbon dioxide and columns packed with 3- and 5-micron particles. J Chromatogr A 2012;1250:105–14.
[22] Berger TA. Instrument modifications that produced reduced plate heights < 2 with sub-2 μm particles and 95% of theoretical efficiency at $k = 2$ in supercritical fluid chromatography. J Chromatogr A 2016;1444:129–44.
[23] Vajda P, Stankovich JJ, Guiochon G. Determination of the average volumetric flow rate in supercritical fluid chromatography. J Chromatogr A 2014;1339:168–73.
[24] De Pauw R, et al. Effect of reference conditions on flow rate, modifier fraction and retention in supercritical fluid chromatography. J Chromatogr A 2016;1459:129–35.
[25] Grand-Guillaume Perrenoud A, Veuthey J-L, Guillarme D. The use of columns packed with sub-2 μm particles in supercritical fluid chromatography. TrAC Trends Anal Chem 2014;63:44–54.
[26] Jentoft RE, Gouw TH. Pressure-programmed supercritical fluid chromatography of wide molecular weight range mixtures. J Chromatogr Sci 1970;8(3):138–42.
[27] Novotny M, Bertsch W, Zlatkis A. Temperature and pressure effects in supercritical-fluid chromatography. J Chromatogr A 1971;61:17–28.
[28] Novotny M, et al. Capillary supercritical fluid chromatography. Anal Chem 1981;53(3):407A–14A.
[29] Gritti F, Guiochon G. The current revolution in column technology: how it began, where is it going? J Chromatogr A 2012;1228:2–19.
[30] Cabrera K. Applications of silica-based monolithic HPLC columns. J Sep Sci 2004;27(10–11):843–52.
[31] Hormann K, et al. Morphology and separation efficiency of a new generation of analytical silica monoliths. J Chromatogr A 2012;1222:46–58.
[32] Lesellier E, West C, Tchapla A. Advantages of the use of monolithic stationary phases for modelling the retention in sub/supercritical chromatography: Application to cis/trans-β-carotene separation. J Chromatogr A 2003;1018(2):225–32.

[33] Sarazin C, et al. Feasibility of ultra high performance supercritical neat carbon dioxide chromatography at conventional pressures. J Sep Sci 2011;34(19):2773–8.
[34] Gritti F, Guiochon G. Mass transfer kinetics, band broadening and column efficiency. J Chromatogr A 2012;1221:2–40.
[35] Gritti F, Guiochon G. The mass transfer kinetics in columns packed with Halo-ES shell particles. J Chromatogr A 2011;1218(7):907–21.
[36] Fekete S, Ganzler K, Fekete J. Facts and myths about columns packed with sub-3 μm and sub-2 μm particles. J Pharm Biomed Anal 2010;51(1):56–64.
[37] Poe DP, et al. Pressure, temperature and density drops along supercritical fluid chromatography columns in different thermal environments. III. Mixtures of carbon dioxide and methanol as the mobile phase. J Chromatogr A 2014;1323:143–56.
[38] Tarafder A, Guiochon G. Use of isopycnic plots in designing operations of supercritical fluid chromatography: I. The critical role of density in determining the characteristics of the mobile phase in supercritical fluid chromatography. J Chromatogr A 2011;1218(28):4569–75.
[39] P. Sandra, A. Medvedovici, A. Kot, and F. David, in *Supercritical Fluid Chromatography with Packed Columns: Techniques and Applications*, C.B. K. Anton, Editor., Marcel Dekker: New York. p. 483.
[40] Lesellier E. Efficiency in supercritical fluid chromatography with different superficially porous and fully porous particles ODS bonded phases. J Chromatogr A 2012; 1228:89–98.
[41] Delahaye S, Lynen F. Implementing stationary-phase optimized selectivity in supercritical fluid chromatography. Anal Chem 2014;86(24):12220–8.
[42] Kirkland JJ, et al. Superficially porous silica microspheres for fast high-performance liquid chromatography of macromolecules. J Chromatogr A 2000;890(1):3–13.
[43] Gritti F, et al. Physical properties and structure of fine core–shell particles used as packing materials for chromatography: relationships between particle characteristics and column performance. J Chromatogr A 2010;1217(24):3819–43.
[44] McCalley DV. Evaluation of the properties of a superficially porous silica stationary phase in hydrophilic interaction chromatography. J Chromatogr A 2008;1193(1–2): 85–91.
[45] Vajda P, Guiochon G. Determination of the column hold-up volume in supercritical fluid chromatography using nitrous-oxide. J Chromatogr A 2013;1309:96–100.
[46] Tarafder A. Metamorphosis of supercritical fluid chromatography to SFC: an overview. TrAC Trends Anal Chem 2016;81:3–10.
[47] Berger TA. Chromatography: Supercritical Fluid | Theory of Supercritical Fluid Chromatography A2-. In: Wilson Ian D, editor. Encyclopedia of Separation Science. Oxford: Academic Press; 2007. p. 1–9.
[48] Vera CM, et al. Contrasting selectivity between HPLC and SFC using phenyl-type stationary phases: a study on linear polynuclear aromatic hydrocarbons. Microchem J 2015;119:40–3.
[49] Glenne E, et al. A closer study of methanol adsorption and its impact on solute retentions in supercritical fluid chromatography. J Chromatogr A 2016;1442:129–39.
[50] Fairchild JN, et al. Chromatographic evidence of silyl ether formation (SEF) in supercritical fluid chromatography. Anal Chem 2015;87(3):1735–42.
[51] Patel MA, et al. Supercritical fluid chromatographic resolution of water soluble isomeric carboxyl/amine terminated peptides facilitated via mobile phase water and ion pair formation. J Chromatogr A 2012;1233:85–90.
[52] Taylor LT. Packed column supercritical fluid chromatography of hydrophilic analytes via water-rich modifiers. J Chromatogr A 2012;1250:196–204.
[53] Zheng J, Taylor LT, Pinkston JD. Elution of cationic species with/without ion pair reagents from polar stationary phases via SFC. Chromatographia 2006;63(5):267–76.

[54] Ashraf-Khorassani M, Taylor LT. Subcritical fluid chromatography of water soluble nucleobases on various polar stationary phases facilitated with alcohol-modified CO_2 and water as the polar additive. J Sep Sci 2010;33(11):1682–91.
[55] Olesik SV. Enhanced-fluidity liquid chromatography: connecting the dots between supercritical fluid chromatography, conventional subcritical fluid chromatography, and HPLC. LCGC Chromatogr 2015;33:24–30.
[56] Hamman C, et al. The use of ammonium hydroxide as an additive in supercritical fluid chromatography for achiral and chiral separations and purifications of small, basic medicinal molecules. J Chromatogr A 2011;1218(43):7886–94.
[57] Ventura M, Murphy B, Goetzinger W. Ammonia as a preferred additive in chiral and achiral applications of supercritical fluid chromatography for small, drug-like molecules. J Chromatogr A 2012;1220:147–55.
[58] Sen A, et al. Analysis of polar urinary metabolites for metabolic phenotyping using supercritical fluid chromatography and mass spectrometry. J Chromatogr A 2016; 1449:141–55.
[59] West C. How good is SFC for polar analytes? Chromatogr Today 2013;22–7. May/June.
[60] Ashraf-Khorassani M, Taylor LT, Seest E. Screening strategies for achiral supercritical fluid chromatography employing hydrophilic interaction liquid chromatography-like parameters. J Chromatogr A 2012;1229:237–48.
[61] Liu J, et al. Extending the range of supercritical fluid chromatography by use of water-rich modifiers. Organ Biomol Chem 2013;11(30):4925–9.
[62] Li JJ, Thurbide KB. A comparison of methanol and isopropanol in alcohol/water/CO2 mobile phases for packed column supercritical fluid chromatography. Can J Anal Sci Spectrosc 2008;53(2):59–65.
[63] Poole CF. Chapter 7-Supercritical Fluid Chromatography, in The Essence of Chromatography. Amsterdam: Elsevier Science; 2003. p. 569–617.
[64] Poole CF, et al. Determination of solute descriptors by chromatographic methods. Anal Chim Acta 2009;652(1–2):32–53.
[65] Park JH, et al. Characterization of some normal-phase liquid chromatographic stationary phases based on linear solvation energy relationships. J Chromatogr A 1998;796(2):249–58.
[66] Vitha M, Carr PW. The chemical interpretation and practice of linear solvation energy relationships in chromatography. J Chromatogr A 2006;1126(1–2): 143–94.
[67] Abraham MH, Ibrahim A, Zissimos AM. Determination of sets of solute descriptors from chromatographic measurements. J Chromatogr A 2004;1037(1–2):29–47.
[68] Bolliet D, Poole CF, Rosés M. Conjoint prediction of the retention of neutral and ionic compounds (phenols) in reversed-phase liquid chromatography using the solvation parameter model. Anal Chim Acta 1998;368(1–2):129–40.
[69] Rosés M, Bolliet D, Poole CF. Comparison of solute descriptors for predicting retention of ionic compounds (phenols) in reversed-phase liquid chromatography using the solvation parameter model. J Chromatogr A 1998;829(1–2):29–40.
[70] Galea C, et al. Is the solvation parameter model or its adaptations adequate to account for ionic interactions when characterizing stationary phases for drug impurity profiling with supercritical fluid chromatography? Anal Chim Acta 2016;924:9–20.
[71] Gohres JL, et al. Spectroscopic investigation of alkylcarbonic acid formation and dissociation in CO_2-expanded alcohols. Ind Eng Chem Res 2009;48(3):1302–6.
[72] Wolrab D, et al. Strong cation exchange-type chiral stationary phase for enantioseparation of chiral amines in subcritical fluid chromatography. J Chromatogr A 2013;1289:94–104.

[73] Wen D, Olesik SV. *Characterization of pH in liquid mixtures of methanol/H_2O/CO_2*. Anal Chem 2000;72(3):475–80.

[74] Lemasson E, et al. Development of an achiral supercritical fluid chromatography method with ultraviolet absorbance and mass spectrometric detection for impurity profiling of drug candidates. Part II. Selection of an orthogonal set of stationary phases. J Chromatogr A 2015;1408:227–35.

[75] West C, et al. An improved classification of stationary phases for ultra-high performance supercritical fluid chromatography. J Chromatogr A 2016;1440:212–28.

[76] West C, et al. An attempt to estimate ionic interactions with phenyl and pentafluorophenyl stationary phases in supercritical fluid chromatography. J Chromatogr A 2015;1412:126–38.

[77] Lesellier E, Latos A, de Oliveira AL. Ultra high efficiency/low pressure supercritical fluid chromatography with superficially porous particles for triglyceride separation. J Chromatogr A 2014;1327:141–8.

[78] West C, Khater S, Lesellier E. Characterization and use of hydrophilic interaction liquid chromatography type stationary phases in supercritical fluid chromatography. J Chromatogr A 2012;1250:182–95.

[79] Berger TA. Characterization of a 2.6 μm Kinetex porous shell hydrophilic interaction liquid chromatography column in supercritical fluid chromatography with a comparison to 3 μm totally porous silica. J Chromatogr A 2011;1218(28):4559–68.

[80] De la Puente ML, Soto-Yarritu PL, Anta C. Placing supercritical fluid chromatography one step ahead of reversed-phase high performance liquid chromatography in the achiral purification arena: a hydrophilic interaction chromatography cross-linked diol chemistry as a new generic stationary phase. J Chromatogr A 2012;1250:172–81.

[81] Aurigemma CM, et al. Automated approach for the rapid identification of purification conditions using a unified, walk-up high performance liquid chromatography/supercritical fluid chromatography/mass spectrometry screening system. J Chromatogr A 2012;1229:260–7.

[82] Perrenoud AG-G, et al. Evaluation of stationary phases packed with superficially porous particles for the analysis of pharmaceutical compounds using supercritical fluid chromatography. J Chromatogr A 2014;1360:275–87.

[83] De la Puente ML, Lopez Soto-Yarritu P, Burnett J. Supercritical fluid chromatography in research laboratories: design, development and implementation of an efficient generic screening for exploiting this technique in the achiral environment. J Chromatogr A 2011;1218(47):8551–60.

[84] Galea C, Mangelings D, Heyden YV. Method development for impurity profiling in SFC: the selection of a dissimilar set of stationary phases. J Pharm Biomed Anal 2015;111:333–43.

[85] Gyllenhaal O. 2.10. Supercritical fluid chromatography (SFC). In: Sándor G, editor. Progress in Pharmaceutical and Biomedical Analysis. Elsevier, Amsterdam; 2000. p. 382–95.

[86] Bui H, et al. Investigation of retention behavior of drug molecules in supercritical fluid chromatography using linear solvation energy relationships. J Chromatogr A 2008;1206(2):186–95.

[87] Shen Y, Lee ML. Silica surface interactions of diol-bonded phases in packed capillary column supercritical fluid chromatography. J Microcolumn Sep 1996;8(6):413–20.

[88] Romand S, Rudaz S, Guillarme D. Separation of substrates and closely related glucuronide metabolites using various chromatographic modes. J Chromatogr A 2016; 1435:54–65.

[89] Desfontaine V, et al. Supercritical fluid chromatography in pharmaceutical analysis. J Pharm Biomed Anal 2015;113:56–71.

[90] Dispas A, et al. Screening study of SFC critical method parameters for the determination of pharmaceutical compounds. J Pharm Biomed Anal 2016;125:339–54.
[91] Willett P. Similarity-based virtual screening using 2D fingerprints. Drug Discov Today 2006;11(23–24):1046–53.
[92] Dunkle M, et al. Synthesis of stationary phases containing pyridine, phenol, aniline and morpholine via click chemistry and their characterization and evaluation in supercritical fluid chromatography. Sci Chromatogr 2014;6(2):85–103.
[93] Aurigemma CM, Farrell WP, Farrell KG. Separation of pharmaceuticals by SFC using mono- and di-hydroxy substituted phenyl stationary phases. Chromatogr Today 2014;7(4):34–7.
[94] Cheng H, et al. Structure-based design, SAR analysis and antitumor activity of PI3K/mTOR dual inhibitors from 4-methylpyridopyrimidinone series. Bioorgan Med Chem Lett 2013;23(9):2787–92.
[95] Goetz GH, et al. High throughput method for the indirect detection of intramolecular hydrogen bonding. J Med Chem 2014;57(7):2920–9.
[96] Goetz GH, Philippe L, Shapiro MJ. EPSA: a novel supercritical fluid chromatography technique enabling the design of permeable cyclic peptides. ACS Med Chem Lett 2014;5(10):1167–72.
[97] Kocic D, et al. Ultra-fast HPLC MS analyses using active flow technology columns. Microchem J 2016;127:160–4.
[98] Shalliker RA, et al. Parallel segmented flow chromatography columns: conventional analytical scale column formats presenting as a 'virtual' narrow bore column. J Chromatogr A 2012;1262:64–9.
[99] Shalliker RA, Ritchie H. Segmented flow and curtain flow chromatography: overcoming the wall effect and heterogeneous bed structures. J Chromatogr A 2014;1335:122–35.
[100] Shalliker RA, Broyles BS, Guiochon G. Physical evidence of two wall effects in liquid chromatography. J Chromatogr A 2000;888(1–2):1–12.
[101] Camenzuli M, et al. Enhanced separation performance using a new column technology: parallel segmented outlet flow. J Chromatogr A 2012;1232:47–51.
[102] Camenzuli M, et al. The design of a new concept chromatography column. Analyst 2011;136(24):5127–30.

CHAPTER

4

Column Characterization

C. West

University of Orleans, Orléans, France

OUTLINE

4.1 INTRODUCTION

The objectives of column characterization in chromatography are multiple.

Firstly, the most frequent reason is to provide some comparison between the numerous stationary phases available to the end-users. In supercritical fluid chromatography (SFC) the need for such comparison

Supercritical Fluid Chromatography.
DOI: http://dx.doi.org/10.1016/B978-0-12-809207-1.00004-5

points has increased significantly as the range of stationary phases available has expanded in the recent years, due to the renewed interest in the technique. Most column manufacturers have selected among their HPLC phases some columns they believed appropriate for SFC use, essentially among polar stationary phases initially designed for normal-phase (NPLC) or hydrophilic interaction liquid chromatography (HILIC), and chiral stationary phases (CSP). But a few manufacturers have proceeded to design stationary phases purposely for SFC use. A comparison method may be desired by column manufacturers, to assess the quality of newly designed stationary phases, and how similar they may be to previously existing phases. But this information is also requested by chromatographers who are looking for high-quality columns and some guidance for selecting a column for a particular application.

Secondly, some prevision of the outcome of the separation is sometimes desired, particularly in the field of enantioselective separations. It is then expected that column characterization could provide some insight into the separation, prior to the experiment. This would essentially accelerate method development in selecting the best chromatographic system without the need to screen for different columns. In this respect, the search for a versatile column that would solve most separation issues is also intense, and may be facilitated by column characterization.

In the present chapter, we will review methods for stationary phase characterization. This is not intended to be a comprehensive review. In particular, several older references will not be cited here, because they report the characterization of columns in operating conditions that are significantly different from the current practice of packed column SFC. This excludes reports employing solely neat carbon dioxide as a mobile phase, and some reports employing very small proportions of cosolvent (typically less than 5%). Besides, as column technology was described in a previous chapter, we have chosen to focus our attention on stationary phases that are commercially available. As a result, original stationary phases that have yet to be developed into commercial products are not reported here.

4.2 NONCHROMATOGRAPHIC METHODS

First, a stationary phase may be characterized through nonchromatographic methods.

Particle size (in μm) is measured with scanning electron microscopy. It also allows the observation of stationary phase morphology like the regularity of the particles and uniformity in particle size. Indeed, the particle size indicated by the column manufacturer is only the average value, while the distribution of particle sizes in bulk silica may vary in width. For instance, superficially porous particles (SPP, also known as core-shell or fused-core particles) are known to have narrower particle size distribution than fully porous particles (FPP). This consistency in particle size is contributing to the improved efficiency observed for SPP particles compared to FPP particles of identical size.

Specific surface area (in m^2/g), pore volume (in cm^3/g), and pore size (in Å) are usually assessed with BET isotherms.

Carbon load (in %) is usually assessed through elemental analysis. However, the method cannot differentiate between carbon atoms from bonded ligands, silanol end-capping groups, or carbon atoms included in hybrid silica supports. Thus the bonding density calculated based on carbon load and specific surface area is not totally accurate.

Spectroscopic methods like Fourrier-transform infrared (FTIR) and nuclear magnetic resonance (NMR) spectroscopy are useful to assess the chemical bonding of ligands to silica supports. FTIR is an inexpensive way to observe residual silanol groups on a bonded silica stationary phase. Besides, when several chemical reactions must be carried out to achieve the desired ligand, completion of these reactions is better assessed with NMR. For instance, many alkyl-type stationary phases comprising polar-embedded groups are usually prepared in two steps: (1) the first step consists in bonding an aminopropyl group to the silica support; (2) the second step is bonding an alkyl chain to the amino group via an amide bond. Spectroscopic techniques can then reveal the completion of the second step.

Although the earlier information may be useful to understand the differences observed in chromatographic behavior (for instance, particle size can be related to efficiency; carbon load may be related to overall retention), their usefulness in guiding the selection of an appropriate stationary phase is rather limited. The characterization of chromatographic behavior is more informative in this respect.

4.3 STABILITY OF STATIONARY PHASES

Column stability is usually evaluated with one or several probe analytes injected repeatedly, either in normal use conditions, or in harsh conditions (higher temperatures than usual, strongly acidic or basic mobile phase, etc.) intended to accelerate the degradation.

The stability of stationary phases employed in SFC was often found to be excellent when used in "classical" operating conditions (avoiding excessive back-pressure and aggressive mobile phase compositions with strong solvents or additives). For instance, de la Puente et al. [1] showed a comparison between a 6-month-old column having been used extensively and a new column. The chromatograms showed only minor differences, which could be due to batch-to-batch variation. Such excellent stability, compared to reversed-phase HPLC (RPLC) columns, can be explained by the general absence of water in the mobile phase composition (except in very small proportions). Indeed water is most aggressive to bonded silica stationary phases causing both hydrolysis of the ligands and silica deterioration.

However, Ebinger et al. [2] observed that retention tended to decrease regularly on polar columns when the mobile phase contained 10 mM ammonium acetate. Some columns were affected more significantly than others. Fairchild et al. [3] attributed this decreasing retention to the formation of silyl ethers due to the reaction of silanol groups with alcohol molecules in the mobile phase, thereby decreasing stationary phase hydrophilicity. This silyl ether formation was however, shown to be reversible, if the column was flushed with water. Note that the simple use of small concentrations of water in the alcohol-modified mobile phase would also be sufficient to avoid this chemical reaction. This observation, however, prompted the synthesis of new stationary phases designed to limit the formation of silyl ethers, which are now commercially available.

4.4 COLUMN EFFICIENCY

Column efficiency is usually assessed with van Deemter plots. It was known for a long time that SFC had an advantage over HPLC owing to higher diffusivity in supercritical fluids, yielding higher overall total efficiency and its retention at higher flow rates. With modern chromatographic columns packed with smaller particles (typically sub-2-μm particles), the van Deemter curve was shown to be even flatter [4], allowing very thin peaks to be maintained at a broad range of mobile phase velocity. Additionally, the SFC mobile phase is much less viscous than liquids employed in HPLC, thus small particles do not cause such a high pressure drop in SFC as in HPLC. Note that such comparisons only make sense now that comparable instruments are available for both techniques with comparable dead volumes.

The kinetic plot method was abundantly discussed in several reports [5–9]. Kinetic plots are intended to determine the kinetic performance of a chromatographic system at its extreme possibilities (highest inlet pressure). The compressibility of the SFC mobile phases causes extra difficulties compared with HPLC. Different approaches were suggested to apply this method, namely the isopycnic method (maintaining constant fluid density in the column) [10] or the conventional method based on calculated viscosities. Whichever method is used, the conclusions are essentially applicable to the set of conditions employed for measurement, and are difficult to extrapolate to other operating conditions.

It should also be noted that the selection of probes to measure column efficiency in SFC is possibly more difficult than for RPLC because of the compressible mobile phase. Indeed when several probes are employed with different retention on the stationary phase under evaluation, they may also have different diffusion coefficients in the selected mobile phase.

Finally, it should be mentioned that efficiency assessment is most useful for identifying operating conditions where the peaks are the thinnest and, thus, resolution is maximized. But modern chromatographic columns for packed column SFC essentially provide comparable efficiency nowadays.

4.5 PROBE ANALYTES

4.5.1 Achiral Probes

Standard chromatographic tests to assess the retentiveness and selectivity of a stationary phase usually comprise a small number of probe analytes. Retention factors and separation factors between pairs of analytes are measured. Several such tests exist for RPLC [11–13], and have been widely adopted by HPLC chromatographers. In that case, for instance, retentiveness of the nonpolar stationary phases (mostly alkylsiloxane-bonded silica) is measured with a hydrophobic analyte (like pentylbenzene in the so-called "Tanaka test" [13]). Hydrophobic selectivity is often assessed through methylene group selectivity, that is to say the separation factor between two analytes differing by one methylene group. Other pairs of analytes may evaluate shape [14] or steric selectivity [15], which is the capability to separate compounds with identical functional groups but different three-dimensional structure. All such tests also include one pair (or more) analytes to measure the accessibility to polar groups on the nonpolar stationary phases, like residual silanol groups [16]. The tested parameters can then be represented through simple two-way plots (for instance with Neue's test [17]) or radar plots (for instance with Tanaka's test [13]). Nearly all such tests are carried out in RPLC aqueous conditions, except for the carotenoid test, which is carried out in SFC conditions.

The carotenoid test [18–22] is a simple test, based on a single analysis of a mixture of carotenoid pigments comprising three probes: *all trans*-β-carotene, 13-*cis*-β-carotene, and zeaxanthin. The first probe is a hydrophobic analyte used to measure hydrophobicity (retention factor of *all trans*-β-carotene). Together with the second probe, it also allows steric selectivity to be assessed (separation factor between *cis* and *trans* isomers) as these two molecules are linear and bent isomers of β-carotene. Together with the third probe, which is a xanthophyll

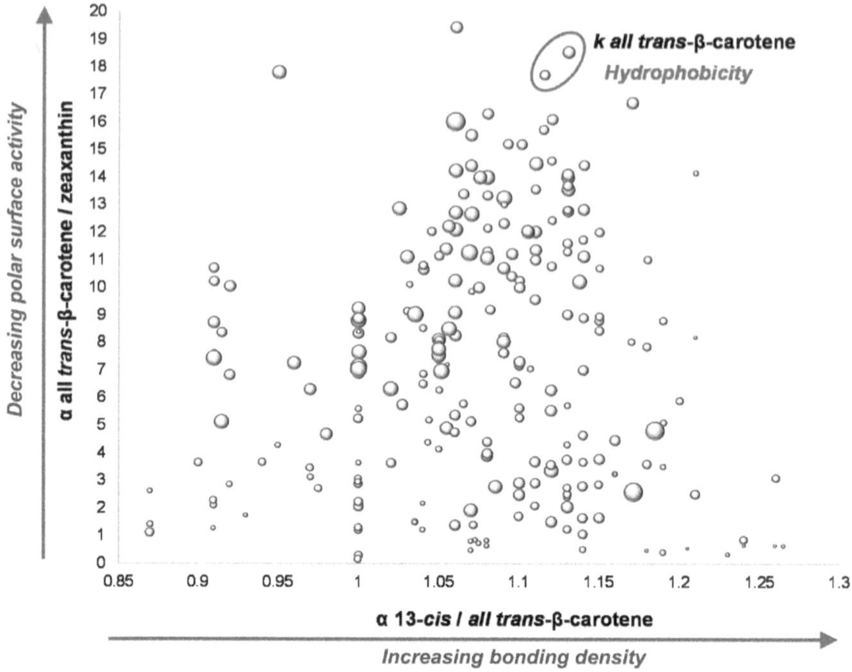

FIGURE 4.1 Classification of octadecylsiloxane-bonded silica stationary phases (carotenoid test) based on the analysis of three carotenoid pigments. Chromatographic conditions: CO_2-MeOH 85:15 (v/v), 25°C, 150 bar, 3 mL/min. Column identification can be found in the related literature.

molecule comprising two hydroxyl groups, it also allows polar surface activity to be evaluated (separation factor between all *trans*-β-carotene and zeaxanthin). Polar surface activity may be due to residual silanol groups or other polar groups present on the silica surface (namely hydrophilic end-capping groups or polar-embedded groups). With these three measurements, each column can be plotted on a simple figure (Fig. 4.1) with *x*-axis representing shape selectivity, *y*-axis polar surface activity, and point size representing hydrophobicity. Individual columns are then easily compared:

- the larger points indicate columns with high retention;
- from left to right, bonding density generally increases, thus columns on the right hand side usually have higher steric resistance, which may be profitable for isomer separations;
- from bottom to top, polar surface activity is reducing, causing less "parasitic" interactions with polar functional groups of the analytes.

While the carotenoid test was initially designed to qualify hydrophobic columns for RPLC, mostly octadecylsiloxane-bonded silica (ODS) stationary phases, it can also be used for SFC separations requiring a nonpolar column, like lipids or polynuclear aromatic hydrocarbons [23–26].

However, a large portion of SFC applications require polar stationary phases like bare silica or silica modified with polar bonded ligands. Most of these phases were initially designed for NPLC or for HILIC. The former were never the object of systematic chromatographic tests like those described earlier for RPLC. However, the recent enthusiasm for HILIC prompted several groups to propose tests to chracterize HILIC columns. As HILIC retention mechanism is a normal-phase process (retention increases with analyte polarity), such tests usually include a probe (or pair of probes) to assess hydrophilicity [27–29]. Structure selectivity may be evaluated also, for instance to assess methylene selectivity [27], shape selectivity, or hydroxyl selectivity (separation factor between two molecules differing by one hydroxyl group) which may be related to the strength of hydrogen bonding. Other features evaluated include ionic interactions [29]. When many such measurements are carried out, the comparison between columns is facilitated by use of radar plots [27], or by subjecting the data to principal component analysis [27,30]. Some of the probes employed in the testing procedures for HILIC columns could be utilized for columns in SFC conditions. However, to the best of our knowledge, no such transfer has been proposed to date.

Most often, the probe analytes employed to evaluate stationary phases for SFC use are drug molecules. This is obviously related to the history of SFC itself: for several years, the pharmaceutical industry was the principal applications area with interest in SFC use, both at the analytical and the preparative scales. In particular, basic molecules, which represent a significant part of small molecules synthesized in the pharmaceutical industry, were (and still are) at the center of attention. For instance, Guillarme, Veuthey and coworkers [31,32] employed 92 basic drugs to evaluate several stationary phases, particularly silica-based stationary phases with 2-ethylpyridine ligands, with particular emphasis on the peak shapes as observed through statistical occurrence of symmetrical peaks. Indeed basic analytes with large values for the acidity constant most often exhibit poor peak shapes both in RPLC and in SFC. While these tailing peaks observed in RPLC are often attributed to a small proportion of so-called "active sites" like residual silanol groups, in SFC they may also be due to poor solubility of highly basic molecules. Adequate probes for stationary phase polar surface activity in SFC must be carefully selected to avoid solubility issues.

The choice of mobile phase to evaluate stationary phase behavior is also a significant point, most notably in relation to basic analytes. Indeed ionizable species (acids or bases) are most often eluted with mobile phase compositions comprising not only carbon dioxide and a cosolvent, but also small concentrations of so-called "additives" (acids, bases, or salts). While such additives are useful in limiting solubility issues for ionizable species, possibly by forming ion pairs with the analytes, they are also adsorbed on the stationary phase. As a result, the stationary phase is no longer "virgin" but covered with the selected additive. In particular, possible "active sites" will be masked. The use of additives in column evaluation is then, to a certain extent, erasing some differences between stationary phases. It is then advisable to combine column evaluation with and without an additive.

McClain and Przybyciel [33] systematically studied 12 polar stationary phases designed for achiral SFC with 60 structurally diverse analytes. Only one mobile phase condition was selected (10% methanol, no additives), with isocratic elution. Based on selected structural families (acids, alcohols, amides, and amines), the columns were ranked based on retentiveness, selectivity, and peak shapes. No single column could be identified with broad utility but selected columns were suggested for particular applications. It was also observed that silanol groups do play an active role in retention mechanisms.

4.5.2 Chiral Probes

CSP have also been evaluated by several research groups using probe analytes. Phinney and Sander [34] proposed a set of five racemates to be used as standard reference material (SRM 877) for enantioselective stationary phases employed in HPLC and SFC. This set was intended for the evaluation of changes in column performance over time and for lot-to-lot variability. For this purpose, enantiomers considered as "likely to be separated" on most CSPs were selected. For instance, warfarin enantiomers are notoriously "easy" to separate. This also illustrates the difficulty in selecting relevant probes for CSP: while a test like this one is preferably based on enantiomers that should be separated on most CSPs, a performance test aiming at column comparison should, on the contrary, also comprise some enantiomer pairs that would not be separated on all columns.

De Klerck extensively tested CSPs in SFC conditions with a large set of 29–57 racemates of pharmaceutical interest [35,36]. The characterization of CSPs is rendered even more complicated by the fact that enantioseparation capabilities, especially on the most famous polysaccharide CSPs, vary significantly depending on mobile phase composition: changing the nature of the alcohol cosolvent (most frequently methanol, ethanol, and isopropanol), may cause a separation to degrade or improve substantially. De Klerck et al. then chose to evaluate the columns with different mobile phase compositions comprising either methanol or isopropanol with basic and acidic additives [37,38]. Such a large data set will result in a large data table, which is unpractical to represent with a simple figure like a radar plot. Statistical treatments are then necessary, and chemometric methods like principal component analysis, hierarchical cluster analysis, and others were applied to extract meaningful factors. A selection of stationary phases to be tested in a preferred order was then proposed, based on the success rate observed for the probe racemates and complementarity of the columns tested [35,39,40].

Hamman et al. [41] also evaluated 25 CSP with a set of 80 racemic compounds with three different mobile phase compositions. This resulted in a proposed set of columns and conditions (ethanol containing ammonium hydroxide) to be screened preferentially to achieve enantioresolution.

Based on the analysis of probe racemates, the comparison of CSPs may be based on enantioselectivity or enantioresolution. The latter is often found more relevant for the true performance of the CSP. Indeed stationary phases based on the same chiral selector but produced by different manufacturers may exhibit comparable enantioselectivity but different column efficiency, resulting in different enantioresolution. However, enantioresolution may also improve through increased retentiveness. It may then be advisable to retain as much data as possible to compare the CSPs based on different criteria.

4.6 QUANTITATIVE STRUCTURE–RETENTION RELATIONSHIPS

4.6.1 Retention Models for Achiral and Neutral Species

Achiral stationary phases have been extensively characterized with quantitative structure–retention relationships (QSRRs). QSRRs are useful for unraveling the interactions contributing to retention and separation mechanisms. QSRRs are based on two sorts of data: retention factors measured in the chromatographic system for a set of probe analytes with structural diversity with known molecular descriptors. For instance, one of the best known molecular descriptor is log $P_{o/w}$, the partition coefficient of a molecule between octanol and water. This descriptor was often used to relate retention in RPLC systems to analyte hydrophobicity. However, in SFC systems log $P_{o/w}$ is usually insufficient: [42] on hydrophobic ODS phases, a positive correlation is observed, similarly to RPLC; on polar stationary phases, a negative correlation is observed (although the fit is not as good as with ODS phases), similarly to NPLC and HILIC; on aromatic phases and ODS phases containing polar groups, no correlation is observed. Therefore other descriptors must be used to achieve adequate description of the retention mechanisms.

To design a QSRR with several descriptors, different approaches have been proposed.

Firstly, a large number of molecular descriptors can be systematically calculated with the help of molecular modeling software, then chemometric methods can help select a subset of relevant descriptors to observe those features that explain retention in the chromatographic system. This approach is usually favored when an accurate description is desired for predicting retention of future analytes. However, it often results in the selection of molecular descriptors that make little sense to chromatographers, and thus is impractical for the interpretation of retention mechanisms. Furthermore, the optimum descriptors may be different for different stationary phases, rendering all comparisons impossible.

A second approach is to select a small number of molecular descriptors, based on some previous knowledge of chromatographic retention behavior. This approach is most often preferred by chromatographers. Although the accuracy of the models obtained may be somewhat inferior to those obtained with the first method, resulting in inferior capabilities for retention prediction, the possibility to interpret the results is a strong motivator for this method [43].

For stationary phases employed in RPLC, the "hydrophobic subtraction model" developed by Snyder and coworkers [44] has found acceptance and is now employed by several column manufacturers for the purpose of column comparison. For stationary phases employed in HILIC, a similar retention model called "hydrophilic subtraction model" [45] was proposed recently. Judging from the diversity of stationary phases employed in SFC, comprising both polar and nonpolar stationary phases, none of these retention models could be applicable to all columns in SFC conditions. A unified characterization method is desirable to allow for the comparison of all usable columns. Furthermore, the methodology requires the choice of a reference analyte and a reference column, which seem both difficult to identify in SFC where reversed-phase or normal-phase retention modes may be observed.

One of the most famous QSRR methods employed to describe a large variety of chromatographic systems, in gaseous phase chromatography (GC), RPLC, NPLC, HILIC, high-performance thin-layer chromatography, or SFC, is the solvation parameter model based on Abraham descriptors [46–48]. It is one form of linear solvation energy relationship (LSER), because it is related to the Gibbs free energy of solvation of the analyte partitioning in a two-phase system. The form usually adopted to describe SFC systems is the following:

$$\log k = c + e\mathrm{E} + s\mathrm{S} + a\mathrm{A} + b\mathrm{B} + v\mathrm{V} \quad (4.1)$$

In Eq. (4.1), capital letters (E, S, A, B, V) are the solute descriptors while lower case letters (c, e, s, a, b, v) are the coefficients obtained when a multiple linear regression is applied to relate retention factors to analyte descriptors. The *c*-term is the model intercept and has little meaning. It comprises the variance remaining unexplained by the other terms, but is also related to phase ratio. The other five terms (*e, s, a, b, v*) are related to the strength of interactions occurring in the chromatographic system. When these coefficients are positive, they indicate that the considered interaction is stronger between analyte and stationary phase than between analyte and mobile phase. When the coefficient is negative, interactions with the mobile phase are stronger. Note that retention factors must be measured in isocratic conditions for the model to be applicable.

Each term of Eq. (4.1) can be easily interpreted in terms of molecular interactions that are familiar to chromatographers.

- The E descriptor measures the capability of an analyte to interact through π and nonbonding electrons. The *e*-term is then somewhat related to dispersive interactions, and to π–π interactions when the stationary phase ligands or mobile phase components also possess polarizable electrons.
- The S descriptor is related to the presence of local dipoles in the analyte. The *s*-term is then related to dipole–dipole interactions.
- The A and B descriptors indicate the capability of the analyte for hydrogen bonding as proton donor (A) or proton acceptor (B). Accordingly, the *a* and *b* terms characterize hydrogen bonding in the chromatographic system.
- The V descriptor is MacGowan's characteristic volume. It can be easily calculated from atom volumes corrected for the bond number and type [48]. The *v*-term may then be related to dispersive interactions, but also to the cavity effect in highly self-associated chromatographic phases as aqueous mobile phases (where water molecules strongly associate via hydrogen bonding) or polar ligands of a stationary phase.

Full details on the construction and interpretation of the solvation parameter model, as well as guidelines of good practice may be found elsewhere [47].

Eq. (4.1) was used extensively over the past 15 years to characterize stationary phases in SFC conditions [42,49–57]. While the results are interesting for comparing stationary phases, they have also brought out a new understanding of retention mechanisms in SFC. Indeed SFC is often described as a normal-phase technique because a large portion of users employ polar stationary phases derived from NPLC. Indeed the retention behaviors observed on an achiral polar stationary phase employed in NPLC [58] or SFC conditions [52,56] are most similar: most polar analytes are highly retained, while hydrophobic analytes with little capability for dipole–dipole interactions or hydrogen

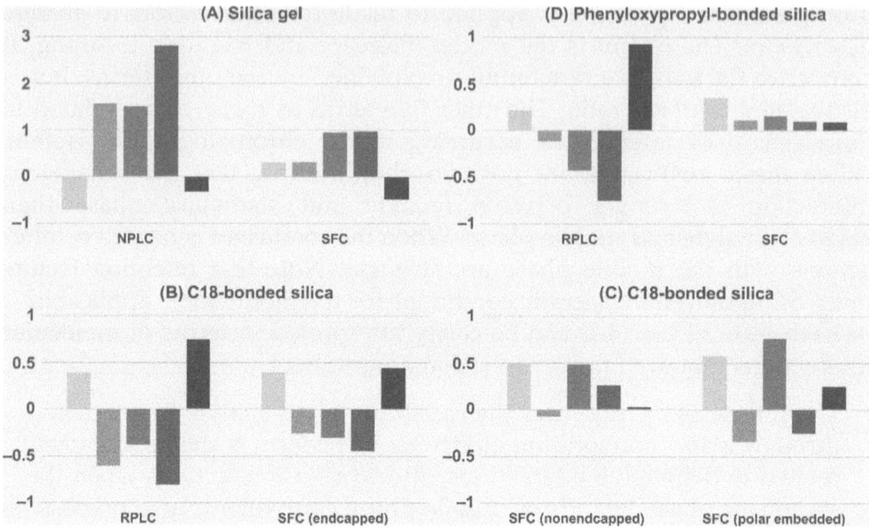

FIGURE 4.2 QSRR models calculated with Eq. (4.1), on different types of stationary phases as indicated in liquid phase and supercritical fluid conditions. SFC conditions: CO_2-MeOH 90:10 (v/v), 25°C, 150 bar, 1 or 3 mL/min depending on column dimensions. RPLC conditions: (b) methanol–water 90:10 (v/v) [59]; (d) methanol–water 70:30 (v/v) [62]. NPLC conditions: hexane-methyl-*t*-butylether 90:10 (v/v) [58].

bonding are little retained. This is illustrated in Fig. 4.2A: polar interactions terms (*e*, *s*, *a*, *b*) are positive, indicating that polar features of the analyte cause favorable interactions with the stationary phase, while the *v*-term is negative, indicating that increased hydrocarbon volume is unfavorable to retention in these systems.

As pointed out in Section 4.5, not only polar stationary phases but also ODS phases employed in RPLC may be useful in SFC. In that case the retention mechanism in SFC conditions [51] is most similar to that observed in RPLC when the aqueous fraction in the mobile phase is small [59], reducing the specific effects of water (hydrophobic effect and strong hydrogen bonding with proton acceptors). Thus polar analytes are little retained, while large, electron-rich molecules with little capability for hydrogen bonding are strongly retained. This is illustrated in Fig. 4.2B: dispersive interactions (*e* and *v* terms) are positive, indicating that increased hydrocarbon volume and number of polarizable electrons cause favorable interactions with the stationary phase, while the polar interaction terms (*s*, *a*, *b*) are negative, indicating that dipole–dipole interactions and hydrogen bonding are stronger with mobile phase components, thus favorable to elution.

While the previous two cases illustrate the possibility to transpose the knowledge acquired from HPLC to SFC, the case of ODS phases can be further expanded to evidence some specificities of SFC behavior. In Fig. 4.2B, a "classical" ODS phase with hydrocarbon ligands and hydrophobic end-capping groups was characterized in the SFC system. However, modern ODS phases may comprise a variety of polar groups. Some of them are considered as "parasitic active sites" like residual silanol groups not successfully covered by the ODS ligands. Other polar groups are intentionally introduced to limit analyte accessibility to residual silanol groups: they are hydrophilic end-capping groups or so-called "polar-embedded" groups, which are polar functions inserted in the alkyl chain, near the silica surface. In RPLC, such polar groups are intended to attract water molecules close to the surface, to create a water shield that the analytes would not cross, thereby preventing them from interacting with silanol groups. In SFC however, water molecules are essentially absent from the mobile phase (except very small proportions 1–5% employed as additives). As a result, the shielding effect does not occur, and analytes can freely interact, not only with alkyl chains but also with polar end-capping groups and polar-embedded groups. Depending on the type of polar group, different polar interactions may occur [57,60,61]. This is illustrated in Fig. 4.2C, with the LSER models obtained on different types of ODS phases. The LSER methodology was thus found to be a good complement to the carotenoid test, to characterize subtle differences between ODS phases [60,61].

Phenyl-bonded stationary phases are also a good illustration of specific behaviors observed in SFC, which are unlike RPLC or NPLC. In Fig. 4.2D, the LSER models obtained in RPLC [62] or SFC [53,63] on such a stationary phase are presented. The RPLC model is in accordance with usual models obtained in reversed-phase systems, comparable to the one in Fig. 4.2B for an ODS phase. The SFC model, on the contrary, is different from all above-described models: it is not similar to polar phases and not similar to ODS phases. Indeed all interaction terms are positive, indicating that all sorts of interactions are favorable to retention on such a column employed in SFC conditions. This could be one form of "mixed-mode" chromatography, where two or more retention mechanisms are superimposed. In the liquid phase, this term is usually employed to describe the combination of reversed-phase and ion-exchange mechanisms, but in the present case it could describe the superimposition of reversed-phase and normal-phase retention mechanisms.

For a given sample where the analytes have identified structures, it is possible to calculate molecular descriptors, then to use the retention models obtained with Eq. (4.1) to predict retention and separation factors on each column previously characterized. It was shown [64] that retention prediction is not totally accurate, but the prediction of selectivity is satisfactory and facilitates the initial choice of stationary phase, prior to optimization of operating conditions.

4.6.2 Retention Models for Achiral Ionizable Species

While Eq. (4.1) has been most useful in many cases, and permits comparison to other chromatographic systems characterized in the gaseous or liquid phase, it has some defects.

First, Eq. (4.1) only applies to neutral molecules. A modified version of the solvation parameter model was thus proposed to take account of ionizable species: [63,65–67]

$$\log k = c + e\mathrm{E} + s\mathrm{S} + a\mathrm{A} + b\mathrm{B} + v\mathrm{V} + d^{-}\mathrm{D}^{-} + d^{+}\mathrm{D}^{+} \qquad (4.2)$$

In Eq. (4.2) D^- represents the negative charge carried by anionic and zwitterionic species, and D^+ represents the positive charge carried by cationic and zwitterionic species, according to the following equations:

$$D^{-} = \frac{10^{(pH^{*}-pK^{*})}}{1+10^{(pH^{*}-pK^{*})}} \qquad (4.3)$$

$$D^{+} = \frac{10^{(pK^{*}-pH^{*})}}{1+10^{(pK^{*}-pH^{*})}} \qquad (4.4)$$

For neutral species, D^- and D^+ are zero so that Eq. (4.2) reverts to Eq. (4.1). Eq. (4.2) was initially developed for HILIC [65] but has since been shown to apply to SFC systems as well, with some approximations made for the pH and pK values [63,66].

Using this modified equation, it was shown that specific interactions due to the presence of ionized moieties for acids or bases could be evaluated. In particular, it was observed that:

- silanol groups (on bare silica stationary phases or as residual silanol groups on bonded silica phases) cause repulsion of anionic species and attraction of cationic species;
- stationary phases well protected against silanophilic interactions indeed do not exhibit significant interactions with charged species;
- pentafluorophenyl-bonded silica phases all exhibit strong retention of cationic species, which could not be explained by silanol interactions and was attributed to the highly electronegative fluorine atoms [63].

Based on the QSRR models obtained, a large number of ultra-high-performance achiral SFC columns were compared and classified. The results obtained with Eq. (4.1) or (4.2) can be simply plotted on a so-called "spider diagram" [42,54,66,68]. In Fig. 4.3, the spider diagram based on Eq. (4.2) is represented, with 35 stationary phases based on sub-2-μm fully porous particles or sub-3-μm superficially porous particles. The figure is easy to read and understand: basically, the horizontal axis represents overall polarity of the stationary phase, with nonpolar

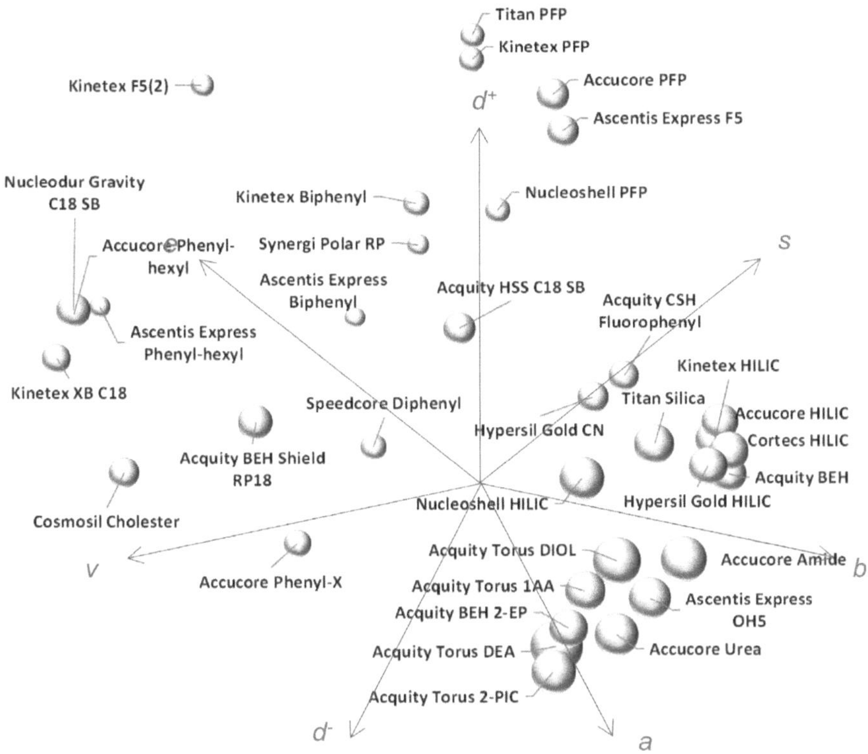

FIGURE 4.3 Classification of achiral stationary phases (spider diagram) based on QSRR models calculated with Eq. (4.2), with retention data measured for 85 neutral species and 24 ionizable species, for the 35 columns in Table 4.1. Chromatographic conditions: CO_2-MeOH 90:10 (v/v), 25°C, 150 bar, 1 or 3 mL/min depending on column dimensions.

stationary phases being plotted on the left (typically end-capped ODS phases) and polar stationary phases (NPLC and HILIC-type phases) being plotted on the right. The scattering of point on the vertical axis is related to varied capabilities for hydrogen bonding and ionic interactions. The bubble size is related to the overall strength of interactions. Stationary phases placed close to each other on this figure have similar capabilities for interactions, thus usually provide similar retention orders and selectivity. When two such columns are compared, the one with the largest bubble size will provide longer retention and larger retention factors. On the contrary, stationary phases placed at some distance in this figure will have different capabilities for interactions with the analytes, thus usually provide different elution orders and selectivity. This classification is constantly updated with new stationary phases and should be available as an open source on the internet in the near future (Table 4.1).

TABLE 4.1 Achiral Stationary Phases Compared in Fig. 3

Column name	Manufacturer	Particles	Bonded ligand	Dimensions (mm)	Particle size (μm)
ACQUITY UPC^2 HSS C18 SB	Waters	FPP	Octadecyl, nonendcapped	100×3.0	1.8
ACQUITY UPC^2 BEH	Waters	FPP	Bare hybrid silica	100×3.0	1.7
ACQUITY UPLC BEH Shield RP18	Waters	FPP	Alkyl with embedded carbamate group	100×3.0	1.7
ACQUITY UPC^2 BEH 2-EP	Waters	FPP	2-Ethylpyridine	100×3.0	1.7
ACQUITY UPC^2 CSH Fluorophenyl	Waters	FPP	Pentafluorophenyl	100×3.0	1.7
ACQUITY UPC^2 Torus 1-AA	Waters	FPP	1-Amino-anthracene	100×3.0	1.7
ACQUITY UPC^2 Torus 2-PIC	Waters	FPP	2-Picolyl-amine	100×3.0	1.7
ACQUITY UPC^2 Torus DEA	Waters	FPP	Diethylamine	100×3.0	1.7
ACQUITY UPC^2 Torus DIOL	Waters	FPP	Propanediol	100×3.0	1.7
Hypersil Gold Silica	Thermo	FPP	Bare silica	100×3.0	1.9
Hypersil Gold CN	Thermo	FPP	Cyanopropyl-bonded silica	100×2.1	1.9
Synergi Polar RP	Phenomenex	FPP	Phenyloxy-propyl	100×3.0	2.5
Titan PFP	Sigma-Aldrich	FPP	Bare silica	100×3.0	1.9
Titan Silica	Sigma-Aldrich	FPP	Pentafluorophenyl	100×3.0	1.9
Accucore HILIC	Thermo	SPP	Bare silica	150×4.6	2.6
Accucore 150-Amide-HILIC	Thermo	SPP	Polyamide	150×4.6	2.6
Accucore Urea-HILIC	Thermo	SPP	Propylurea	150×4.6	2.6
Accucore Phenyl-X	Thermo	SPP	Phenyl-alkyl	150×4.6	2.6

Accucore Phenyl-hexyl	Thermo	SPP	Phenyl-hexyl	150 × 4.6	2.6
Accucore PFP	Thermo	SPP	Pentafluorophenyl	150 × 4.6	2.6
Ascentis Express Biphenyl	Sigma-Aldrich	SPP	Biphenyl	150 × 4.6	2.7
Ascentis Express F5	Sigma-Aldrich	SPP	Pentafluorophenyl	150 × 4.6	2.7
Ascentis Express OH5	Sigma-Aldrich	SPP	Penta-hydroxyl	150 × 4.6	2.7
Ascentis Express Phenyl-Hexyl	Sigma-Aldrich	SPP	Penta-hydroxyl	150 × 4.6	2.7
COSMOCORE Cholester	Nacalai Tesque	SPP	Cholesteryl	150 × 4.6	2.6
Cortecs HILIC	Waters	SPP	Bare silica	150 × 4.6	2.7
Kinetex HILIC	Phenomenex	SPP	Bare silica	150 × 4.6	2.6
Kinetex PFP	Phenomenex	SPP	Pentafluorophenyl	150 × 4.6	2.6
Kinetex F5 (2)	Phenomenex	SPP	Pentafluorophenyl	150 × 4.6	2.6
Kinetex Biphenyl	Phenomenex	SPP	Biphenyl	150 × 4.6	2.6
Kinetex XB C18	Phenomenex	SPP	Octadecyl	150 × 4.6	2.6
Nucleoshell HILIC	Macherey-Nagel	SPP	Sulfobetaine	150 × 3.0	2.7
Nucleoshell PFP	Macherey-Nagel	SPP	Pentafluorophenyl	150 × 3.0	2.7
Speedcore Diphenyl	Fortis Technologies	SPP	Diphenyl-alkyl	150 × 4.6	2.6

FPP, Fully porous particles; SPP, Superficially porous particles.

One important conclusion resulting from the QSRR characterization and classification of achiral stationary phases in SFC was that an efficient screening strategy should preferably comprise stationary phases with different patterns of interactions, thus resulting in complementary selectivity. An adequate short selection of columns could then be chosen based on Fig. 4.3, with columns being well scattered in the selectivity space, rather than columns being all clustered in the same area of the classification.

4.6.3 Retention Models for Chiral Species

Another significant defect of Eq. (4.1) is that it is rather unfit for enantioselective separations. It was used several times to characterize retention propertied of CSPs [69–71], but is incapable of describing enantioselectivity [72,73]. Indeed enantioselectivity has a definite relation to shape and rigidity, in addition to the interactions described previously. For this purpose, a modified version of Eq. (4.1) was proposed to characterize CSPs:

$$\log k = c + e\mathrm{E} + s\mathrm{S} + a\mathrm{A} + b\mathrm{B} + v\mathrm{V} + f\mathrm{F} + g\mathrm{G} \tag{4.5}$$

In Eq. (4.5), F describes analyte flexibility, and G describes globularity or sphericity, as opposed to flat or linear molecules. The flexibility term is usually of little significance with respect to retention, but most significant to explain lack of enantioselectivity. Indeed flexible molecules have more ways to interact with a chiral selector and are usually more difficult to resolve, compared with rigid molecules. The globularity term can be interpreted in terms of steric resistance to insertion in the chiral cavities of a CSP (when considering large analytes). As such, it is generally a negative contributor to retention, as it prevents bulky analytes from fully interacting with polysaccharide CSPs hydrogen bonding sites that are located inside the chiral grooves. However, it may be favorable for enantioselectivity with small molecules, supposedly because small spherical molecules can fit better inside the chiral cavities than flat molecules.

Using Eq. (4.5), many CSPs based on polysaccharide chiral selectors have been characterized and compared in SFC conditions [74–78]. Most notably, the significant changes in hydrogen bonding when the polysaccharide is modified with chlorinated ligands (as opposed to nonhalogenated ligands) was evidenced. However, it must be admitted that the acquired knowledge is still insufficient and has not reached an equivalent level to that of achiral stationary phases. Chromatographers thus still need to rely on systematic screening of CSPs to identify the best chiral selector.

4.7 CONCLUSION

A full column characterization may comprise several steps. Efficiency is significant for high resolution, but is now somewhat comparable for columns packed with sub-2-μm fully porous particles, or sub-3-μm superficially porous particles. Retention and separation properties are best described with the help of probe analytes. Judging from the diversity of retention modes occurring in SFC on a large variety of stationary phase chemistries, it may be preferable to employ a large number of probe analytes with structural diversity, then use chemometric methods to extract meaningful comparisons. The choice of operating conditions is also critical for useful column characterization. Indeed SFC practice has evolved over the years, in particular with mixed mobile phases comprising not only carbon dioxide, but also cosolvents and additives (acids, bases, or salts), or even some portion of water.

References

[1] De la Puente ML, Soto-Yarritu PL, Anta C. Placing supercritical fluid chromatography one step ahead of reversed-phase high performance liquid chromatography in the achiral purification arena: a hydrophilic interaction chromatography cross-linked diol chemistry as a new generic stationary phase. J Chromatogr A 2012;1250:172–81.

[2] Ebinger K, Weller HN. Comparative assessment of achiral stationary phases for high throughput analysis in supercritical fluid chromatography. J Chromatogr A 2014;1332:73–81.

[3] Fairchild JN, Brousmiche DW, Hill JF, Morris MF, Boissel CA, Wyndham KD. chromatographic evidence of Silyl Ether Formation (SEF) in supercritical fluid chromatography. Anal Chem 2015;87(3):1735–42.

[4] Grand-Guillaume Perrenoud A, Veuthey J-L, Guillarme D. Comparison of ultra-high performance supercritical fluid chromatography and ultra-high performance liquid chromatography for the analysis of pharmaceutical compounds. J Chromatogr A 2012;1266:158–67.

[5] Lesellier E. Efficiency in supercritical fluid chromatography with different superficially porous and fully porous particles ODS bonded phases. J Chromatogr A 2012;1228:89–98.

[6] Lesellier E, West C. The many faces of packed column supercritical fluid chromatography—a critical review. J Chromatogr A 2015;1382:2–46.

[7] Delahaye S, Broeckhoven K, Desmet G, Lynen F. Design and evaluation of various methods for the construction of kinetic performance limit plots for supercritical fluid chromatography. J Chromatogr A 2012;1258:152–60.

[8] Lesellier E, Fougere L, Poe DP. Kinetic behaviour in supercritical fluid chromatography with modified mobile phase for 5 μm particle size and varied flow rates. J Chromatogr A 2011;1218(15):2058–64.

[9] Lesellier E, Fougere L. Kinetic behaviour in supercritical fluid chromatography with modified mobile phase for 5 μm particle size. Part II: effect of outlet pressure changes. J Chromatogr A 2014;1373:190–6.

[10] Delahaye S, Broeckhoven K, Desmet G, Lynen F. Application of the isopycnic kinetic plot method for elucidating the potential of sub-2 μm and core–shell particles in SFC. Talanta 2013;116:1105–12.

[11] Euerby MR, Petersson P. Chromatographic classification and comparison of commercially available reversed-phase liquid chromatographic columns using principal component analysis. J Chromatogr A 2003;994(1–2):13–36.
[12] Lesellier E, West C. Description and comparison of chromatographic tests and chemometric methods for packed column classification. J Chromatogr A 2007;1158(1–2):329–60.
[13] Kimata K, Iwaguchi K, Onishi S, Jinno K, Eksteen R, Hosoya K, et al. Chromatographic Characterization of Silica C18 Packing Materials. Correlation between a Preparation Method and Retention Behavior of Stationary Phase. J Chromatogr Sci 1989;27(12):721–8.
[14] Wise SA, Sander LC, Chang H-CK, Markides KE, Lee ML. Shape selectivity in liquid and gas chromatography: polymeric octadecylsilane (C18) and liquid crystalline stationary phases. Chromatographia 1988;25(6):473–9.
[15] Carr PW, Dolan JW, Neue UD, Snyder LR. Contributions to reversed-phase column selectivity. I. Steric interaction. J Chromatogr A 2011;1218(13):1724–42.
[16] Marchand DH, Carr PW, McCalley DV, Neue UD, Dolan JW, Snyder LR. Contributions to reversed-phase column selectivity. II. Cation exchange. J Chromatogr A 2011;1218(40):7110–29.
[17] Neue UD, O'Gara JE, Méndez A. Selectivity in reversed-phase separations: influence of the stationary phase. J Chromatogr A 2006;1127(1–2):161–74.
[18] Lesellier E, Tchapla A. A simple subcritical chromatographic test for an extended ODS high performance liquid chromatography column classification. J Chromatogr A 2005;1100(1):45–59.
[19] Lesellier E, West C, Tchapla A. Classification of special octadecyl-bonded phases by the carotenoid test. J Chromatogr A 2006;1111(1):62–70.
[20] Lesellier E. Extension of the C18 stationary phase knowledge by using the carotenoid test. J Sep Sci 2010;33(19):3097–105.
[21] Lesellier E. Extension of the carotenoid test to superficially porous C18 bonded phases, aromatic ligand types and new classical C18 bonded phases. J Chromatogr A 2012;1266:34–42.
[22] Lesellier E. Additional studies on shape selectivity by using the carotenoid test to classify C18 bonded silica. J Chromatogr A 2011;1218(2):251–7.
[23] Lesellier E, Latos A, De Oliveira AL. Ultra high efficiency/low pressure supercritical fluid chromatography with superficially porous particles for triglyceride separation. J Chromatogr A 2014;1327:141–8.
[24] Lesellier E, Tchapla A. Separation of vegetable oil triglycerides by subcritical fluid chromatography with octadecyl packed columns and CO2/modifier mobile phases. Chromatographia 2000;51(11–12):688–94.
[25] Lesellier E. Analysis of polycyclic aromatic hydrocarbons by supercritical fluid chromatography (SFC). Analusis 1999;27(3):241–8.
[26] Gaudin K, Lesellier E, Chaminade P, Ferrier D, Baillet A, Tchapla A. Retention behaviour of ceramides in sub-critical fluid chromatography in comparison with non-aqueous reversed-phase liquid chromatography. J Chromatogr A 2000;883(1–2):211–22.
[27] Kawachi Y, Ikegami T, Takubo H, Ikegami Y, Miyamoto M, Tanaka N. Chromatographic characterization of hydrophilic interaction liquid chromatography stationary phases: Hydrophilicity, charge effects, structural selectivity, and separation efficiency. J Chromatogr A 2011;1218(35):5903–19.
[28] Ibrahim MEA, Liu Y, Lucy CA. A simple graphical representation of selectivity in hydrophilic interaction liquid chromatography. J Chromatogr A 2012;1260:126–31.
[29] Iverson CD, Gu X, Lucy CA. The hydrophilicity vs. ion interaction selectivity plot revisited: The effect of mobile phase pH and buffer concentration on hydrophilic

interaction liquid chromatography selectivity behavior. J Chromatogr A 2016;1458:82–9.
[30] Dinh NP, Jonsson T, Irgum K. Probing the interaction mode in hydrophilic interaction chromatography. J Chromatogr A 2011;1218(35):5880–91.
[31] Grand-Guillaume Perrenoud A, Boccard J, Veuthey J-L, Guillarme D. Analysis of basic compounds by supercritical fluid chromatography: attempts to improve peak shape and maintain mass spectrometry compatibility. J Chromatogr A 2012;1262:205–13.
[32] Desfontaine V, Veuthey J-L, Guillarme D. Evaluation of innovative stationary phase ligand chemistries and analytical conditions for the analysis of basic drugs by supercritical fluid chromatography. J Chromatogr A 2016;1438:244–53.
[33] McClain RT, Przybyciel M. A systematic study of achiral stationary phases using analytes selected with a molecular diversity model. LCGC N Am 2011;29 (10):894–906.
[34] Phinney K, Sander L. Preliminary evaluation of a standard reference material for chiral stationary phases used in liquid and supercritical fluid chromatography. Anal Bioanal Chem 2002;372(1):101–8.
[35] De Klerck K, Vander Heyden Y, Mangelings D. Pharmaceutical-enantiomers resolution using immobilized polysaccharide-based chiral stationary phases in supercritical fluid chromatography. J Chromatogr A 2014;1328:85–97.
[36] De Klerck K, Parewyck G, Mangelings D, Vander Heyden Y. Enantioselectivity of polysaccharide-based chiral stationary phases in supercritical fluid chromatography using methanol-containing carbon dioxide mobile phases. J Chromatogr A 2012;1269:336–45.
[37] De Klerck K, Tistaert C, Mangelings D, Vander Heyden Y. Updating a generic screening approach in sub- or supercritical fluid chromatography for the enantioresolution of pharmaceuticals. J Supercrit Fluids 2013;80:50–9.
[38] De Klerck K, Mangelings D, Clicq D, De Boever F, Vander Heyden Y. Combined use of isopropylamine and trifluoroacetic acid in methanol-containing mobile phases for chiral supercritical fluid chromatography. J Chromatogr A 2012;1234:72–9.
[39] De Klerck K, Vander Heyden Y, Mangelings D. Exploratory data analysis as a tool for similarity assessment and clustering of chiral polysaccharide-based systems used to separate pharmaceuticals in supercritical fluid chromatography. J Chromatogr A 2014;1326:110–24.
[40] De Klerck K, Vander Heyden Y, Mangelings D. Generic chiral method development in supercritical fluid chromatography and ultra-performance supercritical fluid chromatography. J Chromatogr A 2014;1363:311–22.
[41] Hamman C, Wong M, Aliagas I, Ortwine DF, Pease J, Schmidt Jr DE, et al. The evaluation of 25 chiral stationary phases and the utilization of sub-2.0 μm coated polysaccharide chiral stationary phases via supercritical fluid chromatography. J Chromatogr A 2013;1305:310–19.
[42] West C, Lesellier E. A unified classification of stationary phases for packed column supercritical fluid chromatography. J Chromatogr A 2008;1191(1–2):21–39.
[43] Heaton DM, Bartle KD, Clifford AA, Klee MS, Berger TA. Retention prediction based on molecular interactions in packed-column supercritical fluid chromatography. Anal Chem 1994;66(23):4253–7.
[44] Snyder LR, Dolan JW, Carr PW. The hydrophobic-subtraction model of reversed-phase column selectivity. J Chromatogr A 2004;1060(1–2):77–116.
[45] Wang J, Guo Z, Shen A, Yu L, Xiao Y, Xue X, et al. Hydrophilic-subtraction model for the characterization and comparison of hydrophilic interaction liquid chromatography columns. J Chromatogr A juin 2015;1398:29–46.
[46] Abraham MH, Ibrahim A, Zissimos AM. Determination of sets of solute descriptors from chromatographic measurements. J Chromatogr A 2004;1037(1–2):29–47.

[47] Vitha M, Carr PW. The chemical interpretation and practice of linear solvation energy relationships in chromatography. J Chromatogr A 2006;1126(1–2):143–94.
[48] Poole CF, Poole SK. Column selectivity from the perspective of the solvation parameter model. J Chromatogr A 2002;965(1–2):263–99.
[49] Bui H, Masquelin T, Perun T, Castle T, Dage J, Kuo M-S. Investigation of retention behavior of drug molecules in supercritical fluid chromatography using linear solvation energy relationships. J Chromatogr A 2008;1206(2):186–95.
[50] Weckwerth JD, Carr PW. Study of interactions in supercritical fluids and supercritical fluid chromatography by solvatochromic linear solvation energy relationships. Anal Chem 1998;70(7):1404–11.
[51] West C, Lesellier E. Characterization of stationary phases in subcritical fluid chromatography by the solvation parameter model: I. Alkylsiloxane-bonded stationary phases. J Chromatogr A 2006;1110(1–2):181–90.
[52] West C, Lesellier E. Characterisation of stationary phases in subcritical fluid chromatography with the solvation parameter model: III. Polar stationary phases. J Chromatogr A 2006;1110(1–2):200–13.
[53] West C, Lesellier E. Characterisation of stationary phases in subcritical fluid chromatography with the solvation parameter model IV: aromatic stationary phases. J Chromatogr A 2006;1115(1–2):233–45.
[54] West C, Lesellier E. Orthogonal screening system of columns for supercritical fluid chromatography. J Chromatogr A 2008;1203(1):105–13.
[55] West C, Lesellier E. Characterization of stationary phases in supercritical fluid chromatography with the solvation parameter model. In: Grushka E, Grinberg N, editors. Advances in Chromatography. CRC Press; 2010. p. 195–203.
[56] West C, Khater S, Lesellier E. Characterization and use of hydrophilic interaction liquid chromatography type stationary phases in supercritical fluid chromatography. J Chromatogr A 2012;1250:182–95.
[57] Khater S, West C, Lesellier E. Characterization of five chemistries and three particle sizes of stationary phases used in supercritical fluid chromatography. J Chromatogr A 2013;1319:148–59.
[58] Kiridena W, Poole CF. Influence of solute size and site-specific surface interactions on the prediction of retention in liquid chromatography using the solvation parameter model. The Analyst 1998;123(6):1265–70.
[59] Vonk EC, Lewandowska K, Claessens HA, Kaliszan R, Cramers CA. Quantitative structure-retention relationships in reversed-phase liquid chromatography using several stationary and mobile phases. J Sep Sci 2003;26(9–10):777–92.
[60] Lesellier E, West C. Combined supercritical fluid chromatographic methods for the characterization of octadecylsiloxane-bonded stationary phases. J Chromatogr A 2007;1149(2):345–57.
[61] West C, Fougère L, Lesellier E. Combined supercritical fluid chromatographic tests to improve the classification of numerous stationary phases used in reversed-phase liquid chromatography. J Chromatogr A 2008;1189(1–2):227–44.
[62] Atapattu SN, Poole CF. Factors affecting the interpretation of selectivity on synergi reversed-phase columns. Chromatographia 2010;71(3–4):185–93.
[63] West C, Lemasson E, Khater S, Lesellier E. An attempt to estimate ionic interactions with phenyl and pentafluorophenyl stationary phases in supercritical fluid chromatography. J Chromatogr A 2015;1412:126–38.
[64] West C, Ogden J, Lesellier E. Possibility of predicting separations in supercritical fluid chromatography with the solvation parameter model. J Chromatogr A 2009;1216(29):5600–7.

[65] Chirita R-I, West C, Zubrzycki S, Finaru A-L, Elfakir C. Investigations on the chromatographic behaviour of zwitterionic stationary phases used in hydrophilic interaction chromatography. J Chromatogr A 2011;1218(35):5939–63.
[66] West C, Lemasson E, Bertin S, Hennig P, Lesellier E. An improved classification of stationary phases for ultra-high performance supercritical fluid chromatography. J Chromatogr A 2016;1440:212–28.
[67] Galea C, West C, Mangelings D, Vander Heyden Y. Is the solvation parameter model or its adaptations adequate to account for ionic interactions when characterizing stationary phases for drug impurity profiling with supercritical fluid chromatography. Anal Chim Acta 2016;924:9–20.
[68] West C, Lesellier E. Characterisation of stationary phases in subcritical fluid chromatography by the solvation parameter model: II. Comparison tools. J Chromatogr A 2006;1110(1–2):191–9.
[69] Vozka J, Kalíková K, Roussel C, Armstrong DW, Tesařová E. An insight into the use of dimethylphenyl carbamate cyclofructan 7 chiral stationary phase in supercritical fluid chromatography: the basic comparison with HPLC: liquid Chromatography. J Sep Sci 2013;36(11):1711–19.
[70] Janečková L, Kalíková K, Vozka J, Armstrong DW, Bosáková Z, Tesařová E. Characterization of cyclofructan-based chiral stationary phases by linear free energy relationship. J Sep Sci 2011;34(19):2639–44.
[71] Lokajová J, Tesařová E, Armstrong DW. Comparative study of three teicoplanin-based chiral stationary phases using the linear free energy relationship model. J Chromatogr A 2005;1088(1–2):57–66.
[72] Berthod A, Mitchell CR, Armstrong DW. Could linear solvation energy relationships give insights into chiral recognition mechanisms? J Chromatogr A 2007;1166 (1–2):61–9.
[73] Mitchell CR, Armstrong DW, Berthod A. Could linear solvation energy relationships give insights into chiral recognition mechanisms? J Chromatogr A 2007;1166 (1–2):70–8.
[74] West C, Zhang Y, Morin-Allory L. Insights into chiral recognition mechanisms in supercritical fluid chromatography. I. Non-enantiospecific interactions contributing to the retention on tris-(3,5-dimethylphenylcarbamate) amylose and cellulose stationary phases. J Chromatogr A 2011;1218(15):2019–32.
[75] West C, Guenegou G, Zhang Y, Morin-Allory L. Insights into chiral recognition mechanisms in supercritical fluid chromatography. II. Factors contributing to enantiomer separation on tris-(3,5-dimethylphenylcarbamate) of amylose and cellulose stationary phases. J Chromatogr A 2011;1218(15):2033–57.
[76] Khater S, Zhang Y, West C. In-depth characterization of six cellulose tris-(3,5-dimethylphenylcarbamate) chiral stationary phases in supercritical fluid chromatography. J Chromatogr A 2013;1303:83–93.
[77] Khater S, Zhang Y, West C. Insights into chiral recognition mechanism in supercritical fluid chromatography III. Non-halogenated polysaccharide stationary phases. J Chromatogr A 2014;1363:278–93.
[78] Khater S, Zhang Y, West C. Insights into chiral recognition mechanism in supercritical fluid chromatography IV. Chlorinated polysaccharide stationary phases. J Chromatogr A 2014;1363:294–310.

CHAPTER

5

Method Development in Supercritical Fluid Chromatography

M. Ashraf-Khorassani[1] and M. Combs[2]

[1]Virginia Tech, Blacksburg, VA, United States [2]Celanese Corporation, Narrows, VA, United States

OUTLINE

Supercritical Fluid Chromatography.
DOI: http://dx.doi.org/10.1016/B978-0-12-809207-1.00005-7

5.1 INTRODUCTION

Everyday analysts use different chromatographic techniques to develop separation methods. These methods can be used either in a simple separation using isothermal, isocratic, and isobaric conditions or a more complex separation with a thermal gradient, solvent gradient, or pressure gradient. For each approach, a suitable column, temperature, solvent, and pressure are required. However, the physical and chemical properties of the sample often determine which chromatographic technique and which type of detector are best suited for separation, identification, and quantification of the analytes. Therefore having more information about the sample allows the process of method developing to be simplified. Furthermore, a strong understanding of the fundamentals of chromatographic separation and experience with the techniques are helpful in a systematic approach to developing methods.

There are several things to do when developing a method in chromatography:

1. Select a chromatographic technique such as (but not limited to) gas chromatography (GC), High-Performance Liquid Chromatography (HPLC), or Supercritical Fluid Chromatography (SFC)
2. Select an appropriate detector
3. Select a method for the initial system (column type, mobile phase, temperature, etc.)
4. Select initial conditions (column size, solvent types and relative composition, flow rates, etc.)
5. Optimize the selectivity (stationary phase, solvent, additive, pH, etc.)
6. Optimize the system (solvent gradient program, temperature gradient program, flow rate, pressure, etc.).

Depending on the overall requirements and sample type, one can determine which chromatographic technique is suitable for an analysis. For example, if one is seeking to analyze a mixture of gases, it can be easily determined that GC is the most appropriate technique. If one is trying to analyze water-soluble compounds without derivatization, HPLC is a better choice. However, there are many samples that can be analyzed by either GC, SFC, or HPLC. For these samples, the aforementioned experience and understanding of chromatography is the basis for choosing the best separation technique. It is well documented that analyses can be done most easily, quickly, and efficiently with better sensitivity using GC rather than SFC and HPLC. There are many samples (thermally labile, high mass analytes, polar compounds, etc.) that cannot be analyzed by GC without derivatization. SFC or HPLC are the most viable choices for these analytes. For example there are many compounds (low to intermediate mass polymers, chiral compounds,

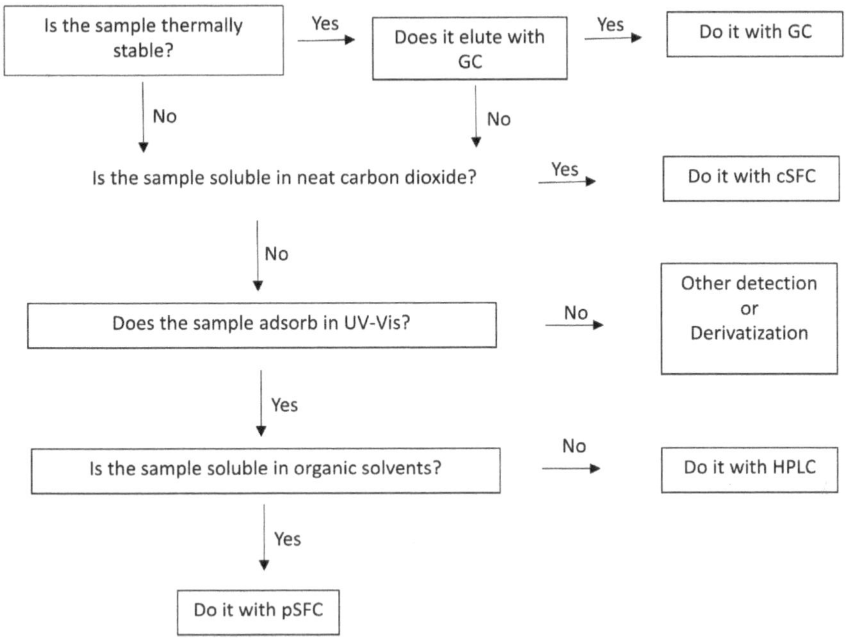

FIGURE 5.1 General strategy for supercritical fluid chromatography (SFC).

and pharmaceutical compounds) that are suited for SFC or HPLC, but due to greater efficiency, faster separation, and ease of method development, SFC would be a better choice than HPLC. Also there are many compounds (e.g., sugars, high mass polymers, proteins and peptides, carbohydrates, etc.) that are more easily analyzed by HPLC without sample preparation or derivatization.

Many publications have demonstrated a general scheme for method development in GC and HPLC. In this chapter our focus is SFC method development. Fig. 5.1 demonstrates the general strategy recommended for using SFC [1]. However, there are many factors that need to be considered for method development in SFC. Over the last two decades, interest has focused more in packed column SFC (pSFC) than in capillary SFC (cSFC) within multiple industries, particularly the pharmaceutical industry. Therefore our focus will be method development using packed column SFC.

5.2 SAMPLE TYPE

Before beginning method development, it is important to review the sample information. This is a common practice that is necessary for all

TABLE 5.1 Information regarding sample composition and properties

A.	Chemical structure
B.	Sample solubility
C.	number of analytes
D.	pKa values
E.	Molecular weight
F.	Concentration range
G.	Analyte UV spectra

chromatographic techniques. Table 5.1 summarizes sample information that is generally helpful for method development. This information can aid in determining the initial parameters for developing a robust method.

In SFC, much like HPLC, analyte solubility in the mobile phase is very important. Usually, the compounds of interest must be soluble in CO_2 or CO_2 that is modified with one of the common organic solvents (e.g., methanol, acetonitrile, isopropyl alcohol, dichloromethane, or chloroform) to be efficiently separated. If the analyte is highly polar and only soluble in water, then there is a high probability that the compound cannot be easily separated using SFC, since water has very poor solubility ($<0.3\%$) in CO_2 [2]. The addition of excess water to CO_2 can cause phase separation resulting in UV detector baseline instability. In another instance, if the compounds of interest are of high mass ($>10,000$ Da) and are polar, SFC is likely to be an unsuitable separation technique [3] for these compounds due to their poor solubility in pure or modified supercritical fluid CO_2.

As with all chromatography techniques, the goal of the separation must be identified. For example, if one is seeking to separate polymer additives from a polymer matrix, it is important to know the properties of the additives, especially the ones that will affect the SFC separation. Also it is important to consider the mass and solubility of the polymer matrix and as a potential interference with the elution of the polymer additives.

Yet another situation is the analysis of free fatty acids (FFAs) from a dietary supplement. The primary interest in this case is the analysis of the FFAs that will affect the SFC separation. However, one has to consider other components (e.g., fatty acid methyl esters, triglyceride mixtures, etc.) that are also present in these supplements and may interference with the SFC separation. Therefore sample preparation may be required to isolate the FFA or choose SFC conditions with specific detectors that allow for the separation and detection of FFAs without interference from other ingredients [4]. In each case, it is important to have standards or related compounds to develop the SFC separation before starting the analysis with a real-world matrix.

The physical and chemical properties of the sample can provide critical information for selection of the initial SFC condition. The choice of initial conditions can be garnered from prior experience with similar compounds, literature information, such as application notes, and a combination of both.

5.3 DETECTORS

Detector selection is an important component of method development for SFC. The following questions should be answered prior to commencing method development:

A. What detection limit is necessary?
B. Do analytes have UV chromophores suitable for detection?
C. Can universal detectors such as MS, ELSD, or FID be used?

Most state-of-the-art packed column SFC systems come with a UV detector installed. It can be used with most supercritical fluids and common solvent modifiers. Also it is compatible with most additives that are commonly used in SFC (e.g., formic acid, isopropylamine, trifluoroacetic acid, ammonium acetate, etc.). In most cases for SFC, a UV detector is the best choice so long as the analytes have a UV chromophore and are of sufficient concentration to be detected. Unlike HPLC, in SFC the back pressure must be controlled after the UV detector in order to maintain the mobile phase density in the flow cell. A change in density of the mobile phase within the flow cell affects the refractive index and results in increased background noise. In order to keep the noise as low as possible, it is important to have a stable pumping system with a reliable back pressure system to reduce pressure fluctuations. All UV flow cells in SFC must withstand high pressure and be pressure rated at least to 400 atm [5].

There are other near-universal detectors such as evaporative light scattering detectors (ELSD) or mass spectrometry (MS) which are often utilized in SFC. However, these detectors have their limitations. Both ELSD and MS require splitting of outlet flow from the SFC column in order to achieve a reasonable level of sensitivity and detection. Also, due to the lack of controlled flow into these detectors, quantitative analysis is often difficult. For both detectors, it is recommended to add make-up flow after the depressurization point to prevent the precipitation of analytes or ice formation, which can cause flow fluctuations. However, packed column SFC separations are typically performed using modifiers, meaning that in these cases, the make-up flow is not necessary unless the modifier percentage is below 5% [6].

For MS the commercially available atmospheric pressure ionization sources (both electrospray ionization (ESI) and atmospheric pressure chemical ionization (APCI)) that are used with LC-MS are also suitable for SFC-MS [7,8]. The addition of make-up solvent is required for most SFC-MS systems. The make-up solvent not only enhances analyte ionization (when modifier does not contain a basic or acidic additive), but it also provides stable flow for analytes eluting early in a separation with a mobile phase containing a low percentages of modifier (typical for gradient separations).

GC detectors, such as flame ionization (FID) [9], nitrogen-phosphorous (NPD) [10], electron-capture (ECD) [11], sulfur and nitrogen chemiluminescence detectors (CLSD or CLND) [12,13] have been used with SFC as well. However, they are mostly limited to pure CO_2 or very low percentages ($<2-3\%$) of modified CO_2. Again quantitative results may be difficult to obtain with these detectors due to poor flow control after the column.

5.4 COLUMNS

Initially capillary columns or polymeric coated packed capillary column stationary phases were used for SFC separations with pure CO_2 and flame ionization detection [14]. These columns were usually coated with nonpolar to mid-polar stationary phases, which limited analyte retention. With these columns, method development was usually focused on pressure, density, or temperature programming. Changing the stationary phase could provide some change in selectivity, but usually the analyte solubility in the mobile phase or its vapor pressure determined the success of the separation. Usually moderately polar compounds eluted with long retention times or with severe tailing, which was attributed to poor solubility in the mobile phase and a limited scope for further method development.

Microbore and analytical scale HPLC packed columns are another option for SFC. Initially, the poor performance of packed columns was attributed to active silanol groups [15]. DColumn manufacturers were able to use various chemistries to deactivate columns. However, poor reproducibility and inherent instability of these stationary phases exacerbated problems due to activity [16]. One of the most common methods to minimize the activity of the silanol groups was the addition of polar modifiers (e.g., methanol, ethanol, isopropanol, etc.) [17]. Later it was found that modifiers not only deactivate the stationary phase active sites but also increase analyte solubility in the mobile phase and increase interactions between analytes and the stationary phase [18]. In the past two decades, the column of choice for pSFC has typically been HPLC packed columns. The increased interest in pSFC in recent years has induced column manufacturers to introduce columns specifically designed for SFC [19]. These columns are more reproducible and robust and compatible with different modifiers, additives, and higher operating temperatures.

Typically both reversed-phase (RP) and normal-phase (NP) columns have been used in SFC since the solvent does not restrict the type of stationary phase as for HPLC. Therefore the most important strategy in achieving a separation by SFC is choosing an appropriate column rather than changing the mobile phase selectivity. Of course selectivity may be improved by varying the modifier type or the modifier gradient. Selectivity can also be enhanced by using additives in the modifier, changing the mobile phase outlet pressure, or varying column temperature as well. Typically, however, the column determines the success of a separation. It is important to emphasize that most pSFC separations, especially those of polar compounds, cannot be achieved without the presence of a modifier.

5.4.1 Reversed-Phase Columns

Reversed-phase (RP) packed columns (C_1, C_4, cyclohexyl, phenyl, C_8, C_{18}, C_{18} endcapped, C_{18} polar endcapped, and C_{18} polar embedded) have been evaluated for the separation of both polar and nonpolar analytes [20,21]. Usually these columns have shown good selectivity for separations of most nonpolar and some intermediate polar compounds by SFC. West et al. [22] characterized more than 170 RP columns using the solvation parameter model, which uses five descriptors (E, S, A, B, V) related to the interactions (electron lone pair, dipole-type, hydrogen bond acceptor, hydrogen bond donor, dispersion, respectively) of the analytes to investigate column separation capabilities. They used identical SFC conditions to acquire retention data for a series of carotenoids and 29 aromatic compounds. Eight-axis "spider" interaction diagrams were created by combining the results from both test mixtures (Fig. 5.2A). Columns located in the same area display similar selectivity. It was established that columns that are located far to the left of the "spider" are the least polar and those located at the bottom far right exhibit the most polar activity. In order to evaluate their classification, two test mixture were used. The first one was a mixture of alkylbenzenes and the second a combination of polar compounds (aspirin, sulindac, and phenylbutazone). The results for alkylbenzenes (nonpolar compounds) with carbon numbers in the alkyl chain from 11 to 15 is

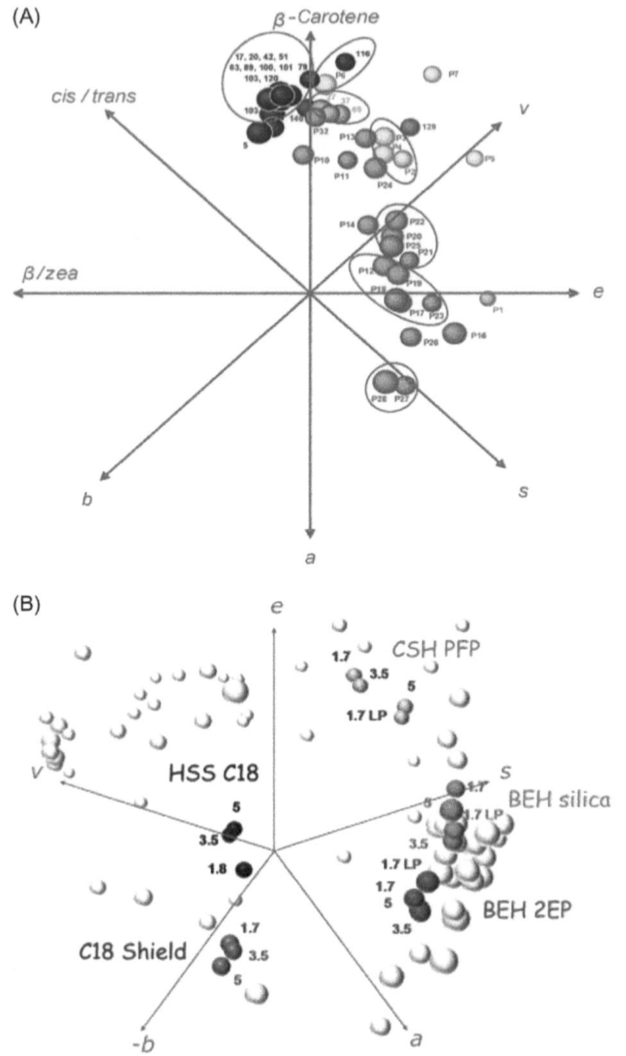

FIGURE 5.2 (A) "Spider" diagram for an eight-dimensional representation of stationary phases evaluated with both the solvation parameter model and the carotenoid test. (B) Spider diagram for five additional stationary phases (see text). *Source: Reprinted from reference West C, Fougére L, Lesellier E. Combined supercritical fluid chromatographic tests to improve the classification of numerous stationary phases used in reversed-phase liquid chromatography. J Chromatogr A 2008;1189:227–44 [22]. Khater S, West C, Lesellier E. Characterization of five chemistries and three particle sizes of stationary phases used in supercritical fluid chromatography. J Chromatogr A 2013;1319:148–59 [23].*

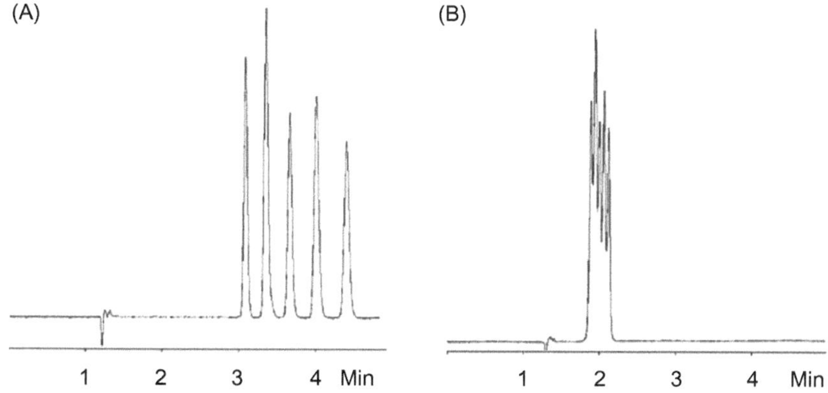

FIGURE 5.3 Separation of alkylbenzenes with alkyl chain with a carbon number 11–15. (A) Uptisphere ODB No. 51, and (B) Uptisphere PLP No.P26. West C, Fougére L, Lesellier E. Combined supercritical fluid chromatographic tests to improve the classification of numerous stationary phases used in reversed-phase liquid chromatography. J Chromatogr A 2008;1189:227–44.

shown in Fig. 5.3A. The column hydrophobicity decreases from the top left (least polar column, stationary phase (SP) 51) to the bottom right (most polar column, SP26, Fig. 5.3B) of Fig. 5.2A. For the polar compounds, it was concluded that the classical endcapped C_{18} columns (SP51 in Fig. 5.2) with a low concentration of accessible silanol groups interacted less with the acidic polar test compounds and were eluted earlier while phenylbutazone eluted last (Fig. 5.4A). For the nonendcapped C_{18} columns (e.g., Uptisphere NEC SP27 in Fig. 5.2A), the acidic compounds (aspirin and sulindac) were retained longer on the column, a phenomenon which demonstrated the importance of the stronger interactions of acidic analyte functional groups with the residual silanol groups on the stationary phase (Fig. 5.4B). This interaction was even stronger for the C_{18} column with polar endcapped phases (Aquasil C_{18}, SP P2 in Fig. 5.2A), since sulindac was eluted at a later retention time than on the nonendcapped columns (Fig. 5.4C). This interaction is stronger for columns with polar embedded phases (Uptisphere PLP SP P26 in Fig. 5.2A), since acidic compounds were retained longer and eluted after phenylbutazone (Fig. 5.3D). These results indicate that RP columns, which are most suitable for the separation of polar compounds in HPLC, are not suitable for the separation of polar compounds in pSFC. Here, the secondary polar interactions of the columns have become the driving force for the separation rather than hydrophobic interactions.

In a similar study, Khater et al. [23] evaluated five stationary phases with a range of polarity for separations in pSFC. Two of the stationary phases, which will be discussed in this section, were identified as HSS

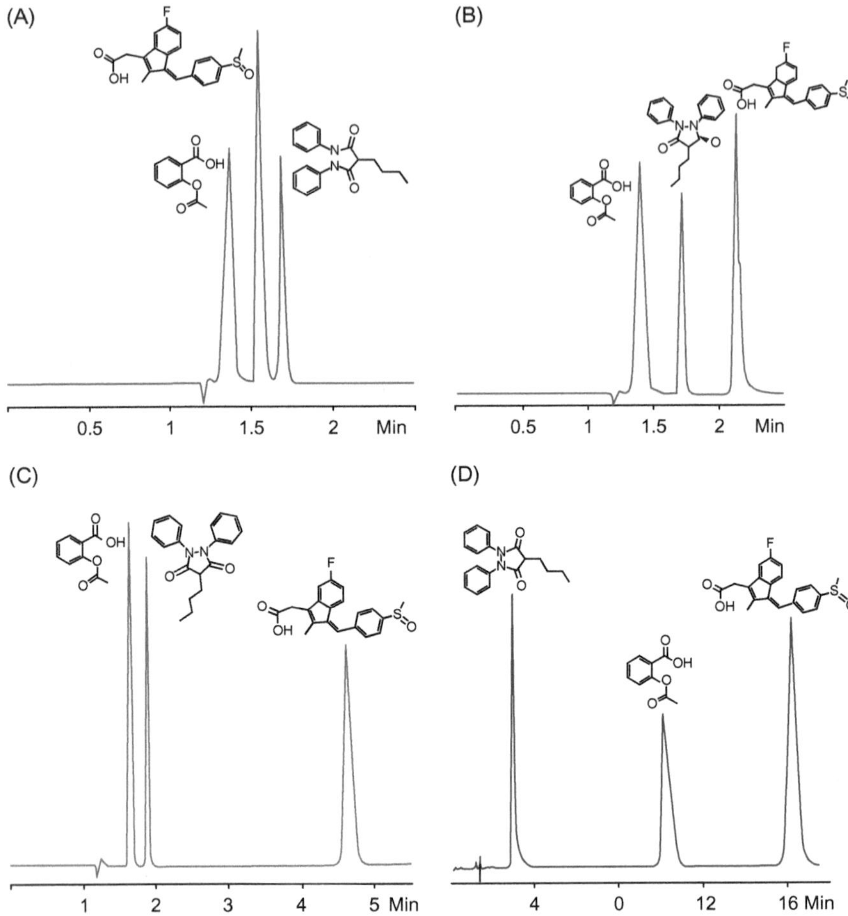

FIGURE 5.4 Separations of nonsteroidal antiinflammatory drugs (aspirin, phenylbutazone, and sulindac). (A) Uptisphere ODB No. 51, (B) Uptisphere NEC No. 27, (C) Aquasil C_{18} No.P2, and (D) Uptisphere PLP No.P26. *Source: Reprinted from reference West C, Fougére L, Lesellier E. Combined supercritical fluid chromatographic tests to improve the classification of numerous stationary phases used in reversed-phase liquid chromatography. J Chromatogr A 2008;1189:227–44 [22].*

C_{18} SB and BEH Shield RP C_{18}. The other three columns that have polar (BEH and BEH 2-EP) and intermediate polar stationary phases (CHS Fluorophenyl) will be discussed in the NP column section. The BEH shield RP C_{18} column has an embedded polar carbamate group and HSS C_{18} SB is a classical nonendcapped C_{18}-bonded phase. They used the solvation parameter model and more than one hundred compounds to characterize these columns. Their findings were added to two other studies additionally reported by West et al. [24,25], which characterized many polar and nonpolar columns using the solvation parameter model, as described previously. All data from these three projects were plotted on a single five axis "spider" diagram (Fig. 5.2B). They concluded that the BEH Shield RP C_{18} column was consistent with polar embedded columns at the bottom left of the "spider" interaction diagram. However, the HSS C_{18} SB column was located at the center of the "spider" and not in the same place as for classical C_{18} phases (upper left corner). This depicted the HSS C_{18} SB as a column with more polar character than classical C_{18} columns. In order to qualify their finding, different mixtures (pesticides and propionic acids drugs) were also tested. Fig. 5.5A shows a separation of propionic acid drugs and pesticides using both HSS C_{18} SB and BEH Shield RP C_{18} columns. For the pesticides mixture (Linuron, Isoproturon, and Diuron), all three compounds eluted faster from the BEH Shield RP C_{18} column compared to the HSS C_{18} SB column (see Fig. 5.5B); while the propionic acid drug mixture was separated with reasonable resolution on the HSS C_{18} SB column and only poorly separated on the BEH Shield RP C_{18} column (see Fig. 5.5A). It was suggested that the poor separation of the acidic drug mixture using the BEH Shield RP C_{18} column was caused by hydrogen bonding of the acidic functional groups with the carbamate oxygen and nitrogen atoms of the stationary phase. The near baseline separation of the acidic drugs on the HSS C_{18} SB column was due to the strong interaction of the analytes with polar active sites (silanol groups) of the stationary phase. Ashraf-Khorassani et al. [4] compared the same two columns (BEH Shield RP C_{18} and HSS C_{18} SB) and three other polar columns (BEH, BEH 2-EP, and HSS CN) to separate 31 FFAs. Of all tested columns, the HSS C_{18} SB column showed the highest selectivity for separation of FFAs while the same mixture eluted from the BEH Shield RP C_{18} column without separation. Fig. 5.6 shows the separation of these FFAs in the HSS C_{18} SB column in less than 7 minutes. This stresses the importance of polar interactions in the separation of compounds in SFC even when these interactions are secondary effects.

There are other studies by West et al. [26], which used the solvation parameter model to characterize additional columns covering a wide polarity range. The results were similar to previous findings presented in Fig. 5.2. Galea et al. [27], characterized 27 different stationary phases (nonpolar and polar) for SFC. They used 64 polar compounds (mostly nitrogen containing compounds) instead of nonpolar to intermediate polar compounds used in earlier studies. When the results were plotted in a "spider" diagram, once again, nonpolar classical C_{18} columns were bunched in one cluster at the top far left corner of the "spider" with most polar columns clustered at the lower right corner of the "spider." Chemometric techniques were then used to select six dissimilar stationary phases (ODS, NH_2, Silica, Phenyl, Hilic, and CN) for comparison.

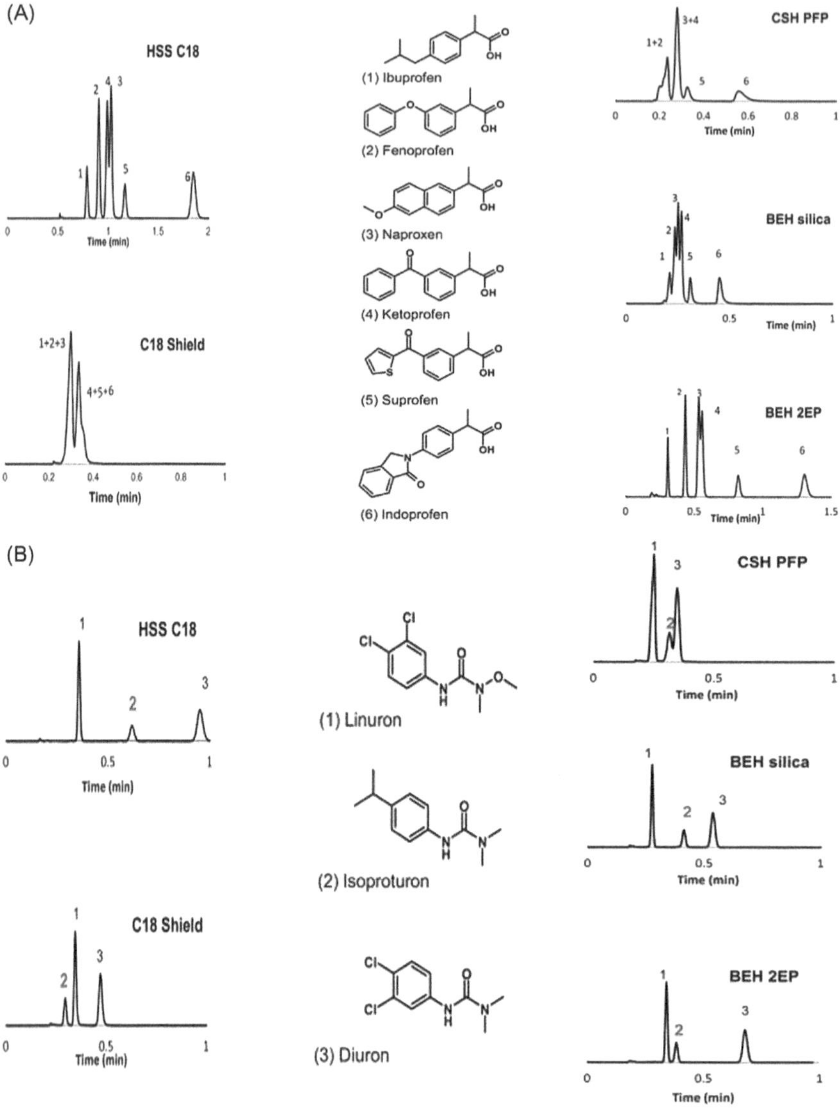

FIGURE 5.5 Separation of (A) Propionic acid drugs and (B) Three phenylurea pesticides on the five different stationary phase. Column dimensions: 100 × 3.0 mm, particle size 3.5 μm, flow rate 3 mL/min, $T = 25°C$, mobile phase: CO_2–MeOH 90:10 (v/v), outlet pressure: 150 bars, UV detection 210 nm, and injection volume: 1 μL. *Source: Reprinted from reference Khater S, West C, Lesellier E. Characterization of five chemistries and three particle sizes of stationary phases used in supercritical fluid chromatography. J Chromatogr A 2013;1319:148–59 [23].*

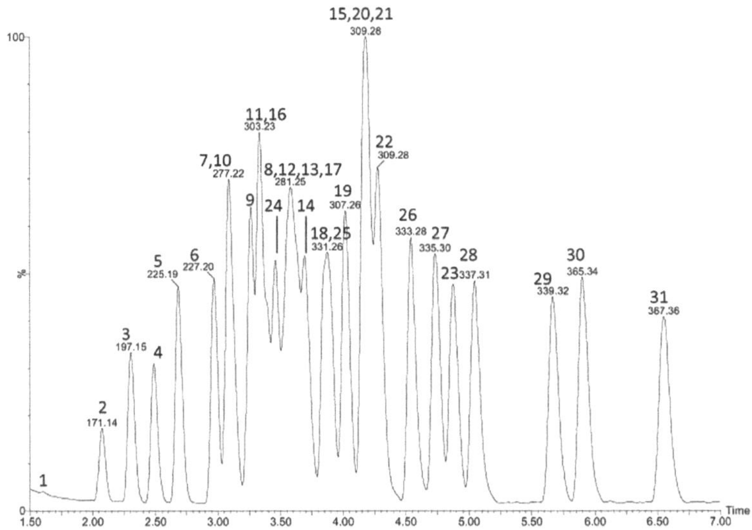

FIGURE 5.6 UHPSFC/MS of GLC411 FFA's mixture using HSS C18 SB column – chromatography conditions: time (min) 0, 98/2 (A/B), time 0.5, 98/2, time 8, 96/4, time 9, 80/20, time 11, 80/20, time 11, 98/2, time 12,98/2. A = CO_2, B = MeOH with 0.1% (v/v) formic acid, flow: 1 mL/min, oven temp.: 25°C, ABPR 1500 psi, column HSS C18 150 × 3.0 mm, 1.8 m, makeup IPA flow rate: 0.2 mL/min, mixture: GLC411 FFA's, conc. 0.003mg/mL per compound. *Source: Reprinted from reference Ashraf-Khorassani M, Isaac G, Rainville P, Fountain K, Taylor LT. Study of ultrahigh performance supercritical fluid chromatography to measure free fatty acids with out fatty acid ester preparation. J Chromatogr B 2015;997:45–55 [4].*

To show the disparity between these stationary phases, two mixtures—one a combination of basic compounds and the other a combination of acidic compounds—were evaluated on these six columns. Fig. 5.7A shows the separation of basic compounds while Fig. 5.7B shows the separation of the acidic compounds on all columns. It is observed that the Sunfire C_{18} (chromatogram B in Fig. 5.7)—the only nonpolar column selected—did not exhibit any selectivity for the separation of basic or acidic compounds. The results from this study confirms that "spider" diagrams can be used to select appropriate columns for a separation when there is sufficient information for analyte polarity and their solubility in the mobile phase.

For RP columns, it can be concluded that hydrophobic stationary phases improve the separation of nonpolar compounds and provide fewer opportunities for the separation of polar compounds if there are no polar or active silanol group on the stationary phase. Also the addition of modifiers will decrease the separation time and enhance peak shapes for late-eluting compounds. However, the addition of an additive will not affect the separation of nonpolar compounds on nonpolar columns.

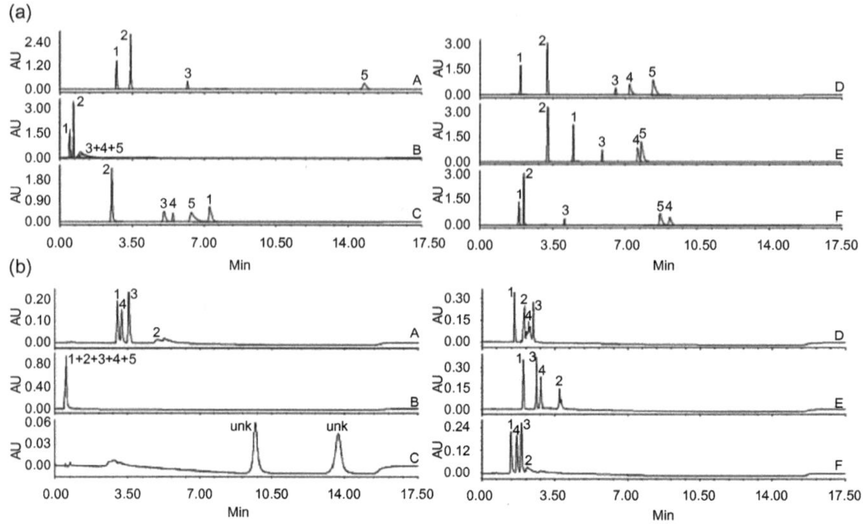

FIGURE 5.7 (a) Separation of five drugs, resorcine (1), carbamazepine (2), digitoxine (3), ranitidine (4), and antazoline (5) using Luna Silica (A); Sunfire C_{18} (B); Luna NH_2 (C); Inertsil Phenyl-3 (D); Luna HILIC (E); Luna CN (F). SFC conditions: 150 bar, 25°C, gradient elution with MeOH containing 0.1% IPAm in CO_2 (5–40% over 10 min), flow rate 3 mL/min, UV detection wavelength: 220 nm. (b) Separation of acetyl salicylic acid (1), salicylic acid (2), acetylsalicylsalicylic acid (3), and salicylsalicylic acid (4). using Luna Silica (A); Sunfire C_{18} (B); Luna NH_2 (C); Inertsil Phenyl-3 (D); Luna HILIC (E); Luna CN (F). SFC conditions: 150 bar, 25°C, gradient elution with MeOH containing 0.1% IPAm in CO_2 (5–40% over 10 min), flow rate 3 mL/min, UV detection wavelength: 220 nm. *Source: Reprinted from reference Galea C, Mangelings D, Heyden Y. Method development for impurity profiling in SFC: the selection of a dissimilar set of stationary phases. J Pharm Biomed Anal 2015;111:333–43 [27].*

5.4.2 Normal-Phase Columns

Most of the methods developed for the separation of achiral compounds have used normal-phase (NP) columns in pSFC. The addition of modifiers and additives to the mobile phase is integral to pSFC for the separation of polar compounds on polar columns.

Since the early days of pSFC, NP columns (e.g., NH_2, Diol, Cyano, silica, 2-ethylpyridine) have been more popular for the separation of polar compounds than reversed-phase columns. Usually if pure CO_2 is used to elute compounds containing basic or acidic functional groups on NP columns, there is a reasonable chance that most will not elute or elute with strong tailing [28,29]. In the early 1980s [30–32] the separation of polar compounds on NP HPLC columns using CO_2 modified with an organic solvent (e.g., methanol, ethanol, etc.) was largely developed. Modifiers not only cover active sites, but also increase solvent strength and the solubility of polar compounds in the mobile phase [19]. In many cases, to reduce the strong interactions of analytes with the stationary phase or active sites on silica, an additive such as trifluoroacetic acid (TFA), formic acid (FA), ammonium acetate (AA), or isopropylamine (IPAm) was added. Additives not only improve peak shape, they also improve selectivity. The use of additives for the separation of complex mixtures was first suggested by Berger [33]. He used tetramethylammonium hydroxide (TMAOH) 0.001% (v/v) added to methanol for the separation of 24 PTH-amino acids. Later, Berger [34–38] used different NP columns (e.g., CN, diol, amino, sulfonic acid, and silica) for the separation of phenolic acids, benzene poly carboxylic acids, benzylamines, phenylenediamines, analines, benzamides, and phenylethylamines with methanol containing different additives (e.g., trifluoroacetic acid, dichloroacetic acid, trichloroacetic acid, acetic acid, chloroacetic acid, IPAm, triethylamine, tetrabutylamine, and diethylamine). In these studies, the effects of additives in common organic modifiers to improve peak shapes and selectivity were conclusively demonstrated using both RP and NP columns. It is notable that all NP columns used in these separations showed some selectivity for the separation of each group of compounds. In two cases, an RP octyl column was tested for the separation of polar compounds. In each case, no selectivity was observed with all compounds eluting with the solvent as a single peak.

By mid-2000, most SFC applications were directed towards pharmaceutical compounds, which were mostly polar compounds. In studies conducted by West et al. [26] and Galea et al. [27], a number of NP columns were studied as well as RP columns. In these studies it was concluded that polar columns cluster at the bottom right of the "spider" interaction diagram and as the polarity of the columns increases, they move towards the bottom right of the "spider" (Fig. 5.2). Galea et al. [27] concluded that for the separation of basic compounds with a series of polar and nonpolar columns (CN, NH_2, HILIC, silica, phenyl, and C_{18}), the polar columns (CN, NH_2, HILIC, and silica) were best suited for the separation. Interestingly all four polar columns exhibited high retention and good selectivity for the separation of the polar mixture (Fig. 5.7A). The phenyl column also showed good selectivity, while no selectivity was observed with the C_{18} column. No optimization was performed on any of the columns. It is important to note that 0.1% IPAm was used as an additive in order to obtain the separations. Furthermore the same polar columns were evaluated for the separation of an acidic test mixture (Fig. 5.7b). Reasonable selectivity was observed with silica, phenyl-3, HILIC, and CN columns using 0.1% IPAm as an additive. However, these acidic analytes were retained much longer on the amino (NH_2) column and did not elute under the same conditions as the other columns. It has been demonstrated by de al Puente et al. [39] that diethylaminopropyl (DEAP), NH_2, and 2-ethylpyridine (2-EP) have the strongest hydrogen bond acceptor values. Therefore it is believed that when polar columns such as DEAP or NH_2 are used in the separation of acidic compounds, it is likely they will be retained for longer or will simply not elute due to strong acid/base interactions.

5.4.3 Chiral Columns

The separation of chiral compounds by SFC is performed using either a chiral stationary phase (CSP) or by addition of a chiral selector to the mobile phase. However, for preparative-scale separations the addition of chiral additives to the mobile phase is less attractive because of the need to separate the analytes from the additive after collection [40,41]. In addition, there are not many chiral selectors with high solubility in SFC mobile phases, limiting their application.

The most common CSP columns are cellulose-based (Chiralcel OD, OJ, OF, OG, OK, Trefoil Cel1, Trefoil Cel2, Lux Cellulose 1, Lux Cellulose 2, etc.) or amylose-based (Chiralpak AD, IC, IA, AS, Trefoil Amy, Lux Amylose-1, etc.), Fig. 5.8. Stationary phases are either chemically bonded (IA, IC, etc.) or physically coated (AD, OD, Cel1, Cel2, Amy, etc.) on a silica-based particle.

Unfortunately there are no standard rules for selecting columns for method development for chiral separations. Several chiral stationary phases are evaluated initially with CO_2 and a modifier (methanol,

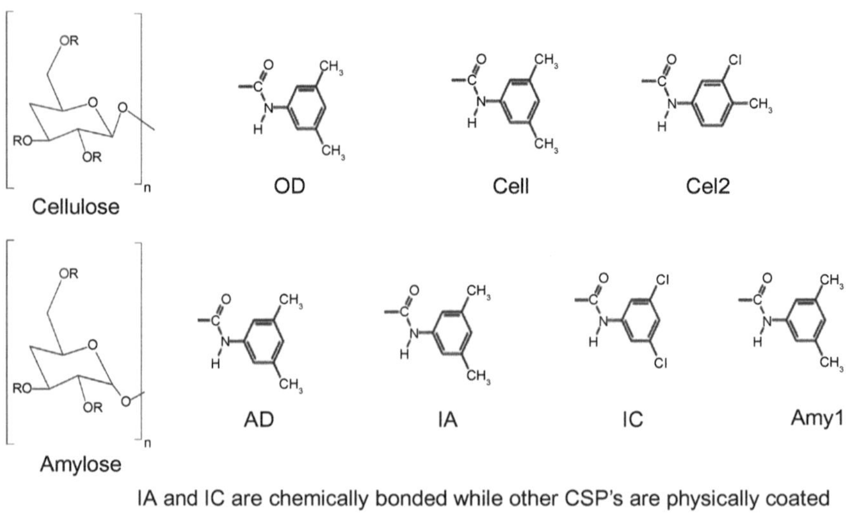

FIGURE 5.8 Structure of some commercially available chiral stationary phases (CSPs).

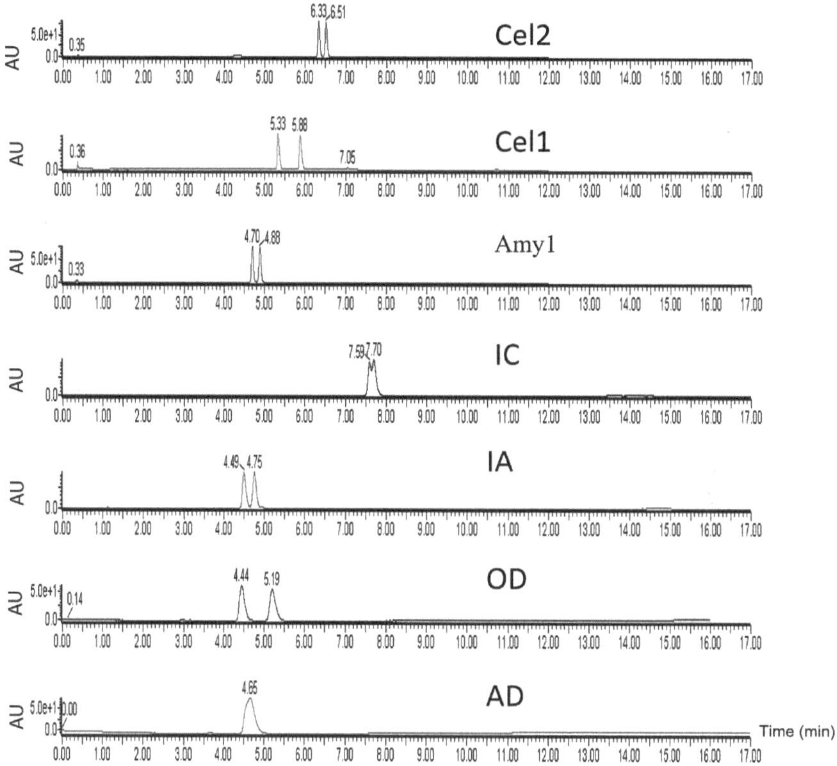

FIGURE 5.9 Separation of propranolol enantiomer with CO_2 modified with MeOH containing 0.25% (v/v) IPAm on different CSPs. *Source: Reprinted from reference Ashraf-Khorassani M, Taylor LT. Supercritical fluid chromatography separation of chiral compounds using 5 and sub 3μm chiral stationary phases. J Chromatogr Sci; submitted for publication [42].*

ethanol, isopropyl alcohol, acetonitrile or their mixtures) with and without additives. Fig. 5.9 shows the separation of propranolol on seven CSPs with cellulose-based (OD, Cel1, and Cel2) and amylose-based (AD, IA, IC, and Amy) columns from two different manufacturers [42]. The separation of propranolol was obtained using all of the cellulose-based CSPs and two of the amylose-based CSPs with methanol as a modifier and 0.25% IPAm as an additive. Several groups have rated the success of chiral columns for their separation capabilities using various modifiers and additives. Villeneuve et al. [43] used three modifiers (methanol, ethanol, and isopropanol) and two additives (trifluoroacetic acid and trimethylamine) for the separation of seven racemic compounds. Based on their observation, the success rate for the separation of the enantiomers was indicated as AD > AS > OJ > OD. Maftouh et al. [44] evaluated more than 40 basic, acidic, and neutral racemic compounds on four columns (Chiralcel OD, OJ, and Chrialpak AD, AS), with two modifiers (10% MeOH and 20% IPA) and two additives (0.5% IPAm for basic compounds and 0.5%TFA for acidic compounds) at 30°C and 200 atm back pressure. The success rate for the separation of enantiomers was indicated as AD > OD > OJ > AS. Ashraf-Khorassani et al. [42] evaluated seven chiral columns with two modifiers and three additives at 40°C and 140 bar. The success rate for varied enantiomers was indicated as IC > AD = IA = AMY > Cel2 > OD > Cel1 stressing the "trial and error" approach to chiral SFC.

It has been demonstrated in many studies that methanol and isopropyl alcohol are the preferred modifiers for the separation of enantiomers. However, this is not true for all compounds. For the separation of ephedrine enantiomers on Chiralpak AS columns using 90% CO_2 and 10% methanol, ethanol, or isopropanol with 0.5% IPAm Maftouh et al. [44] observed that only ethanol was suitable for the separation. Usually additives such as TFA or IPAm are added if the chiral compound is highly ionizable. The addition of an additive usually suppresses the ionization of compounds, which decreases the interaction of the analyte with the CSP. Furthermore additives can adsorb onto the CSP and alter its enantioselectivity [45]. Steric hindrance (IPAm vs TEA) can also effect the separation of chiral compounds. For example in some cases IPAm provided higher selectivity compared with TEA [46].

5.5 MOBILE PHASE FOR pSFC

Carbon dioxide is the most widely used mobile phase for packed column SFC due to its modest critical parameters, low cost, safety, and availability. It has been established that the polarity of CO_2 is more or less similar to hexane [47,48], even at high densities. For a well-behaved separation the polarity of the analytes should be between those of the stationary and mobile phases. Initially more polar fluids (e.g., ammonia and SF_6) were used in SFC for the separation of polar compounds [49,50]. However, due to their high critical parameters, higher cost, corrosiveness, and lack of success they are little used today. Modifiers added to carbon dioxide at low percentages (1–10%) to increase the mobile phase polarity, increase analyte solubility, and decrease analyte interactions with the packing materials became the method of choice to vary mobile phase elution strength and selectivity for the separation of polar compounds. Gere [30] compared 2-methoxyethanol (which is slightly more polar than MeOH) and 2-propanol as modifiers in CO_2 for the separation of caffeine, theophylline, theobromine, and xanthine on a silica column. It was demonstrated that polar analytes can be separated on packed columns using modified SF CO_2 in less than 3 minutes. From the mid-1980s until the early 1990s, different polar solvents (e.g., methanol, dimethyl sulfoxide, tetrahydrofuran, 1-propanol, propylene carbonate, 1-hexanol, dioxane, acetonitrile, methylene chloride, dimethylacetamide, and 2-methoxy ethanol) were evaluated as modifiers in SFC for the separation of polar compounds [17,31,32]. However, in most cases it was demonstrated that alcohols (especially methanol) was the best choice.

In order to obtain the separation of a mixture of wide polarity with high efficiency in a short time, gradient techniques are necessary. In SFC the density gradient (changing pressure, or temperature or a combination of both) is usually applied for the separation of low-polar compounds with pure CO_2 as a mobile phase. Density programming has mostly been used with capillary and micro-packed columns. For polar compounds (neutral or ionic) modifier gradients are often required for successful separations. Modifier gradients usually start with a low percentage of modifier (1–5%) and increase rapidly over a few minutes to 40–50% (ramp rates vary from 5 to 10% per minute). Many research groups have used this type of gradient with a hold at the highest modifier composition for fast SFC separations of polar compounds [39,51–53]. Also this type of gradient is commonly used as a "scouting gradient" to determine if an analyte of interest can be eluted from the column with the selected modifier. If the analytes are eluted with the CO_2-modifier mixture with symmetrical peaks, then changing the gradient program rate may be sufficient to obtain the required separation. Of course, it is important to re-emphasize that the column choice is important at this stage. For complex mixtures, changing the modifier gradient ramp rate is not always effective at separating analytes with similar chemical properties. Changing the selectivity by using a different stationary phase may be more effective. To demonstrate this, Patel et al. [54] compared four columns (2-EP, silica, diol, and cyano) with the same mobile phase gradient (methanol containing 10 mM ammonium acetate mixed with 5% H_2O) for the separation of 10 anionic water-soluble sulfated estrogens. The 2-EP column provided the best separation compared with the other stationary phases, even though all analytes were not fully resolved. Occasionally, lowering the modifier polarity can improve a separation, but in most cases, doing so will result in longer separations. Ashraf-Khorassani et al. [55] compared the separation of four water soluble nucleobases on a diol column using four alcohols as modifiers and the same gradient profile. Peak shapes were sharper with methanol compared with the higher alcohols.

5.6 ADDITIVES

When carbon dioxide with organic modifiers lacks sufficient solvent strength to elute polar compounds with acceptable peak shapes in packed column SFC the addition of organic bases (triethylamine, IPAm, etc.), acids (formic acid, trifluoroacetic acid, etc.) or salts (ammonium acetate, ammonium formate, etc.)—typically secondary modifiers called additives—are recommended. These additives are usually added at low concentrations (0.1–0.5%) into the primary modifier to improve peak shapes. The interactions of the additive with both analytes and the stationary phase are believed to be responsible for the improvement in peak shapes and changes in selectivity. These interactions can be summarized as: (1) Ion suppression of analytes which can increase or decrease the electrostatic interactions of the analytes with the stationary phase, (2) Stationary phase modification to deactivate the active sites on silica which decreases the interactions of polar compounds with silanol groups, or they can adsorb into the stationary phase and create a more selective phase with superior interaction capabilities, allowing for a better separation, (3) Additives can dissociate and form ion-pairs with ionic molecules, which can enhance their solubility in the mobile phase while simultaneously decreasing or increasing interaction with the stationary phase, (4) Ionic additives with the same charge as the analyte can behave similar to an ion-exchange mechanism, (5) Additives can enhance the polarity of modifiers in CO_2 to increase the solubility of the analyte in the mobile phase.

Blackwell et al. [56] studied the effects of acidic and basic additives on the elution of naphthalene derivatives with packed columns using both sub and supercritical fluid conditions. They concluded that the retention of analytes with hydrogen bond donor/acceptor functional groups was more strongly influenced by the presence of additives in the mobile phase than compounds with weak hydrogen bond acceptor properties. They also demonstrated that peak efficiency always improved with the addition of additives under supercritical conditions, more so than under subcritical conditions. Usually the concentration of additives in a modifier depends on the nature of the packed column. The concentrations of basic additives typically vary from 0.25% to 0.5% (V/V) while for acidic additives concentrations of 0.1% to 0.5% (V/V) are more common. Salt concentrations usually range from 10 to 25 mM in modifiers. Since the late 1990s the addition of additives into the modifier has been part of the pSFC method development toolbox for the analysis of all polar compounds [42,57].

Another tool for method development is the simultaneous addition of more than one (typically two) additives to carbon dioxide–modifier mixtures to improve separations. Jones et al. [58] used a mixture of TFA and TEA as additives for the separation of β-agonists. Analytes were eluted with higher efficiency when both additives were combined compared with TEA alone. Ashraf-Khorassani et al. [59] used 10 mM ammonium acetate in methanol containing 5% water for the separation of eight polar compounds. The addition of water enabled the baseline separation of caffeine and theophylline, while the same compounds coeluted in the absence of water in the mobile phase. The addition of a second additive can increase or decrease the interactions of the analytes with the stationary phase, enhancing their resolution.

5.7 DENSITY

In pSFC three factors affect the mobile phase, potentially resulting in a change in both retention and selectivity. These factors are pressure, temperature, and the mobile phase composition. Depending on the pressure and temperature, the density of pure CO_2 can vary usually from 0.2 to 1.1 g/mL. The solubility of analytes depends on the density of the CO_2. Therefore increasing the CO_2 pressure or decreasing the column temperature can increase the solubility of the analyte in CO_2. Thus pressure or temperature gradient/programming can be used to elute analytes with different solubilities in CO_2. However, increasing the temperature can have both positive and negative effects on the solubility of analyte in CO_2. Lou et al. [60] demonstrated that solute solubility can be increased when temperature increases due to the higher vapor pressure of the analyte. At the same time solubility decreases when density is reduced by increasing the temperature. Temperature can also play a major role in solute diffusivity which can enhance column efficiency by decreasing the resistance to mass transfer in the mobile phase. This factor can be critical for certain separations. Both temperature and pressure (i.e., density) play an important role in SFC when CO_2 or CO_2 containing low concentrations (<2%) of modifier is used as a mobile phase. The recommended starting operating density (with pure CO_2 or low concentrations of modifier in CO_2) for analytical separations with an oven temperature of 35–40°C is between 0.7 and 0.75 g/mL (110–140 bar) for stable UV detection. Increasing the temperature at this pressure can cause phase separation in the UV flow cell, which can be a major problem for the UV detector (noisy baseline), but not necessarily for other detectors such as MS or ELSD. Also increasing the temperature can be of concern for thermally labile compounds and certain columns.

Organic modifiers typically have lower compressibility compared with supercritical CO_2. The addition of modifier decreases the compressibility of CO_2 but also increases the density of the mobile phase, which generally has a minimal effect on retention [61]. A low column back pressure of 120–150 bar in separations with a modifier gradient aids in the separation of analytes eluting early in the chromatogram with a low percentages of modifier, while it will not affect the later-eluting analytes at higher modifier concentrations. Increasing the column back pressure to over 200 bar amplifies this process. Therefore increasing the column back pressure is not recommended when using CO_2 with modifier gradient programs. The mobile phase eluent strength is usually controlled by increasing or decreasing the modifier concentration rather than temperature, pressure, or density. In the separation of polar compounds, while temperature and pressure are important, they are not as important as the mobile phase composition. Typically the operating pressure and temperature are set between 120–150 bar and 35–40°C, respectively.

5.8 METHOD DEVELOPMENT IN PRACTICE

In general it can be concluded that in pSFC, RP columns are not usually suitable for the separation of polar compounds which are best separated on polar columns. In addition if the strongest molecular interactions are dipole/induced dipole or dispersion forces pure CO_2 can be used. However, when polar compounds are separated with polar columns, addition of modifiers and potentially additives are necessary to obtain the optimum selectivity and efficiency. The polar columns most used for the separation of polar compounds are 2-EP, CN, Diol, and silica. To start developing a method, the nature of the analytes, whether acidic or basic should be established. Then, one may begin with stationary phases containing 2-EP for bases or Diol for acidic analytes with methanol as a modifier in CO_2. It is best to start with a scouting gradient containing a small percentage of modifier (95:5 CO_2–MeOH) and to increase the modifier concentration to 50:50 CO_2–MeOH over 10 minutes and maintain the final modifier concentration for 5 minutes. Due to the favorable physical properties of CO_2, one can use a high flow rate which can shorten the separation time. If the analytical columns are 250 or 150 × 4.6 mm packed with 3–5 μm particles, the flow rate can be 3–5 mL/minute with a back pressure of 120–150 bar and a column temperature of 35–40°C. Note, however, that the pressure drop across the column is low (10–25 bar) at the start of the separation but as the modifier concentration increases the pressure drop increases and may reach 50–100 bar or even higher depending on the column length and particle size. If the analytes of interest elute with substantial tailing, it is recommended to use an additive that matches the analyte chemistry. Either a basic or acidic additive in the concentration range 0.1–0.5% (V/V) added to the modifier. In some cases addition of a secondary additive (e.g., water) may increase selectivity and improve resolution of partially separated compounds. At the same time the addition of a second additive could have negative effects on other peaks [54].

5.9 CONCLUSION

Packed column SFC is a highly versatile and flexible separation technique typically affording a greater choice of optimization parameters than conventional GC or HPLC. Although GC is the most sought after technique due to its simplicity, pSFC provides the analyst with additional tools for difficult separations. The mobile phase can be altered based on temperature, pressure, and density for nonpolar compounds on RP columns. The polar interactions, even of these RP columns, make separations difficult for polar compounds requiring manipulation of the mobile phase polarity for optimal separations. Polar columns are typically used for the separation of polar compounds to minimize the small secondary effects observed with RP columns. The separation of polar compounds typically requires the use of modifiers and additives (secondary modifiers) in the mobile phase. Often small alcohols are the best modifiers, while matching analyte chemistry to additive properties facilitates the selection of additives.

References

[1] Berger T. Packed column SFC. RSC chromatography monographs. Cambridge: The Royal Society of Chemistry; 1995.
[2] Sabirzyanov AN, Il'in AP, Akhunov AR, Gumerov FM. Solubility of water in supercritical carbon dioxide. High Temp 2003;40:203–6.
[3] Takahashi K. Polymer analysis by supercritical fluid chromatography: review. J Biosci Bioeng 2013;116:133–40.
[4] Ashraf-Khorassani M, Isaac G, Rainville P, Fountain K, Taylor LT. Study of ultrahigh performance supercritical fluid chromatography to measure free fatty acids with out fatty acid ester preparation. J Chromatogr B 2015;997:45–55.
[5] Berger T. Instrumentation for analytical scale supercritical fluid chromatography. L Chromatogr A 2015;1421:171–83.
[6] Ashraf-Khorassani M, Coleman WM, Dube MF, Isaac G, Taylor LT. Synthesis, purification, and quantification of fatty acid ethyl esters after trans-esterification of large batches of tobacco seed oil. Beiträge zur Tabakforschung International 2015;26:205 13.
[7] Chester T, Pinkston DJ. Pressure-regulating fluid interface ande phase behavior consideration in the coupling of packed-columns supercritical fluid chromatography with low-pressure detectors. J Chromatogr A 1998;807:265–73.
[8] Combs MT, Ashraf-Khorassani M, Taylor LT. Packed column supercritical fluid chromatography-mass spectroscopy: review. J Chromatogr A 1997;785:85–100.
[9] Taylor LT, Chang KHC. Packed column development in supercritical fluid chromatography. J Chromatogr Sci 1990;28:357–66.
[10] Ashraf-Khorassani M, Levy JM. Evaluation of thermionic detector with packed capillary supercritical fluid chromatography of nitrogen-containing compounds. Fresenius J Anal Chem 1991;342: 688–371.
[11] Tarver EE, Hill HH. Comparison of a pulsed electron capture detector and a Fourier transform ion mobility detector after capillary supercritical fluid chromatography. Fresenius J Anal Chem 1992;344:453–9.

[12] Shi H, Strode JTB, Taylor LT, Fujinari EM. Feasibility of supercritical fluid chromatography-chemiluminescent nitrogen detection with open tubular columns. J Chromatogr A 1996;734:303–10.
[13] Foreman W, Shellum CL, Briks JW, Sievers RE. Supercritical fluid chromatography with sulfur chemiluminescence detection. J Chromatogr A 1989;465:23–9.
[14] Novotny M, Springston SR, Peaden PA, Fjeldsted JC, Lee ML. Capillary supercritical fluid chromatography. Anal Chem 1981;53:407A–14A.
[15] Ashraf-Khorassani M, Taylor LT. Chromatographic behavior of polar compounds with liquid vs. supercritical mobile phases. J Chromatogr Sci 1988;26:331–5.
[16] Ashraf-Khorassani M, Taylor LT, Henry RA. Packed column SFC comparison of conventional and polymer coated silica bonded phase. Chromatographia 1990;28:569–73.
[17] Blilie AL, Greibrokk T. Modifier effect on retention and peak shape in supercritical fluid chromatography. Anal Chem 1985;57:2239–42.
[18] Ashraf-Khorassani M, Taylor LT. Column efficiency comparison with supercritical fluid carbon dioxide vs. methanol modified carbon dioxide as a mobile phase. Anal Chem 1990;62:1173–8.
[19] Berger T, Berger B, Major RE. A review of column development for supercritical fluid chromatography. LC-GC AM 2010;28:344–51.
[20] Ashraf-Khorassani M, Taylor LT, Henry RA. Packed column supercritical fluid chromatographic comparison of conventional and polymer-coated silica bonded phases. Chromatographia 1989;28:569–73.
[21] Brossard S, Lafosse M, Dreus M. Comparison of ethyoxylated alcohols and polyethylene glycols by high performance liquid chromatography and supercritical fluid chromatography using evaporating light-scattering detection. J Chromatog 1992;591:140–57.
[22] West C, Fougère L, Lesellier E. Combined supercritical fluid chromatographic tests to improve the classification of numerous stationary phases used in reversed-phase liquid chromatography. J Chromatogr A 2008;1189:227–44.
[23] Khater S, West C, Lesellier E. Characterization of five chemistries and three particle sizes of stationary phases used in supercritical fluid chromatography. J Chromatogr A 2013;1319:148–59.
[24] West C, Lesellier E. Characterisation of stationary phases in subcritical fluid chromatography with the solvation parameter model. J Chromatogr A 2006;1110:200–13.
[25] West C, Lesellier E. Chemometric methods to classify stationary phases for achiral packed column supercritical fluid chromatography. J Chemom 2012;26:52–65.
[26] West C, Lesellier E. A unified classification of stationary phases for packed column supercritical fluid chromatography. J Chromatogr A 2008;1191:21–39.
[27] Galea C, Mangelings D, Heyden Y. Method development for impurity profiling in SFC: the selection of a dissimilar set of stationary phases. J Pharm Biomed Anal 2015;111:333–43.
[28] Ashraf-Khorassani M, Taylor LT. Chromatographic behavior of polar compounds with liquid vs. supercritical fluid mobile phases. J Chromatogr Sci 1988;26:331–6.
[29] Schenmakers PJ, Uunk LGM, Janssen H. Comparison of stationary phases for packed-column supercritical fluid chromatography. J Chromatogr 1990;506:563–78.
[30] Gere DR. Supercritical fluid chromatography. Science 1983;222:253–9.
[31] Berry AJ, Games DE, Perkins JR. Supercritical fluid chromatographic and supercritical fluid chromatography-mass spectrometric studies of some polar compounds. J Chromatogr 1986;363:147–58.
[32] Levy JM, Ritchey WM. Investigations of the uses of modifiers in supercritical fluid chromatography. J Chromatogr Sci 1986;24:242–8.
[33] Berger TA, Dye JF, Ashraf-Khorassani M, Taylor LT. Gradient separation of PTH-amino acids employing supercritical CO_2 and modifier. J Chromatogr Sci 1989;27:105–10.

[34] Berger TA, Dye JF. Effect of basic additives on peak shapes of strong bases separated by packed-column supercritical fluid chromatography. J Chromatogr Sci 1991;29:310–17.
[35] Berger TA, Dye JF. Separation of phenols by packed column supercritical fluid chromatography. J Chromatogr Sci 1991;29:54–9.
[36] Berger TA, Dye JF. Separation of hydroxybenzoic acids by packed column supercritical fluid chromatography using modified fluids with very polar additives. J Chromatogr Sci 1991;29:26–30.
[37] Berger TA, Dye JF. Role of additives in packed column supercritical fluid chromatography: suppression of solute ionization. J Chromatogr 1991;547:377–92.
[38] Berger TA, Dye JF. Separation of benzene polycarboxylic acids by packed column supercritical fluid chromatography using methanol-carbon dioxide mixtures with very polar additives. J Chromatogr Sci 1991;29:141–6.
[39] De la Puente ML, Lopez Soto-Yarritu P, Burnett J. Supercritical fluid chromatography in research laboratories: design, development and implementation of an efficient generic screening for exploiting this technique in the achiral environment. J Chromatogr A 2011;1218:8551–60.
[40] Salvador A, Herbreteau B, Dreux M. Preliminary studies of supercritical-fluid chromatography on porous graphitic carbon with methylated cyclodextrin as chiral selector. Chromatographia 2001;53:207–9.
[41] Salvador A, Herbreteau A, Dreux M, Karlsson A, Gyllenhaal O. Chiral supercritical fluid chromatography on porous graphitic carbon using commercial dimethyl β-cyclodextrins as mobile phase additive. J Chromatogr A 2001;929:101–12.
[42] Ashraf-Khorassani M, Taylor LT. Supercritical fluid chromatography separation of chiral compounds using 5 and sub 3 μm chiral stationary phases. (Submitted for publication)
[43] Villeneuve MS, Anderegg RJ. Analytical SFC using fully automated column and modifier selection valves for the rapid development of chiral separations. J Chromatogr A 1998;826:217–25.
[44] Maftouh M, Granier-Loyaux C, Chavana E, Marini J, Pradines A, Vander Heyden Y, et al. Screening approach for chiral separation of pharmaceuticals: part III. Supercritical fluid chromatography for analysis and purification in drug discovery. J Chromatogr A 2005;1088:67–81.
[45] Wang RQ, Ong TT, Ng SC. Chemically bonded cationic β-cyclodextrin derivatives and their applications in supercritical fluid chromatography. J Chromatogr A 2012;1224:97–103.
[46] Zhao Y, Woo G, Thomas S, Semin D, Sandra P. Rapid method development for chiral separation in drug discovery using sample pooling and supercritical fluid chromatography–mass spectrometry. J Chromatogr A 2003;1003:157–66.
[47] Hyatt JA. Liquid and supercritical carbon dioxide as organic solvents. J Org Chem 1984;49:5097–103.
[48] Yonker CR, Frye SL, Kalkwarf DR, Smith RD. Characterization of supercritical fluid solvents using solvatochromic shifts. J Phys Chem 1986;90:3022–7.
[49] Lauer HH, McManigill D, Board RD. Mobile-phase transport properties of liquefied gases in near critical and supercritical fluid chromatography. Anal Chem 1983;55:1370–5.
[50] Leren E, Landmark KE, Greibrokk T. Sulphur dioxide as a mobile phase in supercritical fluid chromatography. Chromatographia 1991;31:535–9.
[51] Perrenoud AG, Boccard J, Venthey JL, Guillarme D. Analysis of basic compounds by supercritical fluid chromatography: attempts to improve peak shape and maintain mass spectrometry compatibility. J chromatogr A 2012;1262:205–13.
[52] Lesellier E, Mith D, Dubrulle I. Method developments approaches in supercritical fluid chromatography applied to the analysis of cosmetics. J Chromatogr A 2015;1423:158–68.

[53] Ebinger K, Weller HN. Comparative assessment of achiral stationary phases for high throughput analysis in supercritical fluid chromatography. J Chromatogr A 2014;1332:73–81.

[54] Patel MA, Hardink MA, Wrisely L, Riley F, Hudalla CJ, Ashraf-Khorassani M, et al. Evolution of strategies to achieve baseline separation of ten anionic, water-soluble sulfated estrogens via achiral packed column supercritical fluid chromatography. J. Chromatogr A 2014;1370:240–5.

[55] Ashraf-Khorassani M, Taylor LT. Subcritical fluid chromatography of water soluble nucleobases on various polar stationary phases facilitated with alcohol-modified CO_2 and water as the polar additive. J Sep Sci 2010;33:1682–91.

[56] Blackwell JA, Stringham RW. Effect of mobile phase additives in packed-column subcritical and supercritical fluid chromatography. Anal Chem 1997;69:409–15.

[57] Cazenave-Gassiot A, Boughtflower R, Caldwell J, Hitzel L, Holyoak C, Lanee S, et al. Effect of increasing concentration of ammonium acetate as an additive in supercritical fluid chromatography using CO_2–methanol mobile phase. J Chromatogr A 2009;1216:6441–9.

[58] Jones DC, Dost K, Davidson G, George MW. The analysis of b-agonists by packed-column supercritical fluid chromatography with ultra-violet and atmospheric pressure chemical ionisation mass spectrometric detection. Analyst 1999;124:827–31.

[59] Ashraf-Khorassani M, Taylor LT, Seest E. Screening strategies for achiral supercritical fluid chromatography employing hydrophilic interaction liquid chromatography-like parameters. J Chromatogr A 2012;1229:238–48.

[60] Lou X, Janssen HG, Cramers CA. Temperature and pressure effects on solubility in supercritical carbon dioxide and retention in supercritical fluid chromatography. J Chromatogr A 1997;785:57–64.

[61] Wu Y. Retention mechanism studies on PCSFC. J Liquid Chromatogr Technol 2004;27:2239–42.

CHAPTER

6

Application of Multiple Column Supercritical Fluid Chromatography

C. Wang, A.A. Tymiak and Y. Zhang

Bristol-Myers Squibb Company, Princeton, NJ, United States

OUTLINE

Supercritical Fluid Chromatography.
DOI: http://dx.doi.org/10.1016/B978-0-12-809207-1.00006-9

6.1 INTRODUCTION

Combining selectivity from multiple columns is a viable approach to resolve multicomponent mixtures [1–4]. A popular way to utilize multiple columns is via multiple-dimensional liquid chromatography (LC), or 2D-LC in its simplest form. In 2D-LC, the effluent from the first dimension (^{1}D) is fractionated and subsequently transferred to the second dimension (^{2}D) for further separation, usually with a different separation mechanism. Alternatively, a tandem column (TC) setup combines different selectivity by connecting two or more columns in series, and the entire eluent is continuously transferred from the front to the back column without fractionation.

2D-LC offers great promise to increase the peak capacity needed for the separation of complex mixtures [1,5]. Since the separation in each dimension can be optimized independently, the overall peak capacity is the product of the ^{1}D and the ^{2}D peak capacities (in the ideal case of comprehensive 2D-LC). In order to ensure the selection of the most appropriate mobile phase for each dimension, separate pumps are needed for each dimension. Recent studies have further demonstrated that diluting the ^{1}D fractions prior to the ^{2}D separation, using an additional pump, can greatly reduce the ^{2}D peak distortion caused by the solvent mismatch between dimensions [6,7], and hence facilitate the coupling of additional separation modes. Additional modifications to the interface could add capabilities to manipulate the ^{1}D fractions online, such as fraction parking and sample trapping. Due to the complexity of the instrumentation for 2D-LC applications are still not common and quite rare outside of research laboratories.

In comparison, TC-LC can be implemented on a regular LC systems equipped with a single pump without too much difficulty. Generally the same mobile phase is used for all serially-connected columns. As a consequence, the columns used in TC-LC often share similar separation mechanisms. In fact the serially-connected columns can be considered as a special multimode single column. Thus TC-LC is also viewed as an approach to optimize separations by tuning column chemistry, as compared to typical mobile phase optimization. In other words, TC-LC is still a one-dimensional technique, and its overall peak capacity, is not the product but the sum of the peak capacity of all serially-connected columns (in an ideal case). Despite its seemingly moderate improvement in selectivity and peak capacity compared with single column LC, TC-LC has been successfully applied to the separations of many complex mixtures that could not be achieved using single column LC [4,8–10].

Supercritical fluid chromatography (SFC) has been widely used for chiral separations, and is increasingly being applied for achiral separations in various industries. SFC is not simply a normal phase technique as it can provide separations based on several retention mechanisms covering the whole spectrum from the normal-phase to reversed-phase modes [11]. Despite the possibility of employing different retention mechanisms, practically all SFC applications today use carbon dioxide (CO_2) with organic solvent modifiers as a mobile phase. Therefore, for series-coupled columns in SFC, mobile phase compatibility is less commonly a concern than for LC applications. In addition, more columns can be coupled in series in SFC on account of the low pressure drop across each column.

In this report we focus on SFC applications that exploit multiple online-coupled columns, including hyphenation between SFC and LC, 2D-SFC, and TC-SFC. Rational approaches to select column combinations and tune TC-SFC separations are presented. Finally the role of pressure on TC-SFC separations is discussed in detail.

6.2 APPLICATIONS OF SFC IN TWO-DIMENSIONAL CHROMATOGRAPHY

Table 6.1 lists the applications of SFC in 2D separations. SFC has been hyphenated with both normal phase and the reversed-phase LC. 2D-SFC has also been reported for both open-tubular (capillary) and packed columns.

6.2.1 Coupling of LC and SFC

SFC has been coupled with LC in several 2D applications [12–15]. The lower viscosity of supercritical and subcritical fluids gives rise to improved mass transfer, facilitates the use of higher flow rates without loss of column efficiency, and produces faster separations as compared with LC. Hence SFC is an ideal choice for ^{2}D separations where fast separation is preferred in order to avoid under-sampling of the ^{1}D separated peaks. For example, Gao et al. employed a 2D NPLC × SFC system for the analysis of extracts of the traditional Chinese medicine, *Ganoderma lucidum* [12]. By using SFC with a monolith column for the ^{2}D separation, a modulation time of 1 minute was achieved and allowed sufficient sampling of the ^{1}D separated peaks to maintain the desired peak capacity. The interface between the two columns was a 2-position 10-port valve with two sampling loops, which is similar to the configuration typically used in the 2D-LC. In addition, a back pressure regulator was added to the 10-port valve waste outlet to balance the ^{1}D and ^{2}D pressures, probably to reduce the baseline disturbance introduced by valve switches.

TABLE 6.1 List of Two-Dimensional (2D) Separations Using Supercritical Fluid Chromatography (SFC)

Separation modes	Interface	Sample	Year and reference
NPLC × SFC	Valve: 2 positions 10 ports; Loop/Trap: 2 × 200 μL loop; Notes: Back pressure restrictor at valve outlet	Traditional Chinese medicine extract	2010 [12]
SFC × RPLC	Valve: 2 positions 10 ports; Loop/Trap: 2 × C18 columns; Notes: Water dilution prior to trapping	Lemon oil	2008 [13]
(SFC − SFC) × RPLC	Valve: 2 positions 10 ports; Loop/Trap: 2 × C18 columns; Notes: Water dilution prior to trapping	Fish oil extract	2009 [14]
RPLC − SFC	Valve: 2 positions 8 ports Agilent 2D valve (G1170A); Loop/Trap: 2 × C18 columns; Notes: Agilent Flexcube (G4227) for multiple peak trapping	Chiral/achiral pharmaceutical mixtures	2016 [15]
SFC − SFC	Flow-switching valve that controls the flow goes to one or two columns; Direct fraction transfer between columns	Coal tar extract, metabolite extract, petroleum distillate	1985 [16]; 1989 [17,18]
SFC − SFC	Valve: 2 × 2 positions 10 ports Loop/Trap: on-column cryogenic trap;	Coal tar extract, polychlorinated biphenyl	1990 [19]
SFC × SFC	Valve: 2 positions 10 ports (nonconventional plumbing); Loop/Trap: 1 × methyl silicone capillary column; Notes: stopping the ^{1}D flow during the ^{2}D injection	Triglycerides in fats and oils	2003 [20]; 2006 [21]

(Continued)

TABLE 6.1 (Continued)

Separation modes	Interface	Sample	Year and reference
SFC × SFC	Valve: 2 positions 10 ports; Loop/Trap: 2 × 45 μL loop; Notes: Splitting flow between the ^{1}D detector and the valve; back pressure regulator at valve outlet	Coal distillate	2012 [22]
SFC − SFC	Valve: 4 × 2 positions 6 ports (Fig. 6.1); Loop/Trap: None; Notes: Direct transfer of the ^{1}D cut to the ^{2}D column head; MS triggered valve switch	Chiral/achiral pharmaceutical mixtures	2011 [23]

In addition to coupling with NPLC, SFC can be coupled directly with RPLC. Francois et al. [13,14] used a SFC × RPLC system for the separation of hydrophobic mixtures such as lemon oils and fish oil extracts. The ^{1}D SFC separation gave results similar to NPLC but with a mobile phase more compatible with the ^{2}D RPLC separation. A 2-position 10-port valve with two C18 trapping columns was installed after the back pressure regulator of the ^{1}D SFC to transfer fractions to the ^{2}D RPLC. At the interface the CO_2 from the ^{1}D column was depressurized and evaporated, producing fractions in organic solvent readily compatible with the RPLC separation. An additional pump was employed to dilute the ^{1}D fractions with water to avoid breakthrough on the C18 trapping columns. The low pressure drop achieved using supercritical and subcritical fluids allowed the series coupling of columns for the ^{1}D separation to increase the peak capacity. The SFC × RPLC system provided high peak capacities due to the orthogonality between the two separation modes.

SFC can also be used for the ^{2}D separation when coupled with RPLC. Venkatramani et al. demonstrated the application of 2D RPLC-SFC for simultaneous achiral and chiral purity analysis of pharmaceuticals [15]. This setup simplified the typical laboratory workflow by reducing the number of single column separations required, but also ensured the unambiguous assignment of chiral purity by excluding the potential coelution of enantiomers with achiral impurities. Due to the inherent dilution of the sample in 2D chromatography, the sensitivity for the ^{2}D SFC separation was about 50% of a conventional single dimensional SFC analysis. Yet the detention limit of 0.1% for chiral impurity can still be achieved by using a C18 trapping-and-focusing column at the interface. In addition, a separation of all eight diastereomers of a drug substance was achieved by combining the selectivity from Sunfire C18 and Chiralpak IC3 columns. The valve interface setup is similar to that used by Francois et al. [15], except for the specially designed Agilent 2D valve with 2 positions and 8 ports.

6.2.2 Two-Dimensional SFC

Multidimensional SFC was employed in the late 1980s for the separation of complex mixtures of high boiling-point hydrocarbons that could not be separated by gas chromatography [16–19]. 2D-SFC using CO_2 or SF_6 as a mobile phase with flame-ionization detection (FID) became the method of choice for such applications. Early practitioners implemented heart-cutting 2D-SFC by using a single pump and flow-switching valves to direct mobile phase flow. The flow can go through a single dimension while pausing the separation in the second dimension. Alternatively the flow can go through two columns in series to transfer the ^{1}D effluent to the ^{2}D column. Both packed capillary and open-tubular columns were evaluated. The coupling of packed capillary columns was demonstrated to afford faster separations and higher loading capacities, whereas the use of two open-tubular columns gave higher efficiency and resolving power [19]. Unique selectivity could be achieved by combining columns with different retention mechanisms through a creative valve configuration [24–26]. However, an attempt to combine a packed capillary column and an open-tubular column with a one-pump system was unsuccessful due to the different optimum flow rates required for the two column types [17]. Juvancz et al. [19] modified the early 2D setup by introducing a second pump to achieve independent flow rate optimizations for each dimension, and adding an on-column cryogenic trap to focus the ^{1}D effluent prior to separation on the ^{2}D column. These modifications were effective at reducing the separation time and producing narrow ^{2}D peaks for a certified coal tar sample. However, open-tubular column SFC has mostly been replaced by packed column SFC since the late 1990s.

The feasibility of packed column SFC × SFC has been demonstrated by two separate research groups. Hirata et al. [20] used a 2-position 10-port valve as an interface. A single capillary column was used to trap the ^{1}D fractions after decompression through a pressure restrictor. The valve was switched briefly to sweep the trapped fraction onto the ^{2}D column for further separation. At the same time, the ^{1}D flow was stopped to avoid loss of fractions. The valve could be switched back to its former position at any time to continue the ^{1}D separation and fractionation. This provided a flexible sampling time for the [1]D separation. This setup was used for the separation of triglycerides in fats and oils using CO_2 as the mobile phase. Compared with single column separations, a useful gain in peak capacity and resolving power was obtained even though the two dimensions used the same C18 columns, which differed only in column length and temperature. This setup was also combined with an online supercritical fluid extractor to achieve the separation of styrene oligomers in polystyrene foams [21]. Guibal et al. [22] built a SFC × SFC interface by modifying the plumbing of a 2-position 10-port valve typically used for 2D-LC. Specifically, the [1]D effluent was split between the [1]D detector and the valve inlet. In addition, a pressure restrictor was installed at the valve outlet to avoid the decompression of fractions in the transfer loop. By coupling C18 with silica columns and using a CO_2 mobile phase, these authors were able to separate a coal distillate containing several hundred thousand components into groups of peaks in the 2D separations space. The authors also noted that further optimization was needed to best utilize the speed and efficiency of the SFC separations.

2D achiral–chiral separations have been reported using SFC [23], similar to the LC-SFC application discussed above. Zeng et al. [23] reported a heart-cutting 2D-SFC system for achiral–chiral separations of pharmaceutical compounds for both analytical and preparative scale separations. The analytical 2D-SFC-MS system (Fig. 6.1) was modified

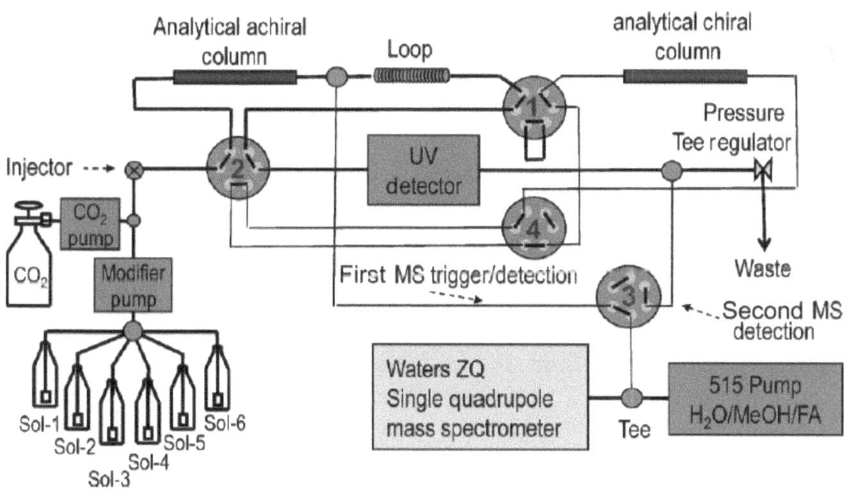

FIGURE 6.1 The system configuration of the analytical heart-cutting 2D-SFC-MS system built by Zeng et al. Four 2-position 6-port valves (1-4) were used to achieve the 2D-SFC separation. *Source: Reprinted from reference Zeng L, Xu R, Zhang Y, Kassel DB. Two-dimensional supercritical fluid chromatography/mass spectrometry for the enantiomeric analysis and purification of pharmaceutical samples. J Chromatogr A 2011;1218:3080–88 [23].*

from a legacy Berger SFC system by adding four 2-position 6-port valves. Valve 1 controlled the ^{1}D effluent to either the ^{1}D UV detector or the ^{2}D column head; valve 2 was used to direct the pump flow to either the achiral or the chiral column; valve 3 was always synchronized with valve 2 to select the inlet of the MS detector; and valve 4 was added as an option to flush the system by bypassing column 2. By adding valve 2 and 3, a single pump with UV and MS detectors was used for the separation in both dimensions. A loop of extra volume was added after the ^{1}D achiral column to delay the effluent's arrival at valve 1 so that the MS signal could be registered in time to trigger the switch of valve 1. No transfer loop was required. The ^{1}D fraction was directly transferred to the ^{2}D column and parked at the column head. The ^{2}D separations were started only after the ^{1}D separation was complete. The preparative 2D system was configured in a similar way, with additional valves to enable the decompression and collection of multiple fractions. One potential limitation of this sophisticated setup is that only one cut from ^{1}D can be transferred for ^{2}D separation. Nevertheless, this setup streamlined the workflow for the targeted chiral purity analysis and for the purification of a single enantiomer in complex mixtures.

6.3 TANDEM COLUMN SFC

TC-SFC is an economic way to combine selectivity from multiple columns. In comparison to the complex instrument setup required for 2D separations, TC-SFC can be conducted using a conventional SFC instrument. An optional zero-dead-volume fitting may be used to connect two columns to reduce the extra-column volume. The data collection and analysis are also identical to a single column analysis. Hence TC-SFC is easier to adopt by analysts with limited experience of operating a designing more complex multidimensional SFC systems. On the other hand, TC is considered a 1D technique. The selectivity of one column can be either improved, unchanged, compromised, or even completely canceled out by the selectivity of another coupled column. Consequently, finding the right column combination for TC-SFC generally takes more time than that for the 2D separations. Nevertheless many successful separations have been demonstrated by connecting columns of different selectivity in series (Table 6.2).

6.3.1 Coupling Columns of the Same Type in Series

Coupling columns of the same type in series is an economic way to "make" longer columns in the laboratory. Longer columns produce a

TABLE 6.2 List of TC-SFC Applications

Types of columns in series	Total column length (cm)	Samples	Year and reference
Silica	220	Lemon oil, PAH, Chimney extract, Gasoline	1993 [27]
Different C18	80	*E-Z* isomers of β-carotene	1999 [28]
Regular and monolith C18	75	*E-Z* isomers of β-carotene	2003 [29]
C18 and silica	50	16-Component PAH standard	1994 [30]
C18, C30, phenyl, cyano	50	15 Steroids	2014 [31]
Silica and diol	75	Wheat lipid extract	2004 [32]
Different achiral and chiral	50	Chiral drug mixture	1998 [33]
AD and OD	50	Diastereomers of a drug	2005 [34]
Silica and AD	30	Chiral reaction product	2006 [35]
Different achiral and chiral	NA	Diastereomers of a drug; chiral reaction product	2007 [36]
Different chiral and achiral	40	Chiral reaction product	2013 [37]
Different chiral	50	Diastereomers of an API; coumachlor degradation mixture	2014 [38]

higher peak capacity and separation efficiency. In practice, the length of a column is often limited by maximum system pressure and separation time constrains. In comparison with LC, SFC uses low-viscosity mobile phases and produces efficient separations at a higher flow rate while generating a lower column pressure drop. Hence longer columns may be used for SFC than for LC separations. For example, Berger and Wilson demonstrated the coupling of 11 columns to achieve an effective column length of 2.2 m, yet the observed pressure drop was only 160 bar [27]. Remarkable gain in peak capacity and resolution were shown for the separation of different complex mixtures. In one application, more than 80 peaks were observed in a 30-minute separation of a chimney soot extract. The plate number and separation time were on par with those of typical GC separations.

Series-coupled columns can be useful for even the separation of two components. In a recent attempt to purify enantiomers of a starting material in our laboratory, a general screening of multiple solvents on

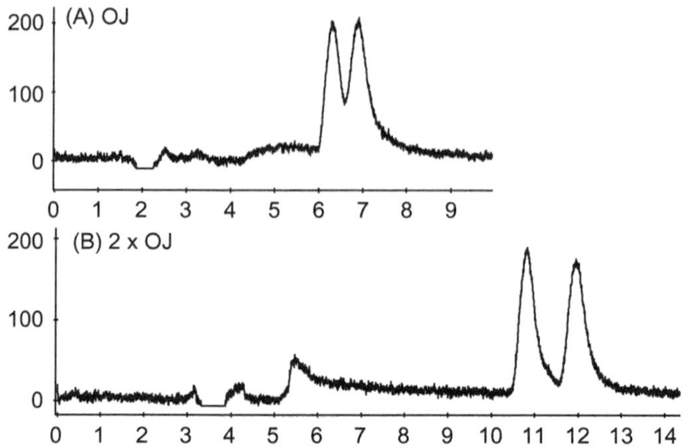

FIGURE 6.2 Enantiomeric purification of a starting material on one (A) and two (B) 250 × 30 mm Chiralcel OJ columns connected in series. The mobile phase is CO_2-isopropanol 95:5 for both separations; all other experimental conditions are identical.

12 commonly used chiral columns only generated a partial separation on one column (Chiralcel OJ column with 5% isopropanol in CO_2 (Fig. 6.2A)). Reducing the organic modifier concentration was not practical due to the requirements for purification. Without spending extra time to exhaust all available chiral columns and mobile phase combinations, we were able to improve the separation (Fig. 6.2B) by coupling two OJ columns for the purification. In fact, increasing the column length by coupling multiple columns is one of the simplest ways to improve a marginal separation. It is probably more frequently practiced than the isolated cases reported in literature.

6.3.2 Coupling Two Achiral Columns of Different Selectivity in Series

When the complete separation of all components on one column type is not achievable, coupling columns with complementary selectivity offers a potential solution. For example, Lesellier et al. [28] systematically studied the separation of the four *E-Z* isomers of β-carotene (all-*trans*, 9-*cis*, 13-*cis* and 15-*cis*) on multiple single columns and found that the Ultrabase UB 225 column could separate the mixture into 3 peaks with the 9-*cis* coeluting with the all-*trans* isomer, and that the Hypersil ODS column generated 4 peaks with a partial coelution of the 13-*cis* and all-*trans* isomers. By coupling one Hypersil ODS and three Ultrabase UB 225 columns in series, a complete separation of all four isomer was achieved. Subsequently, using a similar rationale, the same mixture was separated on a combination of a YMC Pack Pro C18 column and a Chromolith RP 18e column [29].

The series-coupled column approach has also been used for the separation of mixtures with many more components. Sandra et al. [30] reported the first complete SFC separation of the sixteen polycyclic aromatic hydrocarbons in a reference mixture (US Environmental Protection Agency) by coupling two different C18 columns. Delahaye and Lynen evaluated TC-SFC for the separation of a mixture of 15 steroids. They employed a computer-assisted column selection software originally developed for TC-LC for column selection [31]. Two column combinations, each using four column types, were reported to produce separations similar to the predicted separations.

Deschamps et al. studied the SFC separation of five classes of lipids on both silica and diol columns [32]. Both columns produced baseline separation of the mixture with the same elution order, but different relative selectivity between peaks. Combining different numbers and order of columns produced separations with slightly different selectivity, and hence the overall separation of the mixture can be "tuned" too produce an optimum separation with evenly spaced peaks.

6.3.3 Coupling a Chiral Column with a Chiral or Achiral Column in Series

Chiral purity analysis is often conducted on samples in matrixes, such as reaction mixtures, drug formulations, and biological fluids. In some cases, chiral drugs are also analyzed together with coformulated drugs, other drugs on a target list, or internal standards. As a result, the separation of all achiral impurities from the chiral drugs is often mandatory in order to produce definitive chiral purity results. Phinney et al. [33] were the first to combine achiral and chiral columns in TC-SFC to separate a mixture of chiral pharmaceutical compounds. Chiral columns were selected to afford enantiomeric separation for all chiral compounds. Achiral columns were then selected to resolve coeluting drugs and provide complementary selectivity to the chiral column. Racemic mixtures of various chiral drug mixtures were completely resolved by TC-SFC using different combinations of achiral and chiral columns. An example of such a separation of four chiral β-blockers is shown in Fig. 6.3.

In pharmaceutical laboratories, TC-SFC combining one chiral with one achiral column is used mainly for analytical monitoring and preparative purification of reaction mixtures containing both chiral products and starting materials [35–37]. Theoretically, the achiral–chiral column combination should also be able to separate diastereomer mixtures of compounds that have more than one stereogenic center.

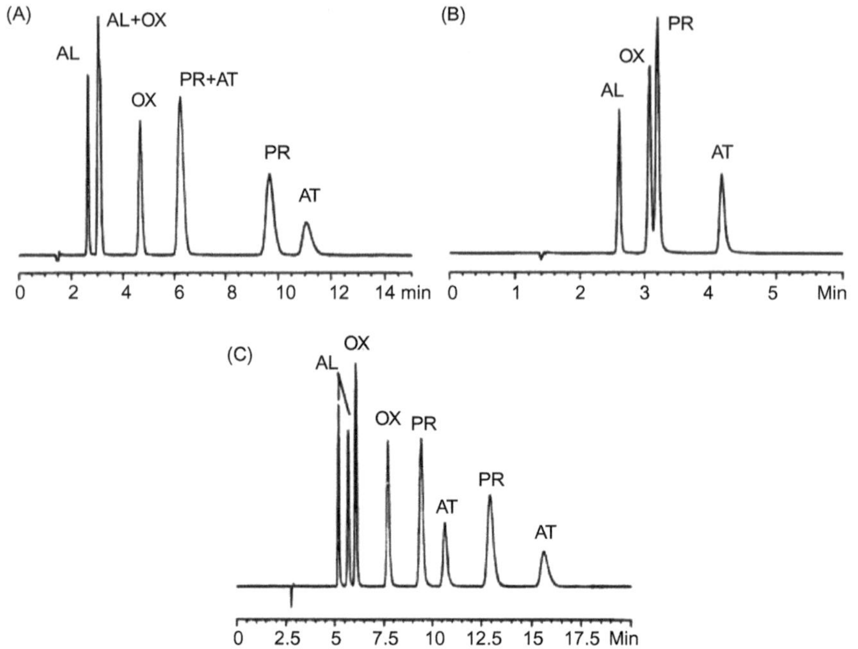

FIGURE 6.3 The separation of chiral β-blockers alprenolol (AL), atenolol (AT), oxprenolol (OX), and propranolol (PR) on (A) the Chiralcel OD column, (B) the achiral cyano column, and (C) the coupled cyano and Chiralcel OD columns. *Source: Reprinted from reference Phinney KW, Sander LC, Wise SA. Coupled achiral/chiral column techniques in subcritical fluid chromatography for the separation of chiral and nonchiral compounds. Anal Chem 1998;70:2331–35 [33].*

In practice, however, Welch et al. noted the generally poor ability of achiral columns to produce substantial separation of diastereomers [36]. This observation was also supported by the fact that most of the reported TC-SFC separations of complex diastereomer mixtures were achieved on two serially coupled chiral columns [34,36,38].

6.4 METHOD DEVELOPMENT IN TANDEM COLUMN SFC

Given that multidimensional chromatography is best suited for extremely complex mixtures, TC-SFC, on the other hand, has great potential for the separation of moderately complex mixtures because it can be easily implemented. One possible challenge is the lack of a systematic approach for column selection and optimization. To select column combinations by hit-or-miss can be inefficient due to the large number of possible combinations. Several approaches for rational method development have been reported. One prerequisite for logically choosing column combinations is the known elution order on the individual columns. The elution orders can be identified by injecting reference standards, comparing relative area percent on different columns, matching online UV or MS spectra, and CD signals.

Lesellier et al. [28,29] focused on combining achiral columns that produced the same elution order but with different critical pairs in a generic screening protocol [28,29]. Prominent column combinations were selected for further mobile phase optimization as typically done in single column SFC. The length of each column was also fine-tuned to afford the best overall selectivity.

Phinney et al. [33] discussed the selection of achiral–chiral column combinations. The chiral separation was developed first to ensure the separation of all enantiomers. A set of achiral columns was then screened under the optimal chiral condition to evaluate their selectivity for the coeluting peaks on the chiral column. The sum of the retention times on the single chiral and achiral columns could also be calculated to estimate TC-SFC retention. The achiral column that produced the best calculated TC-SFC separation was selected for evaluation. Ventura used a similar strategy to identify the optimal isocratic conditions for the chiral separation first [37]. The achiral column selection was based on experimental screening of combinations of the chiral column with different achiral columns.

Selectivity of a TC separation is the retention time weighted average of selectivity from each coupled column [8,39,40], as shown in Eq. 6.1:

$$\alpha_t = \sum f_i \cdot \alpha_i \tag{6.1}$$

where α_t is the overall selectivity of a TC separation, α_i is the selectivity on individual column i, and f_i is the fraction of retention time contribution from column i. Wang et al. studied the TC-SFC separation of a 4 diastereomer mixture using two chiral columns that offer complementary selectivity [38]. Several different approaches were applied to modify α_t by simply changing the f_i while keeping the α_i constant for each coupled columns. The relative retention contribution, f_i, of a column can be effectively reduced by using a relatively shorter or narrower column, and by application of a higher column back pressure. Different elution order and overall selectivity could be optimized simply by modifying these experimental settings while fixing the column chemistries.

Delahaye and Lynen took a similar approach to optimize TC-SFC for achiral mixtures based on Eq. 6.1 [31]. The fine tuning of the f_i, and hence α_t, was achieved by changing the relative column length of five different types of reversed-phase columns. Simply by varying the column lengths, a set of these commercial columns could theoretically offer >3000 different ways to combine the selectivity of the five column chemistries. A computer algorithm was also developed to facilitate the calculation of the overall retention and separation for each possible column combination, and to help pick the most useful column combination. This approach was found to provide reasonably good guidance for column selection for TC-SFC, despite some minor prediction errors that were likely caused by the pressure dependence of supercritical and subcritical fluid density.

Automated screening is also a valuable approach to develop TC-SFC methods. Welch et al. and Ventura installed two multiposition column switching valves in series (Fig. 6.4) to enable the automated screening of different column combinations [36,37]. A bypass line was reserved for each column switching valve so that single columns can also be evaluated using the same setup. For example, using two 6-position valves, a total of 10 single columns and 25 column combinations can be screened for method development.

6.5 THE ROLE OF PRESSURE IN TANDEM COLUMN SFC

Column back pressure has a strong influence on retention in SFC. At higher average column pressure compression of the mobile phase results in an increase in density and in its solvating power, and hence reduces solute retention. On the other hand, pressure was found to barely affect column selectivity [41]. This is probably the main reason why the column back pressure is rarely considered as a parameter to optimize separations in single column SFC.

In TC-SFC, however, pressure is an important parameter for method development [38]. The front column in a TC setup experiences higher pressure than in a conventional single column experiment with the same system back pressure. This elevated effective column pressure leads to a reduced retention contribution from the front column to the TC-SFC separation. As a result, the common practice to use the sum of retention times from single column experiments to predict the TC-SFC retention is generally not accurate, although this simplified prediction is often useful in guiding the selection of column combinations [31,33]. Generally the prediction error increases when the mobile phase is more compressible, the back pressure elevation of the front columns is higher, and the solute is retained longer.

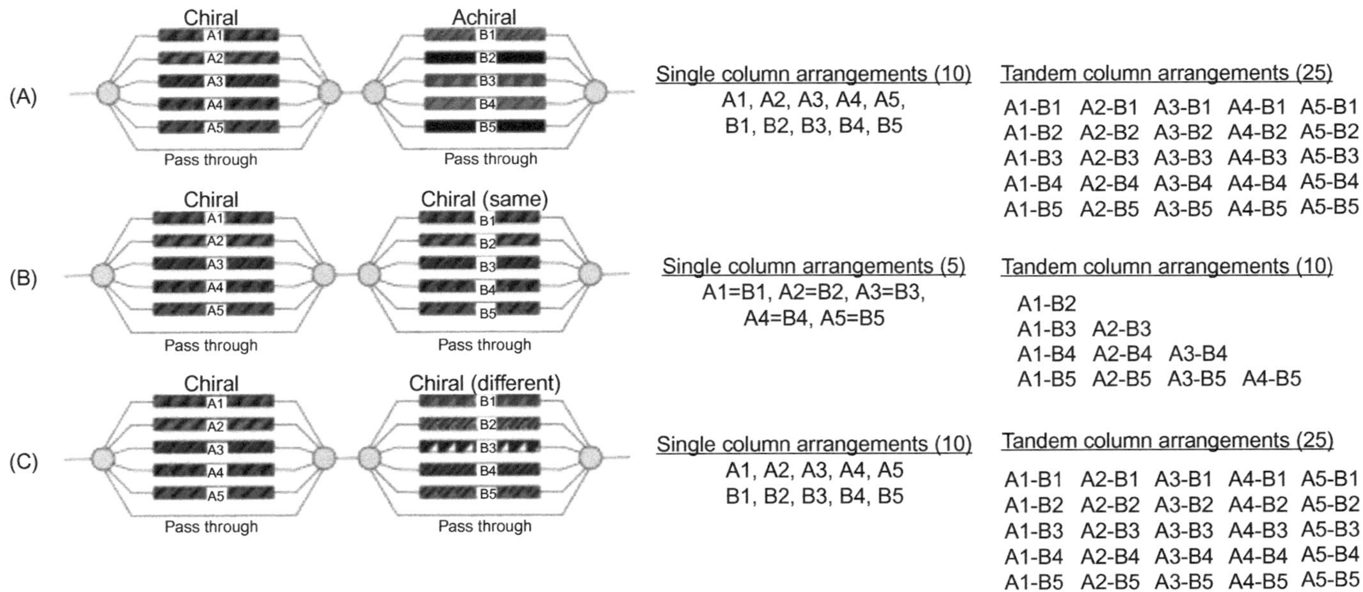

FIGURE 6.4 Examples of employing two column switching valves for automated screening of single and tandem columns (TCs): (A) chiral and achiral; (B) chiral and chiral (same); and (C) chiral and chiral (different). *Source: Reprinted from reference Welch CJ, Biba M, Gouker JR, Kath G, Augustine P, Hosek P. Solving multicomponent chiral separation challenges using a new sfc tandem column screening tool. Chirality 2007;19:184–89 [36].*

To reduce such prediction error originated from neglecting pressure effects, Lesellier et al. used monolith columns, with a low flow resistance producing only a small pressure drop, as downstream columns in TC-SFC [29]. As a result, the front column back pressure was not too different than for the single column experiment, and hence the calculated TC-SFC retention was accurate enough to determine the optimum column length [29]. Alternatively Wang et al. tried to improve the prediction accuracy by using the front column single column retention data at a back pressure identical to that in the TC-SFC setup [41]. To achieve predictions without extra experiments, we extrapolated the single column retention at elevated back pressures from the experimental measurement at a regular back pressure based on an empirical pressure-retention model. This approach was used to successfully predict the selectivity change caused by changing the column coupling order.

Changing the column order in TC-SFC simultaneously changes the back pressure on both columns. As a result, the column order is expected to affect TC-SFC retentions under both gradient and isocratic conditions [36,38,42]. In early studies, however, the column order was found to only slightly affect the overall TC-SFC separation under isocratic conditions. Wang et al. [38] reported a drastic effect of column order on the overall TC-SFC separation of a 4-component diastereomer mixture (Fig. 6.5) [38]. Delahaye and Lynen also showed different selectivity for three critical pairs when changing the column order of a set of achiral columns [31]. The column order effects on TC-SFC separations

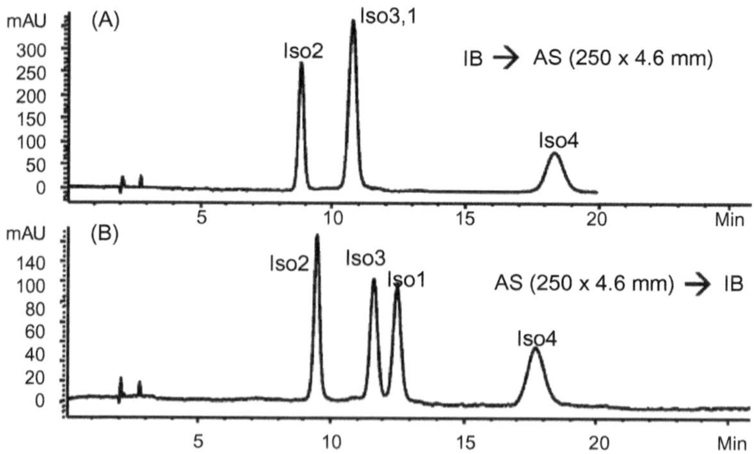

FIGURE 6.5 Different TC-SFC separations obtained using the same AS and IB columns by simply reversing the column coupling order. All experimental settings are identical. *Source: Reprinted from reference Wang C, Tymiak AA, Zhang Y. Optimization and simulation of tandem column supercritical fluid chromatography separations using column back pressure as a unique parameter. Anal Chem 2014;86:4033–40 [38].*

would be more dramatic if the samples were more complex, the difference in selectivity for the coupled columns was greater, or the pressure drop across columns is higher.

Despite its contribution to the added complexity for TC-SFC method development, pressure can also be utilized to optimize a TC-SFC separation. Raising the back pressure of a column would change the overall selectivity α_t by reducing the f_i in Eq. 6.1 because higher pressure generally maintains the selectivity while reducing retention [38,41]. In contrast to conventional parameters that affects f_i, such as column length and inner diameter, the pressure can be changed smoothly to achieve continuous tuning of the α_t of a TC-SFC separation [38]. As an example, in the TC-SFC separation shown in Fig. 6.6, the relative retention

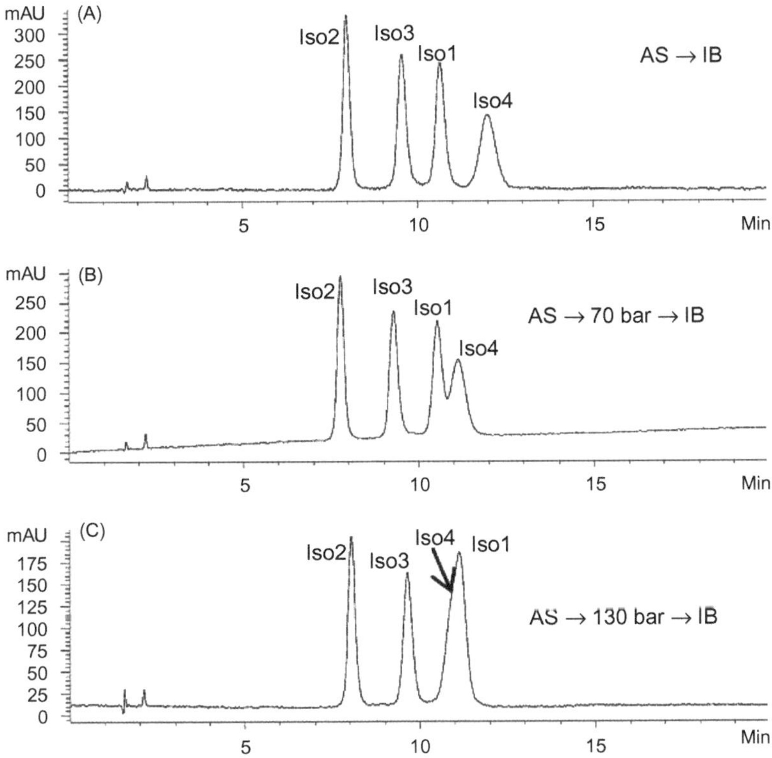

FIGURE 6.6 Tuning the selectivity between *Iso1* and *Iso4* by elevating the column back pressure of the front Chiralpak AS column in a TC-SFC separation. (A) No pressure restrictor is used, (B) 70 bar was added right after the front AS column, and (C) 130 bar was added. The back pressure for back column IB (250 mm×4.6 mm) was kept constant. All separations are achieved under the identical chromatographic conditions. *Source: Reprinted from reference Wang C, Tymiak AA, Zhang Y. Optimization and simulation of tandem column supercritical fluid chromatography separations using column back pressure as a unique parameter. Anal Chem 2014;86:4033–40 [38].*

between *Iso4* and *Iso1* continuously changed when the back pressure of the front column is increased gradually. Although limited by the system pressure, it may be foreseen that *Iso4* and *Iso1* could be baseline resolved again yet with a different elution order from Fig. 6.6A if higher pressure could be applied between the two coupled columns. In another example, by simply changing the front column pressure and the column coupling order, four different elution orders were produced in TC-SFC separations of a 6-component degradation mixture of coumachlor using the same pair of Chiralcel OD and OJ columns [38].

6.6 CONCLUSION

SFC offers unique opportunities in all formats of multiple column separations. Specifically SFC can be coupled with LC to produce complementary selectivity while using compatible mobile phases. The speed and efficiency of the SFC separation make it a desirable choice for ^{2}D separations. Longer columns (achieved by connecting several columns in series) can be used to boost the peak capacity in 2D applications, thanks to the low mobile phase viscosity and pressure drop in SFC. In addition, in TC-SFC applications, a large number of columns of diverse chemistry may be combined due to the low column pressure and compatibility of the mobile phase between different SFC separation modes.

Not surprisingly, most of the reported series-coupled column SFC applications are the analysis or purification of chiral compounds in complex mixtures. Such analysis can be done by both heart-cutting 2D and TC chromatography. The 2D approach requires special instrumentation and software that may be not readily accessible for many laboratories. In comparison, the TC approach uses regular SFC system and can be more widely adopted. In addition, TC-SFC is particularly suitable for purification applications because the stacking injection can still be used to improve productivity. On the other hand, more effort is need to select the right column combinations in TC-SFC than for 2D separations, including selection of column types, dimensions, and coupling order. As a result, the progress in streamlining TC-SFC method development is critical for its broader application to moderately complex mixtures.

References

[1] Stoll DR, Li X, Wang X, Carr PW, Porter SEG, Rutan SC. Fast, comprehensive two-dimensional liquid chromatography. J Chromatogr A 2007;1168:3–43.

[2] Dugo P, Cacciola F, Kumm T, Dugo G, Mondello L. Comprehensive multidimensional liquid chromatography: theory and applications. J Chromatogr A 2008;1184:353–68.

[3] François I, Sandra K, Sandra P. Comprehensive liquid chromatography: fundamental aspects and practical considerations—a review. Anal Chim Acta 2009;641:14–31.

[4] Alvarez-Segura T, Torres-Lapasió JR, Ortiz-Bolsico C, García-Alvarez-Coque MC. Stationary phase modulation in liquid chromatography through the serial coupling of columns: a review. Anal Chim Acta 2016;923:1–23.
[5] Li X, Stoll DR, Carr PW. Equation for peak capacity estimation in two-dimensional liquid chromatography. Anal Chem 2009;81:845–50.
[6] Stoll DR, O'Neill K, Harmes DC. Effects of ph mismatch between the two dimensions of reversed-phase x reversed-phase two-dimensional separations on second dimension separation quality for ionogenic compounds—I. Carboxylic acids. J Chromatogr A 2015;1383:25–34.
[7] Gargano AFG, Duffin M, Navarro P, Schoenmakers PJ. Reducing dilution and analysis time in online comprehensive two-dimensional liquid chromatography by active modulation. Anal Chem 2016;88:1785–93.
[8] Mao Y, Carr PW. Application of the thermally tuned tandem column concept to the separation of several families of environmental toxicants. Anal Chem 2000;72:2788–96.
[9] Gostomski I, Braun R, Huber C. Detection of low-abundance impurities in synthetic thyroid hormones by stationary phase optimized liquid chromatography–mass spectrometry. Anal Bioanal Chem 2008;391:279–88.
[10] Lu J, Ji M, Ludewig R, Scriba GKE, Chen D-Y. Application of phase optimized liquid chromatography to oligopeptide separations. J Pharm Biomed Anal 2010;51:764–7.
[11] Lesellier E. Retention mechanisms in super/subcritical fluid chromatography on packed columns. J Chromatogr A 2009;1216:1881–90.
[12] Gao L, Zhang J, Zhang W, Shan Y, Liang Z, Zhang L, et al. Integration of normal phase liquid chromatography with supercritical fluid chromatography for analysis of fruiting bodies of ganoderma lucidum. J Sep Sci 2010;33:3817–21.
[13] François I, dos Santos Pereira A, Lynen F, Sandra P. Construction of a new interface for comprehensive supercritical fluid chromatography × reversed phase liquid chromatography (SFC × RPLC). J Sep Sci 2008;31:3473–8.
[14] François I, Sandra P. Comprehensive supercritical fluid chromatography × reversed phase liquid chromatography for the analysis of the fatty acids in fish oil. J Chromatogr A 2009;1216:4005–12.
[15] Venkatramani CJ, Al-Sayah M, Li G, Goel M, Girotti J, Zang L, et al. Simultaneous achiral-chiral analysis of pharmaceutical compounds using two-dimensional reversed phase liquid chromatography-supercritical fluid chromatography. Talanta 2016;148:548–55.
[16] Christensen RG. On-line multidimensional chromatography using supercritical carbon dioxide. J High Res Chromatogr 1985;8:824–8.
[17] Davies IL, Xu B, Markides KE, Bartle KD, Lee ML. Multidimensional open-tubular column supercritical fluid chromatography using a flow-switching interface. J Microcolumn Sep 1989;1:71–84.
[18] Payne KM, Davies IL, Bartle KD, Markides KE, Lee ML. Multidimensional packed capillary column supercritical-fluid chromatography using a flow-switching interface. J Chromatogr A 1989;477:161–8.
[19] Juvancz Z, Payne KM, Markides KE, Lee ML. Multidimensional packed capillary coupled to open tubular column supercritical fluid chromatography using a valve-switching interface. Anal Chem 1990;62:1384–8.
[20] Hirata Y, Hashiguchi T, Kawata E. Development of comprehensive two-dimensional packed column supercritical fluid chromatography. J Sep Sci 2003;26:531–5.
[21] Okamoto D, Hirata Y. Development of supercritical fluid extraction coupled to comprehensive two-dimensional supercritical fluid chromatography (sfe-sfcxsfc). Anal Sci 2006;22:1437–40.
[22] Guibal P, Thiébaut D, Sassiat P, Vial J. Feasability of neat carbon dioxide packed column comprehensive two dimensional supercritical fluid chromatography. J Chromatogr A 2012;1255:252–8.

[23] Zeng L, Xu R, Zhang Y, Kassel DB. Two-dimensional supercritical fluid chromatography/mass spectrometry for the enantiomeric analysis and purification of pharmaceutical samples. J Chromatogr A 2011;1218:3080–8.
[24] Lundanes E, Greibrokk T. Group separation of oil residues by supercritical fluid chromatography. J Chromatogr A 1985;349:439–46.
[25] Lundanes E, Iversen B, Greibrokk T. Group separation of petroleum using supercritical fluids: a comparison of supercritical nitrous oxide and carbon dioxide as mobile phases. J Chromatogr A 1986;366:391–5.
[26] Campbell RM, Djordjevic NM, Markides KE, Lee ML. Supercritical fluid chromatographic determination of hydrocarbon groups in gasolines and middle distillate fuels. Anal Chem 1988;60:356–62.
[27] Berger TA, Wilson WH. Packed column supercritical fluid chromatography with 220,000 plates. Anal Chem 1993;65:1451–5.
[28] Lesellier E, Gurdale K, Tchapla A. Separation of cis/trans isomers of β-carotene by supercritical fluid chromatography. J Chromatogr A 1999;844:307–20.
[29] Lesellier E, West C, Tchapla A. Advantages of the use of monolithic stationary phases for modelling the retention in sub/supercritical chromatography: application to cis/trans-β-carotene separation. J Chromatogr A 2003;1018:225–32.
[30] Kot A, Sandra P, David F. Selectivity tuning in packed column sfc separation of the sixteen priority polycyclic aromatic hydrocarbons as an example. J High Res Chromatogr 1994;17:277–9.
[31] Delahaye S, Lynen F. Implementing stationary-phase optimized selectivity in supercritical fluid chromatography. Anal Chem 2014;86:12220–8.
[32] Deschamps FS, Lesellier E, Bleton J, Baillet A, Tchapla A, Chaminade P. Glycolipid class profiling by packed-column subcritical fluid chromatography. J Chromatogr A 2004;1040:115–21.
[33] Phinney KW, Sander LC, Wise SA. Coupled achiral/chiral column techniques in subcritical fluid chromatography for the separation of chiral and nonchiral compounds. Anal Chem 1998;70:2331–5.
[34] Barnhart WW, Gahm KH, Thomas S, Notari S, Semin D, Cheetham J. Supercritical fluid chromatography tandem-column method development in pharmaceutical sciences for a mixture of four stereoisomers. J Sep Sci 2005;28:619–26.
[35] Alexander AJ, Staab A. Use of achiral/chiral sfc/ms for the profiling of isomeric cinnamonitrile/hydrocinnamonitrile products in chiral drug synthesis. Anal Chem 2006;78:3835–8.
[36] Welch CJ, Biba M, Gouker JR, Kath G, Augustine P, Hosek P. Solving multicomponent chiral separation challenges using a new sfc tandem column screening tool. Chirality 2007;19:184–9.
[37] Ventura M. Use of achiral columns coupled with chiral columns in sfc separations to simplify isolation of chemically pure enantiomer products. Am Pharm Rev 2013;16.
[38] Wang C, Tymiak AA, Zhang Y. Optimization and simulation of tandem column supercritical fluid chromatography separations using column back pressure as a unique parameter. Anal Chem 2014;86:4033–40.
[39] Lukulay PH, McGuffin VL. Solvent modulation in liquid chromatography: extension to serially coupled columns. J Chromatogr A 1995;691:171–85.
[40] Mao Y, Carr PW. Adjusting selectivity in liquid chromatography by use of the thermally tuned tandem column concept. Anal Chem 1999;72:110–18.
[41] Wang C, Zhang Y. Effects of column back pressure on supercritical fluid chromatography separations of enantiomers using binary mobile phases on 10 chiral stationary phases. J Chromatogr A 2013;1281:127–34.
[42] Poole CF. Stationary phases for packed-column supercritical fluid chromatography. J Chromatogr A 2012;1250:157–71.

CHAPTER

7

Evolution of Instrumentation for Analytical Scale Supercritical Fluid Chromatography*

T.A. Berger

SFC Solutions Inc., Englewood, FL, United States

OUTLINE

*This chapter draws upon large sections of TA Berger, "Instrumentation for supercritical fluid chromatography," *J. Chromatogr. A*, 1421 (November 2015), 171–183. Reproduced with permission from Elsevier. ©Elsevier 2015.

Supercritical Fluid Chromatography.
DOI: http://dx.doi.org/10.1016/B978-0-12-809207-1.00007-0

7.1 INTRODUCTION

Modern supercritical fluid chromatography (SFC) is a high-resolution chemical separation technique, closely related to high-performance liquid chromatography (HPLC) and, in fact, uses most of the same instrumentation, columns, and software. The primary difference between the two techniques is the replacement of most of the liquid mobile phase used in HPLC with highly compressed, dense carbon dioxide (CO_2), along with relatively small additions of a polar modifier, such as an alcohol. The pressure in the system must be elevated to keep the mobile phase density near those of normal liquids. This is usually accomplished with a back pressure regulator (BPR) mounted after the column/detector.

A generic schematic of an SFC is presented in Fig. 7.1, and indicates the mechanical similarities to HPLC. A binary pump is used to pump CO_2 and modifier. The CO_2 pump head and incoming CO_2 are generally chilled in order for the pump to pump liquid CO_2. The two streams are mixed. An external loop autosampler is used to make injections. The mobile phase is preheated in a heat exchanger, HX, to the column initial temperature. After the column the fluid passes through another HX to condition the fluid temperature to the temperature of the detector flow cell. The most common detector is a diode array UV–Vis detector with a high-pressure flow cell. A BPR is mounted after the detector to main the mobile phase in a dense state. After the pressure is dropped, the liquid waste is collected in an enclosed bottle and the CO_2 vented through a fume hood. The biggest differences between HPLC and SFC are in the inclusion of a new instrumental parameter, outlet pressure, and in the way the pumps are driven by the software.

HPLC is mostly practiced as reversed phase, meaning the separation is based on hydrophobic interaction, where the stationary phase is typically nonpolar, such as C18, with aqueous/organic mobile phases. SFC

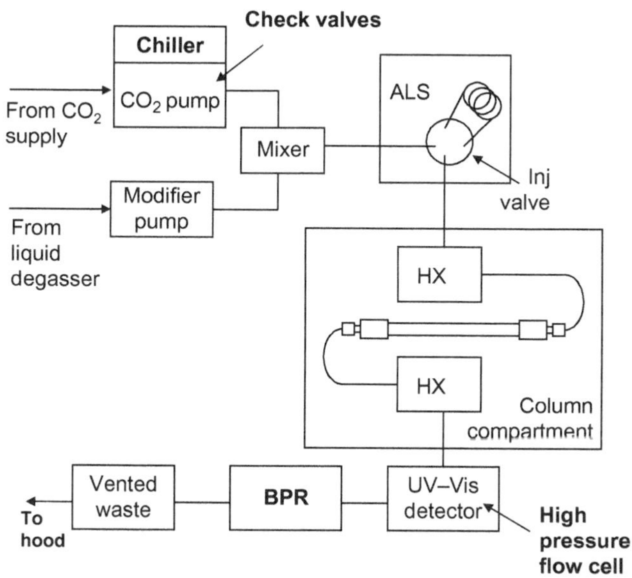

FIGURE 7.1 Generalized scheme of an supercritical fluid chromatography (SFC) instrument. The mechanical differences between an HPLC and an SFC are in bold type. The modifier degasser, the modifier pump, the automatic liquid sampler (ALS), the column compartment ("oven"), and the UV detector can easily be configured to perform either SFC or HPLC. The components most critical for optimal SFC performance include a modified CO_2 pump (different check valves and piston seals), and a back pressure regulator (BPR). The biggest differences are in the software driving the pump and BPR.

is mostly practiced as normal phase, where the stationary phase is polar, such as bare silica, with a relatively nonpolar mobile phase (CO_2 + alcohol). These differences are more perceptual than substantive.

SFC is gaining popularity for a number of practical reasons. The CO_2 is very inexpensive compared to most organic solvents and is recycled from other industries. Thus it is considered to be a "green" technology. The viscosity of SFC mobile phases is much lower than normal liquids, making it inherently higher speed, with lower pressure drops, compared to HPLC, when using columns with similar dimensions. It is often stated that SFC is 3–5 times faster than HPLC on the same-sized particles, and seldom requires pumps capable of more than 400 bar.

7.1.1 Column Developments

The world of HPLC has changed rather dramatically in recent years. The column always dictates instrument design and there have been significant changes in column technology. Sub-2-μm totally porous particles first appeared in the 1980s [1] and promised a significant decrease in analysis times, but they were not commercialized until ≈2000, and instrumentation to fully exploit them was not available until ≈2004. Similarly, porous shell particles were demonstrated long ago but only became available in the same time frame. With both particle types the main objective is much higher speeds.

7.1.1.1 Totally Porous Sub-2-μm Particles

Optimum linear velocity, μ_{opt}, is inversely proportional to particle diameter, d_p. Totally porous 1.8 μm particles require a flow rate ≈2.8 times higher than 5 μm particles and produce the same efficiency with a column 1/2.8 times as long. Thus such particles are ≈8–9 times faster than 5 μm particles. However, sub-2-μm particles are often used to gain even more speed by using much shorter columns, with lower efficiency. It has also been common to use such particles at greater than their optimum velocity to further increase speed. Plots of plate height versus linear velocity are quite flat, resulting in only modest efficiency losses at higher speeds. Sub-1-min chromatograms in HPLC are no longer uncommon. However, this dramatic improvement in speed required a complete redesign of HPLC instrumentation to increase pressure capability, decrease extra-column band broadening, and improve response time.

7.1.1.2 Porous Shell Particles

Another important column improvement has involved the development of porous shell particles. These particles have a solid core, with a thick porous coating. Typically the total particle diameter (d_p) is in the order of 2.5–3 μm, while the solid core is roughly half that diameter. Such particles exhibit efficiencies proportional to the thickness of the porous shell, but pressure drops, ΔP, proportional to d_p^2. Thus they can produce similar efficiency and speed compared to the sub-2-μm totally porous particles, but with as little as one-fourth the pressure drops. Use of these particles still required a somewhat simpler redesign of HPLC instrumentation to minimize extra-column band broadening, and improve response time.

7.1.2 Instrumental Developments

Although HPLC instrumentation to exploit the inherently higher speed of these smaller particles was only introduced in 2004, there have, subsequently, been several newer generations of what is now called "ultra HPLC" or "UHPLC."

HPLC and UHPLC are a more than $4 billion per year industry. While SFC is growing, it is still less than $100 million per year. Since the market size for SFC is no more than a few percentage of HPLC, most instrumental developments in SFC have been later modifications of HPLC instruments. So far, none of the UHPLC's have been converted into SFC's, although it is likely in the future. Nevertheless, some modern SFC's are characterized as having "ultra-high performance," since they are now sometimes used with the modern sub-2-μm and porous shell particles. The use of both sub-2-μm and porous shell particles in SFC has lagged significantly behind HPLC, and has only recently started catching up. This is in spite of the fact that the commercial versions of current SFC instruments are inadequate to achieve full theoretical efficiency of sub-2-μm particles.

7.1.3 Changes in Instrumentation for HPLC (and SFC?) Caused by Using Sub-2-μm Particles

There are significant differences in the physical characteristics of the mobile phases used in HPLC and in SFC. These differences have an impact on instrument design that has not been well defined. While in the past, simply modifying HPLC designs was adequate, it is not necessarily the appropriate approach for advancing SFC or UHPSFC, going forward, since the two techniques will have different optima.

7.1.3.1 Pressure Requirements

7.1.3.1.1 Column ΔP

Higher speed separations come at a price. The most obvious but probably not the most important is the pumping pressure required. The

ΔP across a column is proportional to the viscosity η, length L, the linear velocity μ, and inversely proportional to the square of the particle diameter d_p, and the specific column permeability Θ as indicated in Eq. 7.1.

$$\Delta P = \eta \mu L / \Theta d_p^2 \tag{7.1}$$

Since μ is inversely proportional to d_p, ΔP is actually proportional to d_p^3 (at fixed column length). Even accounting for a shorter column length, when using sub-2-μm particles, ΔP is often 8–9 times higher than when using a column packed with 5 μm particles. Further, the desire to operate at much higher than optimum velocity or to use long columns to quickly generate high efficiencies, further exacerbates ΔP. This has resulted in the development of much higher pressure pumping systems. UHPLC pumps capable of well over 1200 bar are now widely available with much higher pressures likely in the future.

The need for very high-pressure pumps, in HPLC, when using sub-2-μm totally porous particles led to a partial backlash since their adoption would make all older HPLC's, with upper pressure limitations of 400 and even 600 bar, obsolete. This accounts for many tens of billions of dollars in installed base. Porous shell particles offered the possibility of providing similar speeds and efficiencies but with lower pressure drops. There is still a question about the sample capacity of these particles. Their use, however, still requires very low, extra-column band broadening, and very fast data acquisition.

SFC offers higher speeds with lower column pressure drops using either kind of particle. The viscosity, η, of pure CO_2 is as little as 1/20th the viscosity of water. Accounting for the 3–5 times higher flow rate in SFC indicates that the ΔP across an SFC column using pure CO_2 is as little as one-seventh the ΔP in HPLC. However, pure CO_2 is almost never used today. Instead, various amounts of a liquid modifier, such as a short-chain alcohol, are added, typically between 5% and 40% which tends to increase the viscosity of the mixture. For example the kinematic viscosity of 20% methanol at 50°C and 400 bar is 0.141c Stokes, which is still only one-fifth the viscosity of pure methanol (0.69cS) at room temperature and atmospheric pressure. Most HPLC mobile phases are much more viscous than methanol.

The calculated viscosity of mixtures [2] of 5%, 10%, 15%, and 20% methanol in CO_2, at 100 through 400 bar, at 40°C, 50°C, and 60°C, are presented in Table 7.1. Pressure drops, even with totally porous sub-2-μm particles, are typically very low. In fact, using such columns, capable of ≈30,000 isocratic plates, seldom requires a pump pressure as high as 400 bar (at optimum μ), even with an outlet pressure set at 150 bar, and relatively high modifier concentrations.

TABLE 7.1 Kinematic Viscosities of Mixtures of Methanol in CO_2

		Kinematic viscosity($\times 10^{-4}$ cm^2/s)											
MeOH in CO_2		5%			10%			15%			20%		
0.000855	Pressure (bar)	40°C	50°C	60°C	40°C	50°C	60°C	40°C	50°C	60°C	40°C	50°C	60°C
	100	8.55		7.6	9.4	8.4		10.45	9.5	8.55	12.1	10.8	9.6
	150	9.5	8.65	8.05	10.2	9.4	8.75	11.1	10.2	9.45	12.6	11.4	10.1
	200	9.95	9.3	8.75	10.55	10	9.35	11.65	10.85	10.05	13.2	11.9	11.2
	250	10.5	9.75	9.2	11.25	10.5	9.85	12.1	11.33	10.6	13.6	12.4	11.6
	300	10.95	10.38	9.65	11.75	10.95	10.35	12.6	11.8	11.05	14.2	13	12.2
	350	11.45	10.64	10.05	12.1	11.35	10.65	13.05	12.2	11.45	14.7	13.6	12.6
	400	11.77	10.96	10.36	12.45	11.7	11.1	13.3	12.6	11.85	15.3	14.1	13.1

7.1.3.1.2 Extra-Column ΔP

An aspect of observed system pressure drops (difference in pressure between the pumps and the BPR) that is often under-appreciated is the pressure drop in the connector tubing. Sub-2-μm particles require short pieces of smaller ID connector tubing, such as 120 μm, to minimize extra-column band broadening (which is described in more detail later). In one set of experiments [3] system pressure (at pump) was ≈520 bar at 5 mL/min of 20% methanol, with a 3 × 100 mm column packed with 1.8 μm particles, using a back pressure of 150 bar. The optimum flow rate was 1.7 mL/min, so this flow represents nearly 3x optimum. The total ΔP was ≈370 bar. However, removing the column produced a system pressure of 325 bar, indicating that the noncolumn components contributed 175 bar to the total ΔP. Thus the noncolumn components contribute nearly half (47%) of the total pressure drop. Some of this ΔP occurs in front of the column and the rest occurs after the column.

In HPLC, pressure drops in the tubing are basically irrelevant, since retention and selectivity have little or no correlation with pressure. Remember that the flow rate in HPLC is one-third to one-fifth those in SFC for the same column dimensions, partially offsetting the consequences of the higher viscosities in HPLC.

In SFC, the local pressure is rather important, since retention is somewhat related to mobile phase density. The pressure drop in front of the column is not particularly relevant other than limiting the total pressure available for operating long columns. The average pressure/density in a column with small particles can be significantly higher compared to the average pressure/density in a column packed with larger particles. Some suggest that the outlet pressure should be adjusted in order to keep the average pressure, in both cases, constant, to minimize changes in retention and selectivity. The biggest issue with large pressure drops in tubing involves method transfer from larger to smaller particles, particularly if the tubing ID and length is changed.

7.1.3.2 Pressure, Density, Viscosity, and Diffusion Coefficients

In HPLC, diffusion coefficients, and viscosity, are not a significant function of pressure or density. It is assumed that ever smaller particles can be developed, provided that pumps with ever higher pressure capability can be built, in order to create ever faster separations. It is relatively trivial to increase pumping pressure capability. It is somewhat more difficult to decrease detector response time and extra-column band broadening. Much higher pressures may or may not also be useful in SFC, since the effect of pressure, density, and viscosity on solute diffusion coefficients have not been defined.

In SFC, it is often assumed that density increases monotonically with increasing modifier concentration, and that the of ρ and $D_{1,2}$ is a constant (here ρ is density, $D_{1,2}$ is solute binary diffusion coefficient in the mobile phase), at least above $\rho = 0.6$ [4]. However, above 300 bar, the density of CO_2/MeOH mixtures actually decreases significantly with increasing methanol concentration [3]. The diffusion coefficients do not appear to increase with this decrease in density. We know that viscosity increases, with increasing modifier concentration, since ΔP increases, as shown previously. One would assume that at higher pressures, the molecules of the mobile phase would be forced closer together. Further, when using polar modifiers like methanol, one would expect increased hydrogen bonding, with increasing concentration. Both should increase viscosity and decrease $D_{1,2}$. However, there are no measurements of $D_{1,2}$ at high pressures in realistic SFC mobile phase conditions. There is a serious need for accurate measurements of solute diffusion coefficients at very high pressures (like over 400 bar).

7.1.4 Band Broadening and Turbulent Flow

7.1.4.1 Column Variance

The injection of a sample volume into the mobile phase can be idealized as a very narrow square wave of concentration. In the column this square wave is broadened out by diffusion as indicated by a van Deemter–like equation:

$$H = A + BD_{1,2}/\mu + C\mu/D_{1,2} \tag{7.2}$$

The three terms on the right represent eddy diffusion (the A term), axial diffusion (the B term), and the resistance to mass flow (the C term). They all tend to broaden the peak and tend to make it more Gaussian. The use of smaller particles results in much narrower, faster peaks compared to larger particles. The use of smaller diameter columns results in a significant decrease in the volume of the mobile phase containing a peak.

One can easily calculate the variance, σ^2, of a peak based on its standard deviation, σ, obtained from its peak width at half height, $w_{1/2}$, since $w_{1/2} = 2.35\sigma$. A 3 × 100 mm column packed with 1.8 μm particles has 1/12 the column variance of a 4.6 × 150 mm column packed with 3.5 μm particles. A 2.1 × 100 mm column, packed with 1.8 μm particles would have 1/50th the variance of the 4.6 mm column packed with 3.5 μm particles.

7.1.4.2 Extra-Column Variance

Every other component in a chromatograph also tends to broaden out the injection plug. The goal of the instrument designer is to minimize this extra-column band broadening in order to see as much of the inherent efficiency of the column as possible. If the column variance decreases by 50 times, it is likely that the extra-column variance should also be decreased by 50 times.

The components that can contribute to this extra-column variance or dispersion are, the injection volume, injection time, connecting tubing, fittings, detector cell volume, detector electronics, and filter. Each component in a chromatograph has a variance. These variances add. In order to see 95% of the theoretical efficiency of a column, the extra-column variances should be approximately 1/10th the column variance or smaller.

Extra-column variance tends to be dominated by the connector tubing. In HPLC, the flow is generally laminar, and the variance of the tubing can be described by the equation:

$$\sigma^2_{\text{tube}} = \pi r^4 LF/24D_{1,2} \tag{7.3}$$

where r is the radius of the tube, L its length, F is the flow, and $D_{1,2}$ is the binary diffusion coefficient of a solute in the mobile phase. Note that the variance of a tube is proportional to the fourth power of the tube radius and to length. However, this equation is only appropriate for laminar flow.

7.1.4.3 *Turbulent Flow*

The Reynolds Number is a dimensionless variable developed to indicate when laminar flow breaks down, producing turbulence. Turbulent flow is indicated when the Reynolds Number exceeds $\approx$2400. The Reynolds Number is defined by:

$$\text{Re} = \mu d/\eta \tag{7.4}$$

where μ is linear velocity in cm per second, d is the diameter of the tube in cm, and η is the kinematic viscosity in cm^2 per second. Since the viscosity of CO_2-based mobile phases is dramatically less than those used in UHPLC, there is a higher likelihood that flow will be turbulent in the tubing.

Turbulent flow has two effects. The good part is that it decreases the dispersion (extra-column band broadening) caused by the tubing. Thus Eq. 7.3 overestimates the variance of the tubing if the flow is turbulent. Turbulence in the tubing should not have any effect on column efficiency.

The other aspect is that turbulence increases the ΔP across the tubing. With laminar flow the ΔP increases linearly with μ. In turbulent flow ΔP increases with at least μ^2.

Using the viscosities in Table 7.1, it can be shown that between 5% and 20% methanol in CO_2, 100–400 bar, and 40–60°C, flow in 170 μm ID tubing is always turbulent above ≈1 mL/min. Decreasing the tubing ID to 120 μm results in laminar flow at 1 mL/min, but transitioning to turbulent flow starting at 1.5 mL/min. Thus under most conditions used in SFC the flow in the tubing is mostly turbulent, particularly at higher than μ_{opt}. In HPLC, the fluid in the tubing is almost always laminar.

It should be stressed that the flow in the column is always laminar in both techniques, and the ΔP across the columns is a linear function of μ. The ΔP across the SFC column is always substantially less than the ΔP across the HPLC column at the same flow rate.

The ΔP across a SFC chromatographic system sometimes shows a significant nonlinear increase with increasing flow rate, characteristic of increasing turbulent flow (increasing Reynolds number). In one example [3] the ΔP with 71 cm of 120 μm tubing, with a 3 mm, 2 μL flow cell plotted versus flow rate showed such nonlinearity. Extending a straight line from 0 flow and 0 ΔP to asymptotically approach the ΔP curve, indicates what the curve would look like if flow were laminar. Extending this line to 5 mL/min indicted that the actual ΔP was more than four times the expected ΔP if the flow were laminar.

There is at least one example in the literature [5] where the ΔP using pure CO_2 in SFC significantly exceeded the ΔP using pure acetonitrile through the same tubing at the same high flows, caused by turbulent flow with the CO_2 but laminar flow with acetonitrile. This aspect of system design has barely been recognized.

7.1.4.4 Tubing Compromises

Conventional HPLC's have typically used fairly long lengths of 170 μm tubing connecting the injection valve to the column and the column to the detector. Modern UHPLC's use much shorter lengths of 120 μm tubing or smaller (i.e., 75–100 μm) tubing in order to minimize band broadening and, thus, see most of the efficiency of sub-2-μm particles in 2.1 mm ID columns. Commercial SFC's typically use 170 μm tubing to connect the injection valve to the column and the column to the detector, similar to conventional HPLC's. In one design there are 60 cm of 170 μm ID tubing in front of the column and 60 cm after the column.

Most SFC results using columns packed with sub-2-μm particles, and the standard plumbing have found a reduced plate height, h_r of ≈3. In order to see the full efficiency of sub-2-μm particles in 3 mm or smaller ID columns, both the length and ID of the tubing need to be made smaller. It has been shown [3] that 25 cm of 120 μm tubing (including a mini HX) between the injection valve and the column, and 51 cm of 120 μm tubing (including a second mini HX) allows the user to see the full theoretical efficiency of a 3×100 mm column with 1.8 μm particles down to $<k = 2$.

As outlined earlier, the use of sub-2-μm particles has created a dilemma. There is a need to use smaller ID connector tubing for lower dispersion. However, turbulent flow in such tubes makes for very large pressure drops in the tubes. This results in uncertainty as to the actual pressure and density in the column.

7.1.4.5 Thermal Gradients

In the earliest days of HPLC it was predicted [6] that large ΔP's would result in large axial and potentially radial temperature gradients caused by the decompression of the fluid along the length of the column. The axial temperature gradients are not of concern. However, any radial temperature gradient would result in radial gradients of μ, and retention, causing potentially severe efficiency losses. Since recent column developments have emphasized smaller particles for higher speed separations, typical pressure drops, and the potential for thermal gradients, have dramatically increased. Research is aimed at another significant decrease in particle diameter to sub-1-μm, which is likely to require another three- to fourfold increase in ΔP.

The potential for severe efficiency losses, due to such thermal gradients has resulted in several significant developments. It has been recognized that the column should not be directly heated or cooled but rather isolated in a still air oven with the mobile phase preconditioned to the initial temperature by a precolumn HX. Any heat exchange between the column and its surroundings would result in the formation of a radial gradient. If there is no heat exchange with the surroundings, the axial temperature gradient will form but not the radial gradient.

In HPLC the mobile phase always warms during decompression. The larger the pressure drop, the more extensive the warming. In SFC the situation can be very different. At low percent modifier, CO_2 actually cools fairly dramatically with expansion. For instance pure CO_2 cools by ≈16°C when it decompresses from 500 to 100 bar. At 5% modifier, with a similar decompression, the temperature drop is roughly halved. At ≈15% methanol, there is neither cooling nor warming accompanying the decompression. Above 30% there is noticeable warming. Since the pressure changes tend to be much smaller in SFC, compared to HPLC, and the average modifier concentration is ≈20%, where there is minimal warming, there tends to be much less heating or cooling. As a result, somewhat larger ID columns, with sub-2-μm packings tend to be used in SFC and have been shown to deliver full theoretical efficiency.

7.1.4.6 Column ID and Flow Rate

Analytical scale HPLC and SFC columns were traditionally 4.6 mm ID. However, in HPLC it is now common to use smaller column ID's than is traditional in order to minimize the radial distance and, hence, the extent of any thermal gradients caused by the decompression of the mobile phase. In fact 2.1 mm ID has become a widely available defacto standard in HPLC.

The use of 2.1 mm columns dictates serious compromises by the instrument designer. No mechanical device performs truly exceptionally over a dynamic range of >100. The optimum flow rate for a 2.1 mm ID column in HPLC is on the order of 0.3–0.5 mL/min. Thus the flow of the second channel at 5% modifier is on the order of 0.015–0.025 mL/min. This flow must have very high reproducibility. At a flow of 0.3 mL/min, with a set point of 5% modifier, a variation in the modifier flow of ± 0.001mL/min means the delivered percent modifier is only somewhere between 4.67% and 5.33% (±6/6% of the set point). Some state-of-the-art UHPLC's limit the maximum flow rate to 1 mL/min in order to provide better reproducibility at low modifier concentrations when using 2.1 mm or smaller ID columns.

In SFC, the flow rate is always higher than in HPLC, on the same-sized column, due to the higher solute $D_{1,2}$ in CO_2-based mobile phases. A typical SFC flow rate for 2.1 mm ID columns, with sub-2-μm particles is on the order of 0.7–1.0 mL/min. However, since there are minimal problems with thermal gradients in SFC, smaller column ID's are more difficult to pack efficiently, and the mobile phase is inexpensive, there is not much incentive for the use of such small ID columns in SFC.

At the other extreme, the optimum flow rate for a 4.6 mm ID column packed with sub-2-μm particles can be >5 mL/min. Most current SFC's have a typical maximum flow rate of 4 or 5 mL/min, too low to even achieve optimum. It was already shown that higher flows cause serious ΔP's in the connector tubing. Consequently, a good compromise is to use smaller ID columns packed with sub-2-μm particles. Columns with 3 mm in diameter have an optimum flow rate in the range of 1.7–2 mL/min, similar to the optimum of a 4.6-mm ID column packed with 5 μm particles. Since the van Deemter curves for these types of columns are very flat, higher flow rates allow faster analysis with little loss in efficiency. The larger ID makes them easier to pack. The larger size is also more compatible with the flow characteristics of current pumps.

7.2 PUMPING CO_2

New users of SFC typically start with steel cylinders, mounted outside the lab. The pressure in such a cylinder is approximately 55 bar at 20°C, but can change significantly with temperature. In the cylinder, the CO_2 exists as a high-density liquid phase in equilibrium with a somewhat lower density gaseous headspace (~0.8 vs ~0.15 g/cm^3 at 20°C). There is no pressure regulator, just a reducing fitting.

7.2.1 Compression Ratio

The compression ratio of a reciprocating pump is the ratio of the (pump) cylinder volume with the piston fully retracted (full), divided by the volume with the piston fully extended into the cylinder ("empty"). Most modern SFC pumps are reciprocating pumps with relatively low compression ratios. Pistons are typically ~1 mm from the pump cylinder walls, and never approach the end of the cylinder. In one widely used pump, the piston is 3.2 mm OD, with a maximum stroke length of 1.25 cm, while the cylinder is 5.0 mm ID. Ignoring the gap between the end of the piston and the bottom of the cylinder, the empty volume is ≈100 μL, while the full volume is ≈245 μL. Thus the compression ratio of this typical pump is less than 2.45.

7.2.2 Effect of Fluid Temperature on Pumping

Any increase in the temperature or decrease in the pressure of the CO_2, after it leaves the cylinder, results in vaporization of some of the fluid, which can potentially cause cavitation of the pump and, subsequently, no flow. The internal temperature of HPLC pumps often reaches 30°C and higher due to the heat dissipated by the electronics and motor. If the supply cylinder were at 20°C but the pump were at 30°C, the density in the pump during the refill stroke would be dictated by the cylinder pressure of ≈55 bar. Under such conditions, the density of the CO_2 in the pump would only be ≈0.152 g/cm^3 (see Table 7.2).

TABLE 7.2 Density of Pure CO_2 at Various Temperatures and Pressures [7]

	Temperature (°C)				
Pressure (bar)	**−30**	**−20**	**5**	**10**	**30**
20	1.078	1.032	0.0443	0.0430	0.0389
30	1.081	1.036	0.0742	0.0711	0.0623
40			2 phase		
55			0.912	0.876	0.152
100			0.948	0.921	0.771
400			1.066	1.052	0.989
600			1.1113		

In order to increase the pressure of the CO_2 to 100 bar, where the density would be 0.771 g/cm^3 (ignoring any heating due to compression), the pump would need a compression ratio of >5 (0.771/0.152). Pumping to 400 bar would require a compression ratio of >6.7. Obviously a typical reciprocating pump, with a low compression ratio, cannot be used to pump CO_2 from a high pressure steel cylinder without modification.

If, however, the fluid and the pump head were chilled to just 10°C the density of the CO_2 entering the pump cylinder during the refill stroke would be on the order of 0.876 g/cm^3, requiring a compression ratio only 1.05 to reach 100 bar, and 1.2 to reach 400 bar. Almost all approaches to making an SFC CO_2 pump have chilled the pump head to 4–5°C for additional margin. Many SFC's withdraw the liquid phase, using a DIP or eductor tube, mounted on the valve inside the high pressure cylinder, to try to minimize the amount of chilling power required to convert all the fluid to a dense liquid. However, some SFC's withdraw the vapor phase, which is effectively distilled just before use, allowing lower grade CO_2 to be used.

7.2.3 Larger CO_2 Flow Requirements

With high usage cylinders become inconvenient. At 2 mL/min, a 22-kg cylinder lasts ≈7 days if operated 24 h/day. Spare cylinders need to be stored and exchanged. Organizations with many analytical scale and/or semipreparative scale SFC's generally move away from the use of high pressure (atmospheric temperature) supply cylinders and switch to a set of connected Dewars or a larger bulk tank, some holding many tons. The whole facility is sometimes plumbed with welded supply lines distributing the CO_2 throughout multiple labs. This infrastructure development can require significant planning and construction time.

Bulk tanks are more convenient than Dewars since there is no significant handling or need for a lift truck. A large vendor truck simply refills the tank periodically. CO_2 in cylinders and Dewars tends to cost on the neighborhood of \$1 kg^{-1}, whereas the cost from a bulk tank can be as low as \$0.10 kg^{-1}. This low cost is very favorable compared to the cost of heptane or acetonitrile which can cost >> \$80 L^{-1}.

In Dewars and bulk tanks the CO_2 pressure is ≈20 bar, and is chilled to −20°C to −30°C by semicontinuously boiling off of some of the fluid. Any reciprocating pump would need to be chilled to below these temperatures (below −20°C to −30°C) to keep the fluid in a liquid state to allow pumping with a low compression ratio reciprocating pump. CO_2 is a very poor lubricant which affects seal wear, particularly at very low temperatures. The piston seals used in CO_2 pumps are different from those used in reversed-phase HPLC pumps. Pump seals are typically made of ultra-high-molecular-weight polyethylene. Such seals cannot survive long at $-20°C$ to $-30°C$.

A compressor is used to boost the pressure up to that found in a steel, high pressure, cylinder, or higher (+/55 bar). A compressor typically has a much higher compression ratio than a standard reciprocating pump and may not need to be chilled. A number of commercial gas delivery systems (GDS) have appeared. Commercial GDS's typically boost the pressure of the feed from 20–30 bar to between 75 and 100 bar. The metering pump head is still usually chilled.

Older GDS's were large air driven amplifier pumps that were very noisy and inefficient. Large volumes of relatively low pressure ($\approx$7 bar) air were used to generate much smaller volumes of much higher pressure CO_2. Thus a very large air compressor was needed to produce adequate air flow to drive such amplifier pumps, which in turn supplied CO_2 to the metering pump (three pumps in series!!!). However, suitable electrically driven compressors with very high compression ratios are commercially available that are much less noisy, more efficient, and do not require a chiller. Some can raise the pressure from bulk tanks at 20 bar to over 400 bar, without a chiller.

7.2.4 Compressibility

As already noted, chilling the pump head and prechilling the fluid significantly decreases the compression ratio needed to pump CO_2. However, accurately metering the flow of CO_2 is much, more complicated than simply chilling the fluid and the pump head, although in the past this was often (wrongly) assumed to be adequate.

The reciprocal motion of a pump can be divided into a refill stroke, a compression stroke, and a delivery stroke. The compression and delivery stroke are both part of the forward motion of the piston. The compression stroke precompresses the fluid from the source pressure to the column head pressure. No flow leaves the pump during the compression stroke so it tends to be performed relatively quickly in order to try to maintain constant flow versus time. Modern pumps almost always use two or more pump heads, for each fluid, to try to minimize or eliminate the dead spots where no flow emerges. Thus one pump head is typically delivering when the other is refilling.

The fraction of the total forward stroke length used up in compressing the fluid depends on the compressibility of the fluid. If the compression stroke is too short, the pump fails to reach the column head pressure during the compression stroke. Part of the delivery stroke is then wasted to complete the compression. No flow actually emerges during this part of the "delivery" stroke. The pump delivers less flow/stroke than the set point. Alternately if the compression stroke is too long, the pump delivers more flow than the set point. In either case there is usually a significant pressure perturbation and potential composition and flow inaccuracy, along with pressure pulse induced UV noise. Using two or more pump heads typically does NOT eliminate this problem.

Various pumping strategies have evolved that try to eliminate errors in compressibility compensation. These pump developments are at the heart of SFC evolution over the last 25–30 years. The normal liquids used in HPLC have compressibility between 45×10^{-6} per bar (water) and 150×10^{-6} per bar (heptane), and are relatively independent of pressure. These numbers are the change in density associated with each change in pressure of 1 bar. Thus the density of water increases by $\approx 4.5\%$ between 1 and 1000 bar, while the density of heptane increases by 15% over the same pressure range. The pumps deliver flow based on volumetric displacement per unit time during the delivery stroke. If the compressibility compensation is inadequate the actual delivered volume will deviate from the set point. The composition and total flow will both be inaccurate and the degree of inaccuracy will depend on the pump pressure. This is true for both HPLC and SFC, only more exaggerated in SFC.

The compressibility of CO_2, even when chilled to 4–5°C varies widely, depending on pressure. Near cylinder pressure ($\approx$55 bar), the compressibility of CO_2 can be more than 20 times the compressibility of water. At high pressures (i.e., 400–600 bar) the compressibility is much less, approaching that of normal liquids, like heptane.

7.2.4.1 Isothermal versus Adiabatic Compression

There is an additional complication when pumping CO_2. A number of articles [see 8,9] have recently appeared regarding axial and radial temperature gradients that form when CO_2 decompresses across the column pressure drop. With pure CO_2, the decompression results in extensive cooling [8]. With modified fluids, there may be either cooling or warming [9], depending on composition. While this is generally understood, most users do not realize that there can be an even more dramatic extensive warming of the CO_2 during compression from source pressure ($\approx$55 bar), to column head pressure, within the pump. Compressing from 55 to 600 bar can raise the temperature of the CO_2 more than 35°C.

If the compression stroke occurs slowly, there is opportunity for heat to flow from the fluid into the walls of the pump. At one extreme, all the heat generated could be transferred to the pump hardware during compression and the compression would be isothermal. However, both stainless steel and CO_2 are poor heat conductors, making isothermal operation less likely.

If the compression stroke were to occur very rapidly, the compression would be essentially adiabatic, meaning minimal heat would be transferred into the walls of the pump cylinder during compression and the fluid would warm up by as much as 35°C. After the compression stroke, the pump slows to deliver the CO_2 to the column. During this period some heat will flow from the fluid to the colder pump walls. The fluid subsequently cools, and its density increases during the delivery stroke. Mass flow will vary during the delivery stroke, even though volumetric flow (the rate of volumetric displacement) is constant. This tends to cause (reproducible) short-term fluctuations in flow and composition within each stroke.

The theoretical temperature [10] increase accompanying an adiabatic compression of CO_2 from 50 to 600 bar starting at an initial temperature of 5°C is presented as the upper curve in Fig. 7.2. This is compared to an isothermal compression of CO_2 and an adiabatic compression of H_2O, represented by the lower two curves in the Figure.

Compressibility values calculated [10] assuming isothermal compression are much larger than those calculated for adiabatic compressibility, as shown in Fig. 7.3. At 50 bar, the isothermal compressibility of CO_2 is

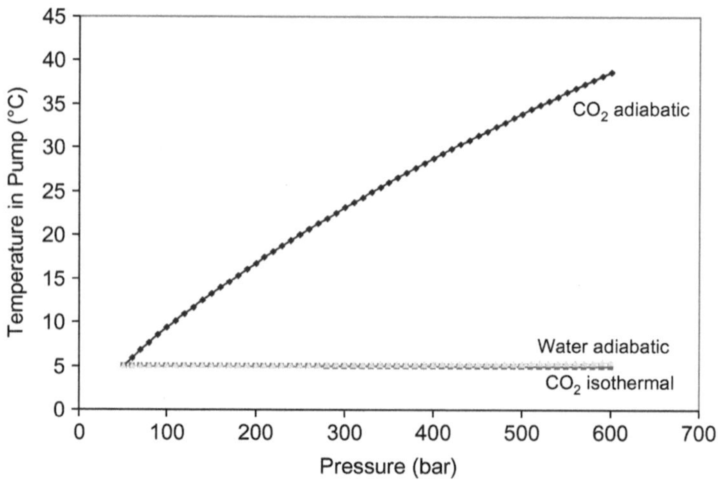

FIGURE 7.2 The potential increase in temperature if CO_2 were compressed from 50 to 600 bar. The upper curve is for adiabatic compression. Most people assume that setting the pump head temperature to some value, such as 5°C guarantees this temperature is maintained. The other two curves show the potential change in temperature for both CO_2 with an isothermal compression and water with an adiabatic compression over the same pressure range. Real compressions are probably somewhere between adiabatic and isothermal.

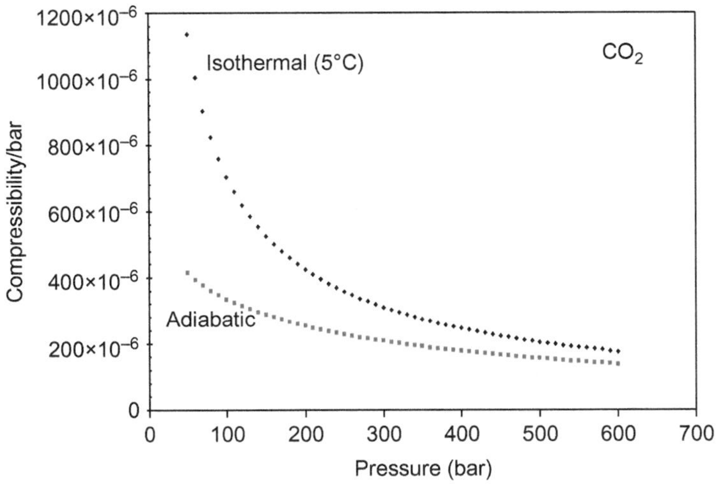

FIGURE 7.3 Calculated adiabatic and isothermal compressibility of CO_2 starting from 5°C and 50 bar (roughly high pressure cylinder pressure). Actual compressibility is somewhere in-between, depending on how much heat is lost to the pump head cylinder walls.

near 1200×10^{-6} per bar (≈27 times water!), while the adiabatic compressibility is a little over 400×10^{-6} per bar (≈9 times water). At 520 bar, the adiabatic compressibility is reduced to $\sim 150 \times 10^{-6}$ per bar, which is similar to the compressibility of heptane. However, the isothermal compressibility remains at 176×10^{-6} per bar, higher than heptane, even at 600 bar.

Real pumps probably operate somewhere in between truly isothermal and truly adiabatic, and depend on flow rate, specific design characteristics, and materials of construction of each pump. High flow rates are likely to favor more nearly adiabatic operation. In any case pumping CO_2 requires extended compressibility compensation, compared to normal liquids.

7.2.4.2 Compression Stroke Length

Even when the pump head and fluid are prechilled to 5°C an inordinate fraction of the total pump stroke length is required in order to reach high pressures, compared to HPLC or UHPLC. The percentage of the total stroke that is required to compress the fluid versus pressure is shown in Fig. 7.4, for isothermal and adiabatic compression of CO_2 and for isothermal compression of water (worst case). The isothermal compression of CO_2 from 50 bar, requires over 18% of the total stroke length to reach 600 bar. An adiabatic compression would require less of the stroke length at ≈10%, while the isothermal (worst case) compression of water requires no more than 2.5% of the total stroke length.

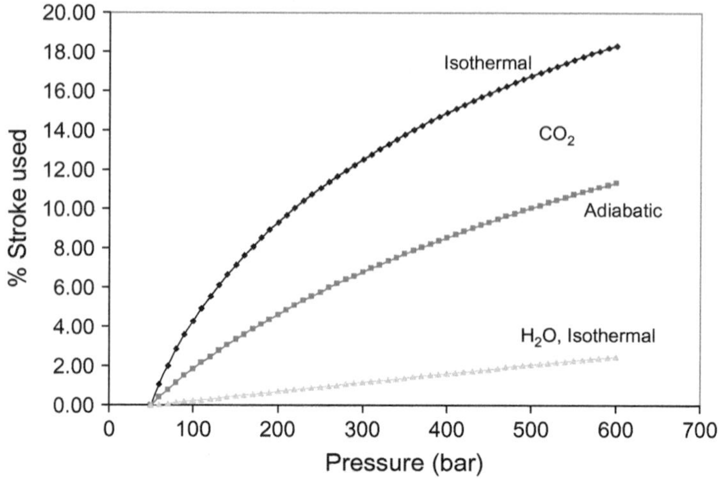

FIGURE 7.4 Fraction of the total stroke length used to compress CO_2, adiabatically or isothermally from 50 to 600 bar, even when the fluid is prechilled to 5°C. These values are compared to water under similar circumstances.

7.2.5 Effect of Pressure on Composition

Even with perfect compressibility compensation, the density of the fluid in the pump varies with pressure. Since the density of each fluid varies with pressure uniquely, constant volumetric displacement will result in variations in the mole ratio between CO_2 and the modifier, as shown in Table 7.3. With a volume/volume set point of 20% methanol in CO_2, the mole% varies from 22.6 Mole% at 100 bar, to 21.2 Mole% at 500 bar. The results were calculated using a compressibility of 120×10^{-6} per bar for methanol, and with CO_2 density calculated using the REFPROP software from NIST.

7.3 PUMP DRIVES

Modern SFC's use two pumps, one to pump CO_2 and the other to pump a modifier. Each pump typically has multiple pistons operated asynchronously. The modifier pumps are identical to HPLC pumps, and will not be discussed. However, there are a number of different approaches to pumping CO_2. In all cases, CO_2 pumps are incompatible with active inlet check valves that allow mixing of multiple modifiers during each stroke (quaternary gradient valves). Such valves are not compatible with the high pressure of the supply CO_2, but could be used on the modifier pump.

TABLE 7.3 The Change in Mole% When the Pump Pressure Changes

	Pump head							
	At 80 vol%			At 20 vol%				
	Temperature, 5°C	CO_2		Total moles/cm^3	Mole% MeOH		Methanol	
Pressure (bar)		moles/cm^3				Moles MeOH	moles/cm^3	density
	Density		Moles CO_2/cm^3 mix			0.00495	0.02475	0.792
100	0.94825	0.02155	0.01724	0.02227	22.60%	0.00503	0.02513	0.804
150	0.97738	0.02221	0.01777	0.02283	22.16%	0.00506	0.02531	0.81
200	1.0007	0.02274	0.01819	0.02329	21.90%	0.0051	0.0255	0.816
250	1.0203	0.02319	0.01855	0.02369	21.70%	0.00514	0.02569	0.822
300	1.0374	0.02358	0.01886	0.02404	21.55%	0.00518	0.02588	0.828
350	1.0526	0.02392	0.1914	0.02435	21.40%	0.00521	0.02606	0.834
400	1.0663	0.02423	0.01938	0.02463	21.32%	0.00525	0.02625	0.84
450	1.0789	0.02452	0.01962	0.02491	21.24%	0.00529	0.02644	0.846
500	1.0905	0.02478	0.01982	0.02515	21.19%	0.00533	0.02663	0.852
	Using 44 g/mole and 32 g/mole							
	Density of MeOH calculated assuming a compressibility of 120×10^{-6} per bar							

The density of each fluid in the pumps changes uniquely as column head pressure changes. Thus with constant volumetric displacement pumps the mole% changes.

7.3.1 Cams

Many older pumps employed two or more asymmetric cams mounted on a shaft driven by a single, constant speed motor. Piston assemblies track the edge of each cam, which translates the elliptical rotation of the cams into a linear reciprocal motion of the pistons in the pump cylinders, as shown schematically in Fig. 7.5. With two cams, one is mounted 180 degrees opposite the other, so that one piston is being withdrawn, refilling with CO_2, while the other is moving forward delivering fluid at the column head pressure.

Check valves allow CO_2 at the source pressure to flow into the pump during the refill stroke, and out of the pump during the delivery stroke. At the end of the refill stroke, the piston changes direction, as shown in Fig. 7.6. Each cam is shaped such that the piston is rapidly accelerated to compress the fluid to the column head pressure. With such cams, the amount of compression is fixed. Thus such pumps only properly compress the fluid at one column head pressure. At all other pressures the pump delivers either more or less fluid than expected, as shown

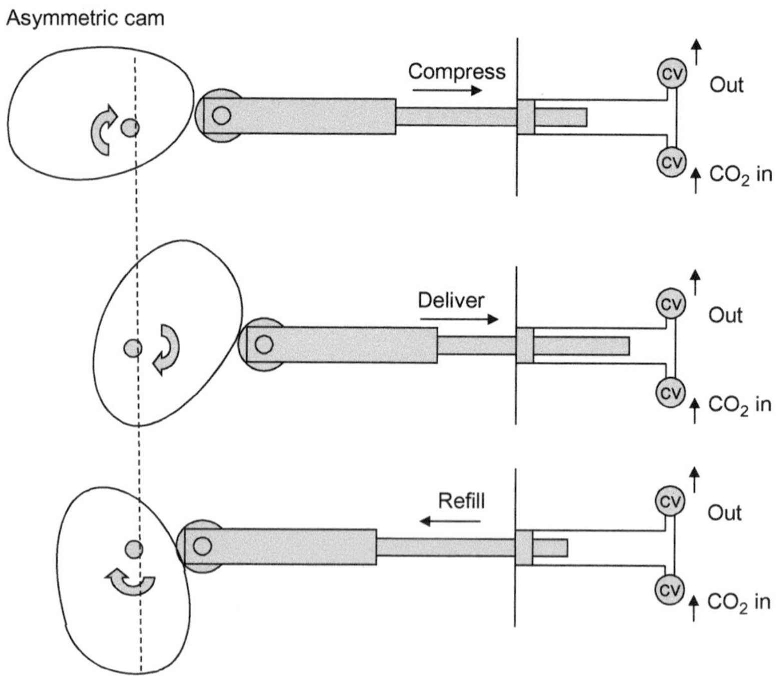

FIGURE 7.5 Schematic diagram of a single cam piston. Most pumps would use two such cams, on a single shaft, mounted 180 degrees apart, to pump each fluid (four cams, two motors for a binary mixture). Not to scale.

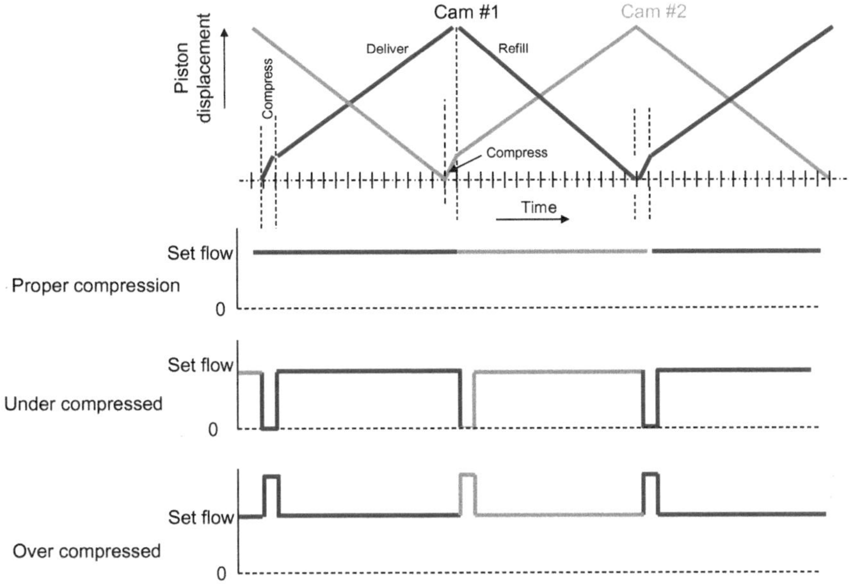

FIGURE 7.6 Top: Piston displacement versus time with two cams on a single shaft/motor, to pump a single fluid, and Bottom: Showing the effect of proper compressibility compensation, under compensation, or over compensating. Such dual cam designs on a single shaft ONLY properly compensate for compressibility at ONE column head pressure. At ALL other column head pressures, the pump either under compensates or over compensates, resulting in inaccurate flow and composition and excessive UV pressure induced noise. When the compressibility compensation is incorrect a mixer is needed to dampen flow/pressure perturbations, BUT the flow and composition will be inaccurate.

schematically at the bottom of Fig. 7.6. This impacts total flow and composition, as well as creating large pressure fluctuations, and UV detector noise. This approach is seriously flawed, and should be avoided, compared to more modern approaches.

7.3.2 Ball Screws

Ball screws or linear actuators have largely replaced cams at least in more expensive systems. Ball screws driven by asynchronous motors with a position sensor, allows the system to control the length and speed of the compression stroke, allowing adjustment of compressibility over a wide range.

At least one, no longer available, commercial system calculated compressibility, based on the delivery pressure and pump head set temperature, using an equation of state (EOS). The piston was then advanced to just compensate for the theoretical compressibility of the fluid. Since the actual fluid temperature tends to differ from the set temperature (due to the extent of isothermal or adiabatic compression), the theoretical optimum stroke length could be incorrect. Subsequently the length of the "next" compression stroke was varied slightly and the amplitude of the resulting pressure spike was measured. The system further dithered the length of the subsequent compression strokes to be both longer and shorter than the theoretical value to empirically find the compression stroke length that minimized the pressure perturbations (which corresponds to optimum compressibility compensation).

Some systems use two ball screws geared together and driven by a single motor as shown schematically in Fig. 7.7. The pistons are plumbed in series instead of parallel. The first, primary, piston had a stroke double the secondary pistons stroke volume and moved twice as fast, as indicated at the top of Fig. 7.8. In this arrangement, there are only two check valves needed (per fluid), on the inlet and outlet of the primary piston. Only the primary piston performs compression. When the primary piston is delivering fluid to the column, the secondary piston is refilling by taking half the flow from the primary. When the primary is refilling, the secondary is delivering.

A problem with this approach is that, during the compression stroke, the secondary piston is withdrawing fluid from downstream of the pump, causing a momentary negative flow to the column. Since CO_2 is much more compressible than normal liquids this effect is more

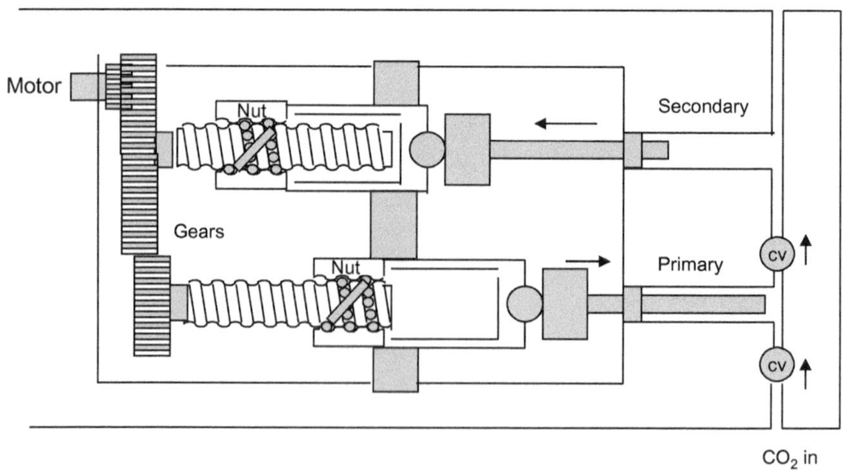

FIGURE 7.7 Schematic diagram of ball screws ganged in series, as used in SFC for more than 20 years. As the shafts turn, the balls force the nuts (and pistons) forward and backwards depending on the direction of rotation of the motor, converting circular motion into linear motion (return springs not shown). The primary piston moves twice as fast, and twice as far, as the secondary piston (gear ratio). This arrangement allows use of only two check valves.

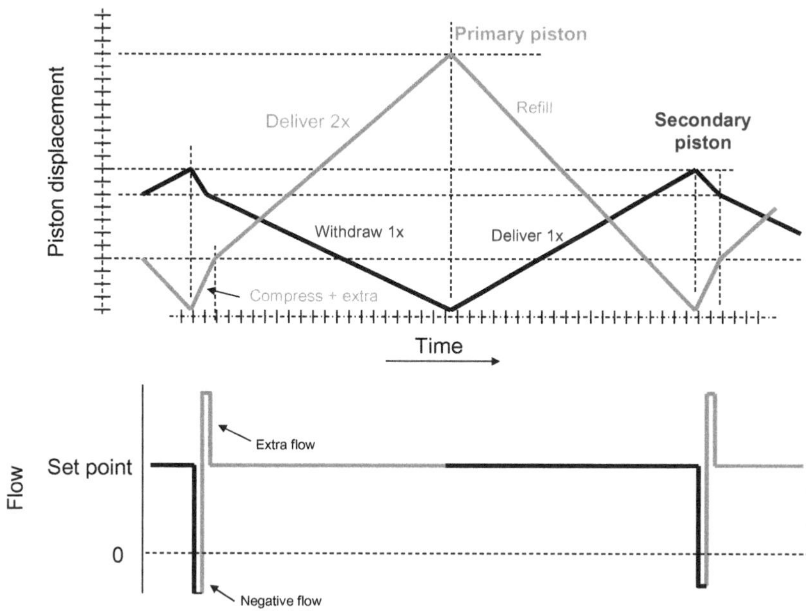

FIGURE 7.8 Top: Piston movement using two ball screws ganged in series. The primary piston moves twice as fast and twice as far as the secondary. Only the primary has a compression stroke. Bottom: Flow versus time. The secondary piston refills by taking half the primary flow, during primary delivery. Note the negative flow during compression, and the extra compensating flow. These flow perturbations must be averaged out using a mixer with a large volume. Never-the-less, flow and composition are accurate.

pronounced in SFC than in HPLC, involving a much larger fraction of the total stroke length (see Fig. 7.4). To compensate, after the compression stroke, the primary piston injects extra fluid to make up for the lack of flow during compression, as indicated schematically at the bottom of Fig. 7.8. This compensation maintains accuracy of flow and composition, but creates significant short-term perturbations in flow and pressure that require relatively large mixing volumes to average out, to minimize UV noise. Large mixing volumes mean long gradient delays. This is significantly accentuated when using smaller ID columns, such as 1 or even 2.1 mm columns that require very low flow rates. Large mixing volumes at low flow rates translate into long gradient delays using such columns, with such pumps.

7.3.3 Multiple Independent Pistons/Motors

Many pumps have used a single motor to drive multiple pistons, largely because the motors and drive electronics were expensive. Recent analytical HPLC's, and several SFC's, use a dedicated motor to

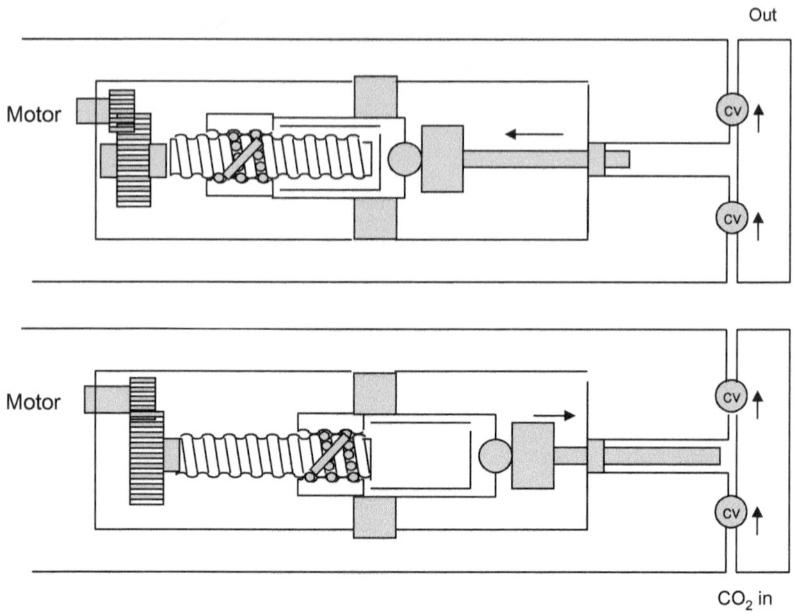

FIGURE 7.9 Independent motors allow maximum versatility. When one pump is filling the other pump is delivering mobile phase. After filling, it precompresses the CO_2 to near the column head pressure, then waits. When the delivering piston nears the end of its stroke it slows down while the other speeds up. Since there is no time of negative flow, there is minimal pressure or composition variation and less need for a mixer.

drive each piston, since the costs of motors and electronics has fallen significantly, while the required analytical throughput has increased significantly. Performance improvements are necessary when using smaller particle packings, and smaller column ID's. Such an approach is shown schematically in Fig. 7.9. In some cases each pump head has an independent pressure transducer. In others measuring current can be used to determine the torque needed to achieve a similar pressure. While one piston is delivering, the other refills and precompresses the fluid to just under the column head pressure, then stops. When the first piston nears the end of its delivery stroke, it slows, while the second piston speeds up and takes over delivery, as suggested in the top of Fig. 7.10. This approach effectively compensates for the compressibility of the fluid and separates compression from metering.

Such an approach largely eliminates the pressure/flow spikes and minimizes the mixing volume required. Surprisingly, this approach was pioneered decades ago (1982), with a pump originally designed to produce uniform flow of a coating agent for photographic film production. [see 11,12]. Many years ago, Rainin developed this concept into a

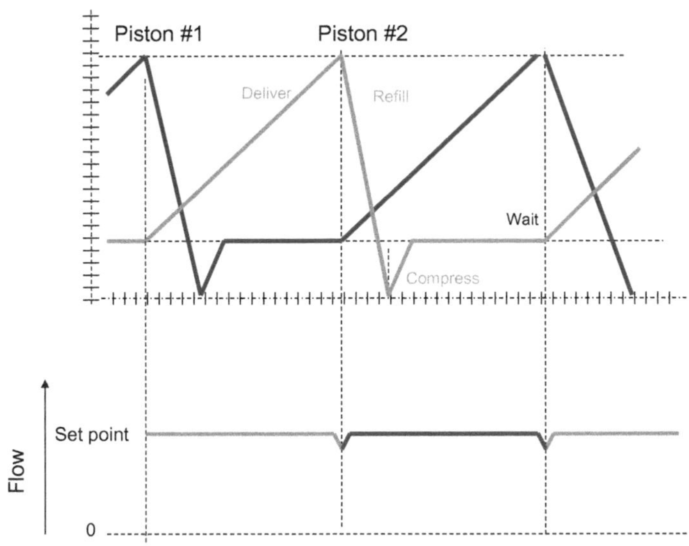

FIGURE 7.10 Top: Piston movement using two independent motors and ball screws, and. Bottom: Flow versus time. This approach is the most nearly ideal for removing the effects of compressibility compensation from both the CO_2 and the modifier flow streams.

semipreparative HPLC pump (the now obsolete Varian/Agilent SD-1). Most UHPLC's now use this approach, which is likely to be the standard in both HPLC and SFC in the future.

7.3.4 Precompression Booster

Another recently developed method for precompressing the CO_2 involves the use of a small booster pump (a compressor) in front of the CO_2 metering pump. Unlike a GDS that precompresses the fluid to a constant, relatively low pressure, such as 75–100 bar, this booster senses the column head pressure and precompresses the CO_2 to a few bar below that pressure. The CO_2 metering pump compressibility is set to 0. The main pump does not compress the CO_2, but only meters its flow. This eliminates the flow and pressure fluctuations common with pumps that both compress and meter. The modifier pump still performs both compression and metering and thus acts similar to the pump described in Fig. 7.8, but with much shorter compression stroke lengths. Piston movement of the CO_2 and modifier pumps are shown schematically at the top of Fig. 7.11, while CO_2, modifier and total flow are illustrated at the bottom of the figure. This approach allows almost any HPLC pump to be used as an SFC, provided the compressibility of the CO_2 pump can be set to zero. In fact the use of a high compression

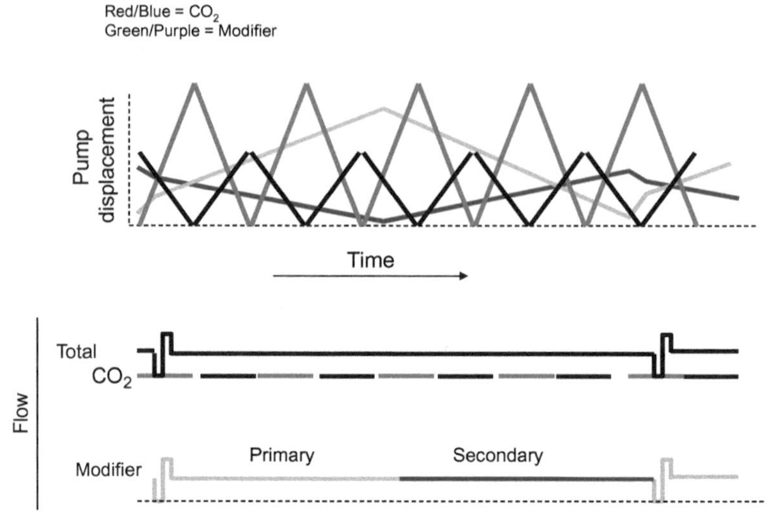

FIGURE 7.11 When a compressor is used to precompress the CO_2, there is no compression stroke on the CO_2 pump. However, there is a compression stroke by the modifier pump. Since the modifier is much less compressible, perturbations in flow and composition are less pronounced and occur at the modifier pump frequency. However, the approach outlined in Fig. 7.10, is slightly superior.

ratio pressure booster (compressor) with a downstream metering pump could eliminate any need for a chiller.

The fluid was precompressed so there is no adiabatic warming in the metering pump. The fluid in the CO_2 pump is at the pump head temperature which is probably higher than ambient (≈30°C). This may not be too different from the actual CO_2 temperature in a pump head chilled nominally to 4°C or 5°C, if the compression is adiabatic. The temperature and density in the chilled pump heads depends on the degree of warming due to the compression stroke. The density of the fluid in the head will be different, using each approach, even though the pressure will be the column head pressure in all approaches. If densities are different using the same volumetric displacement rate versus time (set flow rate) the result is a difference in mass flows of CO_2 causing differences in %B, and total flow between the various approaches.

7.4 PUMP HEAD CHILLERS

7.4.1 Glycol Circulating Baths

The earliest chillers pumped a cold glycol solution through several meters of sometimes uninsulated tubing to a HX bolted to the CO_2

pump head. The glycol solution was chilled to as low as −20°C. The CO_2 was also prechilled then delivered to the pump with a stainless steel tube. It was found that the temperature of the circulating bath needed to be higher than the freezing point of water to avoid forming ice blocks on the tubing and/or pump head due to the condensation of atmospheric moisture, particularly in summer. However, at 4°C or 5°C, water still condenses, sometimes excessively, and can be problematic, forming large amounts of liquid water that needs to be contained or collected. Such condensation/freezing tends to make the cooling system inefficient, since much of the cooling power is wasted condensing the liquid.

In some instruments, the bolt-on HX interfered with routine maintenance, such as changing check valves, pistons, and piston seals. Some later approaches placed the fluid prechiller near or in the pump head. This approach is still used, particularly on larger systems. Baths with pumped glycol solutions tend to be bulky, messy, and somewhat inconvenient. The potential user should evaluate whether this approach is a problem for routine work.

7.4.2 CO_2 Expansion

People have tried to expand some of the CO_2 into a cavity on the pump head to cool it down using a cryogenic valve and an electronic feedback loop. This does allow a low compression ratio reciprocating pump to deliver CO_2. While this appears to be an attractive, simple, and inexpensive approach the level of temperature control, and therefore flow stability is not very good. Consequently this approach tends to be confined to pumps used in supercritical fluid extraction where flow rate stability is less important.

7.4.3 Peltier Chiller

Thermoelectric Peltier-based chillers generate a temperature difference between their two flat surfaces solely with the passage of an electric current. This eliminates the pumped cooling fluid, and most of the problem with condensation. However, Peltier devices are inherently inefficient and become less efficient as the temperature difference between the two sides increases. In addition, any water condensation on the cold side further limits efficiency and scalability. So far, Peltier chillers have only been used at the analytical scale, due to their poor efficiency.

Peltier-based chillers for SFC pumps were introduced around 1992. The cold side of the Peltier device was bolted to an insulated pump head, while the hot side had a HX with tall fins mounted to it.

A high-flow fan pulled lab air through the HX to remove heat from the fins. In some instruments the large HX and insulation got in the way of routine maintenance.

7.4.4 Combination Chiller

A recent chiller variant combined the two earlier approaches. A cold plate, in contact with the cold side of a Peltier, was mounted between the pump head and an insulating spacer, isolating the pump head from the motor drive. Since the cold plate is behind the pump head, routine maintenance, such as changing check valves and main seals, and even pistons, remains simple.

All were mounted inside a stagnant air, insulated enclosure. The hot side of the Peltier, was mounted outside the enclosure, and was in contact with a second HX. A glycol solution was pumped through this hot-side HX to a radiator with a fan mounted on the back of the instrument. In this configuration, the glycol is always warmer than room temperature, eliminating any condensation and oxidation issues.

7.4.5 Refrigeration Chiller

Yet another approach has used an active refrigeration system to chill both the fluid and pump heads on a semipreparative chromatograph. A refrigerant was compressed and subsequently expanded in an HX mounted on the pump head. Such refrigeration systems are available and scalable. Very small systems are available, which should be suitable for analytical scale, but so far none have appeared. This may be the approach of the future for both smaller and larger systems.

7.5 MIXERS

As should be obvious from the previous discussions on pump design, there are large differences in mixer requirements for the various pumping approaches. The pumps that produce significant pressure/flow perturbations during compression strokes require larger mixer volumes to average out the perturbations. Mixers for such pumps are generally sized to hold at least several pump strokes. Mixers are often a short column packed with millimeter-sized steel balls, as shown in Fig. 7.12. In some cases a pulse dampener is also part of the mixer.

Mixer volume is important in terms of gradient delay. There are relatively low volume mixers that have numerous internal flow paths of different lengths as shown schematically in Fig. 7.13. Composition

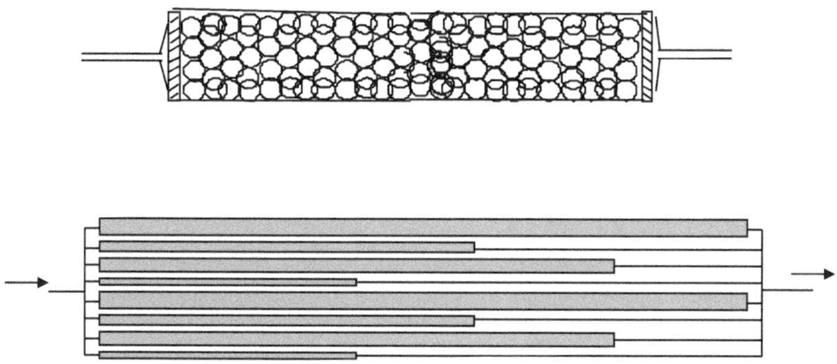

FIGURE 7.12 Mixers. Top: traditional mixer consisting of a small ID chromatographic column filled with large stainless steel balls, to maximize "band broadening" of composition perturbations caused by inadequate pumping strategies. Lower: Etched groves in a metal plate with uniform depth. Longer etched channels have a wider flow path, to match pressure drop of shorter length channels with narrower width, minimize the total volume needed for mixing.

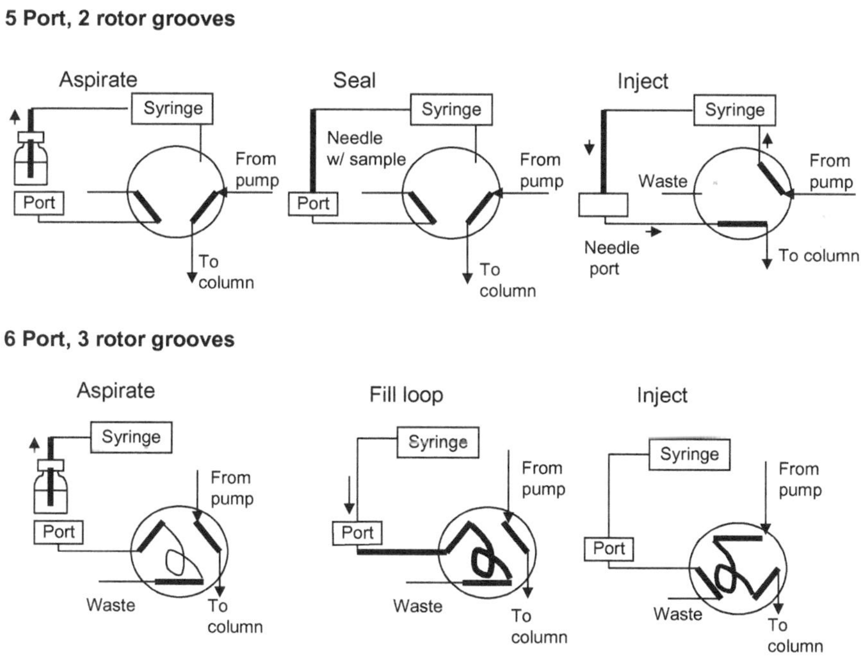

FIGURE 7.13 Some autosamplers common in HPLC (top) use a 2 position 5 port 2 groove rotor with a high-pressure syringe, but are not compatible with SFC. SFC autosamplers employ either a 2-position valve with 6 ports, and a 3-groove rotor with an external loop, or a 2 position internal loop valve.

perturbations travel at different speeds through the flow paths and are thus mixed. For pumps with a motor on each piston, such fluctuations can be minimal and may not require a mixer.

7.6 AUTOSAMPLERS

Almost nobody performs manual injections on a routine basis. Most samples are liquids. As with much of the other equipment used in SFC, the autosamplers (ALS) used are the same as those used in HPLC. However, in some cases alternate plumbing may be required. A noticeable difference between SFC and HPLC is the fact that most polar solutes are dissolved in a polar sample solvent-like methanol. This means that the sample solvent is stronger than the mobile phase, which will result in distortion of early eluting peaks, and/or a loss of efficiency. As a result, injection volumes tend to be smaller in SFC, often less than 5 μL, even on 4.6 mm ID columns.

Some designs move the needle to the samples. Others move the samples to the needle. Reproducibility and carry-over are similar to HPLC.

Some of the earliest ALS's used a simple glass syringe. The sample was aspirated into the syringe and then pushed into an external loop, typically 5 or 10 μL, mounted on a 6 port 2 position injection valve, in the "load" position. Optionally internal loop valves are sometime used to make smaller injection volumes. During sample loading the mobile phase is diverted directly from the pump to the column.

When the injection valve is switched to the "inject" position, the mobile phase is diverted from the pump through the loop and onto the column. Only the loop is placed under high pressure.

Partial loop injections were possible but part of the loop was filled with air. Since the mobile phase is highly compressible, when the valve was switched to the "inject" position, the mobile phase rushed into fill this void, causing a momentary pressure drop. There is no such thing as a bubble in SFC since the mobile phase will expand to fill the space.

The syringe should be left in the injection port when the valve is switched back to the "load" position, although this momentarily subjects the syringe to system pressure. The mobile phase left in the loop will expand roughly 500 times when exposed to the atmosphere. Without the needle in the injection port this mobile phase would expand out the injection port into the lab air. With the needle in place, the fluid can be vented safely out the waste port. In order to prevent carry-over, the syringe is washed multiple times with a solvent.

A better approach uses a larger syringe connected to a two-way valve. The valve connects the syringe to either a solvent wash reservoir or to a flexible tube connected to a needle. The sample is aspirated into the needle, using the syringe, but never reaches the syringe. The sample is then pushed into a loop on a similar 6 port 2 position injection valve. With this approach the loop can be prefilled with a solvent, which can then be partially displaced with sample, for partial loop injections. After the sample is injected the needle and injection port are easily washed with the solvent in the reservoir.

Some modern HPLC autosamplers use an approach in which there is no external loop on a 2 position, 5 port injection valve. Instead a high-pressure needle, and syringe, and the tubing connecting them, form the loop as shown schematically in the top of Fig. 7.13. During sample loading, the mobile phase flow is diverted from the pumps directly to the column. The needle is then withdrawn from a high pressure needle seat which is connected to the injection valve. Sample is aspirated from a sample vial into the needle and sometimes further, into the tubing. The volume of the needle and tubing is typically 100 μL or less. After the sample is withdrawn, the needle is inserted back into the high pressure needle seat. When the injection valve switches, the high pressure mobile phase is diverted from the pump through the syringe, tubing, and needle, and needle seat, through the injection valve, onto the column. When the injection valve is switched back to the load position, the fluid in the needle, tubing, and syringe will decompress to atmospheric pressure. In HPLC, the mobile phase is nearly incompressible, so after decompressing, the needle, tubing and syringe remain filled with the liquid mobile phase. In SFC, the highly compressed mobile phase in the needle, tubing and syringe expands up to 500 times leaving the syringe, tubing and needle filled with a low-density gas. Such an empty syringe cannot aspirate liquid samples.

Such autosamplers must be re-plumbed. The two groove rotor in the injection valve is replaced with a standard three groove rotor and an external loop is installed on the injection valve. The mobile phase is no longer diverted through the syringe, tubing, and needle. The syringe, tube, and needle are all filled with a solvent. The sample is aspirated into the needle and then discharged into the external loop. The syringe, tubing, and needle never see high pressure. A separate, low pressure wash pump is required to flush the system between samples, and leave the syringe needle and tubing filled with the wash solvent.

7.7 OVEN/COLUMN COMPARTMENT

The mobile phase temperature can have a significant effect on solute selectivity. In one example three compounds that coeluted at one temperature were baseline resolved at only 5°C higher temperature. While this is an extreme case, temperature is the second most useful control variable for changing selectivity after modifier concentration and identity. Further the density of CO_2-based mobile phases vary depending on temperature and pressure to a much greater extent than the mobile phases used in HPLC. Consequently SFC's include a thermostated column compartment to carefully control the column temperature using combinations of Peltier elements and resistive heaters. Typical temperature ranges are between 10°C or 20°C and 80°C or 90°C.

Many older SFC instruments employed stirred air ovens, often a gas chromatograph (GC), with much higher upper temperature limits, which were particularly useful for pressure programming with pure CO_2. Many such older applications are incompatible with modern column compartments.

Recent reports [8,9] make it clear that, particularly with sub-2-μm particles, the mobile phase can either warm or cool, depending on composition, during decompression along the column. Axial temperature gradients do not affect efficiency. However, if radial temperature gradients form, there is a strong potential for loss of efficiency since radial gradients in viscosity and linear velocity would also develop. As a result, different elements of the same peak would travel at different linear velocities, distorting the peak shape. Consequently it is advisable to operate columns as adiabatically as possible. By preventing heat loss or gain at the surface of the column, no radial gradient could form. Thus stirred air bath ovens, common in the past, should be avoided.

In a few instances columns have been insulated with foam rubber insulation. However, for most routine work, simply isolating the column from the compartment walls is adequate in a stagnant air bath oven, as suggested schematically in Fig. 7.14 for one such implementation. A flow through HX or other means of heating should be used to preheat the mobile phase to the desired initial temperature, before it enters the column. In the figure a flow through HX designated RHX is shown but other configurations are used. Even with modest thermal gradients, there is little heat flow to or from the stagnant air in the column compartment to/from the column. However, thermal conduction along the stainless steel column tube can potentially cause radial gradients to form.

7.8 UV–VIS DETECTORS

The UV–Vis detectors used in SFC are the same as those used in HPLC, with the exception that the flow cell must be capable of withstanding high pressures. Generally it is advisable to insure the flow cell can withstand the maximum pressure the pump or the BPR can deliver. The flow cell windows should be made of SupraSil, or the like, capable

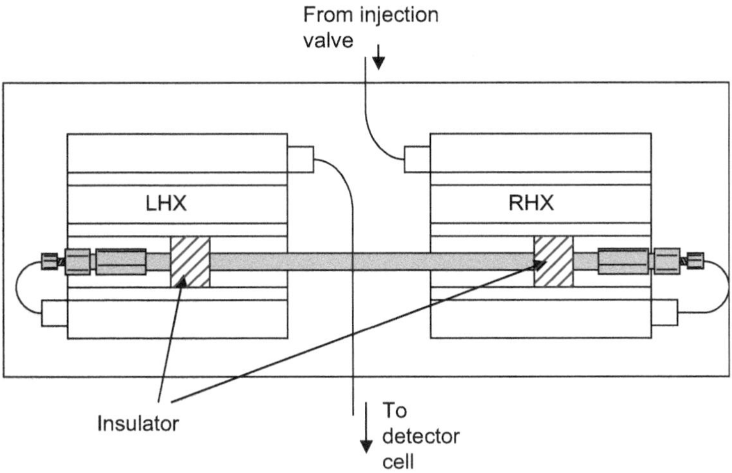

FIGURE 7.14 A common oven or thermostated column compartment configured to preheat the mobile phase to the set-point using the right heat exchanger (RHX), and using the left heat exchanger (LHX) to match the postcolumn fluid temperature to the flow cell temperature.

of passing light down to at least 190 nm. Flow cell windows are typically at least 6 mm thick.

Diode array detectors (DADs or PDAs) are the norm. Modern detectors cover the wavelength range from 190 to well into the visible at 750 nm or higher.

DADs typically are not classic dual beam designs where the light beam from the lamps is split to pass one part through the flow cell and sample to one photo-sensor, while the other is passed directly to a photo-sensor, or perhaps through a flow cell without the sample. This largely eliminates any flicker in the lamp output. Instead the user selects wavelengths where the sample does not absorb light. This reference signal is subtracted from the sample signal. Such an approach largely does the same thing as the dual beam approach since lamps tend to flicker at all wavelengths.

For maximum sensitivity the signal from many diodes should be averaged for both the signal and reference. If the slit width is set-able, it should be set to the same width as the bandwidth of the sample beam. When spectra are collected the bandwidth of the sample beam and the slit width should be as narrow as possible to see any fine structure.

Modern detectors use digital filters. Many different types of digital filters have been used in SFC detectors. For fast chromatography a weighted moving average filter with a Gaussian kernel is probably best. The rise time (time to respond from 10% to 90% of a step function) needs to be appropriate for the sampling rate. Detectors with rise times of 0.025–0.031 s are available.

7.8.1 Postcolumn Temperature Adjustment and Refractive Index

The density, and therefore the refractive index (RI) of CO_2, is strongly dependent on temperature (and pressure). The refractive index of CO_2 varies from 1.06 to 1.24, in the regions used in SFC [13], which is a very wide range. The most common HPLC solvents have RI's between 1.33 and 1.39, a much narrower range, compared to pure CO_2. However, since the CO_2 is typically mixed with a liquid organic solvent-like methanol an even greater range of RI is possible in an SFC detector.

It is often desirable to set the column temperature far from ambient. If the fluid leaving the column is at a substantially different temperature from the detector flow cell, significant UV noise can result. Fluid at one temperature, mixing with fluid at another temperature within the flow cell, causes short-term variations in RI resulting in UV–Vis noise.

Some detector manufacturers actually heat the optical bench to improve ambient rejection (wander and drift). Flow cell temperatures have been measured at between 35°C and 55°C. Thus fluid exiting the column can be either too cold or too warm, compared with the detector flow cell temperature. The temperature of the fluid leaving the column can be adjusted to the detector flow cell temperature using another HX, to match the postcolumn mobile phase temperature to the flow cell temperature, eliminating this potential source of noise. One such arrangement is shown schematically in Fig. 7.14 where the HX is inside the column compartment.

Alternately flow cells have been designed that are insulated from the optical bench. The incoming mobile phase then becomes the dominant force in controlling the flow cell temperature. While this makes the flow cell much less susceptible to noise, it is still advisable to match or nearly match the postcolumn temperature to the inherent detector flow cell temperature.

7.8.2 Phase Separation

In a few unusual situations, such as a detector mounted in the exhaust plume of a mass spectrometer, the internal temperature of the detector could reach excessive temperatures, such as 55°C. At low outlet pressures the mobile phase can break down into two phases effectively destroying the separation or at least causing excessive noise. In such a situation simply raising the pressure 10–20 bar should eliminate the problem.

7.8.3 Conical versus Tapered Flow Cells

It was recognized some years ago that a cylindrical flow path could be a cause of noise induced by changes in RI. Light bent as it enters the cell could reflect off the walls of a cylindrical cell and be lost. By gradually increasing the inner diameter along the length of the cell, any light that is bent significantly does not hit the wall and is not lost. As a consequence, conical shaped flow paths [14] have become the norm. However, the fluids used for SFC all have lower RI than the liquids normally used in HPLC, which results in somewhat less bending of light. The need for a tapered flow cell is less obvious in SFC, compared to HPLC, but since most, if not all, flow cells used in SFC were initially designed for HPLC, most, if not all, have tapered flow paths. This tends to simplify temperature optimization. However, cylindrical flow path flow cells can still be used and, with careful temperature optimization, can produce equivalent low noise [13] compared to a tapered flow path cell.

7.9 BACK PRESSURE REGULATORS

Under atmospheric conditions, CO_2 is a poorly solvating gas. It only exhibits interesting solvent characteristics when it is held at liquid-like densities. This requires minimum pressures, on the order of 80 ± 20 bar (depending on temperature), but in practice pressures between 100 and 150 bar are most commonly used. Fixed restrictors and even needle valves have sometimes been used to elevate the system pressure, but yield poor control and lack versatility. With fixed restrictors, any change in flow or composition results in a change in the outlet pressure. Manual needle valves need to be heated and still tend to deliver sputtering, variable flow, and significant pressure fluctuations, which creates UV–Vis detector noise. Such devices lead to many misconceptions that took years to overcome. Such devices should not be used in SFC.

SFC requires a BPR capable of at least 100–150 bar, mounted after any UV–Vis detector. It is preferable to use one capable of much higher pressure, such as 400 bar. Early BPRs were mechanical but had large poorly swept flow paths. Modern BPRs are all adjustable electronically. This is a throwback to the 1980s and 1990s when pressure programming with pure CO_2 was very common. The column outlet pressure was often programmed from relatively low pressures, like 60–80 bar to the full pressure capability of the pumps (400–600 bar). Most such application also used higher temperatures, such as 100–150°C where the density versus pressure relationship is relatively linear. Such applications have fallen out of favor and are seldom run today, mostly due to hardware, and not application, limitations. However, this capability may be revived in the future since at least one instrument manufacturer provides pressure programming.

The modifier identity and composition is the dominant instrumental means of changing selectivity and retention. When binary or tertiary fluids are used, pressure becomes a secondary control variable [15] and generally has only a small influence on retention or selectivity. Small changes in pressure (<25 bar) tend to have only minimal effect, particularly on selectivity. The only place where pressure has a significant influence is when a method is transferred from a column with large particles operated with standard plumbing to a column with smaller particles with more restrictive plumbing. In that situation, much larger changes in the column outlet pressure could occur, and the actual column outlet pressure becomes unclear. In such a situation, some peaks, particularly early eluting peaks, sometimes merge.

For routine work, one could simply fix the column outlet pressure with an appropriate mechanical BPR and perform all experiments with the same outlet pressure without serious loss in performance. However, when pure CO_2 is used as the mobile phase, pressure/density is a major means of changing solvent strength for relatively nonpolar solutes.

7.9.1 Mechanical Stability

The BPR is probably the most difficult component in an SFC to engineer. This may surprise many readers since it appears to be a rather mundane, well understood device. The problems arise when the mobile phase expands between the walls of an orifice and some sort of pin or diaphragm partially blocking the orifice. The distances involved are microns. The fluid is attempting to expand such that it reaches Mach 1, the speed of sound. A shock wave can attach and detach from various positions in the gap, ripping it apart.

The forces generated are severe and produce degradation both by erosion and corrosion of any susceptible materials in the flow path. Materials must be inert and robust enough to withstand the powerful forces at play. This was the achilles heel of SFC in the 1980s through much of the 1990s, and required tens of engineer years of investment to find a solution. Hundreds of materials were tried. Due to the high costs, the results remain proprietary. Very robust BPRs are now available.

One early work-around employed an electronic BPR in front of a sacrificial mechanical BPR. The electronic BPR only controlled pressure over a fairly narrow range of densities (0.1–0.2 to 1 g/cm^3), while most of the expansion to ambient occurred downstream in the much less expensive mechanical BPR. Most damage occurs at the lower pressures with the much higher velocities.

It is often helpful to place a piece of tubing downstream of the electronic BPR which has adequate pressure drop to keep the outlet of the

BPR above 20–60 bar. At 2–5 mL/min a 20 cm piece of 0.007 in. (170 μm) tubing usually works. In most cases such tubing does not interfere with any post BPR detectors, or peak collection, since the velocity of the mixed fluids is very high, sweeping out the tubing in very short times. Condensation of atmospheric water often occurs on this tubing and needs to be dealt with.

7.9.2 Pressure Noise

As with postcolumn temperature optimization, any pressure noise induced by the BPR will result in RI variations in any UV–Vis detector flow cell, and detector noise [16]. Some modern BPRs can control back pressure to less than ±0.05 bar (0.75 psi) at 200 bar, resulting in remarkably low pressure induced noise.

7.10 MISCELLANEOUS

7.10.1 Waste Collection

The waste stream exiting the BPR rapidly expands up to 500 times to a gas. Entrained in that gas are droplets of modifier and potentially additives and samples. In some cases, aerosols can form that could contain dangerous chemicals. None of these should be vented into the lab air. Consequently the waste stream should be diverted into a closed bottle with an inlet and outlet tube. The liquid droplets will fall out to collect the liquid waste while the gas vents out the top into a fume hood or out a vent to the outside. Bubbling the incoming fluid in the liquid waste helps strip out aerosols. The vented CO_2 does contain up to 3% vaporized modifier.

7.10.2 Leaks

The viscosity of CO_2-based mobile phases is much lower than that of the solvents used in HPLC, as indicated previously. Many HPLC fittings were not designed for such low viscosity fluids. As a consequence it is often necessary to over-tighten fittings to stop gaseous leaks, which tends to shorten the lifetime of the fittings. With the advent of ultra-high-performance LC many fitting designs have emerged that use ferrules made of Vespel or the like which are less prone to leaks. Although PEEK tubing and fused silica lined PEEK tubing has been used, the use of PEEK ferrules is not recommended.

Modern HPLCs have leak sensors that shut down the system when a liquid leak is detected. Such sensors are more or less useless in SFC where leaks are almost always gaseous. The most effective way to find leaks is to apply a few drops of soapy water to each fitting and look for bubbles. Even very small bubbles should be eliminated. Larger leaks can freeze, due to adiabatic cooling, then thaw, repeatedly, creating a noisy UV baseline. Such large leaks are unacceptable.

References

[1] Kirkland JJ. Ultrafast reversed-phase high-performance liquid chromatographic separations: an overview. J Chromatogr Sci 2000;38:535–44.

[2] Tarafder A, Kaczmarski K, Poe DP, Guiochon G. Use of the isopycnic plots in designing operations of supercritical fluid chromatography. V. Pressure and density drops using mixtures of carbon dioxide and methanol as the mobile phase. J Chromatogr A 2012;1258:136–51.

[3] Berger TA. Instrument modifications that produced reduced plate heights <2 with sub-2 μm particles and 95% of theoretical efficiency at k = 2 in supercritical fluid chromatography. J Chromatogr A 2016;1444:129–44.

[4] Poe DP, Martire DE. Plate height theory for compressible mobile phase fluids and its application to gas, liquid and supercritical fluid chromatography. J Chromatogr 1990;517:3–29.

[5] DePauw R, Choikhet K, Desmet G, Broeckhoven K. Occurrence of turbulent flow conditions in supercritical fluid chromatography. J Chromatogr A 2014;1361:277–85.

[6] Halasz I, Endele R, Asshauer J. Ultimate limits in high-pressure liquid chromatography. J Chromatogr 1975;112:37–60.

[7] Encyclopedie de Gaz, Air Liquide. Available online.

[8] Poe DP, Veit D, Ranger M, Kaczmarski K, Tarafder A, Guiochon G. Pressure, temperature and density drops along supercritical fluid chromatography columns in different thermal environments. III. Mixtures of carbon dioxide and methanol as the mobile phase. J Chromatogr A 2014;1323:143–56.

[9] Colgate SO, Berger TA. On axial temperature gradients due to large pressure drops in dense fluid chromatography. J Chromatogr A 2015;1385:94–102.

[10] Lemmon EW, Huber ML, McLinder MO. NIST Reference Database 23: reference fluid thermodynamic and transport properties-REFPROP, Version 9.1. National Institute of Standards and Technology, Standard Reference Data Program, Gaithersburg, 2013.

[11] Eburn, Jr. WH, Kalenik SP. Constant flow pumping apparatus. US Patent 4,321,014, December 31, 1979.

[12] Carpenter CW. Bumpless pump apparatus adjustable to meet slave system requirements. US Patent 4,127,360, December 16, 1976.

[13] Berger TA. Minimizing ultraviolet noise due to mismatches between detector flow cell and post column mobile phase temperatures in supercritical fluid chromatography: effect of flow cell design. J Chromatogr A 2014;1364:249–60.

[14] Nelson KE. Novel photometric system. US Patent 4,011,451, October 24, 1975.

[15] Berger TA. In: Packed column SFC. Smith RM, Series Editor, RSC chromatography monographs. Cambridge, UK, Royal Society of Chemistry; 1995. pp. 145, 166.

[16] Berger TA, Berger BK. Minimizing UV noise in supercritical fluid chromatography. I. Improving back pressure regulator pressure noise. J Chromatogr A 2011;1218: 2320–6.

CHAPTER

8

Hyphenated Detectors: Mass Spectrometry

V. Desfontaine, J.-L. Veuthey and D. Guillarme

School of Pharmaceutical Sciences, University of Geneva, University of Lausanne, Geneva, Switzerland

OUTLINE

Supercritical Fluid Chromatography.
DOI: http://dx.doi.org/10.1016/B978-0-12-809207-1.00008-2

8.1 INTRODUCTION

In the last couple of years, there has been a regain of interest for SFC, mainly driven by the introduction of state-of-the-art systems from major manufacturers of chromatographic instrumentation [1]. In addition, these instruments are now fully compatible with the latest generation of columns packed with sub-2-μm fully porous and sub-3-μm superficially porous particles [2,3]. Then, when combining modern supercritical fluid chromatography (SFC) systems and state-of-the-art column technology, very high kinetic performance can be obtained (at least equivalent or even better than those obtained in ultra-high-performance liquid chromatography, UHPLC), and the previous limitations of SFC, such as poor UV sensitivity, limited reliability of the technique and weak quantitative performance, can be tackled [4,5].

Apart from the excellent complementarity of SFC to the more commonly employed reversed-phase liquid chromatography (RPLC), this chromatographic technique is also compatible with a wide range of detectors [6]. Among them mass spectrometry (MS) is considered as one of the most powerful detectors, thanks to its high sensitivity, selectivity, and quasi universality [7–8]. Then, this detector has become the gold standard for a wide range of applications involving the determination of trace compounds in complex matrices. As an example MS detection is systematically used today in bioanalysis, drug metabolism studies, multiresidue screening, plant extract characterization, and "-omics" approaches such as metabolomics, proteomics, and lipidomics [9,10]. Compared to LC-MS, the potential of supercritical fluid chromatography-mass spectrometry (SFC-MS) has not been fully investigated yet, but looks promising, particularly in terms of alternative selectivity and complementarity to LC-MS. As shown in Fig. 8.1, around 25% of the SFC papers include the use of MS as the detector, and this number is expected to grow in the future, thanks to the recent commercialization of new SFC-MS products and technical solutions.

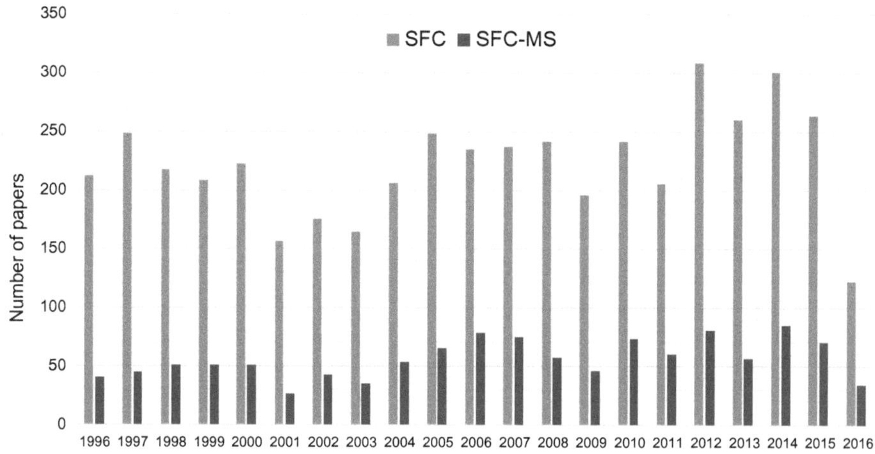

FIGURE 8.1 Number of papers annually published in the field of SFC and SFC-MS, for the last 20 years. For each year, the bars on the left side were obtained with keyword "supercritical fluid chromatography," while the bars on the right side were obtained with an additional filter (keyword "MS"). *Source: Scifinder scholar. Date of information gathering: July 2016.*

The goal of the present review is to summarize the previous and current practice of SFC-MS in terms of instrumentation and applications. For this purpose, the different ionization sources (electron impact, chemical ionization, electrospray ionization (ESI), atmospheric pressure chemical ionization (APCI)...), mass analyzers (quadrupole, ion trap, time-of-flight, tandem and hybrid instruments...), and interfaces (direct coupling, use of make-up pump, postcolumn splitting...) that can be used for SFC-MS operation will be described, with an emphasis on SFC-MS applications. Then, the benefits of SFC-MS versus LC-MS, such as enhanced sensitivity or reduced matrix effects will be critically discussed. Finally a few promising applications of SFC-MS in various fields, such as lipidomics, metabolomics, bioanalysis, food, and environmental analysis will be provided as well.

8.2 IONIZATION SOURCES FOR SFC-MS OPERATION

In the early 1980s, Novotny and Lee introduced the concept of capillary SFC (cSFC) that consists in using capillary or open-tubular columns together with a purely supercritical mobile phase of CO_2 [11–14]. The cSFC instrumentation was very close to GC and therefore, cSFC-MS was performed with GC-type ionization sources, such as electron impact (EI) and chemical ionization (CI). Nowadays SFC is exclusively performed on packed columns (pSFC) using LC-like instrumentation [15,16], and pSFC-MS is carried out with atmospheric pressure ionization (API) sources, commonly used in LC-MS (see Fig. 8.2). Because, the most prevalent ionization source for SFC-MS operation today are the API sources, only this type will be discussed in this chapter. A clear benefit for API sources in SFC is the "self-volatilizing effect" of CO_2 during the nebulization/evaporation process.

Combining SFC and MS analyzer is quite straightforward, since CO_2 is highly volatile and then the SFC effluent can easily be converted into the gas phase during the ionization process. Similarly to LC-MS, the most popular ionization sources for SFC-MS are ESI, APCI, and atmospheric pressure photoionization (APPI). As illustrated in Fig. 8.2, the selection of ionization sources is mainly based on the molecular weight and polarity of the analyzed compounds, but also on their ionizability in solution and the mobile phase flow rate, which can be quite elevated in SFC. However, for a given substance it remains difficult to predict which ionization source (i.e., APPI, APCI, or ESI) will provide the best results. Only preliminary tests will allow the choice of the best ionization source. These three API sources can be used without any further modification in SFC-MS, compared to LC-MS, but some specific conditions for interfacing SFC and MS (see Section 8.3 of this chapter) may be required to enhance ionization and achieve a stable baseline.

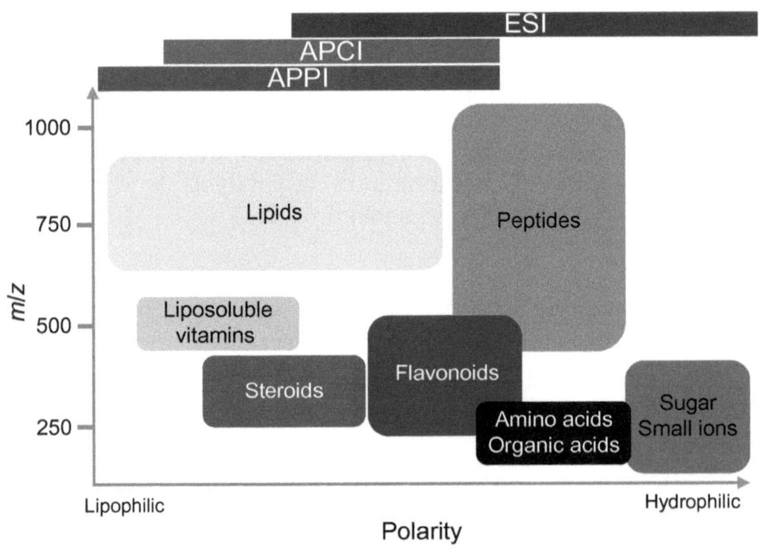

FIGURE 8.2 Application range of atmospheric pressure ionization techniques for selected compound classes in relation to their molecular weights and polarity.

8.2.1 Electrospray

The success of electrospray (ESI) started with the work of Fenn et al., who received the Nobel prize in Chemistry in 2002, and shows that the molecular weight of large proteins can be determined with instruments having a mass limit as low as 2000 Da, due to the presence of multiply charged ions [17,18]. At the beginning, ESI was considered as an ionization source dedicated to proteins analysis. Later its use was extended to the analysis of compounds from medium polarity to ionic species (see Fig. 8.2).

ESI can be considered as the most gentle ionization technique. It is generally performed by forcing a solution of the analyte through a small capillary, so that the fluid sprays into an electric field, thereby generating fine charged droplets. The electric field (several kV) is imposed between the tip of the spraying capillary and a counter electrode. Electrosprayed droplets possess an excess of positive or negative charges, depending on the selected polarity mode. The initial droplets formed at atmospheric pressure in the spray orifice are large (1–10 μm) and diminish in size due to simple evaporation of the solvent in a drying gas at atmospheric pressure. As the droplets shrink, ions (which are not volatile) are retained and their concentration increases. Therefore when the Rayleigh stability limit (the point at which repulsive forces between like charges in an electrolyte solution overcome the cohesive forces of the solvent) is attainted, the droplets undergo disintegration into many smaller droplets of around 100 nm (coulombic explosion). Finally, when the solvent is completely evaporated, the gas phase analyte molecule possesses a residual charge attached and can be directed towards the mass analyzer. In ESI, only volatile additives such as ammonium acetate, ammonium formate, ammonia, formic acid can be added to the SFC mobile phase at relatively low concentration (a maximum of 5–20 mM).

A recent study by Grand-Guillaume Perrenoud et al. showed that the hyphenation of SFC to ESI/MS was more robust than LC-ESI/MS [19]. Indeed the sensitivity decrease was more pronounced in LC-ESI/MS compared to SFC-ESI/MS when the electrospray conditions (temperature and flow of desolvation gas, capillary voltage) were purposely set away from the optimal values. This was probably due to the fact that the LC mobile phase contained water which is more demanding in terms of desolvation.

An important feature of ESI is its response to concentration and not to the total amount of sample injected into the source [20]. In addition, the sensitivity increases when the flow entering the source is reduced to only a few microliters per minute. For a flow rate higher than 500 μL/min, which is generally the case in SFC, the sensitivity is often reduced, but this fact can be compensated by the high volatility of the SFC mobile phase, mostly consisting of CO_2 mixed with a limited proportion of methanol (MeOH).

8.2.2 Atmospheric Pressure Chemical Ionization

APCI is an ionization technique based on gas phase ion–molecule reactions at atmospheric pressure [21]. It is an ionization method close to chemical ionization (CI), which is commonly used in GC-MS, but APCI takes place at atmospheric pressure. The ions are produced by corona discharge on a solvent spray. As illustrated by Fig. 8.2, APCI is mostly applied for the ionization of moderately polar to nonpolar compounds, with relatively low molecular weights (up to 1500 Da), as it gives exclusively mono-charged ions.

In APCI, the analytes in solution are introduced into a pneumatic nebulizer where they are converted into small droplets by a high-speed nitrogen beam. After desolvation of the droplets in a heated vaporization chamber, a corona discharge is applied with an electrode and ionization takes place. The evaporated mobile phase acts as the ionizing gas, and reactant ions are produced from the effect of the corona discharge on the nebulized solvent.

From an historical point of view, APCI was the most widely used ionization source in SFC-MS until the last generation of SFC-MS instruments was commercialized in about 2010. This situation can be explained by the fact that APCI is a mass-flow dependent ionization source, which can accommodate relatively high flow rates (typically up to 2 mL/min) and is better suited for compounds with low to moderate polarity. Coincidently SFC has a higher optimal flow rate than LC (typically 3–5 times higher) and offers suitable retention for compounds of low to moderate polarity. This is why APCI was perceived as the best ionization technique for SFC-MS and was applied for numerous applications [22–24]. Today many SFC-MS applications deal with pharmaceutical compounds, so ESI becomes the most prevalent ionization source (see Fig. 8.3).

8.2.3 Atmospheric Pressure Photoionization

The APPI source is one the most recent atmospheric pressure sources [25,26]. The basic principle consists in using photons to ionize gas-phase molecules. The sample in solution is first vaporized by a heated nebulizer, similarly to APCI. Then, the analyte interacts with photons emitted by a discharge lamp (rather than with electrons produced by the corona needle discharge in APCI), and after a series of gas-phase reactions, the sample molecules are ionized. APPI has the potential to ionize compounds that cannot be ionized in ESI or APCI, particularly the nonpolar substances (as shown in Fig. 8.2). However, the direct ionization of analyte is clearly not sufficiently efficient. Consequently, dopants (toluene and acetone) at relatively high concentration,

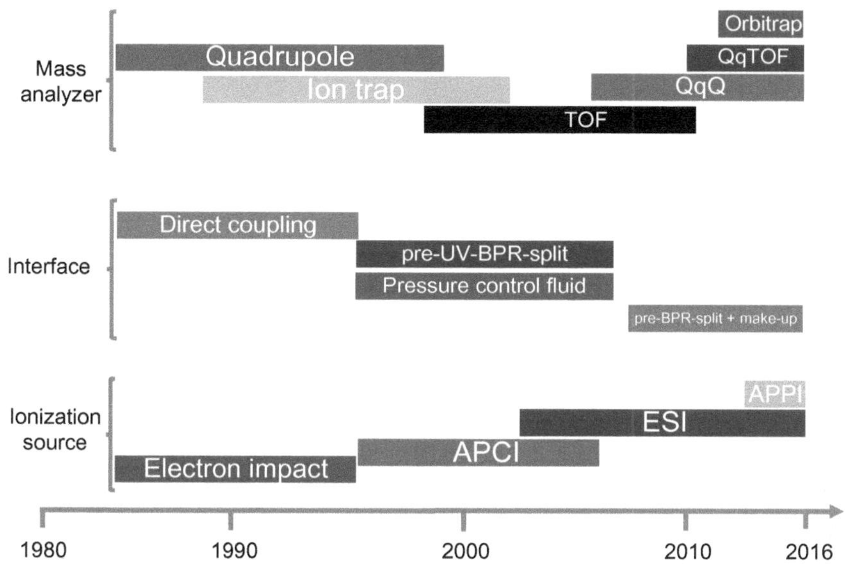

FIGURE 8.3 Evolution of the most widely used ionization sources, mass analyzers, and interfaces in SFC-MS since the 1980s.

compared to the sample, have to be added prior to MS detection, allowing an improvement in the ionization efficiency from 10 to 100 times. Compared to APCI, APPI is more sensitive to the experimental conditions and the properties of solvents, additives, dopants, or buffer components can strongly influence the selectivity and sensitivity of the ionization process. Thus APPI can be considered as a complementary technique to ESI and APCI, useful for nonpolar compounds. APPI was found to be particularly well suited for a few classes of compounds, including flavonoids, steroids, polyaromatic hydrocarbons, liposoluble vitamins, as well as some drugs and their metabolites. To date SFC has been combined only with APPI/MS for the determination of liposoluble vitamins [27] and for compound speciation [28].

8.3 AVAILABLE INTERFACES FOR SFC-MS HYPHENATION

The specially designed interfaces for hyphenating SFC and MS should be able to manage the compressibility of the SFC mobile phase and to preserve as much of the chromatographic separation integrity as possible. With a suitable interface, it is therefore possible to minimize the effect of back pressure modification on retention and selectivity, related to the highly compressible SFC mobile phase. In addition, as the SFC mobile phase elutes from the column, the pressure of the gas is released, the density decreases, the CO_2 can evaporate, and any dissolved compounds can precipitate. It is therefore necessary for the interface to counteract this issue [29]. Finally, the poor ionization yield obtained in electrospray, when the SFC mobile phase is composed of a high proportion of CO_2 (aprotic solvent), can be drastically enhanced with the addition of a sheath liquid.

8.3.1 Available Interfaces for SFC-MS Operation

An interesting review was published in 2005 by Pinkston, highlighting the main advantages and drawbacks of SFC-MS interfaces [30]. Fig. 8.4 illustrates the four main types of interfaces that were successfully used for combining SFC and MS. Although several interfaces can be envisioned, each will have noticeable effects on chromatographic fidelity, flexibility, sensitivity, and user-friendliness [19].

The simplest interface for coupling SFC and MS consists in injecting the total flow from SFC into the MS using a passive back pressure regulator (BPR). In this case, the back pressure is simply controlled with a single piece of restrictor tubing of adequate geometry (length and internal diameter) and directly sent to the MS source to maintain the chromatographic integrity (retention and selectivity). This interface called "direct coupling" is described in Fig. 8.4A. As shown in Fig. 8.3, this interface was used in the early days of SFC-MS hyphenation, using a relatively long fused-silica restrictor (several tens of centimeters) of small ID (25–75 μm). However, with this direct coupling, the ionization efficiency is strongly affected by the mobile phase composition and its flow rate. In addition, this interface suffers from poor flexibility, as the tubing geometry for passive BPR should be changed when altering mobile phase flow rate or composition. Finally, the MS response shows a high variability with this simple interface and it is therefore only recommended for qualitative applications [31].

The second interface is known as "pre-UV-BPR-split." In this configuration, a capillary tube is connected between the SFC column outlet and a zero-dead volume T-union, allowing mobile phase splitting. One part of the flow is directed towards the MS inlet, while the other part is directed towards the back pressure regulator (BPR) through the UV detector (if necessary). A schematic representation of this interface is provided in Fig. 8.4B. Because the splitter is placed under the control of the active BPR, a relatively good flexibility is obtained with this interface, as demonstrated elsewhere [32,33]. As long as the dead volume in the transfer lines is minimized, the chromatographic separation remains intact.

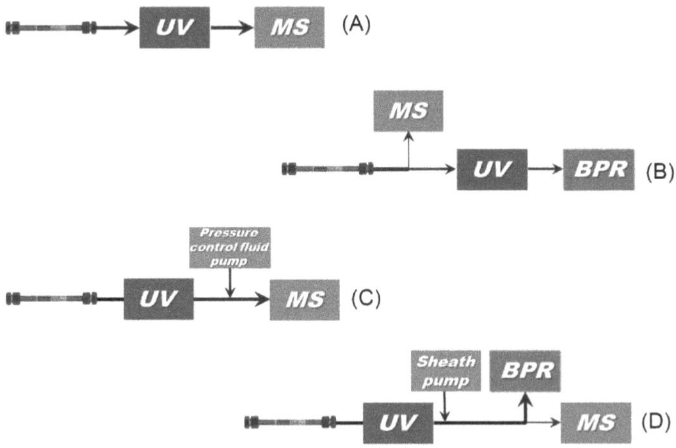

FIGURE 8.4 Schematic representations of the four most common SFC-MS interfaces. (A) Direct coupling interface, (B) Pre-UV and BPR splitter without sheath pump interface, (C) Pressure control fluid interface, and (D) Pre-BPR splitter with sheath pump interface.

In addition, the dimensions of the transfer lines have to be selected to avoid analytes precipitation and phase transition (evaporation). Another source of drawbacks for this interface with regards to quantitation is that the split ratio varies with mobile phase pressure and with any change in the restriction of the transfer line from the split point to the MS inlet. The former is predictable, while the latter is not [30].

The third way to hyphenate SFC and MS consists in using a pressure regulating fluid interface. A low flow of liquid that is miscible with CO_2 (the "pressure regulating fluid") is mixed with the column effluent using a zero-dead volume connector. The miscible liquid works as the pressure regulator and all of the effluent is then pumped into the ionization source. A schematic view of this interface named "pressure control fluid" is provided in Fig. 8.4C. In this interface, the dimensions of the transfer line from the tee to the MS inlet dictate the available ranges of chromatographic effluent flow, pressure regulating fluid flow, and postcolumn pressure [30]. With this design, the pressure is actively controlled near the interface outlet. This avoids the use of long transfer lines with no active pressure control and then minimizes the possibility of phase transition and band broadening. In addition, the entire sample is transferred to the ionization source without any splitting. However, despite these benefits, the pressure regulating fluid interface is not "user friendly" for the following reasons: (1) the dimensions of the transfer line dictate the range of compatible flow rate/pressure combinations and should be occasionally changed, (2) this approach requires an additional pump capable of providing a pulse-free flow, and (3) the additional pressure regulating fluid flow increases the solvent load in the ionization source.

The last interface is the most widely used nowadays on modern SFC-MS instruments and is available from Agilent, Waters, and Shimadzu, for example. In this interface, the UV detector is located just after the column outlet. This interface is composed of two zero-dead volume T-unions located between the UV and MS detectors. The upstream T-union allows the addition of CO_2 miscible make-up liquid delivered by an isocratic pump. This make-up liquid is then mixed with the chromatographic effluent to enhance the ionization yield at low percentages of MeOH and to avoid analyte precipitation during CO_2 decompression. The downstream T-union acts as a flow splitter. Indeed, a fraction of the total flow is directed towards the MS, while the rest of the flow is directed to the BPR. Thanks to the BPR placed downstream of the SFC column, the chromatographic integrity is maintained. This interface is known as pre-BPR splitter with make-up pump interface. A good solution to neutralize the cooling effect of the eluent expansion at the column outlet is to heat the SFC-MS interface up to a temperature of 60°C. Heating of the interface decreases peak broadening observed in SFC-MS, because of the cooling that accompanies decompression of the mobile phase [19]. This option is currently available on the Agilent SFC-MS instruments. This "pre-BPR-split + make-up" interface was reported as the most reliable in terms of retention time and peak area reproducibility. Due to its high sensitivity, linearity and robustness, this configuration is therefore recommended for qualitative and quantitative analyses [31].

8.3.2 Focus on the "Pre-BPR-Split + Make-Up" Interface

The "pre-BPR-split + make-up" interface is the most prevalent today. Grand-Guillaume Perrenoud et al. have recently conducted an exhaustive evaluation of this particular interface [19]. Various types of make-up solvents can be employed in this setup, but a protic solvent like MeOH is often used, since a high proportion of MeOH limits the overall density and eluent strength drop of the mobile phase along the transfer capillary. In addition, sensitivity was also increased in the presence of additional make-up fluid, as MeOH acts as a supplementary proton source, enhancing the ionization process. At the end, the "pre-BPR-split + make-up" interface was found to be the best compromise in terms of chromatographic performance, flexibility, and detection sensitivity.

Due to its design, the operation of the "pre-BPR-split + make-up" interface is directly under the influence of chromatographic parameters. To better understand the principle of operation of this interface, the volume of MeOH content per time unit entering the ionization source has been modeled and the results are reported in Fig. 8.5 for different mobile phase compositions and flow rates (SFC and make-up pumps). As expected, an increase in the make-up pump flow rate or in the

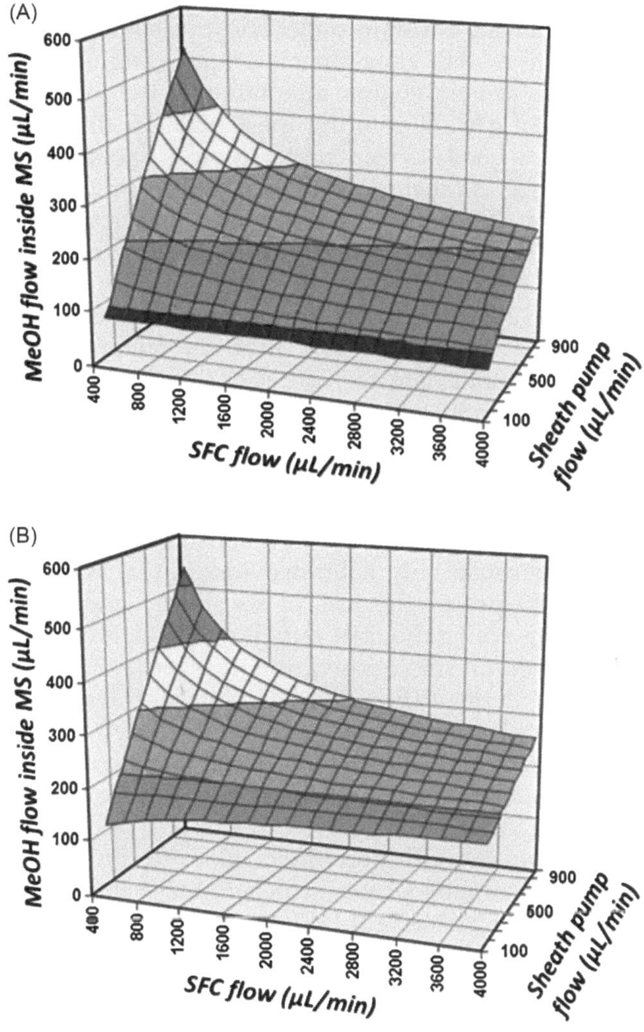

FIGURE 8.5 Calculation of the total MeOH amount entering the ESI probe using the "pre-BPR-split + make-up" interface as a function of the SFC mobile phase flow rate (x-axis) and sheath pump flow rate (y-axis) for two different SFC mobile phase (CO_2/ MeOH) compositions, 95/5 (v/v) for (A) and 80/20 (v/v) for (B), at a backpressure of 150 bar. *Source: Reprinted from Grand-Guillaume Perrenoud A, Veuthey JL, Guillarme D. Coupling state-of-the-art supercritical fluid chromatography and mass spectrometry: from hyphenation interface optimization to high-sensitivity analysis of pharmaceutical compounds. J Chromatogr A 2014;1339:174–84 [Ref. 19]. Copyright 2014, with permission from Elsevier.*

content of MeOH in the SFC mobile phase obviously increased the amount of MeOH directed towards the ionization source. On the other hand, higher SFC mobile phase flow rates with constant MeOH composition led to a reduction of MeOH directed towards the transfer capillary. This counterintuitive behavior is related to the fact that the active BPR always maintains a constant outlet backpressure and therefore, the MS capillary receives less flow. In all of the conditions reported in Fig. 8.5 and thanks to the dynamic adaptation of the split ratio with the BPR, the amount of MeOH entering the ionization source tends to be leveled (from 125 to 300 μL/min under typical operating conditions). This MeOH flow rate is ideal to avoid analytes precipitation inherent to CO_2 decompression, and ensure optimal ESI spray formation and good proton transfer during the ionization process, leading to high sensitivity in SFC-ESI/MS.

Fig. 8.6 illustrates the split ratio (important for a mass-flow dependent detector, like APCI/MS) and dilution factor (relevant for a concentration-dependent detector, like ESI/MS) for various conditions of SFC mobile phase and make-up flows, using the "pre-BPR-split + make-up" interface. As reported in Fig. 8.6A, the split ratio between BPR and MS was comprised between 2 and 8, depending on the flow settings. Regarding MS detection sensitivity, a high split ratio towards the BPR could be critical for a mass-flow sensitive ionization technique (APCI). Indeed, because only a limited amount of analyte would be directed to the MS probe, the loss of sensitivity expected would be proportional to the split ratio, but remains always lower than 1 order of magnitude. Based on this observation, it is clear that the "pre-BPR-split + make-up" interface (Fig. 8.4D) is certainly not the best choice for APCI operation. Indeed, the "direct coupling" (Fig. 8.4A) and "pressure control fluid" (Fig. 8.4C) interfaces would be more appropriate to maximize sensitivity in APCI, as the whole effluent is sent to the MS inlet.

Fig. 8.6B shows the dilution effect due to the make-up fluid adjunction into the mobile phase, which reduces the analyte concentration in the mobile phase. This dilution effect could be critical for ESI since it is a concentration-dependent ionization technique. The dilution factor was calculated for different SFC mobile phase and make-up flow rates. As shown in Fig. 8.6B, the dilution factor was always reasonable with the "pre-BPR-split + make-up" interface and remains below a factor of 2, even under extreme conditions. This result proves that the "pre-BPR-split + make-up" interface is particularly well adapted to SFC-ESI/MS operation. It is also worth mentioning that for a constant make-up flow (common situation), the most retained peaks in gradient elution mode may be more diluted than the less retained peaks.

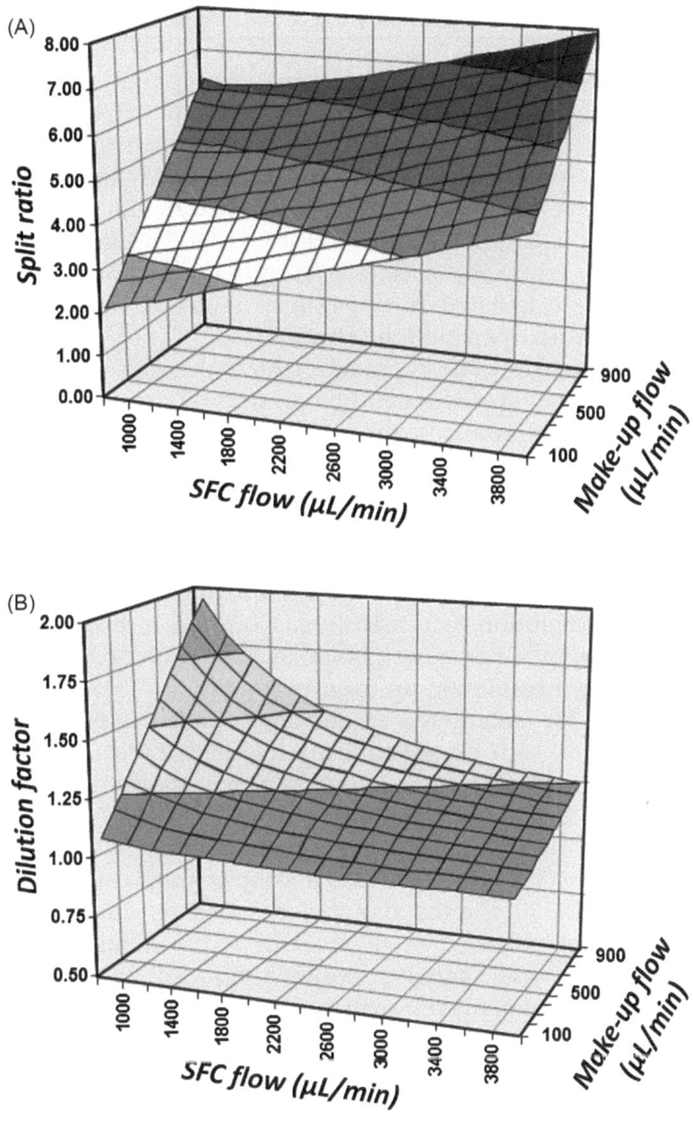

FIGURE 8.6 Model of the split ratio (A) and mobile phase dilution factor caused by the sheath flow (B) for the "pre-BPR-split + make-up" interface as a function of SFC mobile phase flow rate (x-axis) and sheath pump flow rate (y-axis) for a fixed mobile phase (CO_2/MeOH) composition of 90/10 (v/v) and 150 bar backpressure. *Source: Reprinted from Grand-Guillaume Perrenoud A, Veuthey JL, Guillarme D. Coupling state-of-the-art supercritical fluid chromatography and mass spectrometry: from hyphenation interface optimization to high-sensitivity analysis of pharmaceutical compounds. J Chromatogr A 2014;1339:174–84 [Ref. 19]. Copyright 2014, with permission from Elsevier.*

8.4 MASS ANALYZERS FOR SFC-MS OPERATION

The main function of a mass analyzer is to separate ions, produced in the ionization source, based on their mass to charge ratio (m/z). A mass analyzer can be characterized by several features, including its resolving power, mass range, mass accuracy, linear dynamic range, sensitivity, and acquisition speed. The resolving power can be described as the full width at half maximum (FWHM) of a particular peak divided by its m/z value, while the mass accuracy is defined as the difference between the true calculated monoisotopic mass and the experimental one, expressed in part per million (ppm).

Today, there are five main types of analyzer that can be used for SFC-MS operation, namely quadrupole (Q), ion trap (IT), time of flight (TOF), orbitrap, and ion cyclotron resonance (ICR). The analyzers can be classified in two main categories, namely low-resolution and high-resolution analyzers. Low-resolution mass analyzers include both quadrupole and ion trap, offering a resolving power in the range of 1000. This type of MS devices is generally used for targeted analysis (bioanalysis, drug metabolism, multiresidue screening...), when a quantitative determination is required. High-resolution MS instruments include mostly time-of-flight and orbitrap having a resolving power in the range of 10,000–200,000, but also ICR with a resolution up to several millions. These analyzers are principally used for untargeted applications (metabolomics, lipidomics, plant analysis...), which remains mostly qualitative and do not necessitate thorough quantitative evaluation. Fig. 8.3 shows mass analyzers mostly used in SFC-MS at a given time.

To date, ICR mass analyzers were scarcely combined to SFC [34–36], mostly due to the high cost (purchase and operating cost) and complexity, but also due to its limited acquisition speed (an increased acquisition speed can be obtained at the cost of reduced resolving power), incompatible with the narrow peaks produced by modern SFC. Therefore this type of analyzers will not be discussed in more details in this chapter.

8.4.1 Quadrupole Analyzer

A quadrupole analyzer uses a combination of radio frequency alternating current (AC) and direct current (DC) voltages as a mass filter, for separating ions. The quadrupole consists of four parallel rods. The positive DC voltage is applied on two opposite rods, and the same value of the negative DC voltage is applied on the remaining two rods. The AC is connected to all four rods. Combined DC and RF potentials on the quadrupole rods can be set to pass only a selected m/z ratio. All other ions do not have a stable trajectory through the quadrupole mass analyzer and will collide with the quadrupole rods, never reaching the detector.

The single quadrupole is certainly the simplest, cheapest, most robust, and ubiquitous mass analyzer in research and development laboratories, but it suffers from a limited sensitivity, resolving power and mass accuracy. The coupling of SFC with a single quadrupole device was quite common in the early days of SFC-MS (1980s and 1990s), since it was easily controlled by existing software and was easily adaptable to the sources and interfaces described in this chapter [37]. However, this hyphenation is only investigated by few research groups today [38], probably due to the limited selectivity of single quadrupole analyzers and the availability of superior tandem instruments.

8.4.2 Ion Trap Analyzer

On the market, there are two types of ion traps, namely spherical and linear ion traps. Both operate by storing ions in the trap and manipulating them by using DC and RF electrical fields in a series of timed events. Then, the RF and DC potentials can be scanned to eject successive *m/z* ratios from the trap into the detector. The main limitation of spherical ion traps is the space charge effect. Indeed, only a certain maximum number of ions can be stored in a given volume in a spherical ion trap, due to space charge constraints. This results in a loss of resolution with potential *m/z* shift. In comparison, linear ion traps have better ion storage capacity.

The ion trap analyzer is extremely compact. It is able to do multistage mass spectrometry and offers a relatively high sensitivity. However, its quantitative performance and dynamic range are relatively poor and ion trap analyzers remain more difficult to operate than a quadrupole, due to the numerous possible settings (excitation, trapping, and detection conditions). Ion traps were used in the past for SFC-MS operation [39,40], but today they are often replaced by triple quadrupole instruments, which are faster and more sensitive.

8.4.3 Time-of-Flight Analyzer

A TOF analyzer measures the time spent by ions of different masses to move from the ionization source to the detector. There are three main features that explain the high mass accuracy and resolving power achieved with TOF analyzers, namely the reflectron, delayed extraction, and orthogonal acceleration [41,42]. The reflectron (ion mirror) focuses ions with identical *m/z* ratio, but different kinetic energies, thus improving resolving power. A short delay is applied to further reduce kinetic energy spread of individual ions, which is referred to as delayed extraction of ions. The orthogonal acceleration is certainly the best approach to transform a continuous ion beam produced by ESI, APCI, or APPI into a pulsed beam, directly compatible with TOF analyzers.

The inherent characteristics of TOF/MS are its high sensitivity in scan mode (all ions are detected), theoretically unlimited mass range as well as high acquisition speed (the duty cycle of modern instruments can attain 100 Hz). In addition, high-end TOF instruments afford resolving power of 40,000–60,000 and mass accuracies below 2 ppm. The only drawback to TOF analyzers is its limited dynamic range and quantitative performance. TOF analyzers have been employed in SFC-MS [32,43], but tend to be replaced today by QqTOF/MS.

8.4.4 Orbitrap Analyzer

The orbitrap is the most recent mass analyzer, developed by Makarov in 2000 [44,45] and commercialized in 2005. In few words, the orbitrap mass analyzer operates by radially trapping ions around a central electrode. This mass analyzer only uses electrostatic fields (DC) to confine and analyze ions. The m/z values are finally measured from the frequency of harmonic oscillations of ions, along the axis of the electric field.

The orbitrap analyzer offers very high resolving power in the range 100,000–240,000, and excellent mass accuracy below 1 ppm. In addition, it possesses a high space charge capacity which makes this instrument suitable for tackling the most difficult analytical problems. However, this instrument remains quite expensive and difficult to use, at least for beginners. Another drawback of orbitrap analyzers is its low acquisition rate, compared to quadrupole and TOF instruments, which makes them incompatible (at least for the first generation) with the narrow peaks produced in UHPSFC. Up to now, it has only been used in a few SFC-MS applications [46–48], dealing with the determination of pesticides, oil sands profiling, and lipidomics.

8.4.5 Tandem Mass Analyzers

The MS/MS uses two stages of analyzers, to examine selectively the fragmentation of isolated ions. The coupling of two analyzers in series (also known as tandem MS) can be done with the same analyzer (QqQ) or two different analyzers (QqTOF), which are known as hybrid instruments. For dissociation experiments, the most common activation method is collision induced dissociation (CID), where an inert gas is introduced into a collision cell where low energy (10–100 eV) and collisions occur between the precursor ion and the molecules of inert gas.

Today, the most widely used MS/MS instrument for trace analysis in complicated matrices is the triple quadrupole (QqQ), where the first and third quadrupoles are mass selective devices and the second serves as a collision cell. The QqQ is particularly useful for quantitative analysis, due to its ability to work in the highly selective single reaction monitoring (SRM) mode. In the SRM mode, the sensitivity is excellent, and very high acquisition rates (down to <1 ms per transition) can be reached for quantitative analysis of complex multicomponent mixtures.

An alternative to QqQ is the QqTOF where the third quadrupole is replaced with a TOF analyzer. One of the main advantage of QqTOF instruments over QqQ is the high resolving power of TOF, typically in the range 20,000–40,000. As a result, interfering peaks from ions having the same nominal mass can be resolved, thus improving the signal-to-noise ratio.

Tandem mass analyzers (QqQ and QqTOF) have become more popular in recent years and are the most widely used instruments for SFC-MS operation today [49–54].

8.5 PERFORMANCE COMPARISON BETWEEN LC-ESI/MS AND SFC-ESI/MS

8.5.1 Sensitivity

SFC has often been highlighted for its capacity to improve selectivity, throughput or solvent consumption compared to LC. However, one of the main obstacles for its expansion during the last decades was the lack of UV-sensitivity compared to HPLC [55]. Still, when it comes to MS detection, there are some convincing arguments to support SFC. The mobile phase used in SFC is constituted of a large proportion of volatile CO_2 which immediately evaporates in API sources, leaving the analytes in the organic cosolvent fraction (generally MeOH). This is beneficial as the organic solvent alone favors desolvation in the source, compared to the water-containing mobile phase used in RPLC, which are more difficult to evaporate. This phenomenon is clearly in favor of higher sensitivity in SFC-MS compared to RPLC-MS. However, a drawback of SFC is the difficulty of increasing the injection volume without compromising chromatographic quality. Indeed, while it is quite common to inject up to 10 μL or more in LC for high sensitivity, only 1 or 2 μL can be injected in SFC to avoid peak distortion. In a recent study dealing with doping control analysis [51,52], higher sensitivity was obtained in LC-MS versus SFC-MS, because of its higher loading capacity. In addition, the inner diameter of columns typically used in SFC-MS are generally larger than those used in LC-MS (3.0 mm vs 2.1 mm) due to the difference in the external volume contribution for the chromatographic devices [56]. This difference in column volumes adds a dilution factor in SFC-MS, which might further decrease the sensitivity.

More generally, it is hard to conclude whether the sensitivity could be improved in SFC versus LC, using the same MS detection, as it may be strongly compound and equipment dependent. Some early studies have shown that SFC was equivalent to LC in terms of sensitivity, using ESI-MS/MS [57] or APCI-MS [22], while bringing other advantages (throughput, selectivity). Some other studies showed that a gain in sensitivity was possible in SFC-MS. In 2006, Coe et al. reported a sensitivity improvement for the chiral analysis of warfarin in SFC-APCI-MS/MS, compared to the corresponding LC method [58]. More recently, the advent of state-of-the-art SFC systems has also boosted the performance of SFC and, consequently, the number of comparisons between the two techniques. In a recent study, a modern SFC system was used and the sensitivity improvement enabled the detection and quantitation of analytes which could not be detected in LC at low concentrations [59]. However, there are some contrasting examples for which LC-MS globally displayed higher sensitivity compared to SFC-MS [38,52]. In both cases, the design of the MS source seemed to have a great importance: with more recent devices, the desolvation of water-containing mobile phases was enhanced, which is favorable to LC. As an example, the recent comparison of LC-MS/MS and SFC-MS/MS with two different triple quadrupoles from the same brand demonstrated important differences in the achieved sensitivity [50]. With the old-fashioned device, SFC displayed higher sensitivity for 65% of the compounds, whereas this figure collapsed to 38% with a more recent MS apparatus (see Fig. 8.7).

Several parameters are of great importance to achieve high sensitivity in SFC-MS. As previously mentioned, the addition of a make-up solvent at the interface can strongly enhance the MS signal as early-eluting compounds are eluted with a very small amount of cosolvent after CO_2 decompression in gradient separations, and may precipitate. In addition, the composition and flow rate of this make-up solvent has to be tuned. MeOH or ethanol (EtOH) are generally the most widely used. The nature of the cosolvent and additives in the mobile phase can also impact the MS sensitivity and several ones should be screened during method development [49,60]. MeOH is generally preferred as the cosolvent, because it provides high ionization efficiency, while ammonium salts are often selected for their MS compatibility and the good peak shape they provide. A small proportion of water can also improve

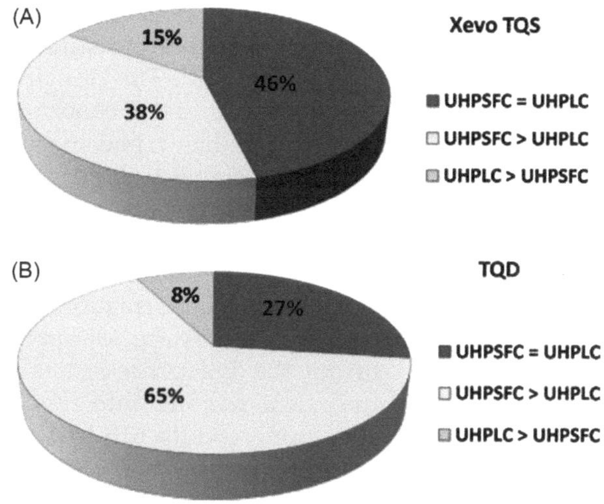

FIGURE 8.7 A comparison of sensitivity between two different triple quadrupole platforms, i.e., Waters Xevo TQ-S (A) and Waters TQD (B) in UHPSFC-MS/MS and UHPLC–MS/MS modes. *Source: Reprinted from Nováková L, Rentsch M, Grand-Guillaume Perrenoud A, Nicoli R, Saugy M, Veuthey JL, Guillarme D. UHPSFC-MS/MS for screening of doping agents. II: analysis of biological samples. Anal Chim Acta 2015;853:647–59 [Ref. 50]. Copyright 2015, with permission from Elsevier.*

MS sensitivity. Finally, MS parameters have to be adapted to the SFC column eluent [60], even if optimal MS parameters for SFC-MS are generally very similar to the ones of LC-MS [19].

8.5.2 Matrix Effects

The apparent reliability of quantitative results obtained with MS detection can actually be altered by the so-called matrix effects (ME). The "matrix" refers to all components in the sample other than the analyte(s) of interest, which could interfere. The concept of ME was first introduced by Kebarle et al. in 1993 and is defined as the impact of the coeluting compounds endogenously present in the matrix on the ionization efficiency and the reproducibility of the ionization source [61]. In other words, when a component from the matrix elutes at the same time as the analyte, it may induce an increase (enhancement) or a decrease (suppression) in the ionization of the analyte, in comparison with the analyte eluting alone [62]. The important influence of endogenous phospholipids present in biological fluids on small molecules MS signal is a common example in the field of bioanalysis. Such a phenomenon may harm method accuracy, precision, and sensitivity, and have to be addressed.

Several methodologies have been developed to measure ME. The postextraction analyte spiking is a quantitative measure and has been introduced by Matuszewski et al. in 2003 [63]. The principle is to compare the MS signal obtained by spiking a given concentration of the analyte in the extracted matrix on the one hand, and in pure solvent or mobile phase on the other hand. If the signal is higher in the solvent than in the extracted matrix, ion suppression exists.

Different strategies can be contemplated to overcome matrix effects. The first one is to switch the ionization source, when possible. As matrix interferences can occur during ion formation in the source and/or desolvation, some ionization sources are less prone to matrix effects than others. Indeed, during the ESI process, analytes have to acquire a charge in solution and then transfer into the gas phase as ions. This is why, ESI is the most vulnerable to ME [64] and switching to APCI or APPI, for which ionization occurs in the gas phase, can be a suitable alternative to tackle this issue. However, ME still happens in the gas phase with APCI and APPI, and some studies even revealed that ME also depend on the source design and that, in some case, APCI can be more sensitive to ME than ESI [65]. In conclusion, ionization source switching is a good alternative to overcome ME but without guarantee, as the ME process will differ from one analyte to another.

A second possibility is to change or improve the sample treatment prior to injection. It is quite obvious that if the sample is cleared from a large part of its endogenous constituents, ME have a good chance to be reduced. Several sample preparation techniques, such as solid-phase extraction (SPE), liquid–liquid extraction (LLE), protein precipitation (PPT), or supported liquid extraction (SLE), are nowadays very widespread in different fields of applications [66]. Switching from one to another can allow better removal of endogenous compounds creating the interference from the sample. Nevertheless, most of these techniques generally induce a preconcentration of the analyte, which is also applied to remaining matrix constituents. In consequence, this can have an opposite effect and increase ME because of a higher concentration of interfering compounds. This is why the right sample treatment should be chosen according to the analyte structure and the type of matrix.

Thirdly, one of the best ways to minimize matrix effects is to use an internal standard. The internal standard is supposed to correct for analyte losses during sample pretreatment and matrix-related signal alteration [67]. More precisely, it has been demonstrated that stable isotope labeled internal standards (compounds where several atoms of the analyte have been exchanged with their stable isotope) are the most appropriate with MS detection [68,69]. The main drawback of these internal standards is their limited availability and cost. In addition, some studies have revealed that an analyte and its deuterated analog can be affected differently by ME, leading to incorrect quantification [70].

Finally, it is possible to overcome the ME by changing the chromatographic separation. Indeed, using a different chromatographic technique may allow the analyte to be separated from the interfering matrix compounds, leading to better ionization of the compound of interest. This has already been investigated with hydrophilic interaction liquid chromatography (HILIC) that is a good alternative to RPLC, because of differences in retention mechanisms [71,72]. On the contrary, ME in SFC have scarcely been evaluated and compared to LC. Only a few papers present a systematic comparison of ME between different techniques [50–52]. In these studies, the ME were generally lower in SFC versus RPLC. This could be explained by the fact that the retention mechanism in RPLC is similar to that of most extraction techniques. As a consequence, extracted impurities from the matrix are prone to elute close to the analyte during the chromatographic step. This is not the case for SFC, which relies on different mechanism (polar retention). More generally, SFC may be appropriate to reduce the ME of polar compounds analyzed in RPLC. Indeed, the major part of ME occur in the first minutes of the analysis as many contaminants are polar, especially in bioanalysis, and poorly-retained analytes are more prone to ME. The improved retention of these analytes in SFC [73] may separate them from the interferences.

8.6 APPLICATIONS OF SFC-MS

8.6.1 Chiral SFC-MS

Enantioselective separations have been the first springboard for the development of SFC. Normal phase HPLC has long been the reference technique for separating enantiomers, despite its numerous drawbacks (toxic solvents, long equilibration...). The first chiral SFC separation was achieved in 1985 [74]. Since then, the interest in chiral SFC has never stopped growing and today, SFC is considered a first choice for enantioseparation, principally because of the important reduction of run time compared to LC and the improvements in terms of throughput and costs when scaling up to preparative chromatography. Moreover, an additional advantage of SFC is the possible hyphenation with MS detectors, which remains difficult with normal phase HPLC. Baker and Pinkston were among the first to investigate the coupling of chiral SFC with MS detection [75].

Nowadays, most chiral SFC-MS applications are found in the pharmaceutical industry, and particularly in the drug discovery departments [76]. It is also widely used during enantioselective drug synthesis to follow up on enantiomeric excess [77].

The purpose of using MS detection with chiral separation is not to increase the selectivity between the enantiomers as the two molecules have the same mass. The main advantages are: (1) increasing the laboratory throughput by being able to inject several pairs of enantiomers at the same time (sample pooling strategy) and to distinguish them by using MS selectivity [33], (2) enhancing analyte detectability compared with UV detection, especially for low concentration levels in biological fluids during pharmacokinetic studies [57,58,78,79], (3) undoubtedly separating enantiomers from other achiral impurities in complex mixtures [80,81]. For chiral screening, the system is generally automated to screen different columns and mobile phases simultaneously [33,82,83].

8.6.2 Bioanalysis and Lipidomics

The measurement of drug concentrations in biological fluids, such as plasma, serum, blood, urine, or bile, is one of the important aspects of drug development. As an example, it is massively performed during pharmacokinetic and metabolic studies, clinical and preclinical trials, and doping control analysis. Biological samples are complex matrices, as they generally contain the compound(s) of interest at very low concentration (ng/mL scale) along with numerous endogenous substances which may create interferences. Thus MS detection has been used in this field predominantly due to its high selectivity and sensitivity.

SFC-MS has been scarcely used for the analysis of complex biological matrices compared to LC-MS, which remains the gold standard [84,85]. However, there has been a growing interest in the application of SFC-MS to bioanalysis in the recent years. Various exhaustive reviews have listed the different bioanalytical [5,86,87] and metabolic [88,89] applications performed with SFC-MS. Despite high MS selectivity, a sample pretreatment is generally required prior to separation, as described in Section 5.2. An organic solvent is usually used during the last stage of these extraction procedures (i.e., MeOH or acetonitrile for SPE, and ethyl acetate, hexane, or heptane for LLE). These types of solvents are suitable for direct injection in SFC. In consequence, the reconstitution step commonly performed prior to LC-Ms injection can be skipped to reduce the sample preparation time.

Some recent papers show a trend in the diversification of MS detectors coupled to SFC: hyphenation of SFC with high resolution MS is encountered more frequently for the analysis of biological fluids [47,90,91] and ion mobility was recently added as an additional dimension to SFC-MS for the separation of nonsteroidal selective androgen receptor modulators [92].

Another field in which SFC has a strong presence is lipidomics. Lipids were among the first compounds to be analyzed during the early days of SFC; because of their high hydrophobicity, they are readily soluble in supercritical CO_2 mobile phases used in cSFC [92]. Nowadays, their analysis in biological samples using pSFC has beeen extensively studied by Bamba et al. [47,94–97]. Another team also recently published a method able to separate 24 lipid classes containing 436 lipid species in a single run [98] (see Fig. 8.8). A review dealing with lipidomics in SFC was published recently [99].

There are also a few applications of SFC-MS in the fields of metabolomics [91,100] and doping control analysis [49–52]. For the latter, more than 200 target compounds were successfully extracted from urine and analyzed using SFC-MS/MS. For a specific class of doping agents (androgenic anabolic steroids), the global performance of SFC-MS/MS was compared to LC-MS/MS and GC-MS/MS [52]. The three methods displayed very different results in terms of selectivity (see Fig. 8.9), sensitivity, and matrix effects.

8.6.3 Achiral SFC-MS for Pharmaceuticals Analysis

Even if chiral separation has been the workhorse of SFC during the last decades, there is also a growing interest in achiral SFC. Pinkston et al. were among the first to illustrate the potential of SFC-MS compared to LC-MS by screening a large library of pharmaceutical compounds [37]. The aim of the study was to prove that the number of "hits" (compounds eluted and detected) was very similar between LC-MS and SFC-MS, even when dealing with polar compounds. Since then, achiral SFC has become increasingly popular, but mainly with UV detection. Some applications using single quadrupoles have been reported but remain rare [37,38,101]. However, the recent commercialization of inexpensive and compact simple quadrupoles may be of interest for analytical development laboratories as a replacement for UV detection, as an example for impurity profiling of drug candidates [102,103].

MS detection can also be applied to preparative SFC. Based on the same principle as mass-directed LC purification, these system are particularly appreciated in drug discovery laboratories [104,105].

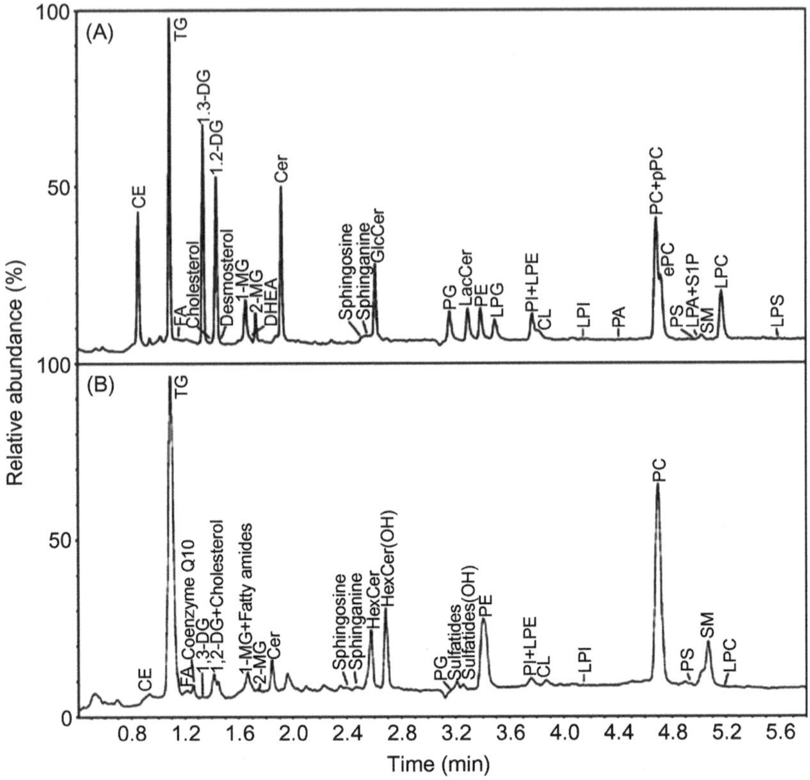

FIGURE 8.8 Positive-ion UHPSFC/ESI-MS chromatograms of the mixture of lipid class standards (A) and the total lipid extract of porcine brain (B). UHPSFC conditions: Acquity BEH UPC2 column (100 × 3 mm, 1.7 μm, Waters), the flow rate 1.9 mL/min, the column temperature 60°C, the ABPR pressure 1800 psi and the gradient of MeOH–water mixture (99:1, v/v) containing 30 mM of ammonium acetate as the modifier: 0 min, 1%; 5 min, 51%; 6 min, 51%. Peak annotation: *CE*, cholesteryl esters; *TG*, triacylglycerols; *FA*, fatty acids; *DG*, diacylglycerols; *MG*, monoacylglycerols; *DHEA*, dehydroepiandrosterone; *Cer*, ceramides; *GlcCer*, glucosylceramides; *HexCer*, hexosylceramides; *PG*, phosphatidylglycerols; *LacCer*, lactosylceramides; *pPE*, 1-alkenyl-2-acyl phosphatidylethanolamines (plasmalogens); *ePE*, 1-alkyl-2-acyl phosphatidylethanolamines (ethers); *PE*, phosphatidylethanolamines; *LPG*, lysophosphatidylglycerols; *PI*, phosphatidylinositols; *LPE*, lysophosphatidylethanolamines; *CL*, cardiolipins; *LPI*, lysophosphatidylinositols; *PA*, phosphatidic acids; *PC*, phosphatidylcholines; *pPC*, 1-alkenyl-2-acyl phosphatidylcholines; *ePC*, 1-alkyl-2-acyl phosphatidylcholines; *PS*, phosphatidylserines; *LPA*, lysophosphatidic acids; *S1P*, sphingosine-1-phosphate; *SM*, sphingomyelins; *LPC*, lysophosphatidylcholines; *LPS*, lysophosphatidylserines. *Source: Reprinted with permission from Lísa M, Holčapek M. Anal Chem 2015;87:7187–95 [Ref. 98]. Copyright 2014 American Chemical Society.*

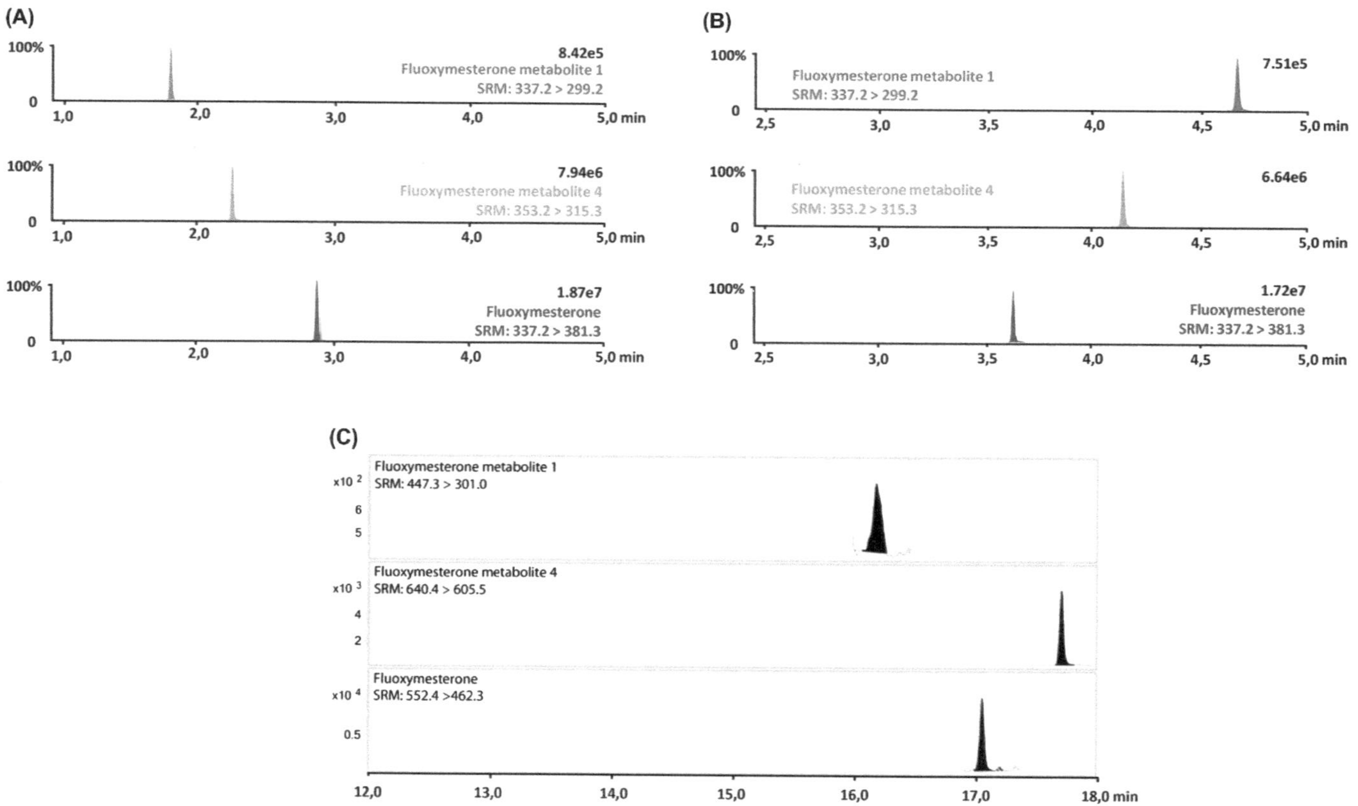

FIGURE 8.9 Chromatograms for injection of urine spiked with fluoxymesterone and two metabolites at 10 ng/mL in UHPLC–MS/MS (A), UHPSFC-MS/MS (B), and GC-MS/MS (C). *Source: Reprinted from Desfontaine V, Nováková L, Ponzetto F, Nicoli R, Saugy M, Veuthey JL, Guillarme D. Liquid chromatography and supercritical fluid chromatography as alternative techniques to gas chromatography for the rapid screening of anabolic agents in urine. J Chromatogr A 2016;1451:145–55 [Ref. 52]. Copyright 2016, with permission from Elsevier.*

8.6.4 Food and Environmental Analysis

The application of SFC-MS to the analysis of food or environmental samples is very recent as most studies from this perspective were conducted after 2010. Once again, lipids were the first reason for using SFC-MS applied to agricultural and food samples [106,107], followed by pesticides [48]. The need for enantiomeric separation of some pesticides also helped in the gain of interest of SFC-MS in this field [108–110].

8.7 CONCLUSION

SFC-UV (and also SFC-MS to a certain extent) has suffered in the past from its too limited sensitivity. Several new instrumental advances in recent years have circumvented this sensitivity issue and stabilized the baseline. Today, SFC-MS can be performed for a wide variety of analyses ranging from normal phase separations of hydrophobic substances, such as lipids, liposoluble vitamins, or polycyclic aromatic hydrocarbons, to reversed phase–type analyses of moderately polar substances, such as drugs, pesticides, or natural products, by using a relatively high amount of MeOH as a mobile phase component (up to 40%) and small amounts of water as an additive. SFC-MS is now entering a more mature phase of development, due to the increasing experience of analysts with this technology. The community using SFC-MS has driven the vendors to produce even more robust and fully integrated systems, which allows equivalent and even sometimes higher performance than LC-MS.

References

[1] Nováková L, Grand-Guillaume Perrenoud A, Francois I, West C, Lesellier E, Guillarme D. Modern analytical SFC using columns packed with sub-2 μm particles: a tutorial. Anal Chim Acta 2014;824:18–35.

[2] Grand-Guillaume Perrenoud A, Veuthey JL, Guillarme D. Comparison of ultra-high performance supercritical fluid chromatography and ultra-high performance liquid chromatography. J Chromatogr A 2012;1266:158–67.

[3] Grand-Guillaume Perrenoud A, Farrell WP, Aurigemma CM, Aurigemma NC, Fekete S, Guillarme D. Evaluation of stationary phases packed with superficially porous particles for the analysis of pharmaceutical compounds using supercritical fluid chromatography. J Chromatogr A 2014;1360:275–87.

[4] Grand-Guillaume Perrenoud A, Veuthey JL, Guillarme D. The use of columns packed with sub-2 μm particles in supercritical fluid chromatography. Trends Anal Chem 2014;63:44–54.

[5] Desfontaine V, Guillarme D, Francotte E, Nováková L. Supercritical fluid chromatography in pharmaceutical analysis. J Pharm Biomed Anal 2015;113:56–71.

[6] Lesellier E, Valarche A, West C, Dreux M. Effects of selected parameters on the response of the evaporative light scattering detector in supercritical fluid chromatography. J Chromatogr A 2012;1250:220–6.
[7] Holčapek M, Jirásko R, Lísa M. Recent developments in liquid chromatography-mass spectrometry and related techniques. J Chromatogr A 2012;1259:3–15.
[8] Rodriguez-Aller M, Gurny R, Veuthey JL, Guillarme D. Coupling UHPLC with MS (/MS): constraints and possible applications. J Chromatogr A 2013;1292:2–18.
[9] Guillarme D, Schappler J, Rudaz S, Veuthey JL. Coupling ultra-high pressure liquid chromatography with mass spectrometry. Trends Anal Chem 2010;29:15–27.
[10] Pitt JJ. Principles and applications of liquid chromatography-mass spectrometry in clinical biochemistry. Clin Biochem Rev 2009;30:19–34.
[11] Novotny M, Springston SR, Peaden PA, Fjeldsted JC, Lee ML. Capillary supercritical fluid chromatography. Anal Chem 1981;53:407A–14A.
[12] Peaden PA, Fjeldsted JC, Lee ML, Springston SR, Novotny M. Instrumental aspects of capillary supercritical fluid chromatography. Anal Chem 1982;54:1090–3.
[13] Springston SR, Novotny M. Kinetic optimization of capillary supercritical fluid chromatography using carbon dioxide as the mobile phase. Chromatographia 1981;14: 679–84.
[14] Peaden PA, Lee ML. Theoretical treatment of resolving power in open tubular column supercritical fluid chromatography. J Chromatogr 1983;259:1–16.
[15] Gere DR, Board R, McManigill D. Supercritical fluid chromatography with small particle diameter packed columns. Anal Chem 1982;54:736–40.
[16] Crowther JB, Henion JD. Supercritical fluid chromatography of polar drugs using small-particle packed columns with mass spectrometric detection. Anal Chem 1985;57:2711–16.
[17] Mann M, Meng CK, Fenn JB. Interpreting mass spectra of multiply charged ions. Anal Chem 1989;61:1702–8.
[18] Fenn JB, Mann M, Meng CK, Wong SF, Whitehouse CM. Electrospray ionization for mass spectrometry of large biomolecules. Science 1989;246(4926):64–71.
[19] Grand-Guillaume Perrenoud A, Veuthey JL, Guillarme D. Coupling state-of-the-art supercritical fluid chromatography and mass spectrometry: from hyphenation interface optimization to high-sensitivity analysis of pharmaceutical compounds. J Chromatogr A 2014;1339:174–84.
[20] Kelly MA, Vestling MM, Fenselau C, Smith PB. Electrospray analysis of proteins: a comparison of positive-ion and negative-ion mass spectra at high and low pH. Org Mass Spectrom 1992;27(10):1143–7.
[21] Dzidic I, Desiderio DM, Wilson MS, Crain PF, McCloskey J. Mass standards for chemical ionization mass spectrometry. Anal Chem 1971;43(13):1877–9.
[22] Anacleto JF, Ramaley L, Boyd RK, Pleasance S, Quilliam MA, Sim PG, et al. Analysis of polycyclic aromatic compounds by supercritical fluid charomatography/mass spectrometry using atmospheric-pressure chemical ionization. Rapid Comm Mass Spectrom 1991;5:149–55.
[23] Ventura MC, Farrell WP, Aurigemma CM, Greig MJ. Packed column supercritical fluid chromatography/mass spectrometry for high-throughput analysis. Part 2. Anal Chem 1999;71(19):4223–31.
[24] Dost K, Davidson G. Development of a packed-column supercritical fluid chromatography/atmospheric pressure chemical-ionisation mass spectrometric technique for the analysis of atropine. J Biochem Biophys Methods 2000;43(1–3):125–34.
[25] Robb DB, Covey TR, Bruins AP. Atmospheric pressure photoionization: an ionization method for liquid chromatography – mass spectrometry. Anal Chem 2000;72(15):3653–9.
[26] Marchi I, Rudaz S, Veuthey JL. Atmospheric pressure photoionization for coupling liquid-chromatography to mass spectrometry: a review. Talanta 2009;78:1–18.

[27] Mejean M, Brunelle A, Touboul D. Quantification of tocopherols and tocotrienols in soybean oil by supercritical-fluid chromatography coupled to high-resolution mass spectrometry. Anal Bioanal Chem 2015;407(17):5133–42.
[28] Cho Y, Choi MH, Kim B, Kim S. Supercritical fluid chromatography coupled with in-source atmospheric pressure ionization hydrogen/deuterium exchange mass spectrometry for compound speciation. J Chromatogr A 2016;1444:123–8.
[29] Kott L. An overview of supercritical fluid chromatography mass spectrometry (SFC-MS) in the pharmaceutical industry. Am Pharm Rev 2013;.
[30] Pinkston JD. Advantages and drawbacks of popular supercritical fluid chromatography/mass interfacing approaches—a user's perspective. Eur J Mass Spectrom 2005;11(2):189–97.
[31] Dunkle M, Vanhoenacker G, David F, Sandra P. Agilent 1260 Infinity SFC/MS Solution—superior sensitivity by seamlessly interfacing to the Agilent 6100 Series LC/MS system. Agilent Appl Note 2011;5990–7972EN.
[32] Bolanos BJ, Ventura MC, Greig MJ. Preserving the chromatographic integrity of high-speed supercritical fluid chromatography separations using time-of-flight mass spectrometry. J Comb Chem 2003;5(4):451–5.
[33] Zhao Y, Woo G, Thomas S, Semin D, Sandra P. Rapid method development for chiral separation in drug discovery using sample pooling and supercritical fluid chromatography–mass spectrometry. J Chromatogr A 2003;1003:157.
[34] Lee ED, Henion JD, Cody RB, Kinsinger JA. Supercritical fluid chromatography/Fourier transform mass spectrometry. Anal Chem 1987;59(16):1309–12.
[35] Laude DA, Pentoney SL, Griffiths PR, Wilkins CL. Supercritical fluid chromatography interface for a differentially pumped dual-cell Fourier transform mass spectrometer. Anal Chem 1987;59(18):2283–8.
[36] Baumeister ER, West CD, Ijames CF, Wilkins CL. Interface for SFC and FTMS. Anal Chem 1991;63:251–5.
[37] Pinkston JD, Wen D, Morand KL, Tirey DA, Stanton DT. Comparison of LC/MS and SFC/MS for screening of a large and diverse library of pharmaceutically relevant compounds. Anal Chem 2006;78(21):7467–72.
[38] Spaggiari D, Mehl F, Desfontaine V, Grand-Guillaume Perrenoud A, Fekete S, Rudaz S, et al. Comparison of liquid chromatography and supercritical fluid chromatography coupled to benchtop single quadrupole mass spectrometer for in vitro CYP-mediated metabolism assay. J Chromatogr A 2014;1371:244–56.
[39] Morgan DG, Harbol KL, Kitrinos Jr NP. Optimization of a supercritical fluid chromatograph-atmospheric chemical ionization mass spectrometer interface using an ion trap and two quadrupole mass spectrometers. J Chromatogr A 1998;800:39–49.
[40] Xu X, Roman JM, Veenstra TD, Van Anda J, Ziegler RG, Issaq HJ. Analysis of fifteen estrogen metabolites using packed column supercritical fluid chromatography – mass spectrometry. Anal Chem 2006;78(5):1553–8.
[41] Bristow AWT. Accurate mass measurement for the determination of elemental formula—a tutorial. Mass Spectrom Rev 2006;25(1):99–111.
[42] Jiwan JLH, Wallemacq P, Hérent MF. HPLC-high resolution mass spectrometry in clinical laboratory? Clin Biochem 2011;44(1):136–47.
[43] Klink D, Schmitz J. SFC-APLI-(TOF)MS: hyphenation of supercritical fluid chromatography to atmospheric pressure laser ionization mass spectrometry. Anal Chem 2016;88:1058–64.
[44] Makarov A. Electrostatic axially harmonic orbital trapping: a high-performance technique of mass analysis. Anal Chem 2000;72(6):1156–62.
[45] Hardman M, Makarov A. Interfacing the orbitrap mass analyzer to an electrospray ion source. Anal Chem 2003;75(7):1699–705.

[46] Pereira AS, Martin JW. Exploring the complexity of oil sands process-affected water by high efficiency supercritical fluid chromatography/orbitrap mass spectrometry. Rapid Comm Mass Spec 2015;29(8):735–44.
[47] Yamada T, Uchikata T, Sakamoto S, Yokoi Y, Nishiumi S, Yoshida M, et al. Supercritical fluid chromatography/Orbitrap mass spectrometry based lipidomics platform coupled with automated lipid identification software for accurate lipid profiling. J Chromatogr A 2013;1301:237–42.
[48] Ishibashi M, Izumi Y, Sakai M, Ando T, Fukusaki E, Bamba T. High-throughput simultaneous analysis of pesticides by supercritical fluid chromatography coupled with high-resolution mass spectrometry. J Agric Food Chem 2015;63(18):4457–63.
[49] Nováková L, Grand-Guillaume Perrenoud A, Nicoli R, Saugy M, Veuthey JL, Guillarme D. UHPSFC-MS/MS for screening of doping agents. I: investigation of mobile phase and MS conditions. Anal Chim Acta 2015;853:637–46.
[50] Nováková L, Rentsch M, Grand-Guillaume Perrenoud A, Nicoli R, Saugy M, Veuthey JL, et al. UHPSFC-MS/MS for screening of doping agents. II: analysis of biological samples. Anal Chim Acta 2015;853:647–59.
[51] Nováková L, Desfontaine V, Ponzetto F, Nicoli R, Saugy M, Veuthey JL, et al. Fast and sensitive supercritical fluid chromatography—tandem mass spectrometry multi-class screening method for the determination of doping agents in urine. Anal Chim Acta 2016;915:102–10.
[52] Desfontaine V, Nováková L, Ponzetto F, Nicoli R, Saugy M, Veuthey JL, et al. Liquid chromatography and supercritical fluid chromatography as alternative techniques to gas chromatography for the rapid screening of anabolic agents in urine. J Chromatogr A 2016;1451:145–55.
[53] Hedrick J, VanAnda J, Brand T. Qualitative analysis of fish oil triglycerides with supercritical fluid chromatography and Q-TOF MS. Agilent Appl Note 2014; 5991–5183EN.
[54] Grand-Guillaume Perrenoud A, Guillarme D, Veuthey JL, Barron D, Moco S. Ultra-high performance supercritical fluid chromatography coupled with quadrupole-time-of-flight mass spectrometry as a performing tool for bioactive analysis. J Chromatogr A 2016;1450:101–11.
[55] Dispas A, Lebrun P, Ziemons E, Marini R, Rozet E, Hubert P. Evaluation of the quantitative performances of supercritical fluid chromatography: from method development to validation. J Chromatogr A 2014;1353:78–88.
[56] Grand-Guillaume Perrenoud A, Hamman C, Goel M, Veuthey J-L, Guillarme D, Fekete S. Maximizing kinetic performance in supercritical fluid chromatography using state-of-the-art instruments. J Chromatogr A 2013;1314:288–97.
[57] Hoke SH, Pinkston D, Bailey RE, Tanguay SL, Eichhold TH. Comparison of packed-column supercritical fluid chromatography-tandem mass spectrometry with liquid chromatography-tandem mass spectrometry for bioanalytical determination of (R)- and (S)-ketoprofen in human plasma following automated 96-Well solid-phase extraction. Anal Chem 2000;72:4235–41.
[58] Coe RA, Rathe JO, Lee JW. Supercritical fluid chromatography–tandem mass spectrometry for fast bioanalysis of R/S-warfarin in human plasma. J Pharm Biomed Anal 2006;42(5):573–80.
[59] Doué M, Dervilly-Pinel G, Pouponneau K, Monteau F, Le Bizec B. Analysis of glucuronide and sulfate steroids in urine by ultra-high-performance supercritical-fluid chromatography hyphenated tandem mass spectrometry. Anal Bioanal Chem 2015; 407:4473–84.
[60] Hsieh Y, Favreau L, Schwerdt J, Cheng KC. Supercritical fluid chromatography/tandem mass spectrometric method for analysis of pharmaceutical compounds in metabolic stability samples. J Pharm Biomed Anal 2006;40:799–804.

[61] Kebarle P, Tang L. From ions in solution to ions in the gas phase—the mechanism of electrospray mass spectrometry. Anal Chem 1993;65(22):972A–86A.

[62] Hall TG, Smukste I, Bresciano KR, Wang Y, McKearn D, Savage RE. Identifying and Overcoming Matrix Effects in Drug Discovery and Development. In: Prasain JK, editor. Tandem mass spectrometry: applications and principles. InTech; 2012. p. 389–420.

[63] Matuszewski BK, Constanzer ML, Chavez-Eng CM. Strategies for the assessment of matrix effect in quantitative bioanalytical methods based on HPLC-MS/MS. Anal Chem 2003;75(13):3019–30.

[64] King R, Bonfiglio R, Fernandez-Metzler C, Miller-Stein C, Olah T. Mechanistic investigation of ionization suppression in electrospray ionization. J Am Soc Mass Spectrom 2000;11:942–50.

[65] Mei H, Hsieh Y, Nardo C, Xu X, Wang S, Ng K, et al. Investigation of matrix effects in bioanalytical high-performance liquid chromatography/tandem mass spectrometric assays: application to drug discovery. Rapid Commn. Mass Spectrom 2003;17(1): 97–103.

[66] Nováková L, Vlcková H. A revwangiew of current trends and advances in modern bio-analytical methods: chromatography and sample preparation. Anal Chim Acta 2009;656(1–2):8–35.

[67] Xu RN, Fan L, Rieser MJ, El-Shourbagy TA. Recent advances in high-throughput quantitative bioanalysis by LC–MS/MS. J Pharm Biomed Anal 2007;44(2):342–55.

[68] Matuszewski BK. Standard line slopes as a measure of a relative matrix effect in quantitative HPLC–MS bioanalysis. J Chromatogr B 2006;830:293–300.

[69] Stokvis E, Rosing H, Lopez-Lazaro L, Schellens JH, Beijnen JH. Switching from an analogous to a stable isotopically labeled internal standard for the LC-MS/MS quantitation of the novel anticancer drug Kahalalide F significantly improves assay performance. Biomed Chromatogr 2004;18:400–2.

[70] Wang S, Cyronak M, Yang E. Does a stable isotopically labeled internal standard always correct analyte response?: a matrix effect study on a LC/MS/MS method for the determination of carvedilol enantiomers in human plasma. J Pharm Biomed Anal 2007;43(2):701–7.

[71] Havlíková L, Vlčková H, Solich P, Nováková L. HILIC UHPLC–MS/MS for fast and sensitive bioanalysis: accounting for matrix effects in method development. Bioanalysis 2013;5(19):2345–57.

[72] Periat A, Kohler I, Thomas A, Nicoli R, Boccard J, Veuthey JL, et al. Systematic evaluation of matrix effects in hydrophilic interaction chromatography versus reversed phase liquid chromatography coupled to mass spectrometry. J Chromatogr A 2016;1439:42–53.

[73] Periat A, Grand-Guillaume Perrenoud A, Guillarme D. Evaluation of various chromatographic approaches for the retention of hydrophilic compounds and MS compatibility. J Sep Sci 2013;36:3141–51.

[74] Mourier PA, Eliot E, Caude MH, Rosset RH, Tambute AG. Supercritical and subcritical fluid chromatography on a chiral stationary phase for the resolution of phosphine oxide enantiomers. Anal Chem 1985;57:2819–23.

[75] Baker TR, Pinkston JD. Development and application of packed column supercritical fluid chromatography/pneumatically assisted electrospray mass spectrometry. J Am Soc Mass Spectrom 1998;9(5):498–509.

[76] Bolaños B, Greig M, Ventura M, Farrell W, Aurigemma CM, Li H, et al. SFC/MS in drug discovery at Pfizer, La Jolla. Int J Mass Spectrom 2004;238(2):85–97.

[77] Alexander AJ, Staab A. Use of achiral/chiral sfc/ms for the profiling of isomeric cinnamonitrile/hydrocinnamonitrile products in chiral drug synthesis. Anal Chem 2006;78(11):3835–8.

[78] Chen J, Hsieh Y, Cook J, Morrison R, Korfmacher WA. Supercritical fluid chromatography-tandem mass spectrometry for the enantioselective determination of propranolol and pindolol in mouse blood by serial sampling. Anal Chem 2006;78 (4):1212–17.

[79] Yang Z, Xu X, Sun L, Zhao X, Wang H, Fawcett JP, et al. Development and validation of an enantioselective SFC-MS/MS method for simultaneous separation and quantification of oxcarbazepine and its chiral metabolites in beagle dog plasma. J Chromatogr B 2016;1020:36–42.

[80] Brondz E, Ekeberg D, Bell DS, Annino AR, Hustad JA, Svendsen R, et al. Nature of the main contaminant in the drug primaquine diphosphate: SFC and SFC–MS methods of analysis. J Pharm Biomed Anal 2007;43:937–44.

[81] Garzotti M, Hamdan M. Supercritical fluid chromatography coupled to electrospray mass spectrometry: a powerful tool for the analysis of chiral mixtures. J Chromatogr B 2002;770:53–61.

[82] Zeng L, Xu R, Laskar DB, Kassel DB. Parallel supercritical fluid chromatography/mass spectrometry system for high-throughput enantioselective optimization and separation. J Chromatogr A 2007;1169:193–204.

[83] Laskar DB, Zeng L, Xu R, Kassel DB. Parallel SFC/MS-MUX screening to assess enantiomeric purity. Chirality 2008;20(8):885–95.

[84] Desfontaine V, Nováková L, Guillarme D. SFC-MS versus RPLC-MS for drug analysis in biological samples. Bioanalysis 2015;7(10):1193–5.

[85] Nováková L. Challenges in the development of bioanalytical liquid chromatography–mass spectrometry method with emphasis on fast analysis. J Chromatogr A 2013;1292:25–37.

[86] Abbott E, Veenstra TD, Issaq HJ. Clinical and pharmaceutical applications of packed column supercritical fluid chromatography. J Sep Sci 2008;31:1223–30.

[87] Ríos A, Zougagh M, de Andrés F. Bioanalytical applications using supercritical fluid techniques. Bioanalysis 2010;2(1):9–25.

[88] Matsubara A, Fukusaki E, Bamba T. Metabolite analysis by supercritical fluid chromatography. Bioanalysis 2010;2(1):27–34.

[89] Taguchi K, Fukusaki E, Bamba T. Supercritical fluid chromatography/mass spectrometry in metabolite analysis. Bioanalysis 2014;6(12):1679–89.

[90] Jumaah F, Larsson S, Essén S, Cunico LP, Holm C, Turner C, et al. A rapid method for the separation of vitamin D and its metabolites by ultra-high performance supercritical fluid chromatography–mass spectrometry. J Chromatogr A 2016; 1440:191–200.

[91] Jones MD, Rainville PD, Isaac G, Wilson ID, Smith NW, Plumb RS. Ultra high resolution SFC–MS as a high throughput platform for metabolic phenotyping: application to metabolic profiling of rat and dog bile. J Chromatogr B 2014;966:200–7.

[92] Beucher L, Dervilly-Pinel G, Cesbron N, Penot M, Gicquiau A, Monteau F, et al. Specific characterization of non-steroidal selective androgen peceptor modulators using supercritical fluid chromatography coupled to ion-mobility mass spectrometry: application to the detection of enobosarm in bovine urine. Drug Test Anal 2016;. Available from: http://dx.doi.org/10.1002/dta.1951.

[93] Giron D, Link R, Bouissel S. Analysis of mono-, di- and triglycerides in pharmaceutical excipients by capillary supercritical fluid chromatography. J Pharm Biomed Anal 1992;10:821–30.

[94] Matsubara A, Uchikata T, Shinohara M, Nishiumi S, Yoshida M, Fukusaki E, et al. Highly sensitive and rapid profiling method for carotenoids and their epoxidized products using supercritical fluid chromatography coupled with electrospray ionization-triple quadrupole mass spectrometry. J Biosci Bioeng 2012;113 (6):782–7.

[95] Uchikata T, Matsubara A, Fukusaki E, Bamba T. High-throughput phospholipid profiling system based on supercritical fluid extraction–supercritical fluid chromatography/mass spectrometry for dried plasma spot analysis. J Chromatogr A 2012;1250:69–75.
[96] Lee JW, Nishiumi S, Yoshida M, Fukusaki E, Bamba T. Simultaneous profiling of polar lipids by supercritical fluid chromatography/tandem mass spectrometry with methylation. J Chromatogr A 2013;1279:98–107.
[97] Bamba T, Lee JW, Matsubara A, Fukusaki E. Metabolic profiling of lipids by supercritical fluid chromatography/mass spectrometry. J Chromatogr A 2012;1250:212–19.
[98] Lísa M, Holčapek M. High-throughput and comprehensive lipidomic analysis using ultrahigh-performance supercritical fluid chromatography-mass spectrometry. Anal Chem 2015;87(14):7187–95.
[99] Laboureur L, Ollero M, Touboul D. Lipidomics by supercritical fluid chromatography. Int J Mol Sci 2015;16(6):13868–84.
[100] Sen A, Knappy C, Lewis MR, Plumb RS, Wilson ID, Nicholson JK, et al. Analysis of polar urinary metabolites for metabolic phenotyping using supercritical fluid chromatography and mass spectrometry. J Chromatogr A 2016;1249:141–55.
[101] Huang Y, Zhang T, Zhou H, Feng Y, Fan C, Chen W, et al. Fast separation of triterpenoid saponins using supercritical fluid chromatography coupled with single quadrupole mass spectrometry. J Pharm Biomed Anal 2016;121:22–9.
[102] Lemasson E, Bertin S, Hennig P, Boiteux H, Lesellier E, West C. Development of an achiral supercritical fluid chromatography method with ultraviolet absorbance and mass spectrometric detection for impurity profiling of drug candidates. Part I: optimization of mobile phase composition. J Chromatogr A 2015;1408:217–26.
[103] Lemasson E, Bertin S, Hennig P, Boiteux H, Lesellier E, West C. Development of an achiral supercritical fluid chromatography method with ultraviolet absorbance and mass spectrometric detection for impurity profiling of drug candidates. Part II. Selection of an orthogonal set of stationary phases. J Chromatogr A 2015;1408:227–35.
[104] Van Anda J. Use of SFC/MS in the purification of achiral pharmaceutical compounds. Am Pharm Rev 2010;13(6):111–15.
[105] Ebinger K, Weller HN, Kiplinger J, Lefebvre P. Evaluation of a new preparative supercritical fluid chromatography system for compound library purification: the TharSFC SFC-MS Prep-100 system. J Lab Autom 2011;16(3):241–9.
[106] Lee JW, Uchikata T, Matsubara A, Nakamura T, Fukusaki E, Bamba T. Application of supercritical fluid chromatography/mass spectrometry to lipid profiling of soybean. J Biosci Bioeng 2012;113(2):262–8.
[107] Tu A, Du Z, Qu S. Rapid profiling of triacylglycerols for identifying authenticity of edible oils using supercritical fluid chromatography-quadruple time-of-flight mass spectrometry combined with chemometric tools. Anal Methods 2016;8:4226–38.
[108] Liu N, Dong F, Xu J, Liu X, Chen Z, Tao Y, et al. Stereoselective determination of tebuconazole in water and zebrafish by supercritical fluid chromatography tandem mass spectrometry. J Agric Food Chem 2015;63(28):6297–303.
[109] Chen X, Dong F, Xu J, Liu X, Chen Z, Liu N, et al. Enantioseparation and determination of isofenphos-methyl enantiomers in wheat, corn, peanut and soil with supercritical fluid chromatography/tandem mass spectrometric method. J Chromatogr B Analyt Technol Biomed Life Sci 2016;1015–1016:13–21.
[110] Pan X, Dong F, Xu J, Liu X, Chen Z, Zheng Y. Stereoselective analysis of novel chiral fungicide pyrisoxazole in cucumber, tomato and soil under different application methods with supercritical fluid chromatography/tandem mass spectrometry. J Hazard Mater 2016;311:115–24.

CHAPTER

9

Theories for Preparative SFC

A. Tarafder

Waters Corporation, Milford, MA, United States

OUTLINE

9.1 INTRODUCTION

In this section and the next some general concepts on preparative supercritical fluid chromatography (SFC) will be briefly discussed to support the discussions presented in the subsequent sections. We hope

Supercritical Fluid Chromatography.
DOI: http://dx.doi.org/10.1016/B978-0-12-809207-1.00009-4

that the material provided here will facilitate a more informed design of separation methods using preparative SFC systems.

9.1.1 Layout of a Preparative-Scale SFC System

Fig. 9.1 [1] shows a layout of a preparative SFC system. Two different pumps are used for delivering the mobile phase, one for liquid CO_2 and the other for an organic cosolvent, to the system. To ensure pumping fluid in the liquid state, the temperature of CO_2 at the pump inlet line is maintained at colder temperatures, e.g., 4°C. There is no industrial standard set-point for this temperature, which can be different from vendor to vendor. After the two liquids are mixed, the mobile phase moves to the injector. Feed from the injector is then driven to the column, and then to the UV detector and the automated back pressure regulator (ABPR), respectively. Eluted bands are collected as different fractions at the outlet of the ABPR in the fraction collection system. Note that this is a general layout. The injection and the fraction collection mechanisms can be different for different vendors. Also note that for a preparative system all the components are the same as in an analytical setup, except the fraction collection mechanism, which is not a regular feature of analytical systems.

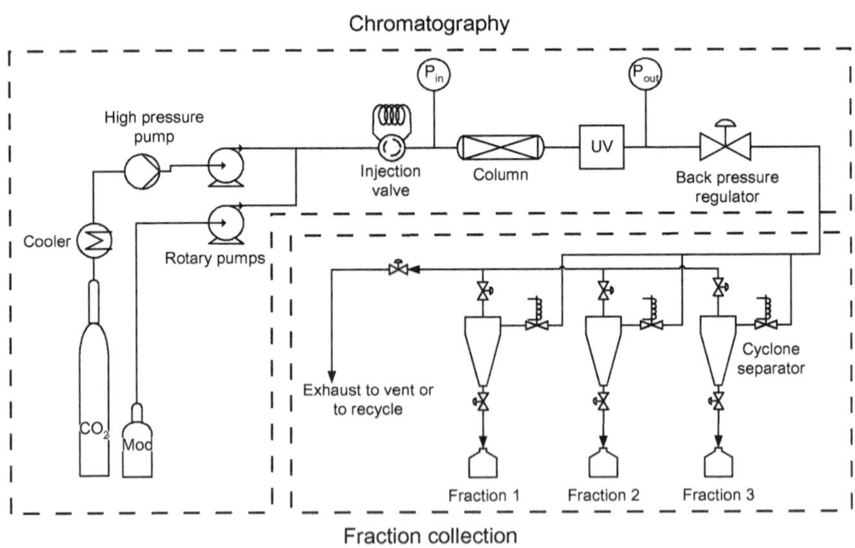

FIGURE 9.1 A preparative-scale SFC system. *Source: Reprint from reference Rajendran A. Design of preparative-supercritical fluid chromatography. J Chromatogr A 2012;1250:227–49.*

9.1.2 Analytical and Preparative-Scale Chromatography

9.1.2.1 Analytical and Preparative-Scale Instruments

Although all the major components of the analytical and preparative instruments have one-to-one correspondences in their respective roles, the volume handled by the systems can be significantly different. For example, standard volumetric flow rates in an analytical SFC system are from 0.5 to 3 mL/min. For preparative systems, on the other hand, flow rates from hundreds to thousands of milliliters per minute are typically employed. From instrument perspective all components are scaled accordingly, although the precision in the preparative systems are usually lower than for analytical systems.

9.1.2.2 Analytical and Preparative Separation

The main difference between analytical and preparative separations, either LC or SFC, is the scale and also the objective of the separation. For preparative scale maximum utilization of the stationary phase is a strong focus, but it is not a priority for analytical-scale separations. For analytical separations the sample is injected in an extremely diluted condition and usually with a low injection volume to ensure adequate resolution for all peaks. Preparative separations, on the other hand, focus on the separation and collection of one or few peaks and with a goal to maximize the utilization of both time and material. This leads to injecting highly concentrated feed solution and/or high injection volumes which overloads the stationary phase, leading to the phenomenon demonstrated in Fig. 9.2. As the analytical feed is dilute, access of sample molecule to the stationary phase surface is not influenced by the other sample molecules (see Fig. 9.2A). For preparative systems, however, the solution is concentrated and access of one solute to the stationary phase is strongly influenced by the other solute molecules. For the example shown in Fig. 9.2B, as the molecule's access to the surface is hindered by other molecules, it moves to a lesser crowded region before being retained. Or, it may displace a retained molecule which then moves further before being retained again. This phenomenon, called "competition effect," leads to a broader band with a sharp concentration front followed by a decreasing concentration tail. Note that, on the time axis the orientation of the band reverses. More on this topic will be discussed in Section 9.3.4.

9.1.3 Preparative-Scale LC and SFC

The main difference between an LC and an SFC system arises from the higher compressibility of the mobile phase in SFC. Preparative LC mobile phases can be compressible too—e.g., pure methanol has an

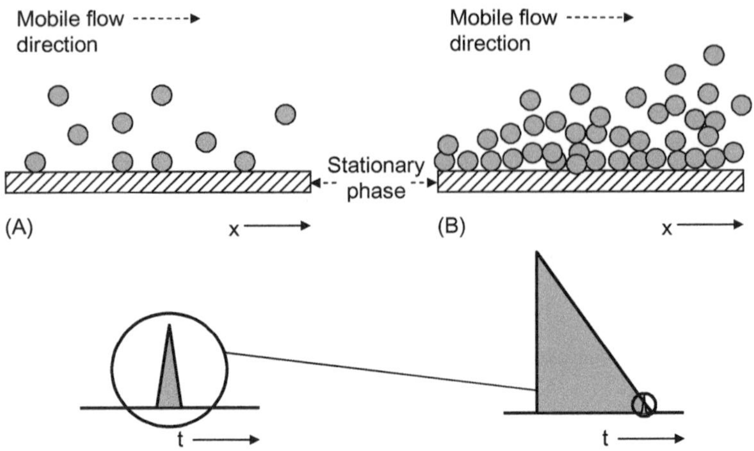

FIGURE 9.2 Schematic representation of solute molecule movement over the stationary phase surface in (A) an analytical system and (B) a preparative system. Diluted solution gives rise to a Gaussian peak, but overloaded solution leads to band broadening and unequal distribution of solute concentration inside the band.

isothermal compressibility of 8.0763×10^{-6} (1/psi) at 40°C and 2000 psi. But compressibility of CO_2-based mobile phases are much higher—e.g., even with 50 mol% of methanol, a CO_2 + methanol mixture has a compressibility of 1.4065×10^{-5} (1/psi) at the same conditions. Compressibility of the mobile phase gives rise to some design challenges that will be discussed in Section 9.3.

9.2 FACTORS INFLUENCING PREPARATIVE SFC PERFORMANCE

In general the elements that directly influence the performance of a preparative SFC system are (1) the solvent delivery system, (2) the feed solubility and injection mechanism, and (3) the separation conditions. All these elements are influential in a general chromatography system as well, but for preparative SFC, the two background factors discussed earlier—the large volume and/or high concentration injection, and compressibility of the mobile phase, influence all these elements and lead to the design and operational challenges typical to preparative SFC. A schematic representation of all the factors that control preparative SFC design and operation are shown in Fig. 9.3. The role played by large injection sizes and compressibility on the four elements are discussed later.

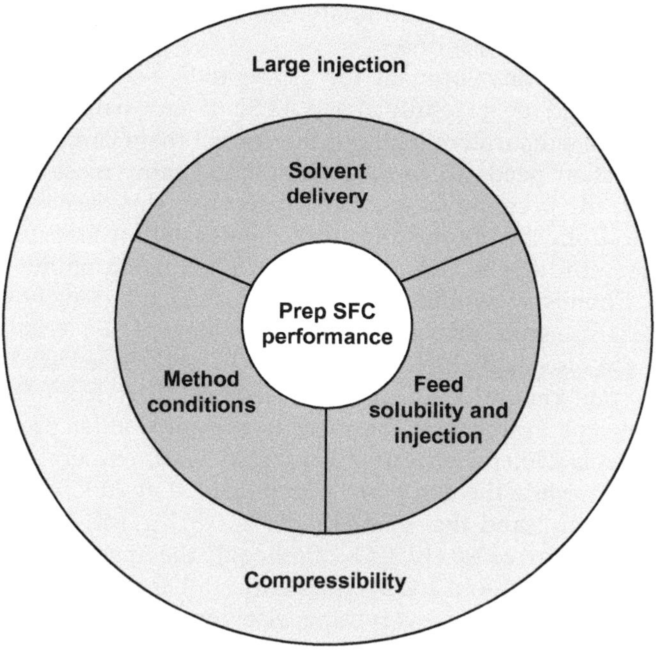

FIGURE 9.3 Factors directly controlling a preparative SFC performance (1) solvent delivery system, (2) feed solubility and injection, and (3) method conditions. The main background factors that influence all these elements are the large injection volume/concentration and the compressibility of the mobile phase.

9.2.1 Solvent Delivery System

The solvent delivery system determines the flow rate as well as the composition of the mobile phase. It can lead to considerable differences in the performance of a preparative SFC system mainly because of the CO_2 compressibility. It is extremely difficult to deliver CO_2 at a certain pressure while maintaining the mass of CO_2 that it is supposed to deliver. That is why most of the CO_2 pumps for preparative scale are controlled by a mass flow meter. The main problem, however, arises from the fact that system design of a preparative system is guided by separation methods designed in an analytical system (more on this topic will be covered in Section 9.3), and that commercially available analytical instruments work with volumetric flow rates and composition. In the world of analytical chromatography flow rate or composition always denotes volumetric. The task of working with a chromatographic instrument that works with mass flow rate and mass composition could make the already misunderstood technology more alienated, which could be the reason why vendors for analytical instruments chose to stay within the volumetric regime.

During method development for a preparative-scale separation this distinction should be carefully noted [2,3]. If an analytical method is transferred to a preparative method, the correct mass composition of the analytical system needs to be known and the same mass composition employed in the preparative system. To conduct this task we first measure or note from the chromatographic data system (CDS) the temperature and pressure of the CO_2 pump head. Then using online resources, e.g., NIST Chemistry WebBook [4], we calculate the density of CO_2 at that temperature and pressure. Density values of the cosolvent e.g., methanol, are also available online from NIST [4]. Based on these density values one can calculate the mass composition from the volume composition. For example, if the measured pump outlet pressure (system pressure) is 4500 psi and the CO_2 pump head temperature is maintained at 13°C, while the cosolvent pump head is at 20°C, the density of CO_2 is 1.013 g/mL and that of methanol is 0.817 g/mL [4]. For a volumetric composition of 90/10 CO_2/methanol, the mass composition at these conditions is 91.8/8.2 CO_2/methanol. But if the CO_2 pump head temperature is −4°C and everything else remains the same, the CO_2 density is 1.07 g/mL and the mass composition becomes 92.2/7.8 CO_2/methanol. Such variation in mass composition may not be important if the target compound is adequately resolved, but could be important if there is a closely eluting impurity, and should be taken care of.

9.2.2 Selection of the Feed Solvent and Injection Mechanism

The SFC injection mechanism has a typical problem—the mismatch between the feed solvent and the mobile phase. Unlike LC, where the feed is typically prepared with the mobile phase solvent, the compressibility of CO_2 prevents the feed solution from being prepared in the mobile phase unless an extraction injection mechanism is employed. The general solution is to dissolve the feed mixture in the cosolvent. In many cases the feed is more soluble in a different solvent, which is not a component of the mobile phase composition. In all cases, the injected volume, just after the switch, creates a solvent mismatch at both sides of the solute plug. This mode of injection, which is typical of LC as well, is a mixed-stream injection. Although this mechanism is routinely employed, it can lead to severe peak distortion, especially for large injection volume [1]. As noted by Rajendran [1], during injection, the feed mixture is present inside a plug of the feed solvent. When injected the compounds of the feed mixture as well as the feed solvent, all migrate along the column, driven by the mobile phase. Depending on the relative retention behavior of the feed mixture and the feed solvent, the retention characteristics can be categorized under three situations [1]

- Case 1: The feed mixture is more strongly retained compared to the feed solvent.
- Case 2: The feed mixture and solvent have similar retention behavior.
- Case 3: The feed mixture is weakly retained compared to the feed solvent.

Note that here, the feed mixture represents the mixture of compounds in the feed that require separation. For those cases where the feed solvent is either more, or less strongly retained, we may assume that it is relative to all the components in the feed. For the case where they are identical, comparison can be made with the feed solvent and the target compound in the feed.

Rajendran [1], calculated the effect of solvent mismatch on the feed band shape using the equilibrium theory of chromatography. Fig. 9.4 demonstrates this effect for two compounds, one more strongly retained than the other, but both are more retained than the feed solvent. For a certain injection volume, the rear part of the solute band travels within the mobile phase while the front part travels, at least for a certain distance, within the band of the feed solvent (See Fig. 9.4). Because the retention of the solute is lower inside the feed solvent, compared to that in the mobile phase, the solute band profiles are broadened due to this tag-along effect. When the injection volume is large enough, it can be noted from Fig. 9.4C that the front of a solute band may not exit the feed solvent band at all. Such situations cause distortion of the elution profile. Instruments using the so-called "modifier-stream" injections, where the feed is injected in the cosolvent stream before it is mixed with the CO_2 stream, can solve this problem to a considerable extent.

9.2.3 Selection of Method Conditions

Selection of the method conditions has the strongest influence on performance of preparative SFC. Both the factors-large injection and compressibility should be addressed to reach more desirable separation conditions. This topic will be addressed in detail covering the entire next section.

9.3 DESIGNING A PREPARATIVE SFC METHOD

Two different approaches for designing a preparative SFC system are described in this section. The first approach, the "rule-based

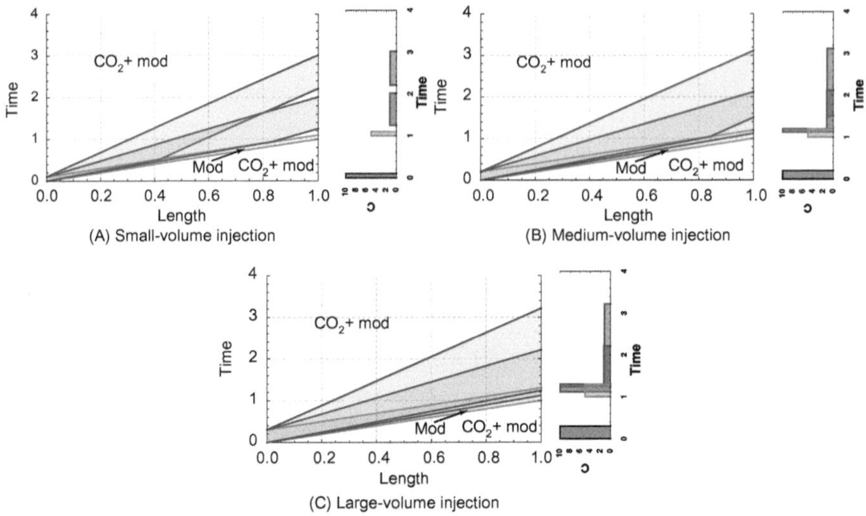

FIGURE 9.4 Demonstration of how the injection mechanism and injection volume influences the band profile in preparative SFC. The plots show the position and extension of a solute band inside the column as a function of time, and also at the column outlet. The movement of the feed solvent band (first-eluting), a less retained compound (second-eluting), and a more retained compound (third-eluting) are shown. The calculated elution profile are also provided at the right hand side. Effect of the size of the injection volume can be noted by comparing (A), (B), and (C). *Source: Reprint from reference Rajendran A. Design of preparative-supercritical fluid chromatography. J Chromatogr A 2012;1250:227–49.*

approach" has a following in the industry. The second approach, the mathematical "model-based approach" is what several academic groups have been working on with promising results over the years. Both approaches are described in this section. But before moving to that, let us clarify the design task for a preparative SFC method.

9.3.1 Design Task for a Preparative SFC Method

The main challenge in the design of a preparative chromatography separation is that the systems are too large for developing and optimizing separation methods directly. This is why preparative method designs are always preceded by method development and optimization steps on an analytical scale, most often using an analytical instrument. That means the main task of preparative SFC method design is either to directly replicate the analytical method to preparative scale, or to design a preparative method based on the analytical results. Both, the rule-based and the model-based approaches, work with these preconditions.

For the model-based approach, analytical results are used to "characterize" the system at different operating conditions, preferably with pure compounds. For the rule-based approach, it is employed for optimizing the method, which is transferred to the preparative system. Note that, both approaches focus on adjusting only the method conditions to achieve preparative-specific separation. The basic criterion is that the stationary phase should be the same in both the analytical and preparative-scale systems. Both approaches, while taking note of the fact that density varies along the system, and so do all physical properties, assume average conditions to address the variation.

9.3.2 Role of Thermodynamic and Transport Properties

Inside a chromatographic system all the fundamental molecular interactions or movements can be broadly classified under (1) thermodynamic and (2) transport phenomena [5]. Thermodynamic phenomena represent the physico-chemical interactions between the solute and the mobile and stationary phases. For a particular mixture of solvents these interactions are controlled by the three state conditions—the composition, the density, and the temperature. Transport phenomena, on the other hand, represent all the factors that control molecular movement inside the chromatographic bed. They are also controlled by the same state conditions as the thermodynamic phenomena, as well as by the geometry of the particles and the chromatographic bed, and the mobile phase flow rate. During the discussion on the rule-based and the model-based approaches in the following sections, these distinctions in physical phenomena on the effects of chromatography, will be frequently referred to.

9.3.3 Rule-Based Approaches

Here the rule-based approaches are based on a diverse set of rules that can be employed to design a preparative SFC method. Most often these rules generate preparative method parameters from analytical method parameters following a few simple steps, e.g., multiplying by a certain factor. Although most of these rules are purely empirical and are developed and used within different laboratories and rarely published, there are some rules, published in the open domain, that have a theoretical basis. This section will cover the theory behind those rules.

9.3.3.1 Scale-Up Rules for Transferring Isocratic Methods

Scale-up rules for preparative SFC can directly follow the rules suggested for scale-up in preparative LC. For the transfer of isocratic methods, where the mobile phase components and all the state conditions remain the same, the scale-up procedure focuses on mitigating the geometric variations on the transport phenomena.

Based on existing practice, Guillarme et al. [6] compiled a set of rules for method-transfer in RPLC. Since scale-up can be construed as a method-transfer from an analytical to a preparative system, the same rules are applicable for preparative method design as well. Instrument vendors often suggest applying these rules for method-transfer and scale-up in LC systems. The rules can be compiled as:

1. Maintain the same stationary phase (chemistry, ligand, pore size, etc.) in the preparative system.
2. Select the same mass or molar composition for the mobile phase.
3. The ratio between the column length (L) and the particle diameter (d_p), L/d_p, of the preparative system should be the same as the analytical system.
4. The reduced linear velocity $\left(ud_p/D_m\right)$ in both the systems should be the same, where u is the linear velocity and D_m is the solute diffusivity in the mobile phase.
5. The ratio between the injection volume and the column void volume in both systems should be the same.

Rules 1 and 2 are the preconditions for any chromatographic method-transfer or scale-up operation. The origin of Rules 3 and 4 is the van Deemter equation in the reduced form, proposed by Giddings. Note Rules 3 and 4 need to be followed only if the same optimum column efficiency in the preparative and analytical systems is to be maintained [7]. Rule 5 is to ensure the same column overloading by volume in the preparative system. The utility of these rules will be discussed further.

From the above rules we can determine some method parameters for a preparative system. The flow rate of the preparative system can be calculated based on maintaining the same reduced linear velocity between the two systems. Assuming that the particle size and the column diameter is preselected,

$$Q_P = Q_A \frac{d_{p,A}}{d_{p,P}} \left(\frac{D_P}{D_A}\right)^2 \tag{9.1}$$

where the suffixes P and A represent the preparative and analytical systems, respectively. Q represents the volumetric flow rates, d_p the particle diameter, and D the column diameter. Note that in practical situations it is probably more prudent to go for the maximum operating flow rate that the system (including the instrument and the column) would allow to operate with. Setting the flow rate to maintain the same optimum column efficiency may not be required for an analytical to preparative-scale transfer because in the latter system peak broadening results mainly from column overloading, not the column efficiency [5].

The column length for the preparative system can be calculated from

$$L_P = L_A \frac{d_{p,P}}{d_{p,A}} \tag{9.2}$$

where L represents the column length.

Injection volumes for the preparative system is generally calculated to match the increase in column volume for the preparative system.

$$V_{inj,P} = V_{inj,A} \frac{L_P}{L_A} \left(\frac{D_P}{D_A}\right)^2 \tag{9.3}$$

where V_{inj} is the injection volume. Note that for preparative SFC additional criteria need to be considered, which are described further under the effect of large injection volumes.

9.3.3.1.1 Effect of Compressibility

To mitigate the effect of compressibility of the SFC mobile phase, we need to add a further rule:

6. The average density of the mobile phase inside the column should be the same.

An elaborate process to calculate the weighted average density inside a column is presented in Ref. [7]. In most situations, however, a simpler approach, maintaining the same average pressure in both systems, will suffice [8]. In this way there is no need to know or calculate the density of the mobile phase. To match the average pressure in the preparative system, the ABPR pressure should be modulated iteratively. The following simple formula is helpful for estimating the new ABPR pressures for the preparative system [7].

$$(P_{ABPR,P})_{i+1} = (P_{ABPR,P})_i + \left[\left(\frac{P_{sys,A} + P_{ABPR,A}}{2}\right) - \left(\frac{(P_{sys,P})_i + (P_{ABPR,P})_i}{2}\right)\right] \tag{9.4}$$

where $(P_{ABPR,P})_{i+1}$ and $(P_{ABPR,P})_i$ are the ABPR pressures for the $(i+1)^{th}$ and i^{th} iteration, and $(P_{Sys,P})_i$ is the system pressure for the ith iteration, all in the preparative system. $P_{sys,A}$ and $P_{ABPR,A}$ are the system and ABPR pressures, respectively, of the analytical system.

9.3.3.1.2 Effect of Large Injection

Although Rule 5 suggests that the injection volume in the preparative system can be directly scaled-up from the analytical injection volume, this is rather a simplistic view. Two points that need to be considered for determining preparative injection volume are listed below:

- Because of the significant effect of solvent mismatch, discussed earlier, injection volume is controlled by (1) the mode of injection-mixed-stream or modifier-stream, and (2) the retention factor of the feed solvent compared to the feed solutes.
- It is always better to have feed solution at the highest concentration whenever possible. Although in preparative SFC this is limited by the possibility of solute precipitation and crystallization inside the system.

9.3.3.2 Scale-Up Rules for Transferring Gradient Methods

For transferring gradient methods from an analytical to preparative system, the following rules compiled by Guillarme for short columns packed with small particles [9] are also applicable. During gradient transfer, two additional parameters, the isocratic hold time and the gradient duration need to be supplied.

1. The isocratic hold in the preparative system should be scaled on the dwell volume of the analytical system.
2. The segment duration in the preparative system should ensure scaling of the gradient swept volume.

Both rules are used to ensure that the analytical gradient in terms of column volume is recreated in the preparative system. The thermodynamic equilibrium inside the columns are controlled by the mobile phase swept volume. Because the mobile phase flow rates for the analytical and the preparative-scale systems are different and so are the column dimensions, the gradient times should be adjusted to maintain the variation of solvent composition in terms of the swept or column volume.

The following equations can be used to determine the preparative SFC gradient hold time and gradient duration.

$$t_{i,P} = \left(t_{i,A} + \frac{V_{d,A}}{Q_A} \right) \frac{L_P}{L_A} \frac{d_{p,P}}{d_{p,A}} - \frac{V_{d,P}}{Q_P} \tag{9.5}$$

$$t_{g,P} = t_{g,A} \frac{L_P}{L_A} \frac{d_{p,P}}{d_{p,A}} \tag{9.6}$$

Although developed for LC, these rules are equally applicable to preparative SFC as well, as long as the average density of the two systems is maintained the same. Note that the pressure, and hence the density, of the mobile phase dynamically varies during a gradient operation. To match the resultant variations in compressibility between the analytical and the preparative systems, ideally, one should take into account the differences between their average density, as a function of time. This, however, is impractical to implement. One way of addressing this issue is to measure the average pressures at the starting and also the end gradient concentrations. The ABPR pressure of the preparative system should be modulated to ensure that the average of these two averages are the same in the preparative system, as it was in the analytical system.

9.3.4 Theories for Modeling Approaches

9.3.4.1 Model-Based Approach—What It Is?

The main purpose of the model-based approach is (1) to harness the current understanding of the physico-chemical behavior of solute molecules inside a chromatographic column, and (2) to exploit the significant computational power available in a desktop computer, to develop a predictive tool. This virtual or in silico tool, if robust enough, can replace the real experimental setup by predicting the outcome of experimental conditions.

The models are a set of mathematical equations that represent the physico-chemical behavior of solute movement inside the chromatographic bed. For preparative chromatography these equations are most often partial differential equations (PDE). When solved they can estimate the band width and concentration distribution of the bands inside and also at the exit of the column. From these results, one can plot simulated chromatograms or track the entire movement of all the bands inside the column. In addition, the results can be used to analyze the projected performance of a method in terms of recovery (yield), productivity, target purity, etc. In silico experiments can lead to rapid evaluation of the problem at hand and suggest an optimized solution. Additionally, such detailed results of the band movement inside the column can help to generate useful insights into a process. This is not possible, or could be very elaborate, to generate through real experiments.

The model-based approach follows the major steps shown in Fig. 9.5. The first step is to select a mathematical model that can reliably predict the physical intricacies of the problem at hand, and also can be solved, either analytically or numerically. The model then needs to be "characterized," which means we need to estimate the values of some "characteristic parameters" of the model from the data generated by a set of

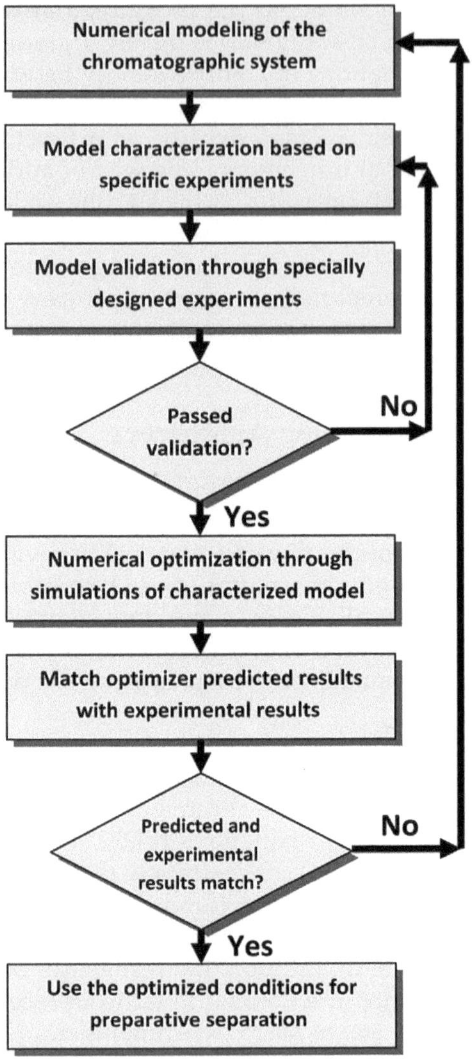

FIGURE 9.5 Flowchart demonstrating the major steps taken in a model-based approach. *Source: Reprint from reference Tarafder A, Hudalla C, Iraneta P, Fountain KJ. A scaling rule in supercritical fluid chromatography. I. Theory for isocratic systems. J Chromatogr A 2014;1362:278–93*

predesigned experiments, performed with a real experimental setup. This is required because the models are not detailed enough to capture the molecular dynamic behavior inside the column. Models with that details are still very compute intensive and are not usually utilized for design purposes. As a compromise we try to estimate the characteristic

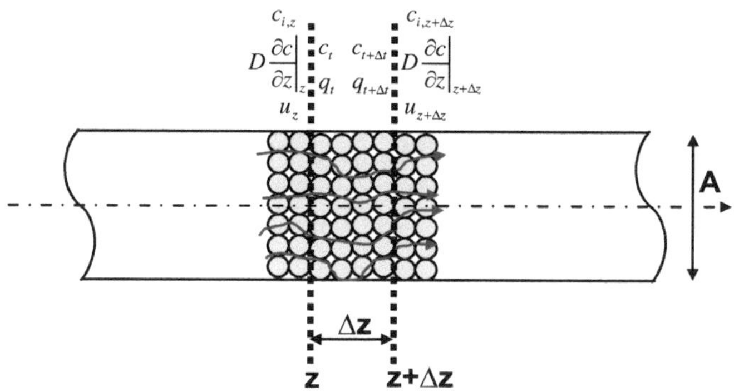

FIGURE 9.6 Schematic diagram of solute movement through an infinitely thin slice of a chromatographic column. The mobile phase along with solute bands enter the slice from the left-hand side and leave from the right hand side. Difference between the rates of mass flowing into the mass flowing out lead to mass accumulation or depletion from that slice.

chromatographic behavior of a solute through a function with a few coefficients. The values for these coefficients are estimated with the experimental results, typically carried out in an analytical system. Once this "characterization" step is done, the equations can be used as a simulator of that experimental setup.

The next step is to detect the optimum operating conditions using this virtual setup. Most often this is done through an optimization program. The optimization program modulates the main operating variables until a predefined optimum condition to run the system is reached. Note that although the model-based approach works with experimental data from analytical systems, it is not limited to designing preparative conditions that reproduce the analytical performance. The model-based approach develops a method, which can be quite different from the analytical method, to meet the objectives specific to preparative separation. Given the level of physical understanding on chromatographic behavior and the availability of computational power, the model-based approach can be very effective. But for wider acceptance it looks that we need more robust and self-supporting programs which ideally need no user intervention.

9.3.4.2 *How the Model Is Developed?*

The model is developed by solving the mass balance equation over a thin cross-section inside the column. Fig. 9.6 represents a chromatographic bed with a cross-sectional area of A. Through the column the mobile phase and the solute bands are moving from the entrance to the exit of the column. To conduct the mass balance, we imagine a slice of an infinitesimally small length (Δz) inside the column at any arbitrarily chosen location. At any moment (say, t) the total mass of a solute entering the infinitesimal volume subtracted from the total mass exiting the volume, will be the net mass accumulation (or depletion if the rate of mass entering is less than the rate of mass exiting the column). Detailed derivation of this equation is available elsewhere [5] and will not be discussed further. The mass balance equation, specific for SFC design is [10]:

$$\frac{\partial c_i}{\partial t} + \frac{(1-\varepsilon)}{\varepsilon}\frac{\partial q_i}{\partial t} + \frac{\partial(c_i u(\rho))}{\partial x} = \frac{\partial}{\partial x}\left(D_{a,i}(\rho)\frac{\partial c_i}{\partial x}\right) \tag{9.7}$$

where

c_i = component i concentration in mobile phase, (g/mL)
q_i = component i concentration on stationary phase, (g/mL)
t = time coordinate (min)
u = interstitial velocity of the mobile phase, (cm/min)
x = space coordinate (cm)
i = component index, $i \rightarrow 1$, *number of components*
ε = void fraction of the column, (-)
ρ = density of the mobile phase, (g/mL)
$D_{a,i}$ = apparent diffusion coefficient of component i, (cm^2/min).

The equation indicates that the mobile phase velocity and the apparent diffusion coefficient are not constant along the column, rather they vary as a function of the density. If the density profile along the column can be estimated as a function of x, u can be expressed as $u(x) = G/\rho(x)$, where $\rho(x)$ is the density profile along the column and G is the mass flow rate. Similarly $D_{a,i}$ could also be expressed as a function of x.

Although such detailed consideration makes the equation more accurate for most conditions, the pressure as well as the density profiles along the column are approximately linear. And a linear average of the end conditions can approximate the overall chromatographic conditions inside the column. The ranges of validity of this simple assumption covers a wide area of the operating conditions for modern SFC [8], and hence can be applied. This assumption transforms Eq. 9.7 to a simpler form:

$$\frac{\partial c_i}{\partial t} + \frac{(1-\varepsilon)}{\varepsilon}\frac{\partial q_i}{\partial t} + u(\overline{\rho})\frac{\partial c_i}{\partial x} = D_{a,i}(\overline{\rho})\frac{\partial c_i^2}{\partial x^2} \tag{9.8}$$

where,

$\overline{\rho}$ = average density of the mobile phase in the column, (g/mL).

Note that this model assumes infinitely fast mass-transfer inside the pores of the particle, which is approximately valid for SFC because of the very high diffusivity of the mobile phase. Also note that this PDE cannot be solved analytically, so numerical solution methods need to be applied, which will be discussed later in this section.

9.3.4.3 Isotherms—The Relationship Between Bulk and Stationary Phase Concentrations

In Eq. 9.8, both the variables c_i and q_i are functions of both space and time. So solution to Eq. 9.8 can generate concentration data for solute i at any point of time and at any location inside the column.

Now to solve Eq. 9.8, first we need to estimate or calculate the values of the characteristic parameters—ε, $u(\overline{\rho})$, and $D_{a,i}(\overline{\rho})$. These are called so because they represent values that characterizes the system to a definite set of column and mobile phase properties. There are, however, still two sets of variables, c_i and q_i, to solve. This is done by adding another set of equations that relates c_i (bulk concentration) with q_i (solute concentration on the stationary phase). These equations are called equilibrium isotherms because it assumes .. in equilibrium with c_i, under isothermal conditions. A general expression of this equation is:

$$q_i = f(\overline{c}) \tag{9.9}$$

Here $\overline{c}$ represents the array of concentrations of all the solutes in the solution. As described in the introduction, this dependence of the retention of a solute molecule on the concentrations of all the other solutes, is typical to preparative chromatography. Based on literature results, most of the retention behavior in preparative SFC can be identified as a Langmuir or bi-Langmuir isotherm. For example, Lochmüller and Mink [11] measured isotherms for dry ethyl acetate on Partisil-10 silica, with CO_2 at 60°C and pressures between 2000 and 4000 psi, at a flow rate of 3 mL/min. The isotherms followed Langmuir behavior. Jha and Madras [12] measured the adsorption of naphthalene, biphenyl, 2,6-dimethylnaphthalene, anthracene, hexachlorobenzene, pentachlorophenol, salicyclic acid, and DDT on soil, ODS (C18-bonded silica), activated charcoal, and a zeolite over a range of pressures and temperatures in the supercritical region. They found that all the data fitted well to a Langmuir model. In recent times Wenda et al. [13] used the Langmuir isotherm to characterize the adsorption of flurbiprofen on Chiralpak AD-H. Enmark et al. [14] used a bi-Langmuir model to fit the adsorption behavior of antipyrine on a Kromasil silica column. Note that in preparative LC, especially for chiral separations, different and more complicated isotherm behaviors are not uncommon [15]. Most likely in preparative SFC, mass overloading is not sufficiently high that it can lead to such complicated retention behavior, as observed in some LC separations. In this chapter we would focus on describing only the Langmuir and the bi-Langmuir isotherms because of their relevance.

9.3.4.3.1 Langmuir Isotherm

The Langmuir isotherm represents a classical view of column overloading. It assumes that the retained molecules form a monolayer on all the available sites of the stationary phase. Any molecule from the bulk either has to replace a retained molecule to get retained or has to move on to the next available site [5]. This behavior can be represented by the following equation:

$$q(c) = \frac{q_s K c}{1 + K c} \tag{9.10}$$

where q is the solute concentration on the stationary phase, and q_s the same, when the surface is fully saturated. K is the equilibrium constant of the desorption versus the adsorption rate of the solute molecules between the mobile and the stationary phases. The solute concentration on the stationary phase continuously increases with its concentration in the bulk, until the saturation capacity is reached (see Fig. 9.7A). Note that sometimes the factor $q_s K$ is represented as H, where H is the Henry's constant. H also represents the equilibrium relationship between the mobile and stationary phase at infinite dilution as $q = Hc$. Also note that the retention factor k is related to H through the relation $k = HF$ where $F = \frac{1-\varepsilon}{\varepsilon}$ is the phase ratio. Compared to k, H does not have any term related to the column property, e.g., ε. So H is a more thermodynamically consistent term. In other words, under the same physical conditions the value of k can be different for two columns having different ε. H, on the other hand, should not change.

All the isotherm parameters, H, K, or q_s are dependent on the state conditions, T, ρ, and composition. For neat CO_2 Rajendran et al. [16] described Henry's constant as a function of ρ at a constant temperature as:

$$H_i = H_i^0 \left(\frac{\rho^0}{\rho}\right)_i^b \tag{9.11}$$

For solvents with CO_2 and a cosolvent, Henry's constant can be described as [16]:

$$H_i = \frac{1}{a_i c_m + d_i} \left(\frac{\rho^0}{\rho}\right)_i^b \tag{9.12}$$

9.3.4.3.2 bi-Langmuir Isotherm

The bi-Langmuir isotherm is an extension of the Langmuir isotherm. Due to different factors, the stationary phase surface available for separations may not be homogeneous for all the solutes. For example, with chiral stationary phases (CSP) which have chiral ligands (e.g., Pirkle phases) of a moderate density, a portion of the CSP surface could

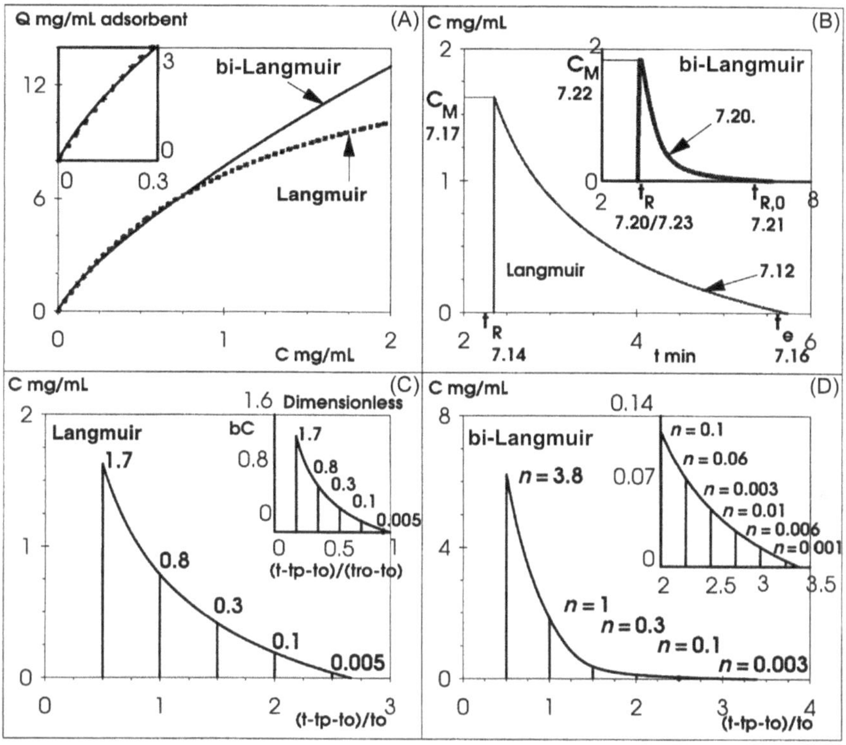

FIGURE 9.7 Different aspects of the Langmuir and bi-Langmuir isotherm. Figure (A) depicts the variation of solute concentration on the stationary phase (q) versus that on the mobile phase (c). Figure (B) depicts the elution profiles or chromatograms of solutes that follow Langmuir and bi-Langmuir (inset) isotherms, respectively. Figures (C and D) depicts the same chromatograms but here the time axis is made dimensionless. *Source: Reprint from reference Guiochon G, Felinger A, Katti AM, Shirazi DG. Fundamentals of preparative and nonlinear chromatography, 2nd ed. Amsterdam, The Netherlands: Elsevier; 2006.*

be covered with the chiral ligands, which results in enantioselective interactions. In many instances, however, the density of the nonselective sites could be significant enough to offer selective retention to one enantiomer over the other [5]. With such a surface that has two different kinds of retention sites, which can behave independently and both following the basic retention behavior of Langmuir model, the bi-Langmuir isotherm can be applied [5]. This behavior can be represented by the following equation:

$$q = \frac{q_{s,a}K_a c}{1 + K_a c} + \frac{q_{s,b}K_b c}{1 + K_b c} \tag{9.13}$$

where the subscripts a and b represent the two types of retention sites. The dependence of the bi-Langmuir isotherm parameters on density and cosolvent concentration should be similar to that for the Langmuir isotherm parameters.

9.3.4.3.3 Multicomponent Isotherms

The two isotherms described previously are single-component isotherms, which means they represent the equilibrium relationship of pure solute only. During preparative separations, there are multiple components and, as mentioned earlier, they influence the retention behavior of each other, especially at higher concentrations. So while modeling a multicomponent retention behavior we have to take into account this effect. It would be tedious to measure and formulate multicomponent isotherms based on experimental results. An easier method is to employ the ideal adsorbed solution (IAS) theory [17] to formulate multicomponent isotherms based on the parameters of single-component isotherms. For example, based on the IAS theory, a binary isotherm, where there are two solutes both following Langmuir isotherms when they are alone, can be written as:

$$q_i = \frac{q_{s,i} K_i c_i}{1 + K_1 c_1 + K_2 c_2} \tag{9.14}$$

where $i = 1,2$ and the subscripts 1 and 2 represent the two components. Note that the isotherm parameters are also characteristic parameters because their values are specific to a set of solute and separations conditions.

9.3.4.4 Solving the Model Equations

The model equation described earlier (see Eq. 9.8), along with suitable isotherm equation(s) need to be solved numerically. There are different types of numerical methods, e.g., finite difference (FD), orthogonal collocation (OC), etc. that have been implemented for this purpose [5]. Here a short description of the finite-difference method is provided.

To solve the PDE using FD we need to discretize the space and the time domain. Reliable and robust solvers to solve initial value problems (IVP) like this are available commercially. For example, the DIVPAG routine from IMSL (International Mathematics and Statistics Library) was used by Wenda et al. [13] to solve the initial value problems for SFC. DIVPAG solves IVPs for ordinary differential equations (ODE) using either Adams-Moulton's or Gear's *backward differentiation formula* (BDF) method.

9.3.4.5 Steps to Characterize the Model–Creating the Simulator

The numerical solvers described above can solve the variables c_i and q_i from the main equation (see Eq. 9.8) and the related isotherm equations, provided, it knows the values of the characteristic parameters, i.e., the bed void fraction (ε), the interstitial velocity (u), the apparent diffusion coefficient ($D_{a,i}$), the saturation capacities of the solutes ($q_{s,i}$), and the equilibrium constants (K_i). The process for estimating the numerical values of these parameters is called model "characterization" because with these values, which are specific to a set of experimental conditions and solute molecules, the equation can simulate the chromatographic behavior specific to those conditions and solutes only. In the rest of this section, methods to estimate these characteristic parameters are described.

9.3.4.5.1 Estimating Bed Void Fraction, ε

Accurate information for the column void volume is necessary to correctly estimate the isotherm parameters from the experimental results. As there is no absolute scale, the retention behavior of the sample mixture is measured and estimated relative to the retention of a nonretained compound. Generally disturbance peaks are used to determine the elution time of an nonretained compound t_0, but Vajda and Guiochon [18] suggested the use of nitrous oxide as the marker when the mobile phase is either pure CO_2 or has a low percent of the cosolvent. Detection of N_2O can be performed at the wavelength range of 195–210 nm. Using the method of peak disturbance to determine t_0 with a compressible mobile phase could be erroneous [18].

9.3.4.5.2 Estimating Interstitial Velocity, $u(\overline{\rho})$

Interstitial velocity through an SFC column is not constant, but varies along the length. This is because the density of the mobile phase varies due to the pressure drop along the column. One can estimate the interstitial velocity profile as a function of density $u(\rho)$, or a more practical way is to measure the velocity with the average density of the column. A convenient way to measure the average density is to measure the pressures at the inlet and the outlet of the column. In most conditions the pressure variation along the column can be approximated as linear—so the linear average of the end pressures can provide the average pressure inside the column [7]. The second assumption is that the conditions inside the column are isothermal. This is approximately true unless the system is operated at conditions of high compressibility [19]. From this pressure and temperature one can measure the density following a reliable Equation of State (EOS). For MeOH/CO_2 mixture NIST published [4] data with the Kunz and Wagner model [20] for calculating mixture densities. Data for other mixtures is not available but approximate values can be estimated employing other EOS.

The second step in the process is to calculate the mass flow rate of the system. Mass flow rate can be assumed to remain constant throughout the system and thus can be measured at any suitable location. The most convenient but expensive way is to install accurate mass flow meters (e.g., Bronkhorst mini CORI-FLOW from Bronkhorst High-Tech B.V., Ruurlo, Netherlands) online and note the data. In their absence, one can measure the temperature and pressure at the pump head of the CO_2 pump. Most often they are supplied by the instrument data system and the value should be close enough to the actual value. Note that although pressure transducer data is more accurate, temperature data may underestimate the temperature. If the pump heat-transfer system is not robust, the actual CO_2 temperature would be higher than that shown by the CDS. From the measured or estimated mass flow rate, and the estimated average density inside the column, one can calculate the average volumetric flow rate as $\overline{Q} = G/\overline{\rho}$, where $\overline{Q}$ is the average volumetric flow rate, G the mass flow rate, and $\overline{\rho}$ the average density inside the column. The average interstitial velocity can be calculated as $u(\overline{\rho}) = \overline{Q}/A\varepsilon$, where A is the cross-sectional area.

9.3.4.5.3 Estimating Apparent Diffusion Coefficient, $D_{a,i}(\overline{\rho})$

The apparent diffusion coefficient at the column average density, $D_{a,i}(\overline{\rho})$, can be estimated from the height equivalent to the theoretical plate (HETP) or plate height data. There are various methods to calculate column HETP from injecting diluted samples in the experimental column. A method that can be easily automated is the method of moments. The second moment calculated from the experimental elution profile of an solute can be expressed as:

$$\sigma_i^2 = \frac{\sum_{j=1}^{n} c_{i,j}(t_{i,j} - \mu_i)^2 \Delta t_{i,j}}{\sum_{j=1}^{n} c_{i,j} \Delta t_{i,j}} \tag{9.15}$$

where σ_i^2 is the peak variance, $t_{i,j}$ and $c_{i,j}$ are the time and the signal of the elution profile, μ_i is the first moment, i.e., the retention time of the peak, and $\Delta t_{i,j}$ is the time interval of the profile. Note that here it is assumed that different solutes in the mixture will have different apparent diffusion coefficients and hence σ_i^2 is measured for individual peaks. In practice the middle peak can be chosen to represent the $D_{a,i}$ for all the solutes.

From σ_i^2, the HETP at that average density can be calculated as:

$$HETP_i(\overline{\rho}) = \frac{\sigma_i^2}{\mu_i^2} L \tag{9.16}$$

where, L is the column length.

The apparent diffusion coefficient $D_{a,i}(\overline{\rho})$ can be calculated from $HETP_i(\overline{\rho})$ as:

$$D_{a,i}(\overline{\rho}) = \frac{HETP_i(\overline{\rho})}{2.0} u(\overline{\rho}) \quad (9.17)$$

9.3.4.5.4 Estimating the Isotherm and the Isotherm Parameters

The equilibrium isotherm and the isotherm parameters can be estimated through various methods. Note that one can measure the concentration distribution between the bulk and the retained concentrations by static methods or by chromatographic (dynamic) methods. Static methods are time intensive and require a separate experimental setup and is excluded from further discussion here. Among the dynamic methods, the Frontal Analysis (FA), Elution by Characteristic Points (ECP), Retention Time Method (RTM), and Inverse Method (IM) had been reported in the preparative SFC literature [14,21,22]. FA is one of the most accurate but on the other hand the most experiment-driven technique. IM, at the other end of the spectrum, needs much lesser experimental efforts and heavily depends on computational power. In this chapter we will provide a description of the FA and the IM methods only.

Frontal Analysis (FA) In FA, to measure the isotherm of a compound, pure solution of the compound is continuously passed through the column until the column is completely saturated and there is a breakthrough of the inlet solute concentration at the column outlet. The time of the front for the breakthrough profile is measured. This operation is carried out with different concentrations, from low to high, to determine the isotherm over a broad range of concentrations. Using FA the stationary phase concentrations of the solute can be measured as a function of the feed concentrations. The method can accurately and quantitatively determine the isotherms, but also requires considerable amount of purified components. In the situations where one has considerable time and enough purified components of the mixture at the beginning, the FA method should be applied to determine the single-component isotherms.

Fig. 9.8 being schematically demonstrates an experimental setup for the determination of equilibrium isotherms in SFC though FA method. The setup has two pumps, one for delivering liquid carbon dioxide (P1) and the other (P2) for delivering the feed sample (S) dissolved in the cosolvent. The two streams from pump outlets are mixed (M) before being delivered to the column (C). Because the feed is continuously delivered to the column mixed with CO_2, no separate injection mechanism is required here. Outlet from the column is monitored in a detector (D) before being led to the waste through the ABPR. Note that

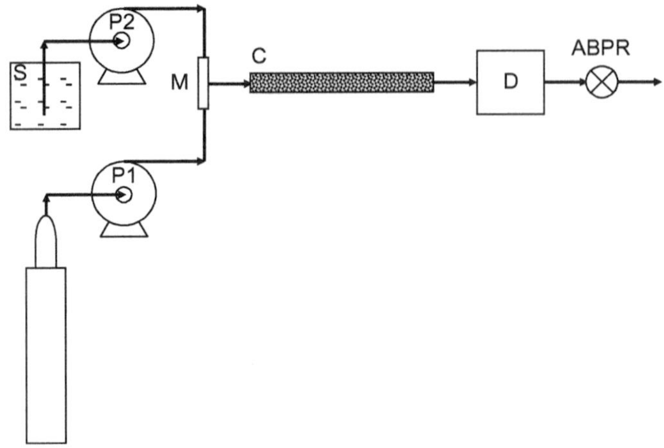

FIGURE 9.8 Experimental setup to determine equilibrium isotherm of a compound through Frontal Analysis (FA) method in SFC. *P1,P2*, pump; *S*, sample; *M*, mixer; *C*, column; *D*, detector; *ABPR*, automated back pressure regulator.

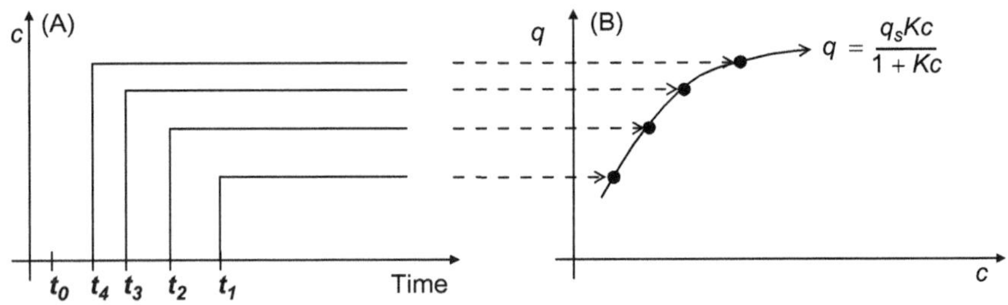

FIGURE 9.9 (A) Schematic representation of breakthrough profiles with increasing feed concentration. The times indicated along the time axis represent the column void volume (t_0) and breakthrough times corresponding to different feed concentrations. (B) Schematic plot of stationary phase concentration versus feed concentration, derived from the breakthrough data. Also shown on Figure (B) a curve joining the experimentally obtained points, which can be fitted with an equation representing Langmuir isotherm.

it is best to carry out the FA procedure with columns of narrow bore, which should offer considerable savings of the amount of valuable pure compounds needed for the breakthrough analysis and also minimize solvent waste. FA is a labor intensive process because it requires a series of solutions of known concentrations. An alternative way is to use the chromatographic setup, with a computer-controlled gradient delivery system, which can accurately deliver step gradients of known concentrations.

The process to estimate the solute concentration inside the column is from the retention times of the breakthrough profiles. Fig. 9.9A schematically demonstrates breakthrough profiles of a pure compound at increasing feed concentrations. For equilibrium behavior following Langmuir isotherm, the front of the breakthrough profile elutes earlier with increasing feed concentration. The breakthrough elution times corresponding to each concentration ($t_1, t_2, t_3 \ldots$) are noted for the determination of the stationary phase concentrations. Note although in Fig. 9.9A the breakthrough profiles are shown as sharp fronts, in reality they are dispersed. So the retention times are measured at the half-height of the plateau. Correction for the extracolumn hold-up time should also be included for correct measurement.

$$q_1 = \frac{V_1 - V_0 - V_{sys}}{V_s} \tag{9.18}$$

Stationary phase concentrations are measured from the breakthrough times following Eq. 9.18. Here V_1 represent the volume of mobile phase that has passed through the column until time t_1. V_0 is the void volume of the column calculated from t_0, V_{sys} the system volume and V_s the volume of the stationary phase. V_1, V_0, and V_{sys} can be calculated from t_1, t_0, and t_{sys}, respectively by multiplying these terms with the average volumetric flow rate ($\overline{Q}$) through the column. Once the values of q are calculated for all the feed concentrations, they are plotted as shown in Fig. 9.9B. To determine the isotherm, these experimental points are fitted with an appropriate equation, which represents the equilibrium isotherm. Note that the equilibrium isotherm can be measured at different percent of the cosolvent with CO_2 by increasing the concentration of the solute in the cosolvent.

A typical problem for FA in SFC arises from the lack of appropriate mechanism to inject the solute feed over an extended period. In LC the feed is directly mixed in the mobile phase solution and pumped to the column. So for an LC FA setup, the components *P*1 and *M* of Fig. 9.8 are not required. Outlet of *P*2 can be directly fed to the column. In SFC the compound mixed in the cosolvent has to be passed through a mixer (M) where it mixes with the CO_2 to prepare the final feed solution. Depending on instruments these mixers can have considerable volume and lead to a dispersed front, resulting in mistakes in the calculation. For example, Kamarei et al. reported a set of isotherm of Naproxen in Ref. [21]. A later publication [22] with a setup having a smaller mixing volume demonstrates the effect of dispersion in the breakthrough fronts. Fig. 9.10 illustrates the difference between the breakthrough profiles obtained through a setup with a small dynamic mixer, and a setup having a large static mixer.

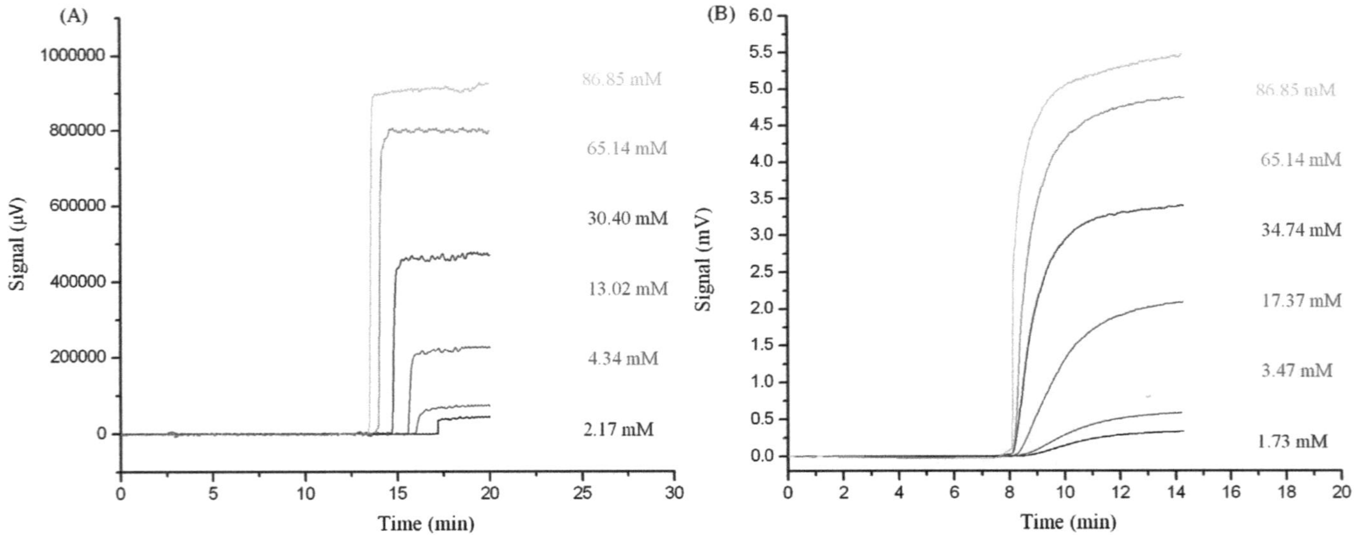

FIGURE 9.10 Results of Frontal Analysis (FA) with two instrumental setup. (A) employing a smaller mixing volume (250 μL), and (B) standard equipment setup having a larger mixing volume (4 mL). *Source: Reprint from reference Kamarei F, Gritti F, Guiochon G, Burchell J. Accurate measurements of frontal analysis for the determination of adsorption isotherms in supercritical fluid chromatography. J Chromatogr A 2014;1329:71–77.*

The Inverse Method FA is accurate but it requires a significant amount of time and purified compound to carry out the measurement. From an industrial perspective, the availability of such resources is not probable. The IM is suitable for such situations because it requires minimum amount of solute and solvent for experimental studies and it is a fast method to obtain isotherm parameters directly from overloaded chromatograms. IM determines the characteristic parameters ($D_{a,i}$, $q_{s,i}$, and K_i) in an iterative way by solving the PDE (see Eq. 9.8), starting with an initial estimation of these parameters and then trying to match the experimental and the simulated profiles. In that way IM solves the problem in an "inverse" way if the direct problem solution is to calculate the band profiles from known isotherm model and parameters.

In the IM the sample mixture is injected at increasing overloading and their elution profiles are recorded. Then using the simulation model and the initial estimations of the parameters described at the beginning of this section the first simulated elution profile is generated. The simulated profiles are then matched with the experimental profiles, from where their difference is calculated. An optimization routine is then used to minimize this difference by supplying different values of these parameters.

The main difficulty with the IM is its dependence on the computational part. While FA is easier to understand and implement in a standard separation laboratory, for IM the user may need fair understanding of the simulation model and the minimization routines, mainly their shortcomings, to obtain reliable results with this method. For example, the user needs to have an idea on the plausibility of the isotherm and plausible values regarding the isotherm parameters. First of all, in IM, the user has to select an isotherm a priori. Although there are guidelines to estimate the isotherm, the choice may not be trivial. Then, during the fitting process, as multiple parameters can lead to a similar effect on the shape of the band, if the optimization process is not robust enough it may lead to estimating improbable values for all the parameters. For example, different combinations of the transport parameter ($D_{a,i}$) and the isotherm parameters ($q_{s,i}$ and K_i) may enable the optimizer to lead to similar error reduction while matching the simulated and experimental profiles. This is called the compensating effect, and during an optimization process it can lead the optimizer to arrive at solutions which, although numerically correct, have no physical relevance.

A way to avoid the compensating effect during optimization is to have an initial estimate of (1) the type of isotherm and (2) the characteristic parameters as closely as possible. Methods to estimate the values of bed void fraction (ε), interstitial velocity (u), and apparent diffusion coefficient ($D_{a,i}$) have been discussed earlier. The same procedure to calculate these factors can be used in the IM study. The user can use the values of $D_{a,i}$; in that case the numbers of variables for the optimizer reduces only to the isotherm parameters, H, K, or q. Accurate values of H can be obtained from the analytical experiments from the retention factor values:

$$H = \left(\frac{\varepsilon}{1-\varepsilon}\right)k \tag{9.19}$$

That leaves only the saturation capacities ($q_{s,i}$) or the equilibrium constants (K_i) to be estimated. This procedure can significantly reduce the compensating effect in parameter estimation.

Regarding determination of equilibrium isotherms, some guidance is available. For example, Fig. 9.7 demonstrates the important difference between the Langmuir and the bi-Langmuir isotherms. When the solute concentration on the stationary phase is plotted against that in the mobile phase (see Fig. 9.7A), the Langmuir isotherm tends to reach saturation more quickly. Note from the figure that insignificant variation of q, vis-a-vis linear increment of c, indicates that the stationary phase is reaching its saturation limit. The chromatogram of the bi-Langmuir isotherm (see Fig. 9.7B) shows this effect with the elongated tail of the band. A quantitative way to determine whether a solute is following Langmuir or bi-Langmuir behavior is to plot the Scatchard plot, which is to plot the values of q/c versus q [5]. If the plot is linear, the solute follows Langmuir isotherm. If the plot is nonlinear it follows bi-Langmuir or another type of isotherm.

The IM also depends on the robustness of the optimization routine for its success. There are many optimization routines that can be used [5]. These routines can be broadly divided into two categories: (1) deterministic methods, e.g., the Marquardt method, the Conjugate gradient algorithm, Nelder-Mead downhill Simplex method, etc., and (2) population-based stochastic techniques, e.g., genetic algorithm, swarm optimization, ant colony optimization, etc. The population-based optimizers are more compute intensive but they are better for detecting the actual solution. The deterministic methods start from a single point estimation and are more prone to error if the initial estimates are not close enough. For further information and implementation of the IM the reader can consult Refs. [1,14,5].

9.4 CONCLUSION

The theory of preparative SFC is explained from two different approaches—the rule-based and the model-based approach. The rule-based approach works with simple algebraic rules to transfer an analytical method directly to preparative scale. The model-based approach, on the other hand, works with more rigorous mathematical equations, mostly partial differential equations, to represent the molecular behavior inside the column. Using the model-based approach one can have an in silico experimental setup which can be employed to determine the optimum operating conditions for a preparative separation. Both approaches discussed here are directly derived from existing preparative LC design approaches. The only factor that differentiates SFC from LC, the effect of compressibility, has been taken into account by expressing the LC model parameters as functions of averaged density.

References

[1] Rajendran A. Design of preparative-supercritical fluid chromatography. J Chromatogr A 2012;1250:227–49.
[2] Enmark M, Asberg D, Leek H, Ohlen K, Klarqvist M, Samuelsson J, et al. Evaluation of scale-up from analytical to preparative supercriticalfluid chromatography. J Chromatogr A 2015;1425:280–6.
[3] Tarafder A, Guiochon G. Accurate measurements of experimental parameters in supercritical fluid chromatography. I. Extent of variations of the mass and volumetric flow rates. J Chromatogr A 2013;1285:148–58.
[4] Lemmon EW, McLinden MO, Friend DG. Thermophysical properties of fluid systems. NIST Chemistry WebBook, NIST Standard Reference 39.
[5] Guiochon G, Felinger A, Katti AM, Shirazi DG. Fundamentals of preparative and nonlinear chromatography. 2nd ed. Amsterdam, The Netherlands: Elsevier; 2006.
[6] Guillarme D, Nguyen DT-T, Rudaz S, Veuthey J-L. Method transfer for fast liquid chromatography in pharmaceutical analysis: Application to short columns packed with small particle. Part I: Isocratic separation. Eur J Pharm Biopharm 2007;66: 475–82.
[7] Tarafder A, Hudalla C, Iraneta P, Fountain KJ. A scaling rule in supercritical fluid chromatography. I. Theory for isocratic systems. J Chromatogr A 2014;1362:278–93.
[8] Tarafder A, Hill J. Scaling rule in SFC. II. A practical rule for isocratic systems. J Chromatogr A 2016; submitted.
[9] Guillarme D, Nguyen DT-T, Rudaz S, Veuthey J-L. Method transfer for fast liquid chromatography in pharmaceutical analysis: application to short columns packed with small particle. Part II: gradient experiments. Eur J Pharm Biopharm 2007;66: 475–82.
[10] Rajendran A, Kräuchi O, Mazzotti M, Morbidelli M. Effect of pressure drop on solute retention and column efficiency in supercritical fluid chromatography. J Chromatogr A 2005;1092:149–60.
[11] Lochmüller CH, Mink LP. Adsorption isotherms of ethyl acetate modifier on silica from supercritical carbon dioxide. J Chromatogr 1987;409:55–60.
[12] Jha SK, Madras G. Modeling of adsorption equilibria in supercritical fluids. J Supercrit Fluids 2004;32:161–6.
[13] Wenda C, Rajendran A. Enantioseparation of flurbiprofen on amylose-derived chiral stationary phase by supercritical fluid chromatography. J Chromatogr A 2009;1216: 8750–8.
[14] Enmark M, Forssen P, Samuelsson J, Fornstedt T. Determination of adsorption isotherms in supercritical fluid chromatography. J Chromatogr A 2013;1312:124–33.

[15] Tarafder A, Mazzotti M. A method for deriving explicit binary isotherms obeying the ideal adsorbed solution theory. Chem Eng Technol 2012;35:102–8.
[16] Rajendran A, Mazzotti M, Morbidelli M. Enantioseparation of 1-phenyl-1-propanol on chiralcel od by supercritical fluid chromatography I. Linear isotherm. J Chromatogr A 2005;1076:183–8.
[17] Myers AL, Prausnitz JM. Thermodynamics of mixed-gas adsorption. AIChE J 1965;11:121.
[18] Vajda P, Guiochon G. Determination of the column hold-up volume in supercritical fluid chromatography using nitrous-oxide. J Chromatogr A 2013;1309:96–100.
[19] Tarafder A, Iraneta P, Guiochon G, Kaczmarski K, Poe DP. Estimations of temperature deviations in chromatographic columns using isenthalpic plots. I. Theory. J Chromatogr A 2014;1366:126–35.
[20] Kunz O, Klimeck R, Wagner W, Jaeschke M. The gerg-2004 wide-range equation of state for natural gases and other mixtures, Tech. rep., GERG Technical Monograph 15. Fortschr-Ber VDI. Dusseldorf: VDI-Verlag; 2007.
[21] Kamarei F, Tarafder A, Gritti F, Vajda P, Guiochon G. Determination of the adsorption isotherm of the naproxenenantiomers on (s,s)-whelk-o1 in supercritical fluid chromatography. J Chromatogr A 2013;1314:276–87.
[22] Kamarei F, Gritti F, Guiochon G, Burchell J. Accurate measurements of frontal analysis for the determination of adsorption isotherms in supercritical fluid chromatography. J Chromatogr A 2014;1329:71–7.

CHAPTER

10

Practical Aspects and Applications of Preparative Supercritical Fluid Chromatography

E.R. Francotte

FrancotteConsulting, Basel, Switzerland

OUTLINE

Supercritical Fluid Chromatography.
DOI: http://dx.doi.org/10.1016/B978-0-12-809207-1.00010-0

10.1 INTRODUCTION

The particular properties of supercritical fluids have fascinated scientists since almost 200 years when Cagniard de la Tour reported for the first time about the critical state phenomenon [1]. However, it is only recently that supercritical fluid chromatography (SFC) has finally been accepted as a reliable and useful separation technique as a complement, or even an alternative, to the other chromatographic methods. Although many substances exhibit a critical point (liquid–vapor critical temperature and pressure) [2], carbon dioxide has been the preferred fluid for chromatographic purposes. There are several practical reasons for this choice: (1) its critical temperature and pressure are compatible with chromatographic instruments and are not destructive for the compounds to be analyzed or purified; (2) it has low reactivity; (3) low toxicity; (4) is nonflammable; and (5) is abundant and inexpensive. The term SFC in principle is not restricted to CO_2, but because most applications of SFC use CO_2 as the supercritical fluid, SFC with CO_2 is implied by most users when they talk about SFC. The advantageous properties of SFC such as low viscosity and high diffusivity have now been fully exploited in analytical SFC using modern instruments with performance similar, or sometimes better, than those of ultra-high performance liquid chromatography (UPLC) [3]. However, for preparative-scale applications the advantages of SFC are even more numerous. This is probably the reason why the rehabilitation of SFC was actually driven by its use for preparative-scale separations, and in particular, for the isolation of enantiomers. As a follow-up of this successful implementation for chiral separations, applications employing packed-column SFC (pSFC) are currently undergoing rapid development in many industrial research organizations.

10.2 SHORT HISTORY OF PREPARATIVE SFC

SFC has been around for a long time. The first reported application by Klesper et al. in 1962 was actually a preparative separation of Nickel etioporphyrin II and Nickel mesoporphyrin dimethylester using chlorofluoromethanes as supercritical fluids [4]. After this first publication there was a long period of about 20 years without significant advances in the field of packed-column SFC. The interest in packed SFC started again in the early 1980s. Several groups were involved in packed capillary SFC but focused on analytical applications. The history of this development was reviewed by Larry Taylor, one of the pioneers of SFC [5]. The situation changed with the introduction of the first commercial preparative instruments under the impulsion of Terry Berger [6] and the development of preparative SFC instruments at ProChrom in France [7]. Between 1980 and 2000, publications appeared regularly reporting applications of preparative SFC but there was no passionate and unanimous adoption of the technology by industrial users.

The pioneer chiral SFC work in the group of Rosset in 1985 and the first preparative chiral SFC separation publication [8] motivated different groups to investigate more deeply the potential of preparative chiral SFC on packed columns. In 1993, in collaboration with ProChrom, we performed one of the first preparative SFC chiral separations with stacked injection for the isolation of the enantiomers of the drug guaifenesine with a cellulose 3,5-dimethylphenylcarbamate column [9]. However, technical problems associated with the formation of aerosol and low recovery were discouraging. Besides the lack of robustness, the instrument costs, and the expensive CO_2 delivery infrastructure, the problem of recovery was certainly a major reason for the slow uptake of the technology in the pharmaceutical industry. Moreover, the lack of concern for environmental considerations at that time was also not in favor of the promotion of the SFC technique.

Finally, improved preparative-scale instruments were introduced spurning a growing interest in the utilization of SFC for the separation of stereoisomers. From the beginning of the new millennium, preparative separations of stereoisomers on chiral stationary phases was gradually adopted by a growing number of pharmaceutical companies and today it is the primary technique for these separations. During this period, the performance and reliability of analytical SFC instruments remained below the standard for HPLC, but it was sufficient for method development and the determination of the optical purity of collected fractions.

Preparative achiral purifications using pSFC was already around before 1990, but the number of applications remained limited and it was almost totally ignored by the pharmaceutical and agrochemical industry. Flash chromatography and reversed-phase HPLC were established as the gold standards. One major limitation of preparative SFC as an alternative to reversed-phase HPLC was the lack of mass-directed fractionation on commercially available instruments. This is almost a must when dealing with complex mixtures containing various impurities. This gap was filled in 2009 with the introduction of the first preparative SFC instrument with mass-directed fractionation. This revolutionized the area of preparative achiral purifications and SFC is now rapidly expanding, especially in the pharmaceutical industry.

10.3 GENERAL CONSIDERATIONS

Today SFC is the method of choice for preparative chiral separations in the drug discovery environment and other life science sectors as it perfectly fits the requirements at this stage of drug development, namely rapid processing of many samples available in relatively small amounts. For very large amounts, tens of kilograms or tons, other techniques like continuous multicolumn chromatography (SMB, varicol) are preferred. Preparative SFC is now also being increasingly adopted for preparative achiral purifications by many research laboratories, which have recognized the numerous advantages of packed-column SFC, such as high diffusivity, low pressure drops, short equilibration times, reduced solvent consumption, and fast solvent removal. In addition to these technical advantages, lower solvent costs, less safety concerns with respect to flammability and toxicity, and reduced impact on the environment make SFC an attractive separation technique.

Whatever the application, chiral or achiral, preparative SFCs offer multiple advantages over HPLC.

10.3.1 Speed

Due to the low viscosity of supercritical CO_2 (near to the gas state), it is possible to work at very high flow rates, which cannot be reached by

HPLC. As a consequence, the run times are significantly shorter and the purifications/separations are on average 3–4 times faster. Another considerable advantage of SFC for preparative applications is related to the recovery process of the purified compounds. Indeed, as CO_2 is removed almost instantaneously, the organic phase which has to be evaporated is much smaller than in an HPLC process.

10.3.2 Safety

Safety is also a positive feature of SFC. CO_2 that constitutes the main component of the mobile phase in SFC is nonflammable and much less toxic than most organic solvents used in HPLC. Alkanes (heptane, hexane) in normal-phase HPLC are neurotoxic and acetonitrile, the most usual organic component in reversed-phase HPLC is highly toxic by inhalation and by contact (skin penetration). Both flammability and toxicity are extremely important in the preparative-scale applications where operators are dealing with large amounts of solvents.

10.3.3 Sample Stability/Quality

In preparative HPLC acidic or basic additives are often used to suppress undesirable peak tailing or distortion effects due to unfavorable interactions with silanol groups on the surface of the stationary phase. As a consequence basic compounds that are most common in the pharmaceutical field are isolated as salts (most frequently as trifluoroacetates). Trifluoroacetic acid is known to be toxic for biological cells and must be removed before testing, necessitating an additional purification step. Using SFC, it is not required to employ additives usually, simplifying the isolation of the target compound(s). Therefore as part of our strategy for preparative chiral or achiral separations/purifications, we try to avoid the use of additives. Moreover, this strategy minimizes the decomposition of acid or base sensitive compounds.

10.3.4 Efficiency

Due to the low pressure drop in SFC (low viscosity), it is possible to work with packing materials of a small particle size, or with longer (tandem) columns, both leading to a significant increase in efficiency. In SFC it is usual to use 5 μm particles for preparative applications. This higher efficiency facilitates good separations even when the selectivity is low.

10.3.5 Costs

Costs are of course a critical factor in preparative chromatography. Replacing organic solvents in normal-phase chromatography by CO_2 allows a saving of 70–90% of the solvent costs. It is slightly less compared with reversed-phase HPLC since water is an inexpensive solvent. However, water removal from isolated fractions is an energetically costly process. According to our estimation it needs seven times more energy compared with the energy required to evaporate small amounts of modifier in SFC. Additionally, waste disposal is considerably reduced in SFC compared to HPLC. This is also a nonnegligible cost factor.

10.3.6 Impact on Environment

SFC is often considered a green technology since it has a positive environmental impact. While often neglected in the past, environmental impact considerations are a major concern in almost all industrial and private sectors today. Here SFC clearly contributes to reducing solvent usage and organic solvent emissions. Moreover, the reduction of waste sent for disposal also indirectly contributes to a decrease in carbon dioxide emissions since organic solvent waste is usually burnt after usage in other chromatographic processes. In this context, it is worthwhile to stress that SFC does not produce CO_2, it just makes use of existing CO_2, mostly obtained as a by-product of large industrial processes.

10.3.7 CO_2 Supply

For preparative-scale applications, the amount of supercritical CO_2 required is obviously much higher than for analytical use. As a consequence the CO_2 supply infrastructure needs to be adapted. This aspect is often underestimated when people embark on preparative SFC and the cost resulting from the setup of a large CO_2 delivery platform is an important part of the total investment for the SFC facility. The most common setups include CO_2 cylinder manifold systems capable of delivering large amounts of CO_2 or large tanks (up to several cubic meters) connected to a booster pump. In both cases CO_2 delivery can occur directly in the liquid state or as a gas, which is recompressed at the point of use (SFC instrument).

10.3.8 Instrumentation

The history and evolution of preparative-scale SFC instruments has been reviewed by Lazarescu et al. [10]. Basically the same preparative SFC systems can be used for either achiral purifications or chiral separations. The instrument choice is determined by the application, such as sample amount and throughput. If mass-directed fractionation is required, the choice is more restricted since not all preparative instruments provide this option.

Today various instrument manufacturers offer preparative-scale SFC systems (Table 10.1). These systems are generally characterized by the maximum flow rate of supercritical fluid they can handle, ranging from a few milliliters for systems combined with an analytical unit to 3 L/min for dedicated large-scale instruments. Small capacity instruments (10–20 mL/min) are used with typical analytical column to isolate micrograms to a few milligrams of substance on columns with inner diameters (i.d.) up to 10 mm. Medium capacity instruments (80–200 mL/min) are usually employed in combination with 15–30 mm i.d. columns. For larger systems (350–3000 mL/min), columns with i.d. between 30 and 50 mm (350–1000 mL/min) and up to 100 mm (3000 mL/min) can be used.

10.3.9 Injection

The stacked injection mode is generally employed in preparative-scale SFC, when multiple injections are required to process the entire sample amount. It consists of injecting the sample solution before completion of the previous run, in order to maximize the throughput and productivity. While it is not commonly applied in LC, it is more or less the norm in SFC. The most common injection modes are mixed-stream and modifier stream. Both methods have been compared. While Miller and Sebastian [11] found that modifier stream injection generally exhibits a higher performance, Cox concluded that there is little difference between the two modes of injection [12,13]. Recently Shaimi and Cox developed a new injection technique using an extractor to dissolve the dry sample in the supercritical fluid prior to injection. This technique has good reproducibility and peak shapes [13]. Another new injection method called dual line injection (DLI) was developed by Jasco [14] and incorporates a bypass line in the sample injector. This method is particularly effective for peaks with short retention times. The influence of injection volume in preparative-scale SFC has also been investigated [15,16] and compared with column overloading [17].

10.3.10 Fraction Collection

For fraction collection, small instruments are usually operated in an open-bed format while larger-scale instruments are typically equipped with a cyclone separator originally developed by Perrut to solve the problem of aerosol formation during depressurization of liquid CO_2 [18].

TABLE 10.1 Preparative SFC Instruments Currently Available in the Market

System	Manufacturer	Optimal column size (i.d. in mm)	Maximum CO_2 flow rate	Modifier flow rate (mL/min)	Detection (standard or optional)	Fractions	CO_2 recycling
CSP-4000 Semi-Prep Analytical and semi-prep	Jasco (Japan)	4–10	1–20 mL/min		UV, PDA, ELSD, MS[a], FID, CD	8 or open bed	
PR-4088 Prep	Jasco (Japan)	10–30	150 mL/min	0–50	UV, PDA, ELSD, MS[a], FID, CD	8	
Supersep 150	Novasep (France)	20–30	30–150 g/min	0–80	UV, ELSD	5 + waste	Optional
Supersep 400	Novasep (France)	30–50	100–400 g/min	0–160	UV, ELSD	5 + waste	Standard
Supersep 1000	Novasep (France)	50–80	250–1000 g/min	0–400	UV, ELSD	5 + waste	Standard
Supersep 3000	Novasep (France)	80–100	600–3000 g/min	0–1000	UV, ELSD	5 + waste	Standard
Hybrid 10–150 Analytical and semi-prep	PicSolution (France)	10–30	150 mL/min	0–50	UV, MS[a]	4 + waste	Optional
PREP 100	PicSolution (France)	10–30	100 mL/min	0–50	UV, MS[a]	6 or open bed	Optional
PREP 150	PicSolution (France)	10–30	150 mL/min	0–50	UV, MS[a]	6 or open bed	Optional
PREP 200	PicSolution (France)	20–30	200 mL/min	0–100	UV, MS[a]	4 + waste	Standard
PREP 400	PicSolution (France)	30–76.5	400 mL/min	0–250	UV	4 + waste	Standard
PREP 600	PicSolution (France)	30–76.5	600 mL/min	0–250	UV	4 + waste	Standard

PREP SFC BASIC Analytical and semi-prep	SepiaTec (Germany)	4–10	20 mL/min	0–13	UV, ELSD, MS[a] (ESI, APCI)	8 or open bed	Optional
PREP SFC 100	SepiaTec (Germany)	10–30	100 mL/min	0–66	UV, ELSD, MS[a] (ESI, APCI)	8 or open bed	Optional
PREP SFC 360	SepiaTec (Germany)	25–50	360 mL/min	0–240	UV, ELSD, MS[a]	8	Standard
Prep 15 SFC Analytical and semi-prep	Waters (USA)	4–20	15 mL/min	0–15	UV, PDA, ELSD, MS[a]	open bed	No
Investigator SFC System	Waters (USA)	4–10	15 mL/min	0–15	UV, PDA, MS	6	No
SFC 80q	Waters (USA)	19	80 g/min	4–70	UV	6 + waste	Optional
SFC 100	Waters (USA)	20–30	100 g/min	5–55	PDA, MS[a]	open bed	Optional
SFC 200q	Waters (USA)	30	150 g/min	4–70	UV	5 + waste	Optional
SFC 350	Waters (USA)	50	300 g/min	20–200	UV	5 + waste	Optional

[a]*With mass-directed fractionation.*

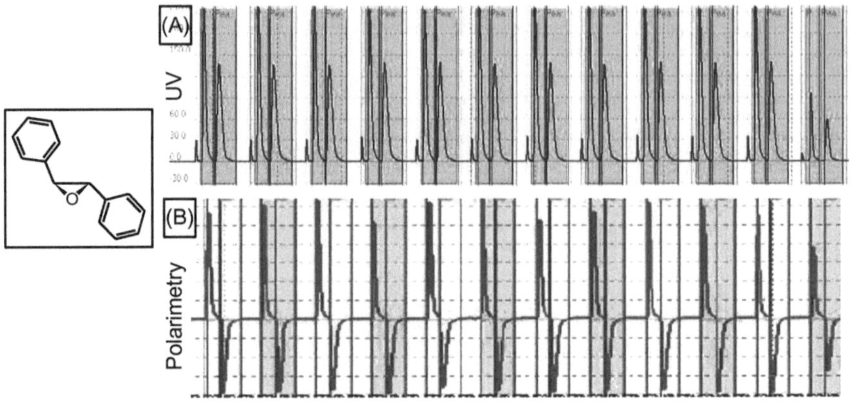

FIGURE 10.1 Resolution of 1 g of racemic *trans*-stilbene oxide by preparative SFC on Chiralpak AD-H (30 × 250 mm). Stacked injections (40 mg loadings) and fraction collection based on UV (A) and polarimetric (B) detection. Conditions: 75 g CO_2 and 16.5 mL MeOH/min. *Source: Adapted from reference Wentz ES. Multimode collections in prep SFC working towards open access SFC Sub-2 um particles for SFC. First International Conference on Packed Column SFC, Pittsburgh, 23-25 September 2007. Slides available from <http://www.greenchemistrygroup.org/past-conferences>.*

10.3.11 Detection

SFC is compatible with almost all detection methods. Beside DAD (UV), the most common detection method, mass spectrometry (MS), light scattering [19], polarimetry [20–21], circular dichroism [22], and chemiluminescence [23] have also been used in preparative SFC. An application of polarimetric detection for a preparative chiral SFC separation is shown in Fig. 10.1.

Coupling MS to preparative SFC instruments has been a great challenge and only became available on commercial instruments a few years ago, although the feasibility and usefulness of semiprepative SFC-MS was demonstrated about 15 years ago [24]. Whereas, it was a home-made setup at that time [24,25], it is now available on several preparative SFC instruments and used routinely, at least at the small and medium scale.

10.4 PREPARATIVE CHIRAL SEPARATION ON PACKED COLUMN SFC

The first chiral preparative-scale separation was reported by Fuchs et al. in 1991 [26]. Only a few further applications were reported during the next 15 years. These include the investigation of the isolation of the enantiomers of guaifenesine by Francotte in collaboration with ProChrom in 1993 [9], a similar study by Miller [27], and a few further applications [28–30]. Nevertheless the revival of pSFC can be attributed to the success of the preparative-scale separation of racemic compounds. When the first bench top preparative-scale SFC instruments became available around 1992 [31], scientists from the pharmaceutical industry quickly realized the potential of the technique in terms of speed, cost saving, and greenness, the latter topic being increasingly relevant to industry at the beginning of the new millennium. Even though the instruments were not as robust as today, their shortcomings were accepted in view of the advantages of SFC. The rapid success of this "niche" application is not surprising, considering that most chiral preparative-scale HPLC separations were performed under normal phase conditions with hydrocarbon solvents as the main component of the mobile phase. This feature greatly facilitated the transfer of methods to SFC.

Today most life sciences companies have fully adopted preparative SFC for chiral separations from milligram to kilogram (mg to kg) scale and it can be considered that chiral separations has rescued the SFC technology.

10.4.1 Chiral Stationary Phases

Enantioselective SFC separations obviously require the use of chiral stationary phases (CSPs). Fortunately, most CSPs applied in HPLC can be used for SFC. There are no chiral stationary phases designed exclusively for SFC. However, by contrast to HPLC which predominantly uses 10 μm particle size, the utilization of smaller particle sizes (usually 5 μm) has become more or less the norm for preparative-scale SFC. The smaller particle size is an advantage in terms of efficiency. Moreover, in order to cope with the risks associated with the properties of compressed gas in SFC, column hardware had to be adapted as well. Both features, small particle size and higher pressure resistance, have led to an adjustment of column properties by column manufacturers to support SFC applications. Although numerous chiral stationary phases (CSPs) are available, only those which exhibit a sufficient loading capacity are useful. This restriction is similar to preparative-scale HPLC. Similar to HPLC, almost all preparative-scale chiral SFC separations (>95%) are performed on polysaccharide-based CSPs (Fig. 10.2). The polysaccharide-based phases were originally introduced by Daicel in Japan, but there are now several manufacturers offering generic CSPs similar to those manufactured by Daicel. A few applications using brush-type phases (Fig. 10.2) have been reported [32]. The new cyclofructan- [33] and zwitterionic quinine-based [34] phase have been used in analytical SFC, but to our knowledge, no preparative practical applications have been reported so far.

A number of polysaccharide-based CSPs are available as either coated materials or in an immobilized form. Among the coated CSPs,

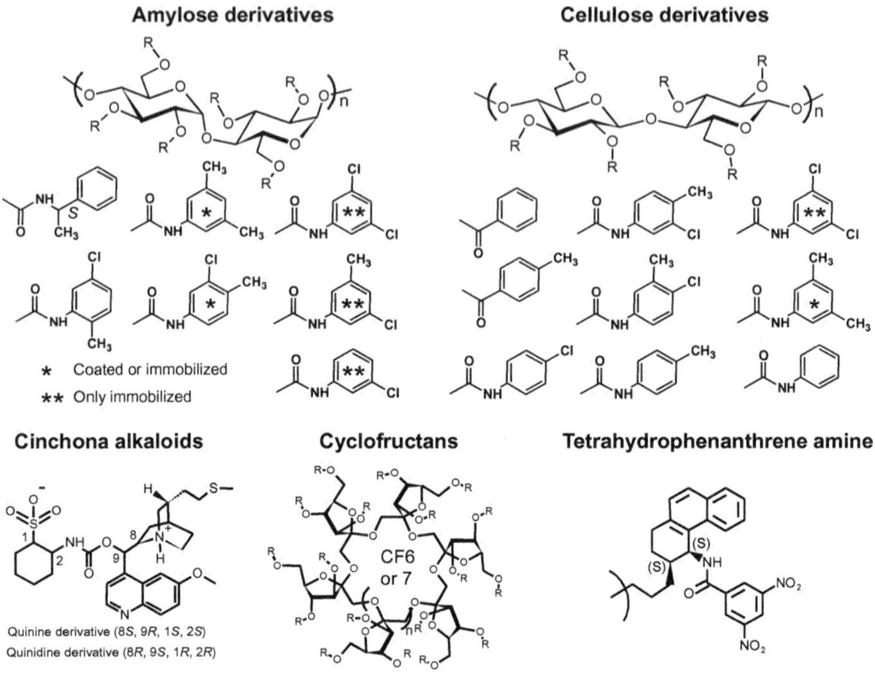

FIGURE 10.2 Structures of chiral stationary phases for preparative enantioselective packed-column SFC separation.

the 3,5-dimethylphenylcarbamate derivative of amylose remains the most popular CSP due to its high success rate. The first coated phase of this series was introduced on the market in 1981 (Daicel) and arose from the work of Yoshio Okamoto and colleagues [35,36]. Most of the commercially available immobilized polysaccharide phases result from a process developed by Francotte and coworkers [37–43]. In particular the immobilized 3,5-dichlorophenylcarbamate derivative of cellulose possesses a broad chiral recognition capability. Other immobilized polysaccharide-based phases have recently been launched and complement the immobilized polysaccharide CSPs [43,44]. Unlike coated phases, the immobilized polysaccharide phases tolerate any kind of solvent, such as chlorinated solvents, ethyl acetate, tetrahydrofuran, dioxane, or dimethoxymethane, for example. This was highlighted by Francotte et al. for chiral HPLC [40–42] and later for chiral SFC [45]. This feature affords additional advantages for optimization of selectivity and/or retention time, and the solubility of target compounds, which is a critical factor for preparative-scale separations [46–52]. In any case, since currently there is no method to predict the separation of

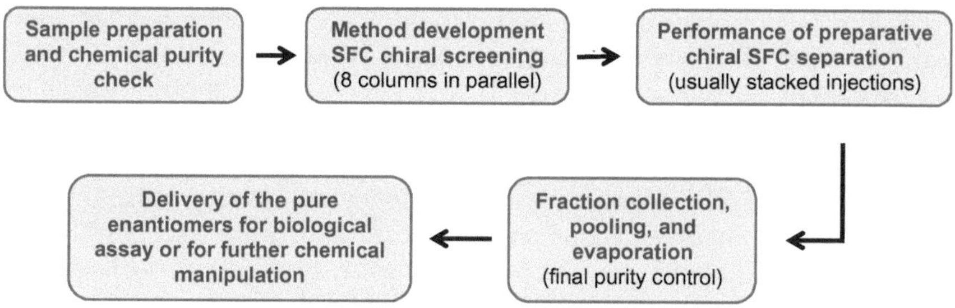

FIGURE 10.3 Process flow for method development and preparative chiral separation.

a racemate on a particular phase or for particular separation conditions, each new application requires a more or less extensive screening.

10.4.2 Process Flow for Preparative Chiral SFC

The standard process for chiral separations is depicted in Fig. 10.3. After submission of the sample, an analytical quantity is subject to screening on a selection of chiral columns with various combinations of modifier, with or without additives. Once the separation conditions are defined, the preparative separation is performed, usually by repetitive injections of small amounts of the sample. After evaporation of the pooled fractions, the optical purity of the isolated single enantiomers is determined. For chiral separations UV detection is usually sufficient but MS might be useful, especially for mixtures containing impurities in addition to the targeted enantiomers.

10.4.3 Stationary Phase Screening and Method Optimization

As usual for preparative-scale chiral separations, the first step is to identify an appropriate CSP and the suitable chromatographic conditions such as modifier, additives, temperature, and flow rate. Subsequently fine tuning is often applied to optimize the separation conditions with respect to selectivity and resolution. As the stacked injection mode is common practice for preparative-scale SFC, it is also usual to optimize retention times by adjusting the flow rate and/or modifier composition. Basic or acidic additives might also help to improve the separation (selectivity) and/or peak shape. The systematic influence of basic and acidic additives in SFC was investigated by Berger and Deye [53] and more recently by Stringham and coworkers [54–56]. Commonly used additives include trifluoroacetic acid, formic acid, ethanesulfonic acid, triethylamine, diethylamine, and ispopropylamine. However, for preparative separations, basic or acidic additive should be used with caution. Unlike analytical separations, the ultimate purpose of preparative separation is to isolate the purified/separated compounds in their intact form. As most additive are reactive, chemical transformations are possible. Moreover as the solute stays in contact with the mobile phase for longer in preparative-scale separations, both in the column and during evaporation of fractions, degradation might be more marked. For example, many carboxylic acids have the tendency to esterify with alcoholic modifiers and can extend the postpurification process. If possible it is recommended to avoid the use of additives in preparative-scale separations. However, in some instances it might be necessary to use small amounts of additives to improve peak shape.

Hamman et al. [57] and Ventura et al. [58] proposed the use of volatile additives, such as ammonium hydroxide and ammonia, respectively, which are easily eliminated by evaporation in order to minimize exposure of the sample to potentially reactive additives and to avoid a high content of residual additives in the purified sample.

Generic screening protocols for chiral method development were proposed by academic and industrial research groups [59–72]. Many pharmaceutical companies have elaborated their own approach based on practical experience. All approaches are similar and the primary screening comprises almost exclusively of polysaccharide-based phases. However, the chiral column field being a dynamic market which is continuously evolving, the setup has to be regularly adapted. Most users have adopted an automated serial column screening setup equipped with switching valves or a parallel system which can screen up to five columns at once. For more than 5 years our standard screening method consists of a high performance instrument capable of simultaneously operating eight columns in parallel [59,73]. This approach allows the rapid identification of optimal conditions (chiral column, mobile phase) for the preparative-scale separation from milligrams to several hundred gram amounts. A typical screenshot of the results from a chiral screening is shown in Fig. 10.4. The racemic compound was screened on eight different columns with six different mobile phases (48 different conditions within 42 minutes) using a commercial SFC instrument. Short columns (150 × 4.6 mm) are used, applying a gradient of 5–40% (v/v) modifier within 5 minutes, at a flow rate of 3 mL/min on each column (7 minutes per run).

With the introduction of immobilized polysaccharide-based phases, the possibility to use strongly solvating modifiers has become almost unlimited [45,59,60]. Typical examples of application of such modifiers are shown in Fig. 10.5. The use of "nonconventional" modifiers for SFC chiral separations on immobilized polysaccharide-based CSP has been exploited by many other users [46–52]. This benefit of the immobilized polysaccharide-based phases is particularly valuable for preparative-scale separations since productivity is often limited by the poor solubility of the racemate in the mobile phase [45]).

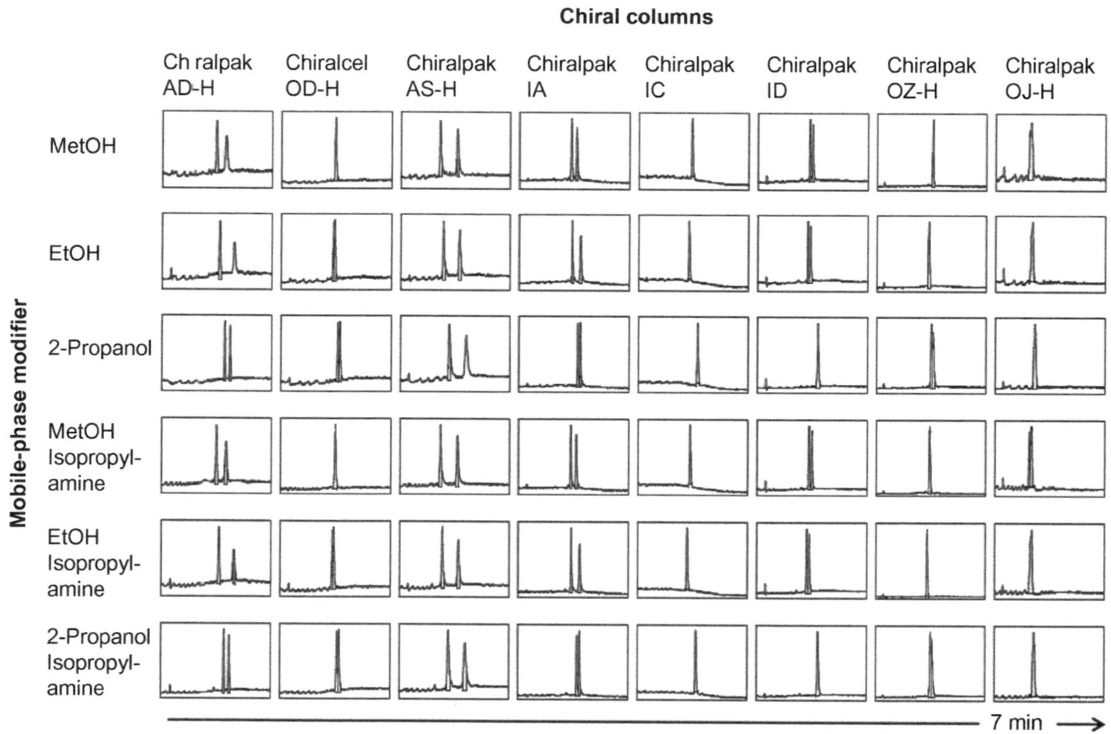

FIGURE 10.4 Typical screenshot of a parallel chiral screening output on a Sepmatix 8× Screening SFC system (Sepiatec, Berlin, Germany). Columns 4.6 mm × 150 mm. CO_2/modifier, gradient 5–40% in 7 min; flow rate, 3 mL/min; injection, 60 μL distributed on the 8 columns.

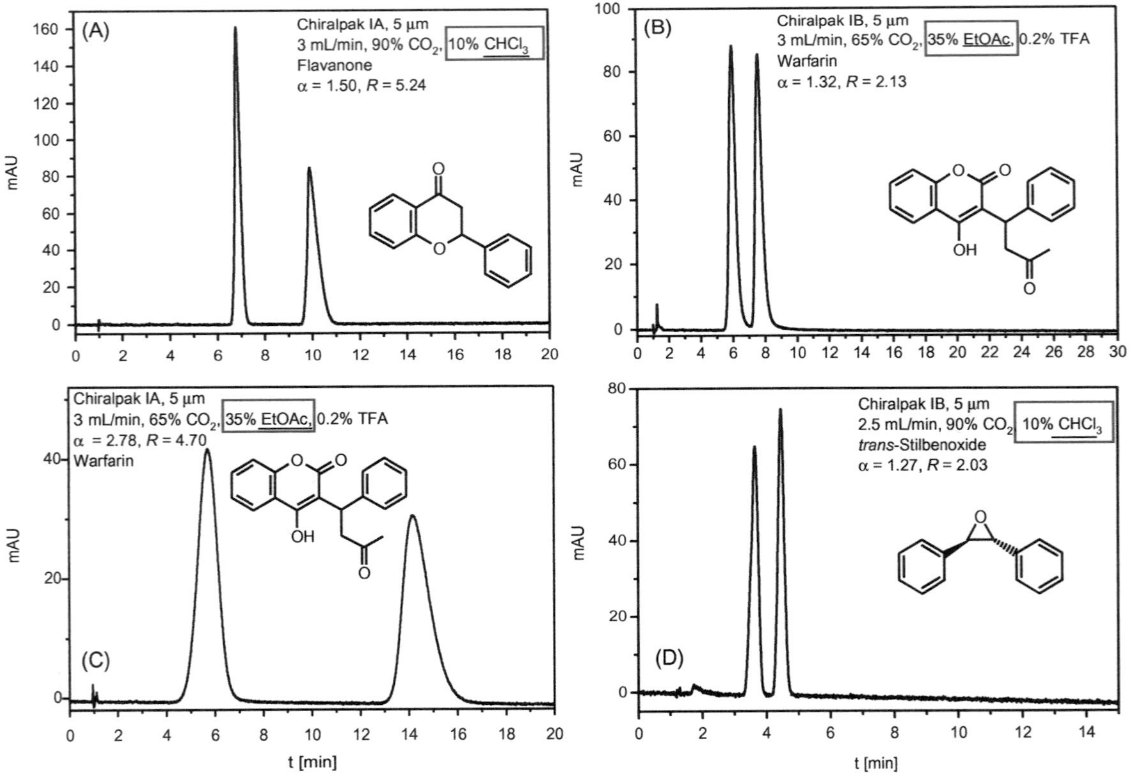

FIGURE 10.5 Chiral SFC separations on immobilized cellulose and amylose-based CSPs with nonclassical modifiers. Column, 4.6 mm × 250 mm, 5 µm; flow, 3 mL/min; (A) flavanone on Chiralpak IA, CO_2/chloroform 90/10; (B) warfarin on Chiralpak IB, CO_2/ethyl acetate/TFA 65/35/0.2; (C) warfarin on Chiralpak IA, CO_2/ethyl acetate/TFA 65/35/0.2; and (D) *trans*-stilbene oxide on Chiralpak IB, CO_2/chloroform 90/10. *Source: Adapted from reference Francotte E, Diehl G. Extending the SFC applicability in the field of enantioselective separations. Lecture presented at the 2nd International Conference on Packed-Column SFC, Zurich, Switzerland, 1–2 October 2008. Slides available from <http://media.wix.com/ugd/2239fc_c0b2afbc6f5c48aeb4fde1a662f05798.pdf>.*

10.4.4 Preparative-Scale Applications (Stacked Injections)

Most pharmaceutical laboratories have fully embraced preparative-scale SFC for chiral separations [27,45,73–81] and the technology is the method of choice in drug discovery as it perfectly fits the requirements at this stage of drug development, namely rapid processing of many samples available in relatively small amounts. The technique is widely applied for amounts ranging from a few grams to kilograms. Preparative chiral SFC instruments having a capacity to operate at flow rates of 10 mL to several liters per minute are available, fulfilling the requirements needed to work at various scales. In contrast to HPLC, it is usual to use short columns and make repetitive injections of small portions of the sample combined with stacked injections and short cycle times to achieve high productivity in SFC. A typical example is shown in Fig. 10.6 [73]. In this example a 50 g chiral drug intermediate was separated isocratically on a Chiralpak AD-H column. Ethanol (5% v/v) was used as the modifier. At each run 208 mg was injected with 32 out of 240 stacked injections shown in Fig. 10.6. The enantiomeric excess of the separated enantiomer was greater than 99%. This application demonstrates the high efficiency of SFC which can reach high productivity up to 2.3 kg racemate per kg CSP per day in this case.

The number of practical preparative-scale chiral SFC applications reported is relatively limited and by far does not reflect the real

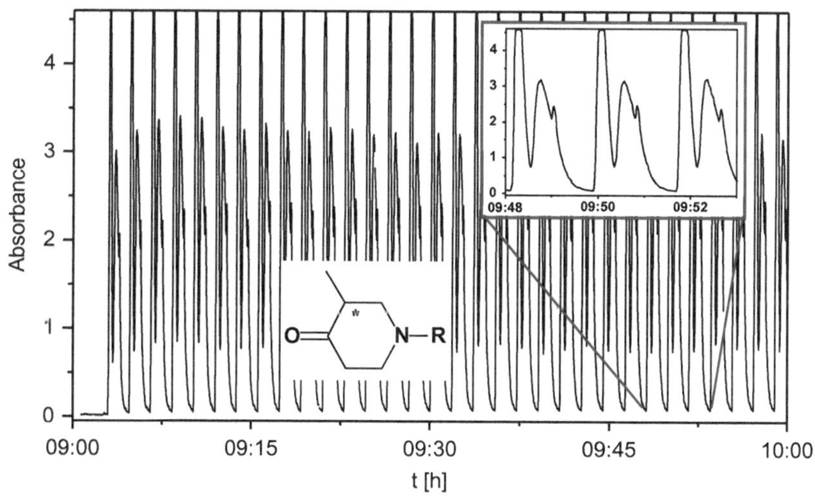

FIGURE 10.6 Preparative chiral SFC resolution of 50 g of a racemic 4-piperidone derivative on Chiralpak AD-H. Column 30 mm × 250 mm; Instrument, Thar SFC 200; Mobile phase, CO_2 containing 5% ethanol (isocratic); Flow rate, 120 mL/min; cycle time 1.8 min. Stacked injection mode, 240 injections (32 injections shown) each 208 mg. Productivity: 6.8 g/h ≈ 2.33 kg/24 h/kg CSP.

utilization of the technique. Considering that most large pharmaceutical companies and many specialized contract research organizations separate the enantiomers of hundreds or thousands of chiral substance every year by SFC, only a very small portion of the applications is in the public domain. The applications cover a wide scale, from a few milligrams to hundreds of grams and kilograms. Several book chapters [82–84] and review articles [75,77–78,85–88] cover the topic. Chiral separations at the pilot scale (>10 kg) have not been reported so far. Among the few recent reports, selected applications include the separation of the enantiomers of the antiviral agent goitrin [89], and of a novel $\sigma 1$ receptor agonist [90], the separation of the enantiomers of rhizopine for determination of its absolute configuration by vibrational circular dichroism [91], the preparation of chiral radiolabeling precursors [92], the resolution of racemic tricyclic indole compound [18 F] GE-180 for evaluation of the enantiomers as positron emission tomography (PET) imaging agents [93], the separation of a pivotal intermediate in the synthesis of glucocorticoid receptor modulators at the kilogram scale on Chiralpak AD [94], and the resolution of racemic resorcinol amide libraries as part of an oncology chemistry program [95].

Besides publications in scientific journals, the website of the "Green Chemistry Group" is a valuable source of information (http://www.greenchemistrygroup.org/). The numerous practical applications found in the compiled and downloadable presentations of the last 9 years attest to the intensive utilization of SFC in the pharmaceutical industry. For intellectual property reasons structures of the separated enantiomers are often missing. Several applications, from a few hundreds of gram and up to kilogram amounts, are reported in this series of conferences.

Comparative studies of SFC and LC have been performed by different pharmaceutical companies and evidently support the understanding that SFC is superior in terms of productivity, speed, and solvent consumption [27,32,59,96–98]. It has been estimated that, depending on the application the organic solvent consumption and running costs in SFC are reduced by 70–95%, productivity is increased by a factor of 7–40, and the processing time is 2–8 times faster. Even though SFC is the method of choice, chiral HPLC remains a complementary technique because there are some compounds which are not separated by SFC or are better separated by HPLC, and for compounds hardly soluble under SFC conditions.

For very large amounts, tens of kilograms or tons, other techniques like continuous multicolumn chromatography (SMB, Varicol) are preferred and afford a higher productivity [98]. It is known that simulated moving bed chromatography (SMB) is particularly efficient with respect to solvent consumption, and moreover, the solvent is almost entirely recycled. While for small-scale preparative separations, CO_2 is usually not recycled, it is the rule for large-scale applications.

10.5 PREPARATIVE-SCALE ACHIRAL SFC PURIFICATIONS ON PACKED COLUMNS

Preparative achiral SFC purifications have been around for more than 30 years, but were mostly applied to lipophilic compounds [5,31]. During this period reversed-phase HPLC became established as a highly effective and generally applicable purification technique. Reversed-phase HPLC had the advantage of using a quasi-universal stationary phase (C_{18}) and a generic mobile phase consisting of mixtures of acetonitrile and water. It is also compatible with MS detection. For these diverse reasons reversed-phase HPLC-MS became the standard approach for the purification of small amounts (a few milligrams to grams) in pharmaceutical research and development and in many other research environments. It has also been employed at the pilot and production scale for various applications. Nevertheless the potential of SFC for purification purposes had been recognized before it was applied to chiral separations. For example in 1977 Hartmann and Klesper reported the preparative-scale separation of styrene oligomers by SFC [99].

10.5.1 Stationary Phases for Achiral SFC Purifications

In early applications of achiral purifications by SFC, a limited number of stationary phases were employed, mainly silica and octadecylsiloxane-bonded silica materials. With the growing interest in analytical and preparative SFC, the situation has changed and the number of stationary phases utilized has expanded spectacularly and continues to expand. This expansion was driven by the need for more efficient and selective stationary phases to cope with a wide variety of chemical substances with different functional groups (e.g., acidic, basic, neutral, small, large, etc.). This feature is important for compounds submitted for biological testing as they are generally obtained from multiple and complex synthetic procedures. Recovery and purity are of first importance in this case. These constraints necessitate the development of a "tailor-made" purification approach for each sample. The same considerations apply to large-scale separations where each operating factor affecting costs is important. The classification of stationary phases for SFC has been proposed by Poole [100] and by West and Lesellier [101–104]. Most of the available phases have been classified according to their polarity, which is a critical parameter for selectivity. This feature provides an opportunity to optimize the selectivity of the separation for individual components of a mixture but it emphasizes also the importance of identifying the right column chemistry for the successful application of achiral SFC. The approach applied is similar to that for chiral compounds and commences with the screening of a group for analytical columns. Column manufacturers have obviously realized the opportunity of this emerging market and the number of columns for achiral purifications by SFC has grown rapidly over the last 5 years.

Nevertheless it is possible that a few columns chemistries will emerge as preferred columns in the future. Fig. 10.7 shows the structures of stationary phases available for preparative achiral purifications. The list is not exhaustive. Some phases from different suppliers might exhibit different properties although they have the same basic structure [101–104]. Several phases are exclusive to a particular manufacturer. Princeton Chromatography offer the widest range of achiral SFC columns. Fig. 10.8 illustrates the separation of a mixture of four compounds on six different column chemistries and clearly demonstrates that selectivity can be modulated by the selection of the column chemistry. The chemical functionality of the phases (Fig. 10.7) is quite diverse (neutral, acidic, basic, polar, apolar) and based on the purification of tens of thousands of samples, our practical experience indicates that some columns are more suitable for basic and others for acidic compounds, and that some accommodate both types [59,60]. A systematic evaluation of currently available achiral stationary phases was performed by different groups [105–107], but as discussed previously, the column chemistry field is dynamic and no universal stationary phase has been found. Moreover evaluations are generally done with a mixture of selected "model compounds" which are supposed to reflect a variety of molecular structures. In reality mixtures to be purified generally consist of structurally related substances generated during the nonselective chemical reactions. We found that a combination of three columns permits to achieve the purification of almost 90% of the mixtures submitted by medicinal chemists. The three phases are the basic phase PEI, the ionic phase C18-WCX, and the BEH silica phase (Fig. 10.7). The 2-ethylpyridine, 4-ethylpyridine, propyl-pyridyl urea, and the diamino phases may also be useful for some substances. In addition to achiral phases, chiral phases, such as the polysaccharide-based phases have also been frequently used for achiral purifications, as they might exhibit useful selectivity. Examples of the separation of regioisomers, diastereoisomers, and metabolites have been shown in Refs. [59,60,73,108,109]. Two examples of achiral separations on a chiral stationary phase are shown in Fig. 10.9.

10.5.2 Preparative-Scale Achiral Purifications

There were ongoing activities in the field of preparative SFC purification especially in France in early 1980s [7,110,111]. Perrut who was a

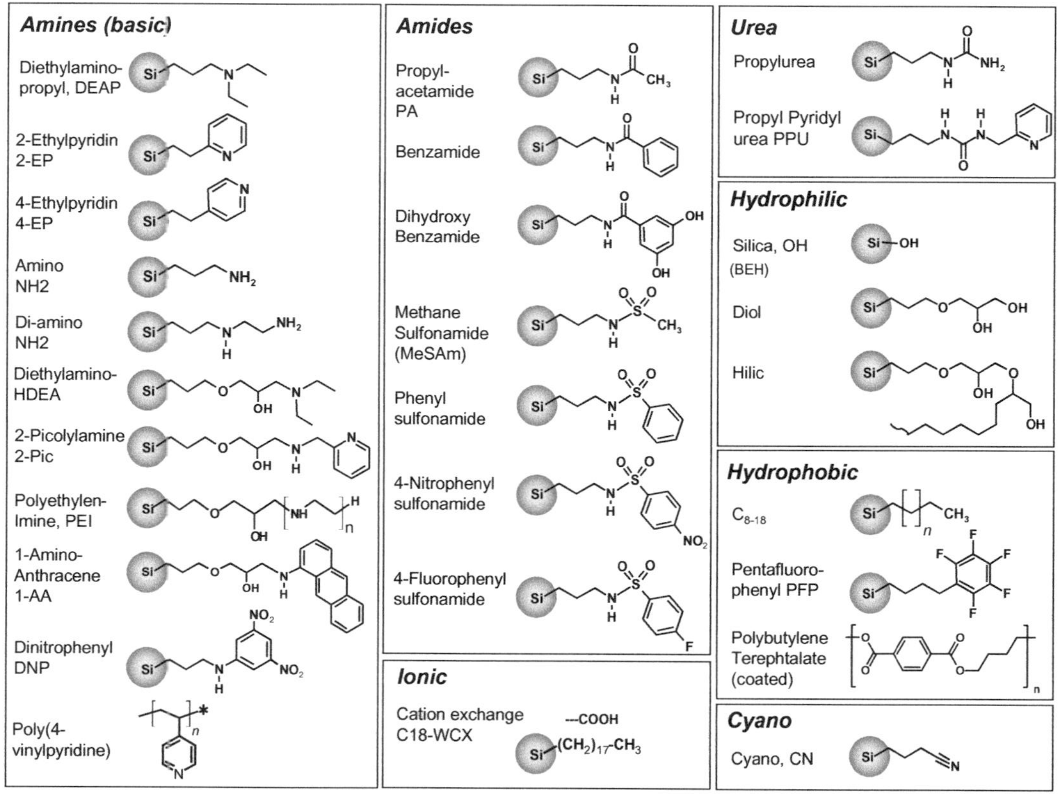

FIGURE 10.7 Chemical functionalities of stationary phases for achiral SFC purifications.

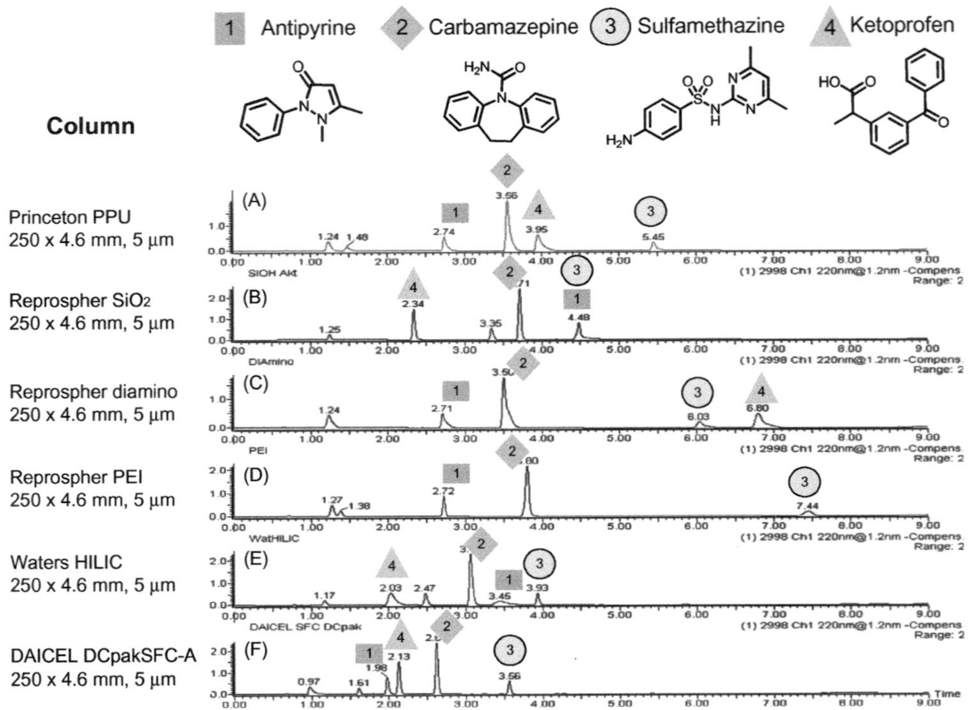

FIGURE 10.8 SFC separation of a mixture of four test compounds on six different column chemistries using a Waters Method Station X5 SFC (Waters Corporation, Milford, MA, USA). Columns 4.6 mm × 250 mm; mobile phase: CO_2/MeOH 90:10; flow rate, 6 mL/min.(A) Princeton Propylphenyl urea PPU, 5 μm, (B) Reprospher (Dr. Maisch) SiO_2, 5 μm,(C) Reprospher (Dr. Maisch) Diamino, 5 μm, (D) Reprospher (Dr. Maisch) Polyethyleneimine PEI, 5 μm, (E) Waters HILIC 5 μm, and (F) DAICEL DCpak SFC-A, 5 μm.

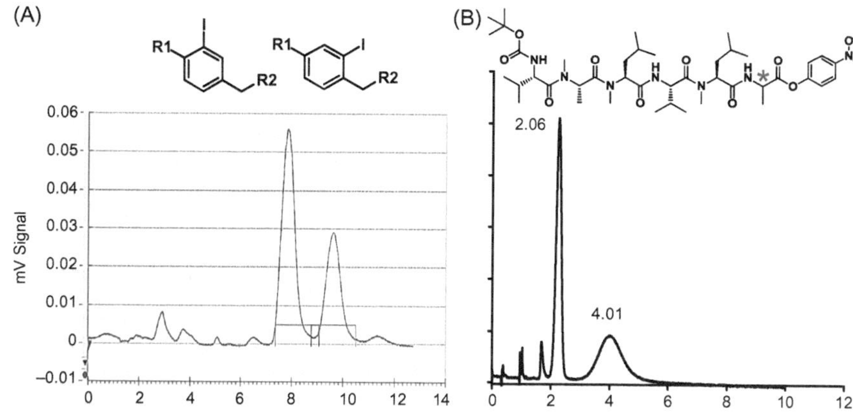

FIGURE 10.9 Achiral separations on chiral stationary phases. (A) Regiosomers of a trisubstituted phenyl derivative. Chiralpak IC; mobile phase, CO_2/2-propanol/2-propylamine 75/25/0.25. (B) Epimers of a small peptide. Chiralpak IA; mobile phase, CO_2/2-propanol 80/20. Flow, 3 mL/min. The asterisk indicates the position of the epimeric center of the molecule.

pioneer in preparative SFC reviewed the status of the technique until 1989 [7]. The major focus at that time was the development of efficient and competitive processes for large-scale applications. Isolation of tocopherol from wheat germ oil [112] and the fractionation of lemon-peel oil [113] by SFC were described in 1990. The isolation and purification of methyl methacrylate oligomers [114] and functional silanes and phosphines [115] were performed during the same period. Most applications targeted lipophilic substances as it was considered that the weak polarity of CO_2 would be more appropriate for the purification of apolar compounds.

Between 1990 and 1999 the interest in using preparative SFC for small-scale applications was relatively limited; the major obstacle was related to technical deficiencies (pressure regulation, injection, aerosol formation, etc.). Nevertheless a number of applications to the purification of achiral mixtures at various scales ranging from milligrams to hundreds of grams was performed. Diverse molecules were purified, e.g., the isolation of steroid hormones [116,117], synthetic oligomers [118], isolation of flavor, and foods constituents [119]. Rocca et al. had already illustrated the potential of preparative-scale SFC for the isolation of minor impurities in chemical mixtures [120]. There was even a report on the isolation of 11C-labeled compounds using ammonia as the supercritical fluid [121]. From 1997 there was growing interest in the purification of docosahexanoic acid (DHA) from fish oil by SFC [122]. DHA is an omega-3 fatty acid, which has an essential biological function in human brain and other organs. Most recent applications include the purification of synthetic intermediates, the isolation of natural products [123–127], the purification of peptides [128], the separation of anomeric derivatized glucosides [129], and the isolation of metabolites or drug related impurities for identification [130–133]. Fig. 10.10 illustrates the separation of natural products [123] and Fig. 10.11 an example of the purification of a synthetic intermediate.

10.5.3 Preparative SFC Purifications With MS-Directed Fractionation

The interest in SFC as a tool to purify polar compounds is relatively recent. The breakthrough came with the introduction of the first preparative SFC instrument equipped with mass-directed fractionation in 2009. Coincidentally there was an issue due to a worldwide acetonitrile shortage at about the same time, which stimulated the exploration of alternative purification methods to replace RP-HPLC. The growing awareness for environmental considerations was also a source of motivation. For years it had been a real challenge to develop a robust and reliable preparative-scale SFC-MS instrument, offering mass-directed fractionation. For the preparative achiral purification of complex mixtures by SFC mass-directed fractionation is a must. A few research groups had successfully developed their own SFC-MS purification system, in particular, in the context of combinatorial chemistry projects [24,25], but these instruments were not accessible to the global community and were limited in their applicability with respect to sample size.

The introduction of the dedicated preparative-scale SFC-MS instrument brought about a dramatic change. Several groups evaluated more or less simultaneously the suitability of the new instrument to replace, at least partially, the "unmatched" reversed-phase purification technique in medicinal chemistry by the SFC technique, although these two chromatographic modes are totally different [73,134–138]. All groups demonstrated that the purification of small synthetic libraries (sample mixtures) by preparative reversed-phase HPLC-MS and by SFC-MS provided equal performance with respect to purity and recovery, but on average SFC was three times faster than RP-HPLC. We found that recovery was better in SFC for some compounds and better in RP-HPLC for others [73]. Later Francotte et al. indicated that in a drug discovery environment of several thousands of samples submitted annually about 80% could be purified by preparative SFC-MS [139]. Recently Ross et al. developed a platform to purify more than 90,000 small-size samples annually using the same SFC-MS instrument and utilizing a generic method [140]. This successful and rapid expansion of achiral preparative SFC was due to the concomitant development of improved SFC instruments and new stationary phases.

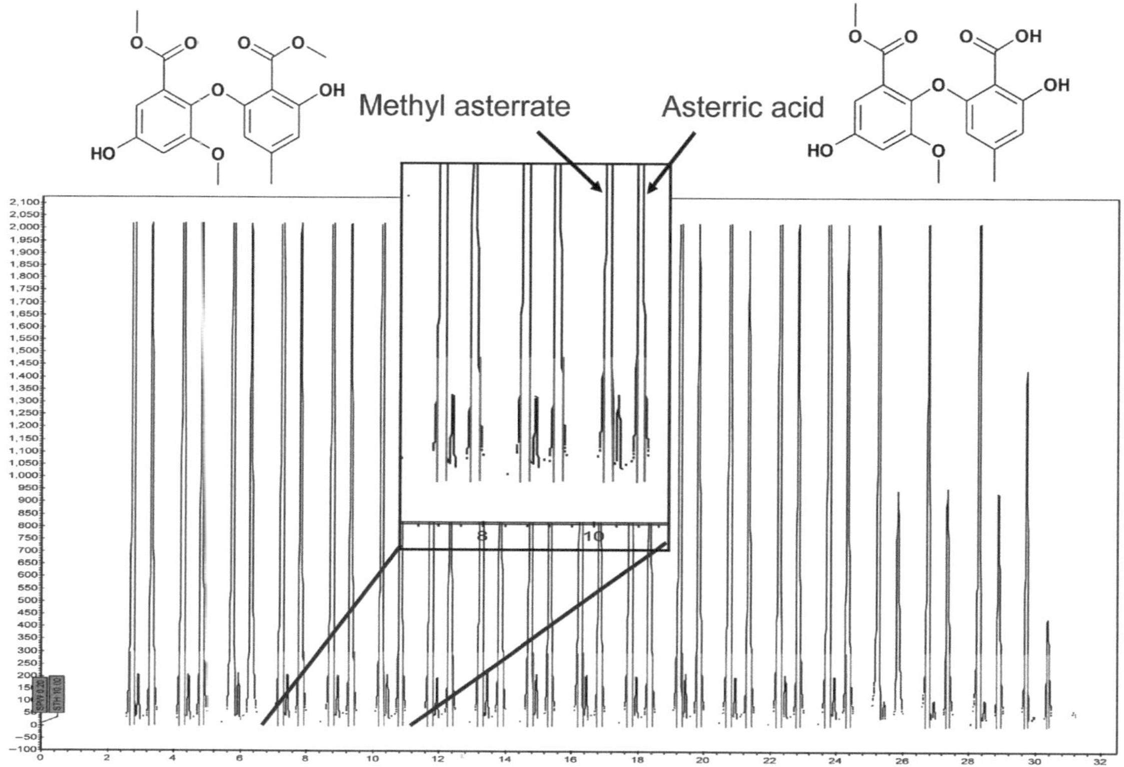

FIGURE 10.10 Separation of fungal metabolites (asterric acid and methyl asterrate on a MiniGram Mettler Berger). Sample 71.8 mg; Column Diol (250 mm × 7.8 mm); Stacked injections (20×); Flow rate 10 mL/min CO_2 with 15% MeOH. *Source: Adapted from reference Wang Y, Stutz G, Roggo S, Diehl G, Francotte E. Development of SFC for high throughput purification in natural product drug discovery. Lecture presented at the 2nd International Conference on Packed-Column SFC, Zurich, Switzerland, 1–2 October 2008. Slides available from <http://media.wix.com/ugd/2239fc_c1832be-f0ef548aba914f87bc86c870d.pdf>.*

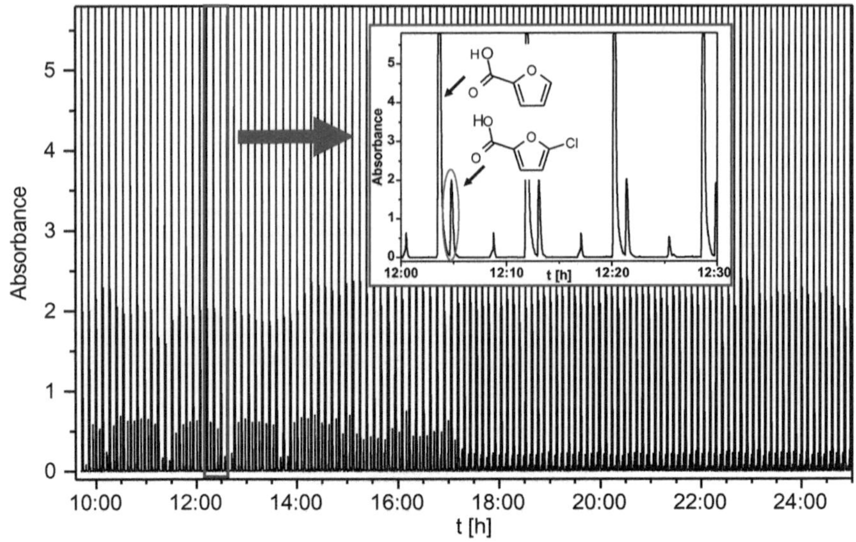

FIGURE 10.11 Separation of 42.9 g of a mixture of 2-furoic acid and 5-chloro-2-furoic acid by preparative SFC. Column: Princeton 2-ethylpyridine, 5 μm, 30 mm × 250 mm, Mobile phase: CO_2/ethanol/TFA 86:14:0.1, 126 g CO_2/min; Cycle time, 7 min, 430 stacked injections at 100 mg (120 consecutive stacked injections shown).

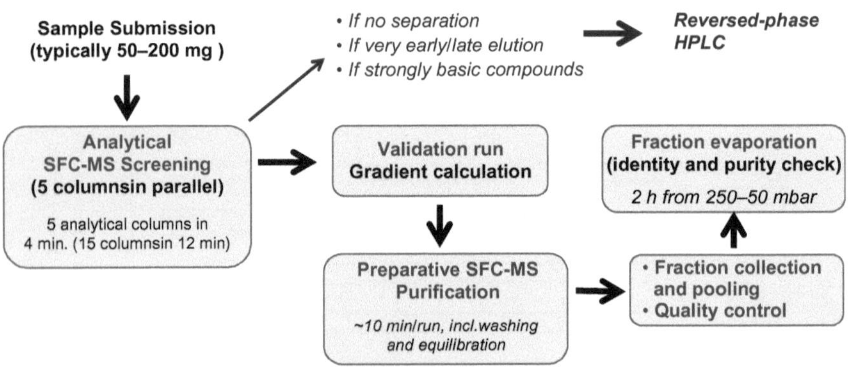

FIGURE 10.12 General workflow for achiral MS-triggered SFC purifications.

10.5.4 Workflow for Medium-Scale Preparative Achiral SFC

Since the general workflow for preparative SFC-MS purification differs only slightly from user to user, the workflow we apply for small-size samples in drug discovery is representative of this approach [59]. The workflow is outlined in Fig. 10.12.

After submission of the sample, typically 50–200 mg, a screening on various column chemistries is performed. Since the retention of the each component of a mixture on a defined stationary phase cannot be predicted, it is important to rely on an efficient screening platform to maximize the quality of the separation with respect to the target substance. The molecular mass of the compound provides critical information for selecting the optimum column and chromatographic conditions. This can be easily performed, using a commercially available SFC-MS system equipped with five columns in parallel. When the bank of preferred five columns is not successful, there is the option to switch to a second or third bank of five columns. After identification of the best column chemistry, a validation run is performed in order to collect the data needed to calculate the optimal gradient for the preparative-scale purification. This gradient calculation is necessary due to the fact that the flow in each column of the parallel screening unit is not necessarily the same as there is no flow control on the currently available SFC-MS parallel screening unit. The gradient calculation procedure has been empirically determined and is specific for each column and size and for a specific instrument [59]. Fig. 10.13 shows a typical application of an achiral SFC purification from the analytical column screening to the final preparative-scale separation. Today achiral SFC replaces RP-HPLC for at least 75% of the purification of small

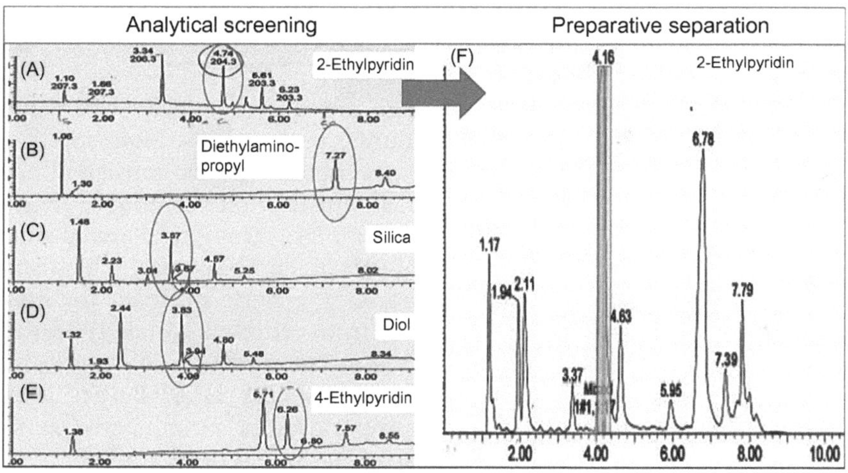

FIGURE 10.13 Method development (SFC-MS) for achiral SFC purification on Waters X5 station (5 columns screened in parallel), target compound is circled in red; columns 250 mm × 4.6 mm; flow rate 20 mL/min; gradient mode 5–50% MeOH within 6 min. (A) on 2-Ethylpyridin; (B) on Diethylamino-propyl; (C) on Silica; (D) on Diol; (E) on 4-Ethylpyridin. (F) Preparative purification on 2-ethylpyridin column, 250 mm × 30 mm. Flow 100 mL/min. *Source: Adapted from reference Francotte E. Why should sfc be the first technology choice for the purification of pharmaceuticals?. Lecture presented at the 8th International Conference on Packed-Column SFC, Basel, Switzerland, 10 October 2014. Slides available from <http://media.wix.com/ugd/2239fc_6eee04127c92436ab62c550d8b8a9317.pdf>.*

molecules in medicinal chemistry. The application range is continuously expanding and covers drug molecules and intermediates, natural products, metabolites, and small linear or cyclopeptides (Fig. 10.14).

Between one and two thousand small-size samples (50–200 mg) can be purified per year on each preparative MS-triggered SFC-100 instrument. The total average turnaround time of 24 hours includes method development, preparative separation, and fraction isolation, evaporation, drying, and quality control, which can be performed by either analytical SFC or RP-HPLC as an orthogonal method. The speed of the purification process, the high purity of the isolate substances, and the absence of TFA which is known to be generally toxic in most biological testing are particularly beneficial in the drug discovery environment.

The shift from RP-HPLC to SFC can be considered as a "green" switch as SFC consumes on average about 20% less organic solvents. Moreover, in RP-HPLC preparative-scale applications, the removal of water is an energy consuming process necessitating about seven times more energy compared to SFC sample evaporation. Basically this approach is not limited to small-size samples. The same workflow is applicable to large-scale preparative-scale SFC-MS instruments.

10.6 HIGH-THROUGHPUT ACHIRAL SFC PURIFICATIONS

In high-throughput synthesis a large number of samples are generally generated, usually in small amounts, and these sample mixtures have to be purified. Given these requirements it was consistent that some scientists were looking for a fast purification technique. In this context utilization of preparative SFC for high-throughput purification attracted a number of groups involved in combinatorial chemistry. Already 17 years ago Coleman proposed a generic protocol for high-throughput purifications of samples from combinatorial chemistry libraries using SFC [141]. High-purity substances could be rapidly isolated up to the tens of milligram level. A year later, Berger et al. reported the development of a new separator for automated fraction collection in semipreparative SFC (up to 50 mg of compound) to avoid the problem of aerosol formation [142]. Kassel and his team improved the system by incorporating mass-directed fractionation [24]. Additional groups developed high-throughput SFC purification platforms with UV-triggered fraction collection [25,143–145]. In almost all cases purifications were performed in the context of hit finding activities where large libraries of potentially new drugs are produced and must be purified quickly using generic methods without column screening and accommodating lower recovery and purity. The

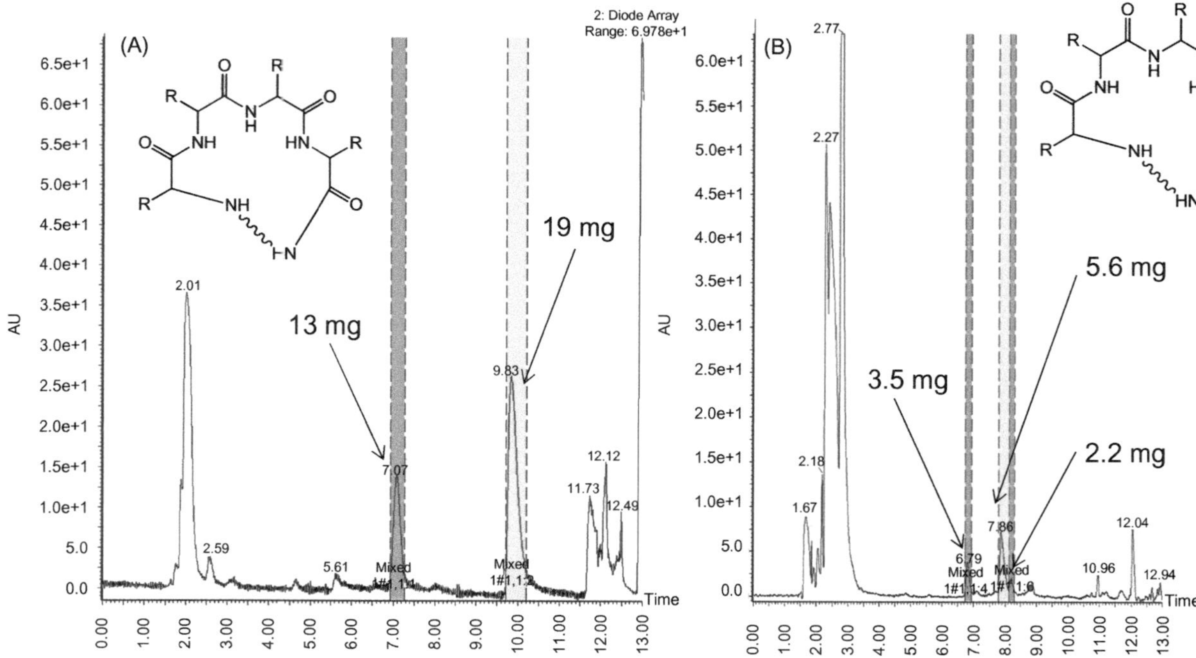

FIGURE 10.14 Separation of the diastereoisomers of two cyclopeptides. (A) Column: Princeton 4-ethylyridine (4EP), 250 mm × 30 mm, 5 μm, gradient 14–24% MeOH in 10 min. Sample (50 mg) dissolved in MeOH/dichloromethane 1:2. (B) Column: Princeton 4-ethylyridine (4EP), 250 mm × 30 mm, 5 μm, gradient 9–14% MeOH in 10 min. Sample (30 mg) dissolved in MeOH/dichloromethane 1:2. *Source: Adapted from reference Francotte E, Adam I, Mann T, Wolf T. SFC evolving as the prevailing purification technique in medicinal chemistry. Lecture at the 6th International Conference on Packed-Column SFC, Brussels, Belgium, October 4–5, (2012). Slides available from <http://greenchemistrygroup.org/pdf/2012/Francotte.pdf>.*

introduction of the new SFC-MS Prep-100 system by Thar (now Waters), set a new standard for this kind of purification. More recently Ross et al. successfully developed a platform to purify more than 90,000 samples (10–20 mg amounts) per year using this MS-based preparative SFC instrument employing a generic protocol [140].

10.7 PREPARATIVE SFC BY SIMULATED MOVING BED CHROMATOGRAPHY (SMB)

As a logical consequence of the successful development of SMB for liquid chromatography, especially for chiral separations, an SFC version was developed by Novasep in France [146,147] and a few years later at the Technical University of Hamburg in Germany [148]. The feasibility of performing supercritical fluid SMB (SF-SMB) was demonstrated by both groups but interest remained low, considering the technical challenge and the fact that the benefit was insignificant compared to LC-SMB in terms of greenness, as the latter technique is highly efficient regarding solvent consumption and solvent recycling (up to 99.5%). A comparison between batch SFC and SMB-SFC was performed by Peper et al. with the conclusion that batch SFC provided higher productivity [149,150]. Nevertheless there was a regain of interest in SFC-SMB at the Shou University in Taiwan in the Liang group. He developed SF-SMB processes for the isolation of substances of pharmaceutical interest, such as sesamin and sesamolin [151] or resveratrol and emodin [152].

10.8 LARGE SCALE AND INDUSTRIAL APPLICATIONS

Even though SFC has been successfully implemented in a number of companies up to the pilot scale for the separation of kilograms of enantiomers, it has not yet been embraced as a purification technique in production. It is possible that some large preparative-scale SFC units have been implemented in industry and have not been reported. Of the few semiindustrial applications reported, the purification of docosahexaenoic acid (DHA) and eicosapentaenoic acid (EPA) is probably the most relevant (Fig. 10.15). Interest in the separation of these high-value compounds started in Japan and a pilot unit was built in 1997 [122]. The technical and economic feasibility of separating DHA-ethyl ester and EPA-ethyl ester from tuna oil by SFC was studied by Akio et al. [153]. Lembke developed a production unit using a column of 1.15 m by 10 cm i.d. and operated at a flow rate of 200 kg/h [154]. This process reduced production costs by 75% compared with HPLC. Additional

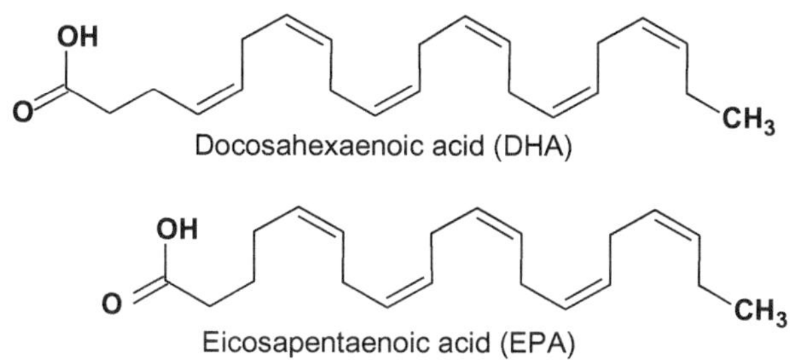

FIGURE 10.15 Chemical structures of Docosahexaenoic acid (DHA) and Eicosapentaenoic acid (EPA).

examples of large-scale SFC separations include the purification of the insecticide piperonyl butoxide and the separation of *cis*- and *trans*-phytol [155]. Productivity of 1.42 kg/day of purified piperonyl butoxide was possible using a 25 cm × 10 cm i.d. column (C18) at a flow rate of 90 kg CO_2 per hour with a cycle time of 2 min. Additional information for large-scale SFC purifications can be found in the review articles by Jusforgues et al. [155] and by Cox [83].

The separation phytonutrients (carotenes, vitamin E, sterols, squalene, α-tocopherol, α-tocotrienol, γ-tocopherol, and γ-tocotrienol) from palm oil at the pilot plant scale was reported by Ng at a 2008 conference in Zurich [156]. A SFC 600 unit was used with a 35 cm i.d. column at a flow rate of 600 kg CO_2/h. Some details of the process were reported recently [157].

10.9 CONCLUSION

For prepafurtherrative chiral separations from the milligrams to kilogram scale, packed-column SFC is now the technique of choice. HPLC remains a complementary technique when the separation requires a mobile phase with high alcohol content or for very large-scale separations (SMB). For achiral purification of small and to some extent medium size molecules, the successful implementation of preparative MS-triggered SFC platforms in many laboratories has initiated a trend towards SFC as the first choice while reversed-phase liquid chromatography (RP-HPLC) is trending to second choice. RP-HPLC it still used but only for samples which are not appropriate for processing by SFC. The success of preparative SFC cannot be estimated by the number of publications since most applications are performed in industry and are not disclosed for intellectual property reasons. For very large-scale applications, there is still a need to explore the scope and limitations of SFC. It can, however, be anticipated that SFC will replace RP-HPLC for further applications. Even though RP-HPLC will remain a complementary technique, SFC is clearly changing the face of the purification world, and if it fits the purpose, SFC should be preferred given the incontestable environmental and economic advantages. In any case, it seems that SFC has now surmounted its hurdles and is unlikely to return to a depression phase, as was regularly the case in the past. This is encouraging in the context of the growing awareness of our society and the big debate about global warming and energy waste, which has triggered a consensus regarding the fact that we have to favor processes which are less energy consuming and more sustainable.

References

[1] Cagniard de la Tour C. Presentation of some results obtained by the combined action of heat and compression on certain liquids, such as water, alcohol, sulfuric ether, and distilled petroleum spirit. Annales de chimie et de physique 1822;21:127–32.

[2] Critical point (thermodynamics). <https://en.wikipedia.org/wiki/Critical_point_(thermodynamics)>.

[3] Grand-Guillaume Perrenoud A, Hamman C, Goel M, Veuthey JL, Guillarme D, Fekete S. Maximizing kinetic performance in supercritical fluid chromatography using state-of-the-art instruments. J Chromatogr A 2013;314:288–97.

[4] Klesper E, Corwin AH, Turner DA. High pressure gas chromatography above critical temperatures. J Org Chem 1962;27:700–1.

[5] Taylor LT. Supercritical fluid chromatography for the 21st century. J Supercrit Fluids 2009;47:566–73.

[6] Berger TA. Packed column SFC. Letchworth, UK: Royal Society of Chemistry; 1995.

[7] Berger C, Perrut M. Preparative supercritical fluid chromatography. J Chromatogr 1990;505(1):37–43.

[8] Mourier PA, Eliot E, Caude MH, Rosset RH, Tambute AG. Supercritical and subcritical fluid chromatography on a chiral stationary phase for the resolution of phosphine oxide enantiomers. Anal Chem 1985;57:2819–23.

[9] Jusforgues P, Shaimi M, Barth D. Preparative supercritical fluid chromatography: grams, kilograms, and tons!. In: Anton K, Berger C, editors. Supercritical fluid chromatography with packed columns. New York: Marcel Dekker, Inc; 1998. p. 417.

[10] Lazarescu V, Mulvihill MJ, Ma L. Achiral preparative supercritical fluid chromatography. In: Webster GK, editor. Advances and applications in pharmaceutical analysis: supercritical fluid chromatography. Singapore: Pan Standford Publishing; 2014. p. 100–8.

[11] Larry M, Ian S. Evaluation of injection conditions for preparative supercritical fluid chromatography. J Chromatogr A 2012;1250:256–63.

[12] Shaimi M, Cox GB. PIC Solution Inc., Application Note 6/13.

[13] Shaimi M, Cox GB. Mixed-stream vs modifier stream injections injection by extraction: a novel sample introduction technique for preparative SFC. Chromatogr Today November/December 2014;42–45.

[14] Horikawa Y, Kamezawa K, Kanomata T, Bounoshita M, Saito M. Novel injection method in preparative supercritical fluid chromatography. Poster presented at the 2nd International Conference on Packed-Column SFC. Zurich, Switzerland, 1–2 October 2008. Available from <http://www.jascoinc.com/docs/application-notes/InjectionMethod_PrepSFC.pdf>.

[15] Dai Y, Li G, Rajendran A. Peak distortions arising from large-volume injections in supercritical fluid chromatography. J Chromatogr A 2015;1392:91–9.
[16] Hirata Y, Kawaguchi Y, Kitano K. Large volume injection for preparative supercritical fluid chromatography. Chromatographia 1995;40(1/2):42–6.
[17] Vajda P, Kamarei F, Felinger A, Guiochon G. Comparison of volume and concentration overloadings in preparative enantio-separations by supercritical fluid chromatography. J Chromatogr A 2014;1341:57–64.
[18] Perrut M. Procédé d'extraction-séparation-fractionnement par fluides supercritiques et dispositif pour sa mise en oeuvre. French Patent 8,510,468, (1985); Eur. Pat. 8,640,139, (1986).
[19] Chen R, Cole J. Feasibility of using ELSD to trigger fraction collection in small-scale purification by SFC. LC-GC Europe 2009;22(7):33.
[20] Yanik GW. Polarimetric detection in supercritical fluid chromatography. In: Webster GK, editor. Advances and applications in pharmaceutical analysis: supercritical fluid chromatography. Singapore: Pan Standford Publishing; 2014. p. 333–46.
[21] Wentz ES. Multimode collections in prep SFC working towards open access SFC Sub-2 um particles for SFC. First International Conference on Packed Column SFC, Pittsburgh, 23-25 September 2007. Slides available from <http://www.greenchemistrygroup.org/past-conferences>.
[22] Kanomata T, Silverman C, HorikawaY, Saito H, Bounoshita M, Saito M. Advantages of Circular Dichroism (CD) Detection for the Determination of Fractionation Timing in Preparative Supercritical Fluid Chromatography (Prep-SFC) for Chiral Separations. Poster presented at First International Conference on Packed Column SFC, Pittsburgh, 23-25 September 2007. Available from <http://www.greenchemistrygroup.org/past-conferences> or LCGC The Peaks, December 2007:7–22.
[23] Mikita I, Masataka M, Toshihiro N. A chemiluminescence reaction analyzer for a supercritical fluid. Bunseki Kagaku 1995;44:91–6.
[24] Wang T, Barber M, Hardt I, Kassel DB. Mass-directed fractionation and isolation of pharmaceutical compounds by packed-column supercritical fluid chromatography/mass spectrometry. Rap Comm Mass Spect 2001;15(22):2067–75.
[25] Zhang X, Towle MH, Felice CE, Flament JH, Goetzinger WK. Development of a mass-directed preparative supercritical fluid chromatography purification system. J Comb Chem 2006;8(5):705–14.
[26] Fuchs G, Doguet L, Perrut M, Tambute A, Le Goff P. Enantiomer fractionation by preparative supercritical fluid chromatography. Proceedings 2nd international symposium on supercritical fluids, Boston (MA). 20–22 May 1991.
[27] Miller L, Potter M, Barnhart W, Gahm K, Eschelbach J. Application of SFC in pharmaceutical discovery and early development. 1st International Conference on Packed Column SFC. September 25, 2007. <http://www.greenchemistrygroup.org/past-conferences>.
[28] Whatley J. Enantiomeric separation by packed column chiral supercritical fluid chromatography. J Chromatogr A 1995;697:251–5.
[29] Geiser F, Schultz M, Betz L, Shaimi M, Lee J, Champion W. Direct, preparative enantioselective chromatography of propranolol hydrochloride and thioridazine hydrochloride using carbon dioxide-based mobile phases. J Chromatogr A 1999;865:227–33.
[30] Saito M, Yamauchi Y, Higashidate S, Okamoto I. Preparative supercritical fluid chromatography and its applications to chiral separation. In: Hatano H, Hanai T, editors. International symposium on chromatography, the 35th anniversary of the research group on liquid chromatography in Japan, Yokohama, January 22-25, 1995;863–866.
[31] Berger TA. The past, present, and future of analytical supercritical fluid chromatography. Chromatogr Today 2014;7(3):26–9.

[32] Wu DR. Chiral supercritical fluid chromatography in drug discovery: from analytical to multigram-scale. 1st International Conference on Packed Column SFC. September 25, 2007. <http://www.greenchemistrygroup.org/past-conferences>.

[33] Sun P, Wang C, Breitbach ZS, Zhang Y, Armstrong D. Development of new HPLC chiral stationary phases based on native and derivatized cyclofructans. Anal Chem 2009;81:10215–26.

[34] Zhang T, Holder E, Franco P, Lindner W. Zwitterionic chiral stationary phases based on cinchona and chiral sulfonic acids for the direct stereoselective separation of amino acids and other amphoteric compounds. J Sep Sci 2014;37:1237–47.

[35] Okamoto Y, Kawashima M, Hatada K. Useful chiral packing materials for high-performance liquid chromatographic resolution of enantiomers: phenylcarbamates of polysaccharides coated on silica gel. J Am Chem Soc 1984;106(18):5357–9.

[36] Okamoto Y, Yashima E. Polysaccharide derivatives for chromatographic separation of enantiomers. Angew Chem Int Ed 1998;37:1020–43.

[37] Francotte E, Zhang T. Patent, PCT Int. Appl. WO 9704011, 1997.

[38] Francotte ER. Enantioselective chromatography: an essential and versatile tool for the analytical and preparative separation of enantiomers. Chimia 1997;5:717–25.

[39] Francotte ER. Enantioselective chromatography as a powerful alternative for the preparation of drug enantiomers. J Chromatogr A 2001;906:379–97.

[40] Francotte E, Huynh D. Immobilized halogeno-phenylcarbamate derivatives of cellulose as novel chiral stationary phases for enantioselective drug analysis. J Pharm Biomed Anal 2002;27:421–9.

[41] Francotte E, Huynh D, Wetli H. *G.I.T* Multi-parallel chiral screening. Screening of chiral Stationary and mobile phases for method development of preparative separations. Lab J Eur 2006;10:46–8.

[42] Wetli H, Francotte E. Automated screening platform with isochronal-parallel analysis and conditioning for rapid method development of preparative chiral separations. J Sep Sci 2007;30:1255–61.

[43] Amoss C, Cox G, Franco P, Zhang T. CHIRALPAK® ICTM—An Immobilized Polysaccharide Chiral Stationary Phase with a Unique Chiral Selector LCGC's The Application Notebook (September 2008).

[44] Zhang T, Nguyena D, Franco P, Isobe Y, Michishita T, Murakami T. Cellulose tris (3,5-dichlorophenylcarbamate) immobilised on silica: a novel chiral stationary phase for resolution of enantiomers. J Pharm Biomed Anal 2008;46:882–91.

[45] Francotte E, Diehl G. Extending the SFC applicability in the field of enantioselective separations. Lecture presented at the 2nd International Conference on Packed-Column SFC, Zurich, Switzerland, 1–2 October 2008. Slides available from <http://media.wix.com/ugd/2239fc_c0b2afbc6f5c48aeb4fde1a662f05798.pdf>.

[46] Cox GB. Solvent and stationary phase selectivity in enantioselective SFC. 1st International Conference on Packed-Column SFC, Pittsburgh, USA. 23–25 September 2007. <http://www.greenchemistrygroup.org/past-conferences>.

[47] Miller L. Evaluation of non-traditional modifiers for analytical and preparative enantioseparations using supercritical fluid chromatography. J Chromatogr A 2012;1256:261–6.

[48] Al-Othman ZA, Ali I, Asim M, Khan TA. Recent trends in chiral separations on immobilized polysaccharides CSPs. Comb Chem High Throughput Screening 2012;15(4):339–46.

[49] Huang J, Yuan M. Separation of substituted phenylpiperazine derivatives with immobilized polysaccharide-based chiral stationary phases by supercritical and subcritical fluid chromatography. J. Chinese Pharm Sci 2013;22(3):244–50.

[50] Miller L. Use of dichloromethane for preparative supercritical fluid chromatographic enantioseparations. J Chromatogr A 2014;1363:323–30.

[51] Lee JT, Watts WL, Barendt J, Yan TQ, Huang Y, Riley F, et al. On the method development of immobilized polysaccharide chiral stationary phases in supercritical fluid chromatography using an extended range of modifiers. J Chromatogr A 2014;1374:238–46.
[52] Da Silva JO, Coes B, Frey L, Mergelsberg I, McClain R, Nogle L, et al. Evaluation of nonconventional polar modifiers on immobilized chiral stationary phases for improved resolution of enantiomers by supercritical fluid chromatography. J Chromatogr A 2014; 1328:98–103.
[53] Berger T, Deye J. Role of additives in packed column supercritical fluid chromatography: suppression of solute ionization. J Chromatogr 1991;547:377–92.
[54] Stringham R. Chiral separation of amines in subcritical fluid chromatography using polysaccharide stationary phases and acidic additives. J Chromatogr A 2005;1070: 163–70.
[55] Ye Y, Lynam K, Stringham R. Effect of amine mobile phase additives on chiral subcritical fluid chromatography using polysaccharide stationary phases. J Chromatogr A 2004;1041:211–17.
[56] Blackwel JA, Stringham RW, Weckwerth JD. Effect of mobile phase additives in packed-column subcritical and supercritical fluid chromatography. Anal Chem 1997;69:409–15.
[57] Hamman C, Schmidt Jr DE, Wong M, Hayes M. The use of ammonium hydroxide as an additive in supercritical fluid chromatography for achiral and chiral separations and purifications of small, basic medicinal molecules. J Chromatogr A 2011;1218: 7886–94.
[58] Ventura M, Murphy B, Goetzinger W. Ammonia as a preferred additive in chiral and achiral applications of supercritical fluid chromatography for small, drug-like molecules. J Chromatogr A 2012;1220:147–55.
[59] Francotte E. Why should sfc be the first technology choice for the purification of pharmaceuticals? Lecture presented at the 8th International Conference on Packed-Column SFC, Basel, Switzerland, 10 October 2014. Slides available from <http://media.wix.com/ugd/2239fc_6eee04127c92436ab62c550d8b8a9317.pdf>.
[60] Francotte E. Practical advances in SFC for the purification of pharmaceutical molecules. LCGC Europe 2016;29(4):194–204.
[61] Woods RM, Breitbach ZS, Armstrong DW. Comparison of enantiomeric separations and screening protocols for chiral primary amines by SFC and HPLC. LCGC Europe 2015;28(1):26–33.
[62] De Klerck K, Vander Heyden Y, Mangelings D. Pharmaceutical-enantiomers resolution using immobilized polysaccharide-based chiral stationary phases in supercritical fluid chromatography. J Chromatogr A 2014;1328:85–97.
[63] Sharp VS, Hicks N, Stafford J. A multimodal liquid and supercritical fluid chromatography chiral separation screening and column maintenance strategy designed to support molecules in pharmaceutical development (Part 1). LCGC Europe 2013;26 (11):608–18.
[64] De Klerck K, Tistaert C, Mangelings D, Vander Heyden Y. Updating a generic screening approach in sub- or supercritical fluid chromatography for the enantioresolution of pharmaceuticals. J Supercrit fluids 2013;80:50–9.
[65] Subbarao L, Wang Z, Chen R. Multi-channel SFC system for fast chiral method development and optimization. LCGC Special Issues on Supercritical Fluid Chromatography (SFC), Application Notebook, 2010 (November), 20–24.
[66] Hamman C, Wong M, Aliagas I, Ortwine DF, Pease J, Schmidt DE, et al. The evaluation of 25 chiral stationary phases and the utilization of sub-2.0 μm coated polysaccharide chiral stationary phases via supercritical fluid chromatography. J Chromatogr A 2013;1305:310–19.

[67] Franco P, Zhang T. Common screening approaches for efficient analytical method development in LC and SFC on columns packed with immobilized polysaccharide-derived stationary phases. In: Scriba GK, editor. Chiral separations: methods and protocols. Methods in Molecular Biology, vol. 970. Springer Science 2013. p. 113–26.
[68] Speybrouck D, Corens D, Argoullon JM. Screening strategy for chiral and achiral separations in supercritical fluid chromatography mode. Curr Topics Med Chem 2012;12(11):1250–63.
[69] Hicks MB, Zhang Y, Moore D, Apedo A. Advancing SFC method development with a multi-column supercritical fluid chromatography with gradient screening. Am Pharm Rev 2011;14(6).
[70] Webster GK, Kott L. Method development for pharmaceutical chiral chromatography. In: Ahuja S, Scypinski S, editors. Separation Science and Technology. Handbook of Modern Pharmaceutical Analysis, vol. 10. Elsevier 2011. p. 251–82.
[71] Kiplinger JP, Lefebvre PM, Kavrakis SK. Toward development of "generic" separation methods for achiral pharmaceutical analysis using SFC. Poster presentation at SFC 2009. <http://media.wix.com/ugd/2239fc_6f38e23017ff49c38ea3e91b01580116.pdf>.
[72] Ning J, Schafer W, Welch C, Gong X. Parallel SFC method development screening for enhanced speed and quality. SFC 2011, July 20–22, 2011, New York City. Slides available from <http://media.wix.com/ugd/2239fc_3942a40defba42e49aa7d93a-da306602.pdf>.
[73] Francotte E, Diehl G. SFC: The universal purification technique? Lecture at the 4th International Conference on Packed-Column SFC, Stockholm, Sweden September 15–16, 2010. Slide available can be loaded from <http://media.wix.com/ugd/2239fc_fc3a01aa76d5458a86c0f0eaeac983e8.pdf>.
[74] Miller L. Pharmaceutical purifications using Preparative Supercritical Fluid Chromatography. Chimica Oggi 2014;32(2):23–6.
[75] Villeneuve M, Schmidt R, Zhao Y. Introduction of unique Analytical and Preparative SFC units. Lecture presented at the 5th International Conference on Packed-Column SFC, New York, USA, 20–22 July 2011. Slide available from <http://media.wix.com/ugd/2239fc_896087ae6931466d8c871ff783b75295.pdf>.
[76] Van Anda J. The use of supercritical fluid chromatography (SFC) for chiral pharmaceutical analytical and preparative separations. Am Pharm Rev 2009;12:48–53.
[77] Miller L, Potter M. Preparative chromatographic resolution of racemates using HPLC and SFC in a pharmaceutical discovery environment. J Chromatogr B 2008;875: 230–6.
[78] Wu DR, Leith L, Balasubramanian B, Palcic T, Wang-Iverson D. The impact of chiral supercritical fluid chromatography in drug discovery: from analytical to multigram scale. Am Laborat 2006;38:24–6.
[79] Seest E, Belvo M, Perun T, McDermott P. Green lab scale chiral chromatography: comparison of purification techniques for cost effective and efficient processing of samples while minimizing environmental impact. Lecture presented at the 5th International Conference on Packed-Column SFC, New York, USA, 20–22 July 2011. Slides available from <http://media.wix.com/ugd/2239fc_d6b5ee0432a5472-da202e1f2d9318ed1.pdf>.
[80] Welch C, Leonard W, DaSilva J, Biba M, Albaneze-Walker J, Henderson D, et al. Preparative chiral SFC as a green technology for rapid access to enantiopurity in pharmaceutical process research. LCGC Europe 2005;18:264–72.
[81] DaSilva JO, Yip HS, Hegde V, Zaks A. Supercritical fluid chromatography (SFC) as a green chromatographic technique for support in rapid development of pharmaceutical candidates. Lecture Presented at the 2nd International Conference on Packed-Column SFC, Zurich, Switzerland, 1–2 October 2008. Slides available from <http://media.wix.com/ugd/2239fc_233026ba94ee41a5b79b463c0de9b66a.pdf>.

[82] Villeneuve MS, Miller LA. Preparative-scale supercritical fluid chromatography. In: Cox GB, editor. Preparative enantioselective chromatography. Blackwell Publishing; 2005. p. 205–23.
[83] Ventura MC. Chiral preparative supercritical fluid chromatography,. In: Webster GK, editor. Advances and applications in pharmaceutical analysis: supercritical fluid chromatography. Singapore: Pan Standford Publishing; 2014. p. 171–93.
[84] Phinney KW, Stringham RW. Chiral separations using supercritical fluid chromatography. In: Subramanian G, editor. Chiral separation techniques. 3rd ed. Weinheim: Wiley-VCH; 2007. p. 135–54.
[85] Plotka JM, Biziuk M, Morrison C, Namiesnik J. Pharmaceutical and forensic drug applications of chiral supercritical fluid chromatography. TrAC, Trends Anal Chem 2014;56:74–89.
[86] Phinney KW. Enantioselective separations by packed column subcritical and supercritical fluid chromatography. Anal Bioanal Chem 2005;382(3):639–45.
[87] Freund E, Abel S, Huthmann E, Lill J. Chiral chromatography in the early phases of pharmaceutical development. Chimica Oggi 2009;27(5):62–4.
[88] Da Silva JO, Yip HS, Hegde V. Supercritical fluid chromatography (SFC) as a green chromatographic technique to support rapid development of pharmaceutical candidates. Am Pharm Rev 2009;12(1):98–104.
[89] Lixing N, Zhong D, Shuangcheng M. Improved chiral separation of (R,S)-goitrin by SFC: an application in traditional Chinese medicine. J Anal Meth Chem 2016; 5782942/1–5782942/5.
[90] Rossi D, Marra A, Rui M, Brambilla S, Juza M, Collina S. "Fit-for-purpose" development of analytical and (semi)preparative enantioselective high performance liquid and supercritical fluid chromatography for the access to a novel σ1 receptor agonist. J Pharm Biomed Anal 2016;118:363–9.
[91] Krief A, Dunkle M, Sandra P, Bahar M, Bultinck P, Herrebout W. Elucidation of the absolute configuration of rhizopine by chiral supercritical fluid chromate ography and vibrational circular dichroism. J Sep Sci 2015;38(14):2545–50.
[92] Shoup TM, McCauley JP, Lee Jr DF, Chen R, Normandin MD, Bonab AA, et al. Synthesis of the dopamine D2/D3 receptor agonist (+)-PHNO via supercritical fluid chromatography: preliminary PET imaging study with [3-11C]-(+)PHNO. Tetrahedron Lett 2014;55(3):682–5.
[93] Chau WF, Black MA, Clarke A, Durrant C, Gausemel I, Khan I, et al. Exploration of the impact of stereochemistry on the identification of the novel translocator protein PET imaging agent [18F]GE-180. Nucl Med Biol 2015;42(9):711–19.
[94] de Mas N, Natalie KJ, Quiroz F, Rosso VW, Chen DC, Conlon DA. A partial classical resolution/preparative chiral supercritical fluid chromatography method for the rapid preparation of the pivotal intermediate in the synthesis of two nonsteroidal glucocorticoid receptor modulators. Org Proc Res Dev 2016;20(5):934–9.
[95] Cho-Schultz S, Patten MJ, Huang B, Elleraas J, Gajiwala KS, Hickey MJ, et al. Solution-phase parallel synthesis of Hsp90 inhibitors. J Comb Chem 2009;11(5): 860–74.
[96] Black R, Esser C, Morris J, Leonard WR Jr, Sajonz P, Helmy R, et al. Chiral SFC in support of drug discovery: basic research through preclinical development. Lecture at the 1st International Conference on Packed Column SFC. September 25, 2007. Slides available <http://www.greenchemistrygroup.org/past-conferences>.
[97] Farrell B, Riley F. Unleashing the power of sfc: a case study in drug discovery and development. Lecture at the 1st International Conference on Packed Column SFC. September 25, 2007. Slides available from <http://www.greenchemistrygroup.org/past-conferences>.

[98] Yan TQ, Orihuela C, Swanson D. The application of preparative batch HPLC, supercritical fluid chromatography, steady-state recycling, and simulated moving bed for the resolution of a racemic pharmaceutical intermediate. Chirality 2008;20 (2):139–46.
[99] Hartmann W, Klesper E. Preparative supercritical fluid chromatography of styrene oligomers. J Polym Sci Polym Lett. 1977;15(12):713–19.
[100] Poole CF. Stationary phases for packed-column supercritical fluid chromatography. J Chromatogr A 2012;1250:157–71.
[101] West C, Lesellier E. Chemometric methods to classify stationary phases for achiral packed column supercritical fluid chromatography. J Chemom 2012;26:52–65.
[102] West C, Lesellier E. A unified classification of stationary phases for packed column supercritical fluid chromatography. J Chromatogr A 2008;1191:21–39.
[103] Khater S, West C, Lesellier E. Characterization of five chemistries and three particle sizes of stationary phases used in supercritical fluid chromatography. J Chromatogr A 2013;1319:148–59.
[104] West C, Khalikova MA, Lesellier E, Heberger K. Sum of ranking differences to rank stationary phases used in packed column supercritical fluid chromatography. J Chromatogr A 2015;1409:241–50.
[105] Korsgren P, Lanborg Weinmann A. Column selection for achiral purification using SFC-MS. Am Pharm Rev 2012;15(4).
[106] McClaina R, Hyunb MH, Lib Y, Welch CJ. Design, synthesis and evaluation of stationary phases for improved achiral supercritical fluid chromatography separations. J Chromatogr A 2013;1302:163–73.
[107] Ebinger K, Weller HN. Comparative assessment of achiral stationary phases for high throughput analysis in supercritical fluid chromatography. J Chromatogr A 2014;1332:73–81.
[108] Regalado EL, Welch CJ. Separation of achiral analytes using supercritical fluid chromatography with chiral stationary. TrAC, Trends Anal Chem 2015;67:74–81.
[109] Shibata T, Shiruka S, Ohnishi A, Mukarami Y, Ueda K. Achiral SFC on polysaccharide phases. Lecture Presented at the 9th International Conference on Packed column SFC, Philadelphia, Pensylvannia 2015. Slides available from <http://media.wix.com/ugd/2239fc_b5a4641c46d24e92a3b2e6fdd3d6c6d1.pdf>.
[110] Jusforgues P, Berger C, Perrut M. New separation process: preparative supercritical fluid chromatography. Chem Ingen Techn 1987;59:666–7.
[111] Jusforgues P. Preparative supercritical fluid chromatography: principles, potential, and development. Spectra 2000 1992;166:53–6.
[112] Saito M, Yamauchi Y. Isolation of tocopherols from wheat germ oil by recycle semi-preparative supercritical fluid chromatography. J Chromatogr 1990;505(1):257–71.
[113] Yamauchi Y, Saito M. Fractionation of lemon-peel oil by semi-preparative supercritical fluid chromatography. J Chromatogr. 1990;505(1):237–46.
[114] Ute K, Miyatake N, Asada T, Hatada K. Stereoregular oligomers of methyl methacrylate. 6. Isolation of isotactic and syndiotactic methyl methacrylate oligomers from 19-mer to 29-mer by preparative supercritical fluid chromatography and their thermal analysis. Polym Bull (Berlin, Germany) 1992;28(5):561.
[115] Fritz G, Feucht G. Analytical and preparative separation of functional carbosilanes and phosphines by means of SFC (supercritical fluid chromatography). Zeit Anorg Allgem Chem 1991;593:69–89.
[116] Hanson M. Small-scale preparative supercritical fluid chromatography of cyproterone acetate. Chromatographia 1995;40(3/4):139–42.
[117] Hanson M. Micropreparative supercritical fluid chromatography as a bench-scale method for steroid purification. LC-GC 1996;14(2) 152, 154, 156, 158.

[118] Ihara E, Tanabe M, Nakayama Y, Nakamura A, Yasuda H. Characterization of lactone oligomers isolated by preparative SFC. Macrom Chem Phys 1999;200(4):758–62.
[119] Flament I, Keller U, Wunsche L. Use of semi-preparative supercritical fluid chromatography for the separation and isolation of flavor and food constituents. In: Rizvi SSH, editor. Supercritical fluid processing of food and biomaterials. New York: Springer; 1994, p. 62–74.
[120] Cretier G, Neffati J, Rocca JL. Preparative LC and preparative SFC: two complementary techniques in the fractionation of an impurity from a major component. J Chromatogr Sci 1994;32(10):449–54.
[121] Jacobson GB, Markides KE, Langstrom B. Supercritical fluid synthesis and online preparative supercritical fluid chromatography of 11C-labeled compounds in supercritical ammonia. Acta Chem Scand 1997;51(3, Suppl):418–25.
[122] Kadota Y, Tanaka I, Ohtsu Y, Yamaguchi M. Separation of polyunsaturated fatty acids by chromatography using a silver-loaded spherical clay. I. Pilot-scale preparation of high-purity docosahexaenoic acid by supercritical fluid chromatography. Nihon Yukagakkaishi 1997;46(4):397–403.
[123] Y Wang, G Stutz, S Roggo, G Diehl, E Francotte. Development of SFC for high throughput purification in natural product drug discovery. Lecture presented at the 2nd International Conference on Packed-Column SFC, Zurich, Switzerland, 1–2 October 2008. Slides available from <http://media.wix.com/ugd/2239fc_c1832be-f0ef548aba914f87bc86c870d.pdf>.
[124] Song W, Qiao X, Liang WF, Ji S, Yang L, Wang Y, et al. Efficient separation of curcumin, demethoxycurcumin, and bisdemethoxycurcumin from turmeric using supercritical fluid chromatography: from analytical to preparative scale. J Sep Sci 2015;38(19):3450–3.
[125] Vicente G, Garcia-Risco MR, Fornari T, Reglero G. Isolation of carsonic acid from rosemary extracts using semi-preparative supercritical fluid chromatography. J Chromatogr A 2013;1286:208–15.
[126] Pokrovskii OI, Krutikova AA, Ustinovich KB, Parenago OO, Moshnin MV, Gonchukov SA, et al. Preparative separation of methoxy derivatives of psoralen using supercritical-fluid chromatography. Russ J Phys Chem B 2013;7(8): 901–15.
[127] Wang Z. Supercritical fluid chromatography—a powerful tool for polymethoxyflavone analysis and isolation. Lecture presented at the 3rd International Conference on Packed-Column SFC. Philadelphia, PA, USA. July 22–23, 2009. Slides available from <http://media.wix.com/ugd/2239fc_180972ec65134882a1a-cad3b88b03557.pdf>.
[128] Patel MA, Riley F, Ashraf-Khorassani M, Taylor LT. Supercritical fluid chromatographic resolution of water soluble isomeric carboxyl/amine terminated peptides facilitated via mobile phase water and ion pair formation. J Chromatogr A 2012;1233:85–90.
[129] Montanes F, Rose P, Tallon S, Shirazi R. Separation of derivatized glucoide anomers using supercritical fluid chromatography. J Chromatogr A 2015;1418:218–23.
[130] Terfloth G. Preparative isolation of impurities. In: Smith RJ, Webb ML, editors. Analysis of drug impurities. Oxford: Blackwell Publishing; 2007. p. 215–34.
[131] Zelesky T. Supercritical fluid chromatography (SFC) as an isolation tool for the identification of drug related impurities. Am Pharm Rev 2008;11(7):56–60, 62.
[132] Buskov S, Hasselstrom J, Olsen CE, Sorensen H, Sorensen JC, Sorensen S. Supercritical fluid chromatography as a method of analysis for the determination of 4-hydroxybenzylglucosinolate degradation products. J Biochem Biophys Meth 2000; 43(1–3):157–74.

[133] Klobcar S, Prosen H. Isolation of oxidative degradation products of atorvastatin with supercritical fluid chromatography. Biomed Chromatogr 2015;29(12):1901–6.

[134] Aurigemma C. Breaking the flow rate barrier: utilization of high flow mass-directed sfc for pharmaceutical applications. Lecture at the 3rd International Conference on Packed-Column SFC. Philadelphia, PA, USA. July 22–23, 2009. Slides available from <http://media.wix.com/ugd/2239fc_af89841a5cc843f4842e24142ce3bec0.pdf>.

[135] Ma L, Lazarescu V, Mulvihill MJ. Toward a "Universal Approach" for mass-directed sfc purification of small molecule compound libraries. Lecture at the 4th International Conference on Packed-Column SFC, September 15–16, 2010 Stockholm, Sweden. Slides available from <http://media.wix.com/ugd/2239fc_764d60b4afcf4f73aec61e83f4918955.pdf>.

[136] Mich A, Matthes B, Chen R, Buehler S. A comparative study on the purification of library compounds in drug discovery using mass-directed preparative SFC and preparative RPLC. LC-GC Europe 2010;12–13.

[137] Van Anda J. Use of SFC/MS in the purification of achiral pharmaceutical compounds. Am Pharm Rev 2010;13:111–15.

[138] Ebinger K, Weller HN, Kiplinger J, Lefebvre P. Evaluation of a new preparative supercritical fluid chromatography system for compound library purification: the TharSFC SFC-MS Prep-100 system. J Lab Automat 2011;16(3):241–9.

[139] Francotte E, Adam I, Mann T, Wolf T. SFC evolving as the prevailing purification technique in medicinal chemistry. Lecture at the 6th International Conference on Packed-Column SFC, Brussels, Belgium, October 4–5, 2012. Slides available from <http://greenchemistrygroup.org/pdf/2012/Francotte.pdf>.

[140] Rosse G. Maximizing efficiency in drug discovery: SFC-MS as the technique of choice for small molecules purification. Lecture at the 9th International Conference on Packed-Column SFC. Philadelphia, PA, USA, July 23, 2015. Slide available from <http://media.wix.com/ugd/2239fc_9c99ce71471742199f2637d360a10cb5.pdf>.

[141] Coleman K. High-throughput preparative separations from combinatorial libraries. Analusis 1999;27(8):719–23.

[142] Berger TA, Fogleman K, Staats T, Bente P, Crocket I, Farrell W, et al. The development of a semi-preparatory scale supercritical-fluid chromatograph for high-throughput purification of 'combi-chem' libraries. J Biochem Biophys Meth 2000;43(1-3):87–111.

[143] Hochlowski J, Olson J, Pan J, Sauer D, Searle P, Sowin T. Purification of HTOS libraries by supercritical fluid chromatography. J Liq Chromatogr Rel Techn 2003;26(3):333–54.

[144] Hochlowski J. High-throughput purification: triage and optimization. In: B. Yan, editor. Analysis and purification methods in combinatorial chemistry. Chemical Analysis, vol. 163. New York: John Wiley & Sons; 2004. p. 281–06.

[145] Ventura M, Farrell W, Aurigemma C, Tivel K, Greig M, Wheatley J, et al. High-throughput preparative process utilizing three complementary chromatographic purification technologies. J Chromatogr A 2004;1036(1):7–13.

[146] Clavier JY, Nicoud RM, Perrut M. A new efficient fractionation process: the simulated moving bed with supercritical eluent. Process Technology Proceedings 1996, 12 (High Pressure Chemical Engineering), p. 429–34.

[147] Clavier JY, Nicoud RM, Perrut M. A New efficient fractionation process: the simulated moving bed with supercritical eluent. In: von Rohr PR, Trepp C, editors. High pressure chemical engineering. London: Elsevier Science; 1996. p. 429–34.

[148] Depta A, Giese T, Johannsen M, Brunner G. Separation of stereoisomers in a simulated moving bed-supercritical fluid chromatography plant. J Chromatogr A 1999;865:175–86.

[149] Peper S, Johannsen M, Brunner G. Preparative chromatography with supercritical fluids. Comparison of simulated moving bed and batch processes. J Chromatogra A 2007;1176(1-2):246–53.

[150] Johanssen M. Supercritical fluid chromatography: from analytical tool to industrial separation processes. Lecture presented the 4th International Conference on Packed-Column SFC, Sept. 15-16, 2010 Stockholm, Sweden. Slides available from <http://media.wix.com/ugd/2239fc_ca204fa5d0e146128a97c132d290d00f.pdf>.

[151] Liang MT, Liang RC, Huang LR, Hsu PH, Wu YH, Yen HE. Separation of sesamin and sesamolin by a supercritical fluid-simulated moving bed. Am J Anal Chem 2012;3:931–8.

[152] Liang MT, Liang RC, Yu S, Yan R. Separation of resveratrol and emodin by supercritical fluid-simulated moving bed chromatography. J Chromatogr Sep Tech 2014;4(3):1–5.

[153] Alkio M, Gonzalez C, Jantti M, Aaltonen O. Purification of polyunsaturated fatty acid esters from tuna oil with supercritical fluid chromatography. J Am Oil Chem Soc 2000;77(3):315–21.

[154] Lemke P. Production of high purity n-3 fatty acid-ethyl esters by process scale supercritical fluid chromatography. In: Anton K, Berger C, editors. Supercritical fluid chromatography with packed columns. New York: Marcel Dekker; 1998. p. 429–43.

[155] Jusforgues P, Shaimi M, Barth D. Preparative supercritical fluid chromatography: grams, kilograms, and tons!. In: Anton K, Berger C, editors. Supercritical fluid chromatography with packed columns. New York: Marcel Dekker; 1998. p. 403–27.

[156] Ng MH, Choo YM, Yahaya H. Pilot scale supercritical fluid chromatography: an experience. Lecture presented at the 2nd International Conference on Packed Column SFC. Zurich, Switzerland, 1–2 October 2008. Slides available from <http://media.wix.com/ugd/2239fc_8387fad969ff45afab2a826ab7712a4a.pdf>.

[157] Ng MH, Choo YM. Chromatography for the analyses and preparative separations of palm oil minor components. Am J Anal Chem 2015;6:645–50.

CHAPTER

11

Validation of Supercritical Fluid Chromatography Methods

A. Dispas[1], *P. Lebrun*[2] *and P. Hubert*[1]

[1]University of Liège, Liège, Belgium [2]Arlenda SA, Louvain-la-Neuve, Belgium

OUTLINE

Supercritical Fluid Chromatography.
DOI: http://dx.doi.org/10.1016/B978-0-12-809207-1.00011-2

11.1 INTRODUCTION

Analytical method validation is one of the main steps in an analytical method life cycle (Fig. 11.1). The United States Pharmacopeial Convention (USP), defines the method validation concept as follows: *"Validation of an analytical procedure is the process by which it is established, by laboratory studies, that the performance characteristics of the procedure meet the requirements for the intended analytical applications"* [1]. The Food and Drug Administration (FDA) provides a more generic definition: *"Analytical method validation is the process of demonstrating that an analytical method is suitable for its intended purpose"* [2]. All the regulatory agencies (FDA, EMA, GMP, etc.) require the validation of analytical

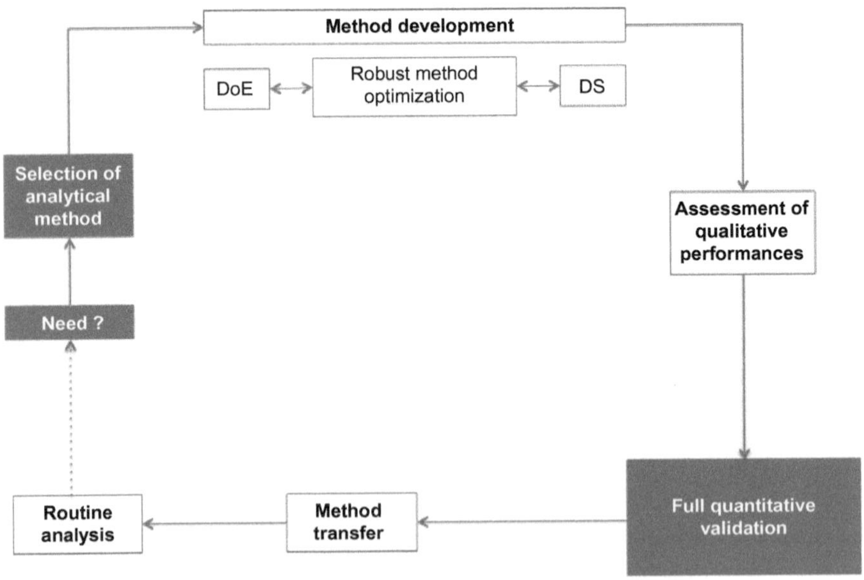

FIGURE 11.1 Analytical method life cycle.

methods but the deep evaluation of all required validation criteria and the conclusion about method acceptance remains confused [1–7]. The first objective of the present chapter is to give an overview of the validation concept and to propose a state-of-the-art validation approach, i.e., total error approach. Secondly, the validation of supercritical fluid chromatography (SFC) methods will be deeply discussed.

11.2 VALIDATION CRITERIA

11.2.1 Specificity

ICH defines specificity as *"the ability to assess unequivocally the analyte in the presence of components which may be expected to be present"* [4]. The components might include matrix compounds, degradation products, impurities, etc. When dealing with chromatographic techniques such as SFC, the term selectivity could be used instead of specificity. Indeed, the specificity testing will ensure that all potential compounds that could be detected are at least baseline separated. Selectivity/specificity of a SFC method will be determined by the chromatographic selectivity and by the specificity of the detection mode (i.e., UV, ELSD, MS, etc.). The specificity is the first validation criterion that has to be evaluated to ensure the identity of the analyte and to ensure the reliability of the quantitative results.

11.2.2 Response Function

The response function (calibration curve) is the relationship between the response (e.g., peak area) and the concentration of analyte in the sample. The response function selection is the first step before any validation data treatment. This step is critical because a significant source of imprecision and bias could be attributed to the choice of the statistical model for the calibration curve.

11.2.3 Precision

According to USP, *"the precision of an analytical procedure is the degree of agreement among individual test results when the procedure is applied repeatedly to multiple samplings of a homogeneous sample."* The precision is an estimation of random error and is measured using the (relative) standard deviation of test results. The precision could be defined considering three levels, depending on the testing conditions:

- Repeatability expresses the precision when the independent results are obtained using the same operating conditions (same analytical method, same samples, same laboratory, same analyst, same equipment) over a short interval of time. Repeatability is also called intraseries precision.

- Intermediate precision expresses the precision when the independent results are obtained by independent samples on different days (and may be different analysts on different equipment), using the same analytical method over a quite short interval of time. The intermediate precision is an estimation of within-laboratory variations. Intermediate precision also includes the repeatability (intraseries precision). In the literature, intermediate precision is wrongly called interseries precision because it encompasses also the repeatability criterion.
- Reproducibility expresses the precision when the independent tests results are obtained from different laboratories working with different analysts and equipment. Reproducibility is a criteria involved in collaborative studies, i.e., for the standardization of a pharmacopeial method.

11.2.4 Trueness

The trueness expresses the closeness of agreement between the average value obtained from a series of test results and the value which is accepted either as a conventional true value or as an accepted reference value (e.g., standard international or standard of a pharmacopeia). The trueness is an estimate of the systematic error and is measured by the bias or the recovery.

The concept of trueness was also called accuracy (see next point). Some confusion exists between the trueness and the accuracy criteria, especially when different nomenclatures are used in regulatory documents. In the present chapter, the term trueness will be used systematically in order to avoid any further confusion, even if the referenced paper mentions accuracy.

11.2.5 Accuracy

The accuracy expresses the closeness of agreement between the value measured and the value that is accepted either as a conventional true value or an accepted reference value. The USP definition of accuracy is more understandable *"the accuracy of an analytical procedure is the closeness of test results obtained by that procedure to the true value"*. As it is impossible to separate the types of errors from an individual result, the accuracy is therefore the resultant of the systematic and random error. Then accuracy could be defined as the expression of the sum of precision and trueness [5].

11.2.6 Linearity

"The linearity of an analytical procedure is its ability to elicit test results that are directly, or by a well-defined mathematical transformation, proportional to the concentration of analyte in samples within a given range" [1]. Consequently the linearity criterion refers to the linearity of the relationship between the concentrations introduced and the observed concentrations considering the whole dosing range. USP provides a general definition dealing with the "proportional" term but to be more precise, introduced and observed concentrations should be close to an identifer line with an intercept equal to 0 and a slope equal to 1. The linearity criterion should not refer to the relationship between the analytical response and the concentration, which is the response function (see Section 11.2.2). In some cases, mathematical transformations are required in order to attain the linearity between concentration and signal. A typical example when dealing with SFC is the evaporative light scattering detector (ELSD), which could require log transformation of both concentration and peak area. ICH requires a minimum of five concentration levels to determine the method linearity.

11.2.7 Range

The range, or dosing range, is established by confirming that the analytical method provides acceptable results for all the validation criteria mentioned (i.e., precision, trueness, accuracy, linearity). Some guidelines are available in the field of pharmaceutical analysis. As an example, the following minimum concentration range should be considered:

- for the assay of a drug substance (raw material) or a finished product: from 80% to 120% of the targeted concentration;
- for the content of uniformity: from 70% to 130% (a wider range should/could be evaluated for some pharmaceutical formulations);
- for a dissolution testing: ±20% over the specified range;
- for the determination of an impurity: from the reporting level to 120% of the specification.

11.2.8 Limit of Quantification

Two limits of quantification could be determined: the lower limit of quantification (LLOQ) and the upper limit of quantification (ULOQ) that are respectively the lower and the upper amount of analyte in a sample that can be determined with an acceptable accuracy. The limits of quantifications define the dosing range (see Section 11.2.7).

11.2.9 Limit of Detection

The limit of detection (LOD) is the smallest concentration that could be detected but not quantified with an acceptable precision and accuracy. LOQs and LOD could be estimated using several protocols [1,4–8].

11.2.10 Robustness

The robustness of an analytical method is a measure of its capacity to remain unaffected by small but deliberate variations in method parameters and provides an indication of its reliability during normal use. Previously the robustness was evaluated during of after the validation step. According to ICH Q8 [9], USP now recommends to test the robustness during method development.

11.3 TOTAL ERROR APPROACH

As explained previously, the validation of an analytical method is well described in the literature. Nevertheless, despite many regulatory documents (GMP, ISO, FDA, ICH, etc.) the validation protocol and the conclusion about method acceptance criteria remain confused. In this context, a SFSTP commission proposed a harmonized approach for the validation of quantitative analytical methods [10–13]. The objective of method validation was defined considering the final objective of an analytical method, i.e., routine analysis: *"the objective of validation is to give to the laboratories and regulatory bodies "guarantees" that every single measure that will be performed in routine analysis will be close enough to the unknown true value of the sample to be analyzed or at least that the difference will be lower than an acceptable limit taking into account the intended use of the method."* This definition offered a new validation perspective focused on the method final objective. Then the SFSTP commission proposed a new validation strategy focused on risk management using a clear decision tool as recommended by ICH [14]. In this context, future versions of USP and ICH guidelines will include the total error approach. This chapter will provide a few statistical explanations; the interested reader is invited to read the specific references [8–13].

11.3.1 Objective of an Analytical Method

The first step is to define the true objective of the method. For quantitative methods, the objective is to be able to provide accurate results. In mathematical words, the difference between the measured value (X) and the unknown "true value" (μ_T) must be lower than an acceptance limit λ:

$$|X - \mu_T| < \lambda \tag{11.1}$$

The acceptance limit can be defined considering the objective of the analytical method and the requirements. Regulatory documents provided some requirements, especially in the pharmaceutical field (e.g., $\pm 2\%$ for API raw material, $\pm 5\%$ for API pharmaceutical formulation, $\pm 10\%$ for preservatives). It is important to notice that these requirements are focused on the product to be analyzed. Consequently the analytical method should present better performances than these requirements in order to be able to characterize this product with an acceptable high quality level.

11.3.2 Total Error

The total error of an analytical method is related to the closeness of agreement between the value found and the value that is accepted either as a conventional true value or an accepted true value. The closeness of agreement observed is based on the sum of the systematic and random errors (i.e., total error). In mathematical words, an observed test result (X_i) can be seen as a composite of the unknown true value (μ_T), the difference between the mean of measurements (X_M) and the true value ($X_M - \mu_T$), and the analytical variability ($\sigma^2 = \Sigma(X_i - X_M)^2/(n-1)$). In other words, the measured value (X_i) is determined by the unknown sample true value μ_T, the method bias, and precision. The total error concept is summarized in Eq. 11.2 and Fig. 11.2.

$$X_i = \mu_T + \text{bias} + \text{precision} \leftrightarrow X_i - \mu_T = \text{total error} \tag{11.2}$$

In this validation approach, validation criteria (i.e., precision, trueness, accuracy) are estimated for each concentration level by statistical analysis of the back-calculated concentrations (also named observed concentrations).

11.3.3 Validation Protocol

As the validation should reflect the future routine analyses, reconstituted samples should be prepared. These solutions are named "validation standards": the matrix is spiked with a well-known amount of analyte. During the data treatment, these validation standards are the unknown samples. The validation protocol could also involve real

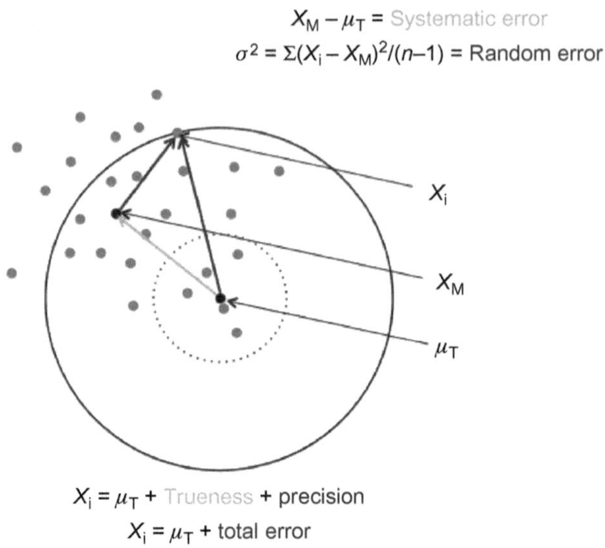

FIGURE 11.2 Schematic representation of total error.

samples previously analyzed by reference methods (i.e., USP monograph). Their concentrations are back-calculated using the response function established for the calibration standards solutions. These back-calculated concentrations are used to calculate all other validation criteria.

It is important to note that the total error validation methodology enables method validation to be performed using one experimental design. Practically, it means that all validation criteria are evaluated simultaneously using an "all-in-one" validation design. Consequently this more powerful approach is not more time and cost consuming than the classical validation protocols involving "one criterion at a time."

11.3.4 Decision Rule

As mentioned previously, one of the main concerns of the validation strategy is to propose a decision tool. Several authors proposed a reliable decision tool based on the β-expectation tolerance interval [10–13,15,16]:

$$E_{\hat{\mu}_M,\hat{\sigma}_M}\left(P_X\left[\hat{\mu}_M - k\hat{\sigma}_M < X < \hat{\mu}_M + k\hat{\sigma}_M \middle| \hat{\mu}_M, \hat{\sigma}_M\right]\right) = \beta \qquad (11.3)$$

where k is a factor determined so that the expected proportion of the population falling within the tolerance interval is equal to β [17]. If the β-expectation tolerance interval is included in the acceptance limits, the expected proportion of results within the same acceptance limits is at least superior than β. The β value is fixed by the analyst, which should manage the risk.

Most of the time, a range of concentrations should be evaluated. For that purpose, during the validation step, the analyst should prepare samples at different concentrations (to cover the whole range investigated) and a β-expectation tolerance interval is computed at each concentration level. Then the accuracy profile is obtained by connecting the upper limits of the β-expectation tolerance interval of all concentration levels and by connecting the lower limits of the β-expectation tolerance interval.

The graphical representation of an accuracy profile is presented in Fig. 11.3. For each concentration level, the β-expectation tolerance interval is plotted. The method is valid for those doses at which the β-expectation tolerance interval in included within the acceptance limits. This validation approach also enables the lower and the upper limits of quantification to be identified, which correspond to the dosing range limits. Moreover, on this accuracy profile, the systematic error is represented by the bias ((red) continuous line) and the random error is represented by the dispersion ((blue) dashed lines) of the individual measurement ((green) points) for each concentration level. Moreover, it is important to keep in mind that, using this unique and simple tool, all the validation criteria required by ICH are fulfilled by this approach.

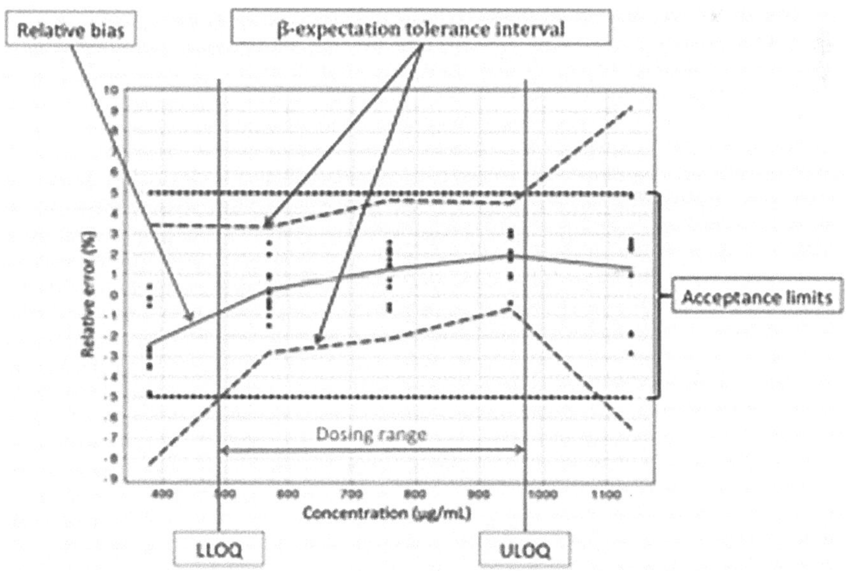

FIGURE 11.3 Schematic representation of an accuracy profile.

Finally, this predictive approach enables to evaluate the β proportion of the future routine results that will fall within the acceptance limits. Consequently this validation approach is focused on the true objective of method validation: giving a guarantee for the ability of the method to provide accurate and reliable results for the future routine analyses.

11.4 ROBUST OPTIMIZATION STRATEGY

The concept of Quality by Design (QbD) is now well established in pharmaceutical development. The QbD is defined by ICH Q8 R2 [18] as *"a systematic approach to development that begins with predefined objectives and emphasizes product and process understanding based on sound science and quality risk management."* Furthermore, the QbD concept was recently introduced in the field of analytical method development and validation [19,20]. Indeed an analytical method can be seen as a process that must have an output of acceptable quality. Borman et al. [21] demonstrated that the QbD concept for manufacturing processes could also be applied to analytical methods. Design of Experiments (DoE) considering risk management by means of error propagation is considered as a keystone to optimize process or method in the QbD environment [22].

In this context, the Design Space (DS) was introduced as a key component of analytical method development [20,23]. The DS is defined as *"the multidimensional combination and interaction of input variables (e.g., material attributes) and process parameters that have been demonstrated to provide assurance of quality"* [18]. Thus the DS is a subspace of the experimental domain of the input variables, in which the assurance of quality has been proven. As previously described [23], the Design Space could be defined as a region of an experimental domain-χ-where the posterior probability that the critical quality attributes (CQAs) are within acceptance criteria Λ, is higher than a specified quality level π, conditionally on the available data.

$$\mathrm{DS} = \{x_0 \in \chi : P(\mathrm{CQAs} \in \Lambda \mid x_0, \mathrm{data}) \geq \pi\} \tag{11.4}$$

CQAs provide some indications about the overall achievement of the analytical method. In chromatography, CQAs may be the resolution (Rs) or the separation (S) of a critical pair of peaks, or other criteria such as the width of peaks or their asymmetry, while the acceptance criteria Λ may be $\mathrm{Rs} > 1.5$ and/or $S > 0$. In this context, a result given as a predictive probability that the CQAs will be within the acceptance interval establishes the assurance of quality. For chromatographic method development, the DS could be defined as the space of

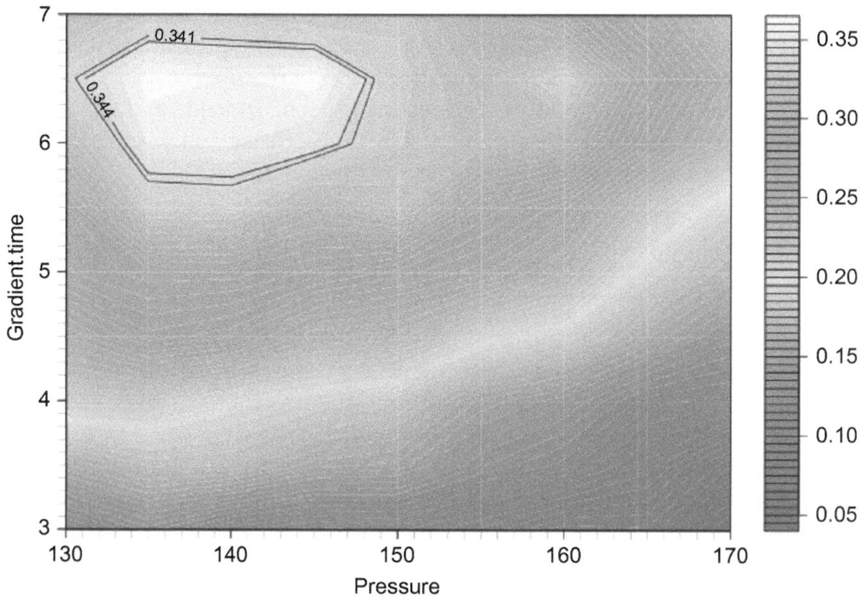

FIGURE 11.4 Schematic representation of design space.

chromatographic conditions that will ensure the quality of the separation. Therefore the method robustness is guaranteed inside the DS limits.

Fig. 11.4 shows an example of analytical design space illustrating the predictive probability for the CQA (i.e., S) to be equal to the acceptance criteria ($S > 0$). In this image a subspace of the experimental domain with two factors studied, i.e., pressure (bar) and gradient time (minute) is presented. The DS (circled in dark) represents the working conditions zone in which the separation of all peaks is guaranteed, and then the DS is the robustness area.

Previously method robustness was evaluated prior to or during the validation step. Only a few examples of SFC robustness studies were published [24]. Nevertheless USP now requires robustness evaluation, before the initiation of method validation, namely during the development/optimization step [1]. The interest in a robust optimization strategy, especially for SFC methods development, was previously described [25–27]. In this context, a robustness study is no longer required if robust method optimization was performed and an acceptable guarantee level π was found. Thus the DS strategy fulfils the USP recommendations and at the same time allows speeding up the analytical life cycle (Fig. 11.1). More recently, Hubert et al. [28] proposed a combined QbD strategy to integrate the robust optimization and validations phases. This strategy enables the whole design space chromatographic conditions to be validated. This first demonstration of method optimization and validation steps is totally in accordance with USP requirements. Further investigations should be performed, especially using SFC, to implement this new integrated approach. Indeed, a deep understanding of the method provided by DS methodology is a guarantee of quality for the future use of the method for routine analysis.

11.5 METHOD TRANSFER

In the pharmaceutical industry, the method development and validation steps are not necessary performed in the laboratory where the method will be employed. Furthermore the production sites could be numerous and the method might well need to be transferred in all quality control laboratories. The transfer of a method from a sending laboratory (e.g., R&D laboratory) to a receiving laboratory (e.g., quality control (QC) laboratory) is an important step in an analytical method life cycle. Generally analytical method transfer is the intermediate step between method validation and routine use. This step should ensure that the results obtained in the receiving laboratory are reliable and close to those obtained by the sending laboratory. This laboratory should provide the guarantee that it is able to implement the method and to get reliable results. The whole method transfer methodologies is well described elsewhere [29,30].

11.6 SFC METHODS VALIDATION

In the 2010s the resurgence of interest in SFC led to an increase in the number of publications. The main work was focused on fundamental studies of the technique (kinetic performances and interaction mechanisms, for example) but more recent work is focused on the quantitative performances of SFC. The present chapter will provide an overview of SFC methods developed for quantitative determinations with and a critical description of the validation protocols described. A summary of the method validations referred to is provided in Table 11.1.

11.6.1 Conventional Approach

11.6.1.1 Bioanalytical Method Validation

Method validation is a large topic covering all the applications domains. Regarding SFC, some interesting bioanalysis studies were reported [31–39]. The first validation criterion to assess before starting

TABLE 11.1 Summary of Supercritical Fluid Chromatography (SFC) Method Validations

References	Sample/matrix	Analytes	Guidelines/ regulatory documents	Validation criteria								
				Selectivity	Trueness	Precision repeatability	Intermediate precision	Linearity	Accuracy	LOD	LOQ	Robustness
METHODOLOGY												
[31]	Mouse liver	Polar lipids	N/A			✓	✓	✓		✓		
[32]	Plasma	Vitamin D and metabolites	N/A			✓	✓	✓		✓		
[33]	Plasma	R/S-warfarin	N/A	✓	✓	✓	✓				✓	✓
[34]	Mouse plasma	Cytarabine	N/A	✓	✓	✓						
[35]	Plasma	Ginkgolides and metabolites	N/A		✓	✓	✓				✓	
[36]	Plasma	Bifendate	N/A	✓	✓	✓	✓					
[37]	Beagle dog plasma	Oxcarbazepine and chiral metabolites	N/A	✓	✓	✓	✓					
[38]	Urine	Doping agents	N/A		✓			✓		✓		
[39]	Urine	Doping agents	N/A		✓	✓						
[40]	Cosmetics	UV filters	N/A			✓						
[41]	Cosmetics	Provitamin B5	ICH	✓		✓	✓	✓			✓	
[42]	Plant products	Synthetic cannabinois	N/A		✓			✓		✓		

(*Continued*)

TABLE 11.1 (Continued)

References	Sample/matrix	Analytes	Guidelines/ regulatory documents	Validation criteria								
				Selectivity	Trueness	Precision repeatability	Intermediate precision	Linearity	Accuracy	LOD	LOQ	Robustness
[43]	Bee products and environmental samples	Dinotefuran and chiral metabolites	N/A		✓	✓		✓		✓	✓	
[44]	Microalgae	Carotenoids	ICH		✓	✓	✓			✓	✓	
[45]	Chili containing sample	Illegal dyes	FDA (bioanalytical method)	✓	✓	✓	✓	✓		✓	✓	
[46]	Raw material	Mometasone furoate and impurities	N/A		✓	✓		✓			✓	
[47]	Raw material	R/S-timolol	Ph. Eur	✓	✓	✓	✓			✓		✓
[48]	Raw material	Chiral drugs	GMP		✓	✓	✓			✓	✓	✓
[49]	Raw material and tablets	agomelatine	ICH		✓	✓	✓	✓		✓	✓	
[50]	Cortisone 10 Plus crème	Hydrocortisone	N/A	✓		✓		✓				
[51]	Pharmaceutical product	Retinoids	N/A	✓	✓	✓		✓		✓	✓	
[52]	Tablets	Chloroxazone, paracetamol, aceclofenac	ICH		✓	✓	✓	✓		✓	✓	

[53]	Pharmaceutical aqueous formulation	Drug compound AZY	ICH, FDA	✓	✓	✓	✓	✓				✓
[54]	Tablets	Isoniazid, pyrizaramide	ICH, Ph. Eur.	✓	✓	✓	✓					✓
[27]	Capsules	Amoxicilline	ICH, SFSTP	✓	✓	✓	✓	✓	✓	✓	✓	✓
[55]	Raw material	Pyroglutaminde detivatives	SFSTP	✓	✓	✓	✓	✓	✓	✓	✓	
[56]	Medical devices	Plasticizers	SFSTP		✓	✓	✓	✓	✓	✓	✓	
[57]	Capsules	Quinine sulfate	ICH, SFSTP	✓	✓	✓	✓	✓	✓	✓	✓	

the true validation protocol is the specificity/selectivity of the method. When dealing with biological samples (e.g., plasma, blood, urine, . . .), it is mandatory to test the absence of interference from the matrix compounds using at least six individual sources of the appropriate blank matrix. Each blank matrix sample is individually analyzed and evaluated for interference. Some guidelines are provided by EMEA [58]: the absence of interfering compounds is accepted where the response is less than 20% of the LLOQ for the analyte and 5% for the internal standard (IS). Unfortunately the specificity criterion was not evaluated systematically for the bioanalytical methods reported.

Generally bioanalyses are performed using chromatographic techniques hyphenated with a mass spectrometric detector. In this context, matrix effect should be investigated using at least six batches of blank matrix from individual donors (pooled matrix should not be used). Indeed, as observed for liquid chromatography-mass spectrometry (LC-MS), some phenomena of ionization suppression or enhancement were described for SFC-MS analysis [38]. For each analyte (and internal standard), the matrix effect (ME) is calculated for each matrix batch as follows:

$$\mathrm{ME}(\%) = \frac{\text{peak area of analyte spiked in matrix}\left(\text{postextraction}\right)}{\text{peak area in standard solution(absence of matrix)}} \times 100 \qquad (11.5)$$

Normalized ME could be calculated by dividing the analyte ME by the IS ME. The determination of ME should be performed at low concentration (close to LLOQ) and high concentration (close to ULOQ) in order to cover the whole dosing range. Then a "relative" matrix effect ($\mathrm{ME}_{\mathrm{REL}}$) could be calculated as $\mathrm{ME}_{\mathrm{REL}} = \mathrm{ME} - 100$. A positive value is synonymous to signal enhancement while a negative value means signal suppression. It is interesting to note that these matrix effect in SFC-MS were studied and compared to LC-MS, the MS response are different for the same compounds analyzed with LC or SFC. Generally speaking the matrix effect observed in SFC was lower than that observed in LC [38].

Usually sample preparation involving an extraction step is required to analyze biological samples. In this context the determination of extraction recovery is mandatory to be able to evaluate properly the method trueness. For each analyte (and internal standard), the extraction recovery (RE) is calculated as follows:

$$\mathrm{RE}(\%) = \frac{\text{Peak area of analyte spiked in matrix}\left(\text{preextraction}\right)}{\text{Peak area of analyte spiked in matrix}\left(\text{postextraction}\right)} \times 100 \qquad (11.6)$$

To properly quantify the analyte, the impact of significant ME and RE should be taken into account. Indeed relevant ME and RE means that the signal observed for identical concentrations in matrix and in standard solution will be different, which means that a bias will be observed. Several approaches could be proposed to control these effects: a correction factor could be calculated [59] to prevent an incomplete extraction recovery or to prevent a known source of bias (i.e., signal enhancement or suppression). The standard addition method or calibration in matrix are experimental approaches that enable the matrix effect to be minimized.

Some reported results are confused considering the validation protocol and validation criteria studied. As an example, Lee et al. [31] proposed a SFC-MS/MS method for the analysis of polar lipids from mouse liver. Unfortunately the validation study was performed using stock solution of the analytes and then real analyses of mouse liver were made. Obviously a validation step using stock solution does not mimic real samples as required. Regarding the first criterion selectivity specification, it was not tested meaning that the method should not be used to analyze real samples. Jumaah et al. [32] recently proposed a method to quantify vitamin D and its metabolites in human plasma. The selectivity was not evaluated but a partial validation was performed utilizing spiked plasma. However, vitamin D and its metabolites are endogenous compounds, which could be present in the plasma used for the study. The plasma samples should be tested without spiking to evaluate the native concentrations of these endogenous compounds. Consequently a positive bias could be observed because spiked plasmas are quantified using calibration solutions in methanol. The trueness criteria and the matrix effect were not evaluated in this study. As mentioned by the authors, this work was only a partial validation and more work is mandatory to quantify properly vitamin D and its metabolites in plasma. Liu et al. [35], reported a pharmacokinetic study for the quantification of gingkolide metabolites without selectivity/specificity evaluation. Moreover, large values for bias (<|20%|) were reported, which could indicate a matrix effect.

In the field of chiral analysis, Coe et al. [33] reported a method to quantify R/S-warfarin in human plasma. The validation protocol was really complete and involved selectivity on 10 plasma batches, matrix effect and extraction recovery. One interesting concept was to test the selectivity against well-known coadministered drugs. They also determined trueness, repeatability, and intermediate precision. The accuracy criterion involving total error was not measured. The method robustness was tested using several operators, two SFC instruments and different APCI sources for MS detection. Yang et al. [37] proposed a chiral separation method for the determination of oxcarbazepine metabolites in dog plasma. The selectivity was evaluated according to EMEA requirements [58], other validation criteria were studied: trueness and precision (intraseries and interseries).

Finally, recent work was proposed to quantify doping agents in urine using SFC-MS/MS [38,39]. The method was not validated properly but a large effort was made to evaluate the method trueness (matrix effect and extraction recovery). This method is a good starting point for a full validation in the SFC-MS bioanalysis field.

11.6.1.2 Method Validation for Various Matrices

SFC is a technique used for numerous sample types. An SFC method was developed for the determination of UV filters in cosmetics [40]. The SFC was hyphenated to ELSD, and a log–log response function was used. Analyses of real samples were performed without any evaluation of selectivity (at least an injection of placebo matrix). Moreover, negative or positive bias (under estimation or higher estimation) was observed during routine analysis. These results highlight that validation is a mandatory step for all analytical methods, even outside the highly regulated pharmaceutical industry. In the same field, an SFC method was optimized for the determination of an enantiomeric purity of provitamin A in cosmetic formulations [41]. The selectivity was properly evaluated using a placebo formulation indicating the absence of a peak at the retention time of the compounds of interest. The linearity of the response (ratio AUC analyte/AUC SI versus the concentration) was determined using a Fischer-Snedecor test indicating the significance of the slope. However, the linearity of the results was not determined as required (see Section 11.2.6). Precision and trueness were also determined.

In the analysis of natural products [42–44] method validation is challenging since a placebo matrix is not always available. Toyo'oka et al. [42] reported a reliable method for the detection of cannabinoids by SFC-MS and demonstrated its application to real samples. The quantitative performances were challenged using herb tea spiked with cannabinoids. The selectivity and matrix effect of actual cannabinoid plants were not truly evaluated. A legal plant from the *cannabaceae* family would be more appropriate to mimic the sample matrix. Some precision tests were performed but a formal validation was not carried out. The conclusion of the study is that the method is reliable and suitable for the qualitative and quantitative analysis but no objective demonstration is available in this report to support this statement. Abrahamsson et al. [44] reported a SFE-SFC method for the determination of carotenoids in microalgae. The method validation was performed following ICH guidelines, except for the selectivity criterion because of the lack of placebo samples. Since the AUC grows when samples are spiked with stock solution of analyte, the signal observed can be assigned to the analyte and then the method can be considered as selective.

Finally, Khalikova et al. [45] developed a SFC method for the determination of illegal dyes in different spice preparations. All validation criteria were evaluated following the FDA requirements. The accuracy (total error) was not calculated. The quantitative performances of UHPSFC were compared to UHPLC, which can be considered as a reference technique. For this application, the performances of LC are superior to SFC considering the precision criterion and LOD. Generally speaking the detectability of SFC-UV is lower than for LC-UV due to the optical properties of the supercritical fluid [27,45].

11.6.1.3 Pharmaceutical Method Validation

As SFC is now presented as a promising technique in the field of pharmaceutical analysis, the number of publications dealing with the determination of pharmaceuticals is growing. Unfortunately the methods proposed are not always properly validated with respect to the requirements of the regulatory authorities.

Several studies proposed SFC methods for the quality control of pharmaceutical raw material. Wang et al. [46] developed a method for the control of mometasone furoate and its related impurities. A partial validation was described involving trueness, repeatability, and LOQ determination. At least, the determination of LOD would be useful in the context of impurities detection. A full validation is still required to use this method for the control of mometasone furoate API. In the framework of chiral separations, Marley et al. [47] proposed an enantio-selective SFC separation as an alternative to a normal-phase LC method described in the pharmacopeia for timolol. The selectivity was demonstrated using a solution containing all related impurities for timolol, which represented the potential interfering compounds. Robustness testing was performed for flow rate, temperature, and back-pressure deliberated variations. Chromatographic parameters (relative retention time, resolution, etc.) and repeatability were selected as criteria to assess the method robustness. One interesting point is the evaluation of the response factor. The objective of the method is to quantify S and R-timolol using a standard of S-timolol. The UV response was not the same for each enantiomer, and then relative response factor (RRF) was calculated using the differences in peak area versus concentration of a standard solution containing identical concentration of S and R-timolol.

In the same way, Hicks et al. [48] developed a method to evaluate the enantiomeric purity of a chiral synthesis process. The acceptance limits were set at RSD ≤ 25% for precision and bias ≤15% for the trueness criterion. This tolerance is quite large considering the raw material. Moreover, if the total error is considered, including precision and trueness, then an acceptance limit of 40% should be set, which is not acceptable in the field of pharmaceutical material GMP analysis as claimed by the authors. Hopefully the results presented are largely lower than the specifications. Finally, a comparative study of state-of-art UHPLC and state-of-art UHPSFC was published for the determination of agomelatine impurities [49]. The UHPLC-UV is still more sensitive than UHPSFC but better precision values were obtained for the SFC method.

The next step of pharmaceutical analysis is the quality control of finished product. The validation of quantitative determination in different pharmaceutical matrices is reported in the literature. De Phillipo et al. [50] proposed a SFC method to determine hydrocortisone in a cream. They showed absence of matrix effect since all excipients were eluted at the dead time. However, evaluation of the trueness criterion is mandatory to confirm this observation. Repeatability and response linearity criteria were studied. The range of concentration investigated was really large, i.e., 80–2000 μg/mL, which is not relevant for the quality control of a pharmaceutical (known target concentration). Méjean et al. [51] studied lipophilic vitamins (retinoids) in oil solution. Evaluation of method linearity was performed for each analyte using the response linearity: peak area versus concentration. However, the response function selected is a least squares linear regression using $1/\text{quantity}^2$ as the weighting factor. In this context, the linearity of the results would be more appropriate to evaluate the adequacy of the model as required by ICH [4]. Note that linearity of the response function is never required to obtain linearity of the results. The linearity of the results should be evaluated by plotting the back-calculated concentrations using the calibration model versus the introduced concentrations. Once again, the values obtained for trueness (relative bias from −20% to 6%) and precision (RSD < 10%) were not totally in agreement with the regulatory requirements for a quantitative analysis. On the basis of these results, the absence of matrix effect was assessed. The bias is the estimation of systematic error, this kind of error could be reduced/avoided if the error source is identified. In the present case, it would be interesting to prepare calibration solutions in matrix to test the trueness criterion, if the bias is reduced, matrix effect can be suspected. The quantitative determination of antiinflammatory drugs in tablets was published by. Desai et al. [52]. Unexpectedly the precision and trueness criteria were not evaluated at the same concentration levels. For the determination of API in a pharmaceutical matrix, the method should be valid in the dosing range 80–120% of the targeted concentration as required by the EMA and ICH. Then, the more pragmatic approach is to evaluate the validation criteria in the dosing range required (or a larger dosing range) using the same concentration levels. Moreover, the RSD values obtained for the interday precision were lower than those for intraday, which is an indication of an incorrect estimation of precision. To avoid these issues, it is generally advised to analyze the inter- and intraday variability in a single combined design of experiments. The validation of an SFC method for the determination of drug in aqueous formulation was proposed by Mukherjee [53]. The trueness was evaluated using standard solutions, which is not representative of actual samples. Finally, Prajapati et al. [54] reported an SFC-MS/MS method for two antitubercular drugs in tablets. Selectivity and trueness were properly estimated but the RSD values presented for intermediate precision criterion (RSD $<$ 0.22%) seemed to be uncommonly low in comparison with the technical specifications of SFC instruments and typical data reported in the literature. A deep study of the validation protocol and the computing of the results is mandatory to properly interpret these findings.

11.6.2 Total Error Approach

The interest in the total error validation methodology is already well documented for several analytical techniques, especially LC. A few reports describe SFC method validation by this approach. For the enantioseparation of a pharmaceutical raw material, Baudelet et al. [55] proposed a comparison of the quantitative performances of LC and SFC using the total error approach (Fig. 11.5). The dosing range was larger for LC, explained by the better sensitivity of LC-UV detector compared with SFC-UV. Consequently the lower level of concentration tested was set five times lower for LC. The SFC method was valid in the dosing range from 0.1 to 0.75 mM. Consequently SFC can be used as a valid technique to determine the enantiomer purity of the raw material tested in this study. The precision is slightly better for LC (mean RSD of 0.90% for repeatability and 1.62% for intermediate precision) than for SFC (mean RSD of 1.34% for repeatability and 1.81% for intermediate precision). For SFC analysis, the authors used a hybrid instrument (analytical

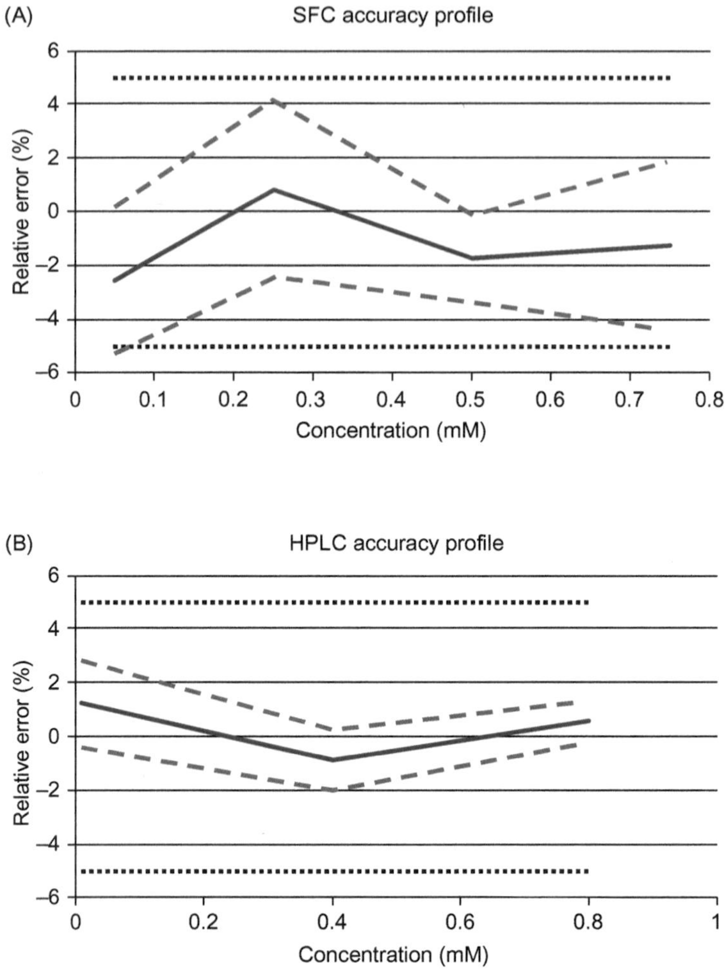

FIGURE 11.5 Accuracy profile of SFC and LC method for the determination of an enantiomer considering a linear regression. The acceptance limits represented by the dotted (black) lines were set at 5%. The relative bias was represented by the plain (red) line and the β-expectation tolerance limits by the dashed (blue) lines. *Source: Reprinted with permission from reference Baudelet D, Schifano-Faux N, Ghinet A, Dezitter X, Barbotin F, Gautret P, et al. Enantioseparation of pyroglutamide derivatives on polysaccharide based chiral stationary phases by ultra-high-performance chromatography and supercritical fluid chromatography: a comparative study. J Chromatogr A 2014;1363:257–69 [55].*

and semiprep with a large injection loop) which cannot be considered as modern SFC. In this context, the variability observed during the validation could be partially explained by the superiority of the injected volume of the LC instrument.

Liquid and supercritical fluid chromatography coupled to ELS detection were also compared for the determination of plasticizers in medical devices [56]. The main issue is the preparation of validation standards using stock solutions. As validation should reflect the future use of the method for routine analysis, the validation standards should be as close as possible to the future samples. In the present study, the plasticizers were extracted from PVC medical devices using chloroform and these samples were analyzed directly using SFC. However, the validation was performed using stock solutions of plasticizers in acetonitrile. Consequently the validation presented is not a demonstration of the ability of the method to determine the analytes in medical devices because the selectivity and all the validation criteria were not evaluated properly.

Finally, the quantification of API in a pharmaceutical formulation was described [27,57]. The first method was optimized using a DS strategy for the separation of several antibiotic drugs. Then the method was fully validated for the quantification of amoxicillin in capsules. The SFC method was valid in the range 80–120% of the target concentration considering the acceptance limits set at ±5% as required by the EMA for pharmaceuticals (Fig. 11.6A). Consequently the SFC method can be used for the quality control of this medicine. Nevertheless this first work showed the superiority of UHPLC-UV, especially in terms of sensitivity and dosing range. The main issue was the solubility of the salt form of the pharmaceutical compounds in the organic solvent mixture generally used in SFC. This first demonstration of method validation for a pharmaceutical using an accuracy profile methodology led to two fundamental questions: some work is mandatory to improve the SFC-UV detectability and a suitable alternative is needed for the preparation of the pharmaceutical compounds. Following these results, the research was focused on studying the critical method parameters for the quantitative determination of the pharmaceutical compounds. The main objective of this work [57] was to propose a protocol to prepare pharmaceutical samples (salt molecules) without using apolar solvents such as *n*-heptane. The results demonstrated the interest in the new generation of SFC stationary phases to obtain narrow peaks in order to improve the method sensitivity. The sample preparation protocol (involving water and a short aliphatic alcohol) and the optimal chromatographic method were validated according to the accuracy profile methodology using quinine sulfate as a model compound. The method was valid in the range 70–130% of the target concentration considering acceptance limits set at ±5% (Fig. 11.5B). These results are really promising for SFC quantitative analysis.

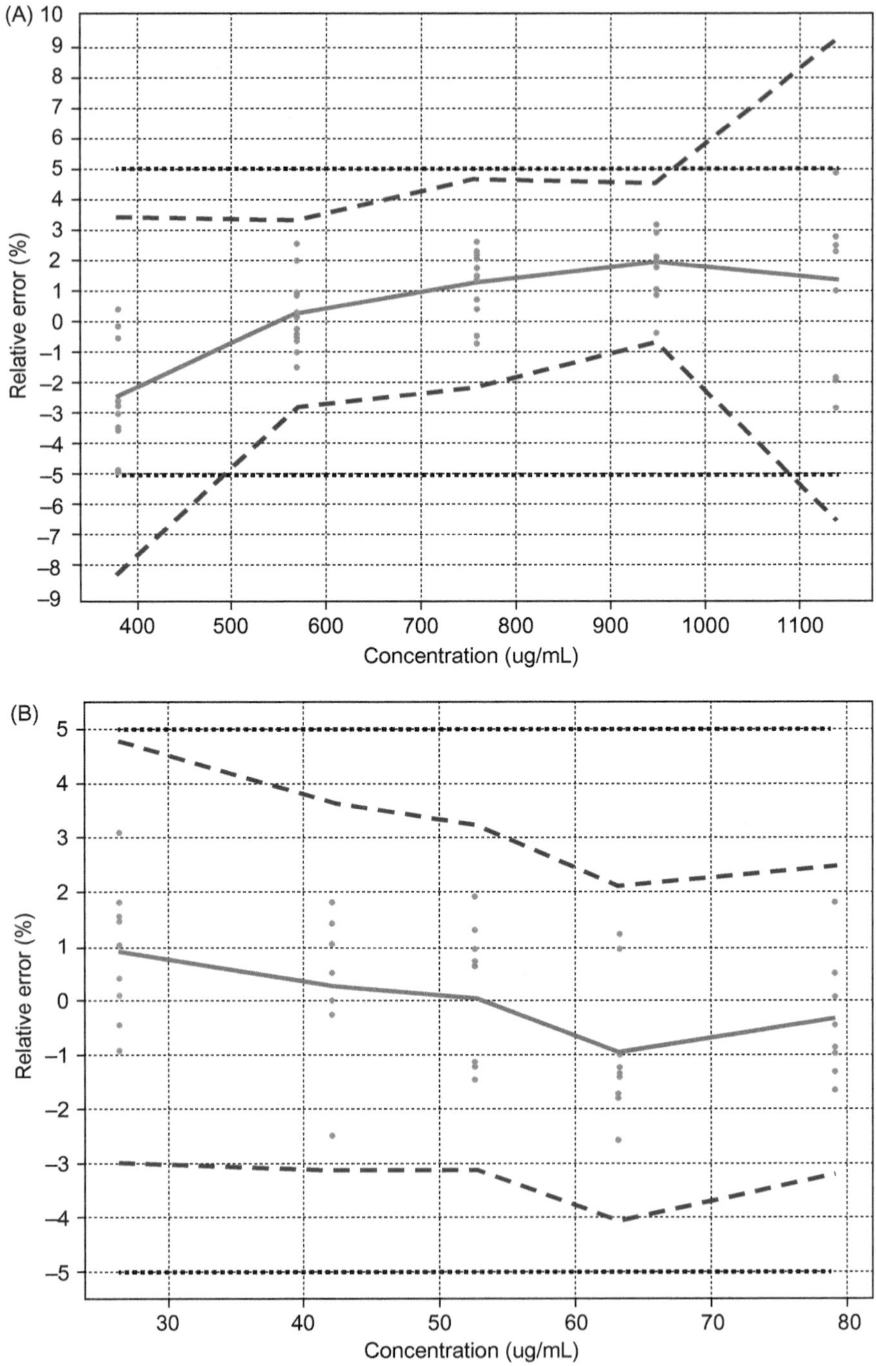

FIGURE 11.6 Accuracy profiles of SFC methods for the determination of amoxicilline in capsules (A—Ref. [27]) and quinine in capsules (B—Ref. [57]) considering linear regression. The acceptance limits represented by the dotted (black) lines were set at 5%. The relative bias was represented by the plain (red) line and the β-expectation tolerance limits by the dashed (blue) lines.

11.7 CONCLUSION

The potential of SFC as a quantitative technique is now well established and was demonstrated by several reports. Considering the technical view, the sample detectability of the SFC-UV detector remains the weakness of this technique due to the optical properties of super and subcritical fluids. Nevertheless chromatographic efficiency improvements have helped producing narrow symmetric peak leading to a higher detector signal. Regarding validation methodology, the definition of acceptance limits and the assessment of an analytical method as valid remains confused for the large part in the field of SFC. One of the main topics to address is the selection of an adequate validation protocol involving reconstituted samples (validation standards) in order to reflect as close as possible the nature of the future samples. Then the selectivity should be evaluated systematically before beginning a quantitative method validation. After this assessment of qualitative performances of the method, the next mandatory step is to select an adequate response function before performing the mathematical assessment of the validation data. Finally, as demonstrated in the literature, the total error approach is the more powerful validation strategy. The use of accuracy profiles as a decision tool should be systematically used for SFC method validation.

References

[1] USP 38, Chapter 1225—Validation of Compendial Procedures.

[2] Analytical procedures and methods validation for drugs and biologics—guidance for industry. U.S. Department of Health and Human Services, Food and Drug Administration; July 2015, <http://www.fda.gov/downloads/drugs/guidancecomplianceregulatoryinformation/guidances/ucm386366.pdf>.

[3] Eudralex Volume 4, Part 1—Chapter 6: Quality control, October 2014, <http://ec.europa.eu/health/files/eudralex/vol-4/2014-11_vol4_chapter_6.pdf>.

[4] International conference on harmonization of technical requirements for registration of pharmaceuticals for human use. Validation of analytical procedures: text and methodology Q2(R1), 2005; <http://www.ich.org/fileadmin/Public_Web_Site/ICH_Products/Guidelines/Quality/Q2_R1/Step4/Q2_R1__Guideline.pdf>.

[5] SO 17025, General requirements for the competence of calibration and testing laboratories, 2005.

[6] Rozet E, Ceccato A, Hubert C, Ziemons E, Oprean R, Rudaz S, et al. Analysis of recent pharmaceutical regulatory documents on analytical method validation. J Chromatogr A 2007;1158:111–25.

[7] Bouabidi A, Rozet E, Fillet M, Ziemons E, Chapuzet E, Mertens B, et al. J Chromatogr A 2010;1217:3180–92.

[8] Vial J, Jardy A. Experimental comparison of the different approaches to estimate LOD and LOQ of an HPLC method. Anal Chem 1999;71:2672–7.

[9] International conference on harmonization of technical requirements for registration of pharmaceuticals for human use, Pharmaceutical development Q8(R1), 2009; <http://www.ich.org/fileadmin/Public_Web_Site/ICH_Products/Guidelines/Quality/Q8_R1/Step4/Q8_R2_Guideline.pdf>.

[10] Hubert P, Nguyen-Huu JJ, Boulanger B, Chapuzet E, Chiap P, Cohen N, et al. Harmonization of strategies for the validation of quantitative analytical procedures—a SFSTP proposal – part I. J Pharm Biomed Anal 2004;36:579–86.

[11] Hubert P, Nguyen-Huu JJ, Boulanger B, Chapuzet E, Chiap P, Cohen N, et al. Harmonization of strategies for the validation of quantitative analytical procedures—a SFSTP proposal – part II. J Pharm Biomed Anal 2007;45:70–81.

[12] Hubert P, Nguyen-Huu J-J, Boulanger B, Chapuzet E, Cohen N, Compagnon P-A, et al. Harmonization of strategies for the validation of quantitative analytical procedures—a SFSTP proposal – part III. J Pharm Biomed Anal 2007;45:82–96.

[13] Hubert P, Nguyen-Huu J-J, Boulanger B, Chapuzet E, Cohen N, Compagnon P-A, et al. Harmonization of strategies for the validation of quantitative analytical procedures—a SFSTP proposal – part IV. J Pharm Biomed Anal 2008;48:760–71.

[14] International conference on harmonization of technical requirements for registration of pharmaceuticals for human use, Quality Risk management Q9, 2005; <http://www.ich.org/fileadmin/Public_Web_Site/ICH_Products/Guidelines/Quality/Q9/Step4/Q9_Guideline.pdf>.

[15] Boulanger B, Devanaryan V, Dewé W, Smith W. Statistical considerations in analytical method validation, in Pharmaceutical Statistics, SAS Press, p. 69–94.

[16] USP 1210, Statistical tools for procedure validation, chapter in process revision.

[17] Mee RW. β-Expectation and β-content tolerance limits for balanced one-way ANOVA pandom model. Technometrics 1984;26:251–4.

[18] International conference on harmonization of technical requirements for registration of pharmaceuticals for human use. Pharmaceutical development Q8(R2), 2009.

[19] Nethercote P, Ermer J. Quality by design for analytical methods: implications for method validation and transfer. Pharm Tech 2012;36:52–5.

[20] Rozet E, Lebrun P, Debrus B, Boulanger B, Hubert P. Design spaces for analytical methods. Trends Anal Chem 2013;42:157–67.

[21] Borman P, Truman K, Thompson D, Nethercote P, Chatfield M. The application of Quality by Design to analytical methods. Pharm Tech 2007;31:142–52.

[22] Peterson JJ. What your ICH Q8 Design Space needs: a multivariate predictive distribution. Pharm Manuf 2010. Date of publication : 25 JUNE 2010.

[23] Lebrun P, Boulanger B, Debrus B, Lambert P, Hubert P. A Bayesian design space for analytical methods based on multivariate models and predictions. J Biopharm Stat 2013;23:1330–51.

[24] Dejaegher B, Heyden YV. Ruggedness and robustness testing. J Chromatogr A 2007;1158:138–57.

[25] Dispas A, Lebrun P, Sassiat P, Ziemons E, Thiébaut D, Vial J, et al. Innovative green supercritical fluid chromatography development for the determination of polar compounds. J Chromatogr A 2012;1256:253–60.

[26] Dispas A, Lebrun P, Andri B, Rozet E, Hubert P. Robust method optimization strategy, a useful tool for method transfer: the case of SFC. J Pharm Biomed Anal 2014;88:519–24.

[27] Dispas A, Lebrun P, Ziemons E, Marini R, Rozet E, Hubert P. Evaluation of the quantitative performances of supercritical fluid chromatography: from method development to validation. J Chromatogr A 2014;1353:78–88.

[28] Hubert C, Houari S, Rozet E, Lebrun P, Hubert P. Towards a full integration of optimization and validation phases: an analytical-quality-by-design approach. J Chromatogr A, Vol. 1395, p. 22 May 2015, 88–98.

[29] Rozet E, Dewé W, Ziemons E, Bouklouze A, Boulanger B, Hubert P. Methodologies for the transfer of analytical methods: a review. J Chromatogr B 2009;877:2214–23.
[30] Rozet E, Mertens B, Dewe W, Ceccato A, Govaerts B, Boulanger B, et al. The transfer of LC-UV method for the determination of fenofibrate and fenofibric acid in Lidoses: use total error as decision criterion. J Pharm Biomed Anal 2006;42:64–70.
[31] Lee JW, Nishiumi S, Yoshida M, Fukusaki E, Bamba T. Simultaneous profiling of polar lipids by supercritical fluid chromatography/tandem mass spectrometry with methylation. J Chromatogr A 2013;1279:98–107.
[32] Jumaah F, Larsson S, Essén S, Cunico LP, Holm C, Turner C, et al. A rapid method for the separation of vitamin D and its metabolites by ultra-high performance supercritical fluid chromatography-mass spectrometry. J Chromatogr A 2016;2440: 191–200.
[33] Coe RA, Rathe JO, Lee JW. Supercritical fluid chromatography-tandem mass spectrometry for fast bioanalysis of R/S-warfarin in human plasma. J Pharm Biomed Anal 2006;42:573–80.
[34] Hsieh Y, Li F, Duncan CJG. Supercritical fluid chromatography and high-performance liquid chromatography/tandem mass spectrometric methods for the determination of cytarabine in mouse plasma. Anal Chem 2007;79:3856–61.
[35] Liu X-G, Qi L-W, Fan Z-Y, Dong X, Guo R-Z, Lou F-C, et al. Accurate analysis of ginkgolides and their hydrolyzed metabolites by analytical supercritical fluid chromatography hybrid tandem mass spectrometry. J Chromatogr A 2015;1388:251–8.
[36] Liu M, Zhao L, Yang D, Ma J, Wang X, Zhang T. Preclinical pharmacokinetic evaluation of a new formulation of bifendate solid dispersion using a supercritical fluid chromatography-tandem mass spectrometry method. J Pharm Biomed Anal 2014;100: 387–92.
[37] Yang Z, Xu X, Sun L, Zhao X, Wang H, Fawcett JP, et al. Development and validation of an enantioselective SFC-MS/MS method for simultaneous separation and quantification of oxcarbazepine and its chiral metabolites in beagle dog plasma. J Chromatogr B; Vol. 1020, p. 1 May 2016, 36–42, <http://dx.doi.org/10.1016/j.jchromb.2016.03.013>.
[38] Novakova L, Rentsch M, Perrenoud AG-G, Nicoli R, Saugy M, Veuthey J-L, et al. Ultra high performance supercritical fluid chromatography coupled with tandem mass spectrometry for screening of doping agents. II: analysis of biological samples. Anal Chim Acta 2015;853:647–59.
[39] Novakova L, Desfontaine V, Ponzetto F, Nicoli R, Saugy M, Veuthey J-L, et al. Fast and sensitive supercritical fluid chromatography—tandem mass spectrometry multi-class screening method for the determination of doping agents in urine. Anal Chim Acta 2016;915:102–10.
[40] Lesellier E, Mith D, Dubrulle I. Method developments approaches in supercritical fluid chromatography applied to the analysis of cosmetics. J Chromatogr A 2015; 1423:158–68.
[41] Khater S, West C. Development and validation of a supercritical fluid chromatography method for the direct determination of enantiomeric purity of provitamin B5 in cosmetic formulations with mass spectrometric detection. J Pharm Biomed Anal 2015;102:321–5.
[42] Toyo'oka T, Kikura-Hanajiri R. A reliable method for the separation and detection of synthetic cannabinoids by supercritical fluid chromatography with mass spectrometry, and its application to plant products. Chem Pharm Bull 2015;63:762–9.
[43] Chen Z, Dong F, Li S, Zheng Z, Xu Y, Xu J, et al. Response surface methodology for the enantionseparation of dinotefuran and its chiral metabolite in bee products and environmental samples by supercritical fluid chromatography/tandem mass spectrometry. J Chromatogr A 2015;1410:181–9.

[44] Abramsson V, Rodriguez-Meizoso I, Turner C. Determination of carotenoids in microalgae using supercritical fluid extraction and chromatography. J Chromatogr A 2012;1250:63–8.
[45] Khalikova MA, Satinsky D, Solich P, Novakova L. Development and validation of ultra-high performance supercritical fluid chromatography method for determination of illegal dyes and comparison to ultra-high performance liquid chromatography method. Anal Chim Acta 2015;874:84–96.
[46] Wang Z, Zhang H, Liu O, Donovan B. Development of an orthogonal method for mometasone furoate impurity analysis using supercritical fluid chromatography. J Chromatogr A 2011;1218:2311–19.
[47] Marley A, Connolly D. Determination of (R)-timolol in (S)-timolol maleate active pharmaceutical ingredient: validation of a new supercritical fluid chromatography method with an established normal phase liquid chromatography method. J Chromatogr A 2014;1325:213–20.
[48] Hicks MB, Regalado EL, Tan F, Gong X, Welch CJ. Supercritical fluid chromatography for GMP analysis in support of pharmaceutical development and manufacturing activities. J Pharm Biomed Anal 2016;117:316–24.
[49] Plachka K, Chrenkova L, Dousa M, Novakova L. Development, validation and comparison of UHPSFC and UHPLC methods for the determinataion of agomelatine and its impurities. J Pharm Biomed Anal 2016;125:376–84.
[50] DePhilippo T, Chen R. Separation and quantitative determination of hydrocortisone in Cortizone 10 Plus crème by supercritical fluid chromatography (SFC). LCGC Europe, 21. 2008. p. 37–8.
[51] Méjean M, Vollmer M, Brunelle A, Touboul D. Quantification of retinoid compounds by supercritical fluid chromatography coupled to ultraviolet diode array detection. Chromatographia 2013;76:1097–105.
[52] Desai PP, Patel NR, Sherikar OD, Mehta PJ. Development and validation of packed column supercritical fluid chromatographic technique for quantification of chloroxazone, paracetamol and aceclofenac in their individual and combined dosage forms. J Chromatogr Sci 2012;50:769–74.
[53] Mukherjee PS. Validation of direct assay of an aqueous formulation of a drug compound AZY by chiral supercritical fluid chromatography (SFC). J Pharm Biomed Anal 2007;43:464–70.
[54] Prajapati P, Agrawal YK. SFC-MS/MS for identification and simultaneous estimation of the isoniazid and pyrazinamide in its dosage form. J supercrit fluids 2014;95: 597–602.
[55] Baudelet D, Schifano-Faux N, Ghinet A, Dezitter X, Barbotin F, Gautret P, et al. Enantioseparation of pyroglutamide derivatives on polysaccharide based chiral stationary phases by ultra-high performance chromatography and supercritical fluid chromatography: a comparative study. J Chromatogr A 2014;1363:257–69.
[56] Lecoeur M, Decaudin B, Guillotin Y, Sautou V, Vaccher C. Comparison of high-performance liquid chromatography and supercritical fluid chromatography using evaporative light scattering detection for the determination of platicizers in medical devices. J Chromatogr A 2015;1417:104–15.
[57] Dispas A, Lebrun P, Sacré P-Y, Hubert P. Screening study of SFC critical method parameters for the determination of pharmaceutical compounds. J Pharm Biomed Anal 2016;125:339–54.
[58] European Medicines Agency. Guideline on bioanalytical method validation; 2011.
[59] AFNOR NF. V03-110 (2010), Protocole de caractérisation en vue de la validation d'une méthode d'analyse quantitative par construction du profil d'exactitude.

CHAPTER

12

Separation of Stereoisomers

C.M. Galea, Y. Vander Heyden and D. Mangelings

Vrije Universiteit Brussel (VUB), Brussels, Belgium

OUTLINE

Supercritical Fluid Chromatography.
DOI: http://dx.doi.org/10.1016/B978-0-12-809207-1.00012-4

12.1 INTRODUCTION

Stereoisomerism is the arrangement of atoms in which isomeric molecules have the same molecular constitution, but a different three-dimensional spatial arrangement. The arrangement around a stereocenter can lead to molecule pairs that are nonsuperimposable mirror images of each other, which are called enantiomers. On the other hand, molecule pairs can also have more than one stereocenter or a carbon double bond with a different arrangement of atoms across the bond, and these are called diastereomers. The existence of nonsuperimposable isomers of a molecule is referred to as chirality.

Chirality is a fundamental feature of living organisms because proteins and carbohydrates are chiral. Therefore living organisms will display different biological responses to enantiomers. Because their three-dimensional spatial arrangement of the atoms is different, enantiomers can interact with receptors in different manners, and consequently bring about different responses. Prior to the 1960s, it was common practice to develop chiral drugs as racemates. However, the thalidomide tragedy raised awareness that different drug enantiomers have different pharmacological activities and that it is beneficial to administer a chiral drug as a single enantiomer [1]. Nowadays enantioselective separation, identification, and quantification methods should be developed for each active pharmaceutical ingredient having chiral properties. Different techniques are available to separate enantiomers, for instance, crystallization, kinetic resolution, membrane-based separations as well as chromatography. The predominant application of chiral separations has been for drug enantiomers in the pharmaceutical industry. In other industries methods are required, e.g., for the separation of enantiomeric fungicides or insecticides in the agricultural industry and chiral amino acids in food supplements or flavors in the food industry. On the other hand, chromatography is the main technique used for diastereomer separation.

High-performance liquid chromatography (HPLC) has been the leading technique for stereoselective separations during many years. Long equilibration and separation times, the need for toxic and flammable solvents as well as rather low efficiencies are all drawbacks that could be encountered when developing HPLC methods. Supercritical fluid chromatography (SFC) is now gaining ground and is slowly becoming the preferred choice for stereoselective separations. This is due to the introduction of improved SFC instruments together with an increased awareness of the potential advantages of the technique [2].

Carbon dioxide (CO_2) is used as the main mobile phase component in SFC because of its low critical temperature and pressure. An additional advantage is that CO_2 evaporates easily when the pressure is reduced leading to decreased waste generation [3]. Moreover, CO_2 can be recycled and purified after use for re-use on a preparative scale, leading to an overall cost reduction. These properties of CO_2 allow SFC to be considered as a green technology [4].

This chapter will discuss the different methods for the separation of enantiomers and diastereomers by SFC. Enantiomers have different properties in a chiral environment and therefore require chiral methods to separate them, while diasteromers have the same properties, making their separation also possible with classical achiral methods. Because diastereomer separations are less studied than enantiomer separations, only a small section of this chapter will be dedicated to applications of diastereomer separations

12.2 STEREOISOMERISM

Stereoisomers can be subdivided into two main groups, enantiomers and diastereomers (Fig. 12.1) [5]. Enantiomers or optical isomers occur due to an atom, usually a carbon, which is bonded to four different atoms or groups of atoms. The central carbon is referred to as the asymmetrical or chiral carbon atom, or as stereogenic center or stereocenter [6,7]. Enantiomers can have one of two configurations (R or S) determined by the priority of the substituents surrounding the stereocentre. The two configurations are mirror images of each other and therefore nonsuperimposable. A mixture of enantiomers in equal proportions is called a racemic mixture and is optically inactive [6,7].

On the other hand, diastereomers are compounds having two or more stereogenic centers, each with its own configuration (R or S) (Fig. 12.1). This would lead to a total of four isomers, since both centers may be (R) or (S) [6,7]. Therefore a compound having two enantiomer pairs would have four diastereomer pairs. Diastereomers are characterized by different physico-chemical properties such as density, solubility in water and melting point and by some differences in chemical behavior toward achiral and chiral reagents.

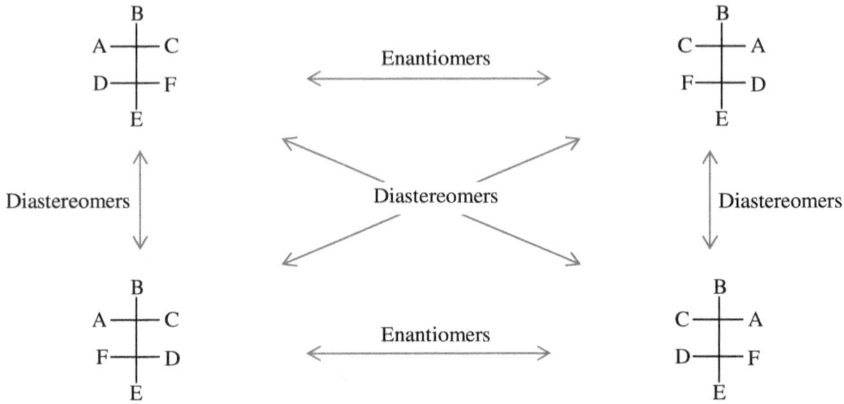

FIGURE 12.1 Stereoisomerism—Scheme showing differences between enantiomers and diastereomers.

12.3 SEPARATIONS OF ENANTIOMERS

Enantiomers cannot be separated by distillation because they possess an identical boiling point and they cannot be separated on an achiral chromatographic stationary phase because they posses the same affinity for the phase. Different methods are available for chiral separation: crystallization, kinetic resolution, membrane-based separations, and chromatographic methods.

Crystallization involves the seeding of a supersaturated solution of the racemic mixture with a crystal of the desired enantiomer. The pure crystals are allowed to grow in a controlled way, and this process is repeated in a cyclic manner resulting in high optical purity in about 90% of the cases [8]. Enantioselective crystallization is used for about one-fifth of chiral separations at the kilogram or larger scale [9]. The ease of the process and the cost of manufacture are the reasons for the popularity of this technique. However, in order to provide sufficient material recoveries, fractional crystallization trains have to be used, which add to the processing time and costs [8,9].

Kinetic resolution uses a chiral catalyst or an enzymatic reagent to promote selective reaction of one enantiomer over the other resulting in a mixture of enantio-enriched starting material and product, from which the desired component has to be isolated [10]. Application of this method is limited due to the long development times and the availability of suitable materials for selective reactions [9].

Synthetic enantioselective membranes, either liquid or solid, can be used for the transport of specific enantiomeric forms. Advantages of membrane-based separation techniques over other techniques are their continuous operation capability, large processing capability, low-energy consumption and their simplicity of use. Liquid membranes contain enantiomer recognizing carriers such as a chiral crown ether or cyclodextrins. These show high enantioselective permeability but a low durability. Conversely solid membranes are more stable facilitating extended use [11,12].

Chromatographic methods are generally easier to develop and the most popular. Both analytical and preparative-scale separations are possible and methods are easy to transfer between laboratories and to upscale. Chromatographic separations of enantiomers can be broadly classified into indirect and direct methods.

The indirect methods consist of derivatization of the enantiomers with an enantiomerically pure reagent to form covalently bonded diastereomeric derivatives followed by their separation on achiral phases, thanks to the different physico-chemical properties of the reaction products [13]. These derivatives can then be separated with different separation techniques, e.g., thin layer chromatography (TLC), HPLC, gas chromatography (GC) as well as SFC.

Indirect chiral separation methods provide a possible improvement for detection limits when forming strongly UV-absorbing or fluorescent derivatives. However, the approach requires reagents with a high stereochemical purity because reagent impurities can lead to the formation of multiple diastereomers for a single enantiomer. Possible derivatization reagents include 9-fluorenylethyl chloroformate (FLEC), 2,2,4,6-tetra-*O*-acetyl-β-D-glucopyranosyl isothiocyanate (GITC), and Marfey's reagent [14]. However, the need for a fast and complete reaction, the possible formation of side-products and the necessity for costly reagents make this approach less attractive [14].

The direct approach consists of creating temporary diastereomeric complexes between the enantiomers and a chiral selector. The direct chiral separations can be further subdivided into those applying either chiral stationary phases (CSPs) or chiral mobile phase additives. The CSPs have the chiral selector coated or bonded onto a stationary phase backbone, which is usually silica-based. In the chiral mobile phase additives approach, the chiral selector is added to the mobile phase and an achiral stationary phase can be used. Although chiral mobile phase additives can be used in SFC, they have two main drawbacks. These additives often possess poor solubility in typical SFC mobile phases [15,16]. In addition, method development is often lengthy and scale-up for preparative purposes is complicated by the need to isolate the enantiomers from the collected fractions [17]. These drawbacks make CSPs the first choice for SFC enantioseparations.

12.4 CHIRAL STATIONARY PHASES

Chiral columns for SFC separations can be initially classified into capillary and packed column phases. The use of capillary columns exhibits a limited sample capacity and is not suitable for preparative-scale applications [18]. In addition, linear velocities of 10–20 times the optimum may be required for realistic separation times. Due to poor reproducibility and a limited application range, capillary SFC is rarely used outside of the petroleum industry today [19].

The use of a fixed restrictor downstream of the capillary columns [20], in packed column SFC is replaced by a back pressure regulator. Packed columns have also overcome the additional problems associated with modifier addition, sample injection, and automation. In the 1990s LC-type CSPs were widely adopted for SFC leading to the dominance of packed column methods [19].

Packed columns for the separation of enantiomers can be further divided into classes based on the type of chiral selector they contain (Table 12.1). The most common CSPs are those based on polysaccharides, which consist of amylose or cellulose derivatives, coated or immobilized on a silica gel support. Other macromolecular selectors, such as synthetic polymers and proteins, are also used. Macrocyclic selectors, such as cyclodextrins and glycopeptides, and low-molecular selectors, such as the Pirkle- or brush-type CSPs, are less commonly used in SFC. A variety of CSPs is important to accommodate the specific nature of enantiomeric separations. An overview of the different classes of chiral selectors and their applications is given later.

12.4.1 Polysaccharide Derivatives

Polysaccharide-based CSPs are the most commonly used stationary phases for the separation of enantiomers by SFC. Their attractiveness is due to a number of characteristics: a wide application range, repeatable results, high loadability and availability [17]. Underivatized cellulose and amylose show only limited enantioselectivity because their helical structure is too dense to allow inclusion and enantiorecognition of many molecules. The introduction of electron donor groups, such as alkyl groups, or electron-withdrawing groups, like halogens or trifluoromethyl groups onto the polysaccharide backbone enhances the range of their enantioselective interactions [21,22]. Several polysaccharide-based phases are now available from different manufacturers (Table 12.1).

Because of the number of binding sites, the chiral recognition mechanisms of derivatized polysaccharides are quite complex and are not yet fully understood. Chiral discrimination can result from

TABLE 12.1 Overview of the Commercially Available Chiral Selectors Used in SFC

Selector types	Commercial CSP names	Selector
Polysaccharide selectors	Lux Cellulose-1 (previously Sepapak 1)	Cellulose tris(3,5-dimethylphenylcarbamate)
	Chiralcel OD-H; Chiralpak IB	
	RegisCell	
	Kromasil CelluCoat	
	Chiralpak AD-H; Chiralpak IA	Amylose tris(3,5-dimethylphenylcarbamate)
	Kromasil AmyCoat	
	RegisPack	
	Lux Cellulose-2 (previously Sepapak 2)	Cellulose tris(3-chloro-4-methylphenylcarbamate)
	Chiralcel OZ-H	
	Chiralcel OX-H	Cellulose tris(4-chloro-3-methylphenylcarbamate)
	Lux Cellulose-4 (previously Sepapak 4)	
	Lux Amylose-2 (previously Sepapak-3)	Amylose tris(5-chloro-2-methylphenylcarbamate)
	Chiralpak AY-H	
	RegisPack CLA-1	
	Chiralcel-OJ-H	Cellulose tris(4-methylbenzoate)
	Lux cellulose-3	
	Sepapak 5	Cellulose tris(3,5-dichlorophenylcarbamate)
	Chiralpak IC	
	Chiralpak AS-H	Amylose tris((S)-α-methylbenzylcarbamate)
	Chiralpak ID	Amylose tris(3-chlorophenylcarbamate)
	Chiralpak IE	Amylose tris(3,5-dichlorophenylcarbamate)
	Chiralpak IF	Amylose tris(3-chloro-4-methylphenylcarbamate)
Glycopeptides	Chirobiotic T	Teicoplanin
	Chirobiotic T2	Teicoplanin
	Chirobiotic R	Ristocetin A

(Continued)

TABLE 12.1 (Continued)

Selector types	Commercial CSP names	Selector
	Chirobiotic V	Vancomycin
	Chirobiotic TAG	Teicoplanin aglycone
Cyclodextrins	Sumichiral OA-7500	Heptakis(2,3,6- tri-*O*-methyl)-betacyclodextrin
	β-Cyclose-OH T	Mono-2-*O*-pentenyl-β-cyclodextrin
	β-Cyclose-6-OH T	Mono-6-*O*-pentenyl-β-cyclodextrin
	β-Cyclose-2-OH	Oxidized mono-2-*O*-pentenyl-β-cyclodextrin
	β-Cyclose-6-OH	Oxidized mono-6-*O*-pentenyl-β-cyclodextrin
	MPCCD	Mono-6-(3-methylimidazolium)-6-deoxyperphenylcarbamoyl-β-cyclodextrin chloride
	MDPCCD	Mono-6-(3-methylimidazolium)-6-deoxyper(3,5-dimethylphenylcarbamoyl)-β-cyclodextrin chloride
	OPCCD	Mono-6-(3-octylimidazolium)-6-deoxyperphenylcarbamoyl-β-cyclodextrin chloride
	ODPCCD	Mono-6-(3-octylimidazolium)-6-deoxyper(3,5-dimethylphenylcarbamoyl)-β-cyclodextrin chloride
	Cyclobond I 2000 RN	(R)-Naphthylethylcarbamoylated-β-Cyclodextrin
	Cyclobond I 2000SN	(S)-Naphthylethylcarbamoylated-β-cyclodextrin
Cyclofructans	Larihc CF7-DMP	3,5-Dimethylphenylcarbamate cyclofructan 7
	Larihc CF6-P	Alkyl derivatized cyclofructan-6
	Larihc CF6-RN	R-naphthylethyl-functionalized cyclofructan-6
Pirkle-type selectors	R,R Whelk-O1 or S,S Whelk-O1	1-(3,5-dinitrobenzamido)-1,2,3,4-tetrahydrophenanthrene
	R,R Whelk-O2	1-(3,5-dinitrobenzamido)-tetrahydrophenanthrene
	Chirex 3005	(R)-1-Naphthylglycine and 3,5-dinitrobenzoic acid
	ChyRoSine-A	3,5-Dinitrobenzoyl tyrosine

(*Continued*)

TABLE 12.1 (Continued)

Selector types	Commercial CSP names	Selector
Ion-exchange selectors	Chiralpak QD-AX	*O*-9-(*tert*-butylcarbamoyl) quinidine
	Chiralpak QN-AX	*O*-9-(*tert*-butylcarbamoyl) quinine
	Chiralpak ZWIX (+)	Quinine combined with (S,S)-*trans*-2-aminocyclohexanesulfonic acid
	Chiralpak ZWIX (−)	Quinidine combined with (R,R)-*trans*-2-aminocyclohexanesulfonic acid

inclusion inside the polysaccharide helical structure, $\pi - \pi$ interactions can take place between aromatic and carbonyl functional groups and the selector, while hydrogen bond donor and acceptor groups may add to the enantioselectivity. Flexibility and globularity of the phases contribute to their enantiorecognition abilities, and to a lesser extent, to the achiral interactions of the polysaccharide phases [23,24].

Most chiral polysaccharide-based phases have been prepared by coating the chiral selector onto the silica support. However, such CSPs are restricted to the use of polar solvents typically used in normal- and reversed-phase conditions. This problem was mitigated by forming immobilized phases with covalent binding of the chiral selector to the silica matrix [25]. These phases are compatible with nonstandard solvents, for instance, dichloromethane, chloroform, ethyl acetate, tetrahydrofuran, dioxane, toluene, and acetone. Exposure of coated CSPs to such solvents would cause swelling, dissolution and elution of the physically adsorbed chiral selector, resulting in the destruction of the column. The use of immobilized phases with atypical solvents may lead to improved analyte solubility, which is critical for preparative-scale separations [26,27]. In addition, these solvents may result in improved efficiency or alternative enantioselectivity and therefore enhanced resolution.

12.4.2 Cyclic Oligosaccharides

12.4.2.1 Cyclodextrins

Cyclodextrins are cyclic structures composed of glucopyranose units. The cyclodextrins have a hydrophobic interior cavity with hydrophilic edges containing hydroxyl groups. The hydrophobic center allows the entrapment of hydrophobic parts of molecules, while the exterior hydroxyl groups allow the selector to interact with analytes through hydrogen bonding and dipole−dipole interactions [28]. Frequently the cyclodextrin is linked to the silica support via a spacer arm [29].

The low polarity of supercritical mobile phases has limited the use of cyclodextrin CSPs, since the native cyclodextrins are not able to form inclusion complexes with the analytes. Physical coating has been used to develop a series of cationic cyclodextrin phases, which show enhanced chiral recognition capabilities. Such phases include mono-6-(3-methylimidazolium)-6-deoxyperphenylcarbamoyl-β-cyclodextrin chloride (MPCCD), mono-6-(3-methylimidazolium)-6-deoxyper(3,5-dimethylphenylcarbamoyl)-β-cyclodextrin chloride (MDPCCD), mono-6-(3-octylimidazolium)-6-deoxyperphenylcarbamoyl-β-cyclodextrin chloride (OPCCD), and mono-6-(3-octylimidazolium)-6-deoxyper(3,5-dimethylphenylcarbamoyl)-β-cyclodextrin chloride (ODPCCD) [30,31]. In addition, a radical copolymerization reaction was also used to create two covalently bonded cationic cyclodextrin phases used for the separation of flavones and amino acid derivatives as well as thiazides [32].

12.4.2.2 *Cyclofructans*

Cyclofructans are cyclic structures with six or more β-2,1 linked D-fructofuranose units (Fig. 12.2). These chiral selectors can incorporate different numbers of fructofuranose units in the macrocyclic ring, with

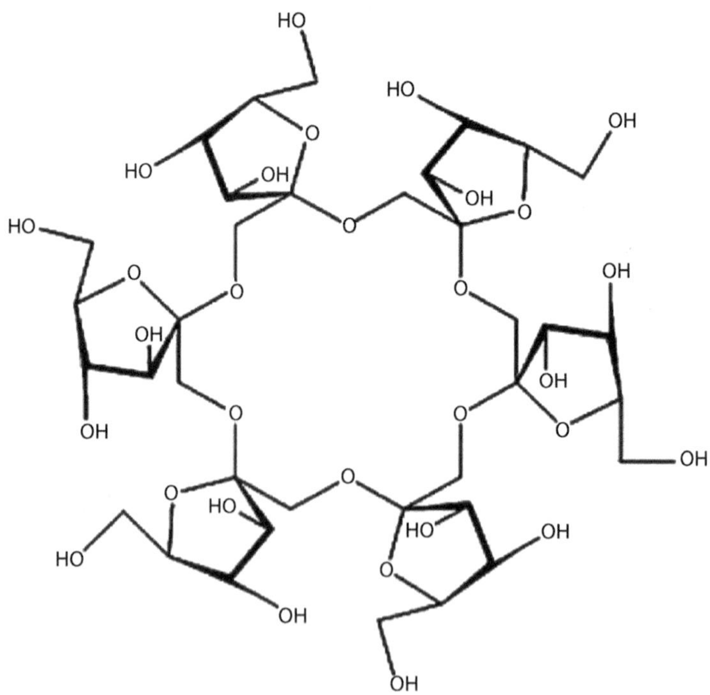

FIGURE 12.2 Molecular structure of cyclofructan-6.

each fructofuranose unit containing four stereogenic centers and three hydroxyl groups, which can be used for derivatization [33]. Derivatized cyclofructan-based chiral phases include isopropyl carbamate CF6 (CF-P), R-naphthylethyl-carbamate CF6 (CF-RN), and dimethylphenyl carbamate CF7 (CF-DMP), where the number reflects the number of D-fructofuranose units. These phases show increased enantioselectivity compared to the native structures.

The dimethylphenyl carbamate CF7 CSP was used in SFC and the results compared with those obtained by HPLC [34]. Linear free energy relationships were built to understand the interactions contributing to retention in different mobile phase compositions. The results show that adsorption of certain components of the mobile phases to the stationary phase plays an more important role in SFC than in HPLC, while dispersion interactions were similar for both techniques. The tendency to interact with n- and/or π-electron pairs was only significant for SFC [34].

In a recent study, cyclofructan chiral phases were used to separate α-aryl ketones with normal-phase liquid chromatography (NPLC) and SFC conditions [35]. The α-arylated carbonyl group is found in many biologically active natural products. Arylation of the ketone group results in the formation of racemic mixtures. Using the cyclofructan phase, baseline separations were observed for 17 of the 21 α-aryl ketones via NPLC and for 10 of the 21 compounds by SFC [35].

12.4.3 Glycopeptides

Macrocyclic antibiotics, for instance vancomycin (Fig. 12.3), teicoplanin, teicoplanin aglycone, and ristocetin, are widely used as CSPs in SFC. These phases contain numerous chiral centers and exhibit a wide range of enantioselectivity. They contain a similar aglycone part with fused macrocyclic glycopeptide rings and linked carbohydrate functionalities. The selectors are always bonded to a silica support and are stable to typical separation conditions. However, long equilibration times are necessary [28,36]. The primary interactions responsible for enantiomer recognition include hydrogen bonding, dipole–dipole and $\pi-\pi$ interactions, hydrophobic interactions, and steric repulsions [37]. These macrocyclic antibiotics have shown excellent selectivity toward diverse chiral compounds, such as amino acids, β-blockers, chiral heterocycles, and pharmaceutical drugs [38].

12.4.4 Low-Molecular Weight Selectors

Pirkle-type or brush-type stationary phases are designed to target specific chiral interactions with the analyte. They consist of single strands with either π-donor or π-acceptor aromatic groups, together

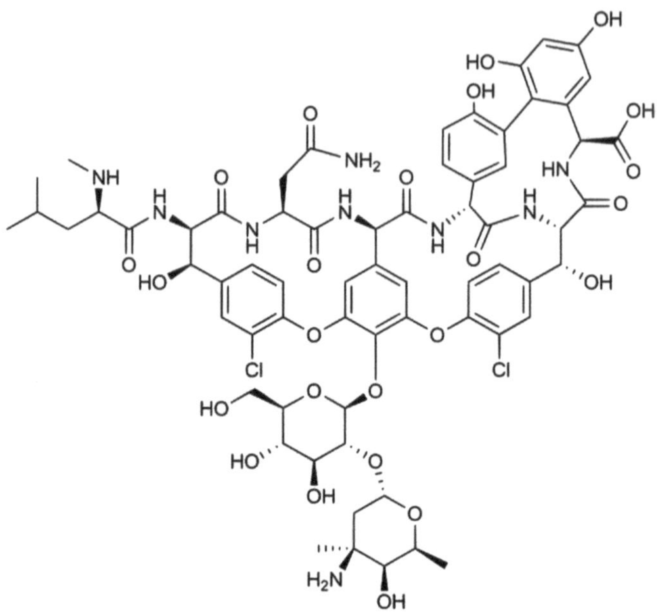

FIGURE 12.3 Molecular structure of vancomycin.

with hydrogen bonding and dipole-stacking inducing functional groups covalently bonded to the silica surface. The interactions that take place with Pirkle-phases require $\pi - \pi$ interactions, hydrogen bonding, and dipole interactions, which are favored in nonpolar solvents, explaining why the first reported packed column SFC enantioseparation used this type of phase [39]. Since then, more Pirkle-type phases have been developed, which show broader selectivity and better efficiency.

Pirkle-type phases have been used successfully for racemate screening in SFC. A chiral (R,R)-Whelk-O1 column and four polysaccharide phases (Chiralcel OD and OJ, Chiralpak AS, and AD-H) were tested under HPLC and SFC conditions for the separation of chiral amine enantiomers having a carbobenzyloxy (cbz) protecting group [40]. All amine cbz derivatives showed favorable resolutions on Chiralpak AD or Chiralpak AD-H columns. Chiralpak AS and Whelk-01 were unable to separate the amine enantiomer derivatives, but only about one-half to baseline. The study concluded that both HPLC and SFC should be investigated in parallel when developing a new method.

12.4.5 Ion-Exchange Phases

Ionizable compounds, for instance acids, are frequently separated on synthetic polymeric chiral phases, derivatized amylose or cellulose

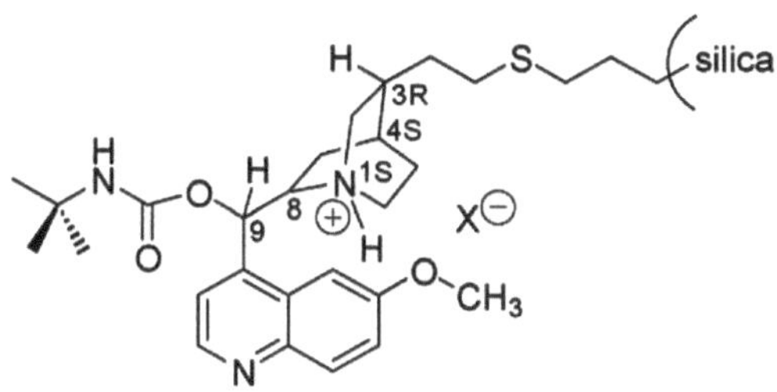

FIGURE 12.4 Quinine and quinidine carbamate-based chiral stationary phases (CSPs). Chiralpak QN-AX: (8S,9R), quinine derived; Chiralpak QD-AX: (8R,9S), quinidine derived [42].

phases, and macrocyclic glycopeptide phases. Quinine (Chiralpak QN-AX) and quinidine (Chiralpak QD-AX) based phases (Fig. 12.4) are weak anion exchangers and have been used to separate chiral acidic compounds [41]. Retention can be tuned by varying the amount and type of co- and counterions (additives) in the modifier, without affecting the enantioselectivity. Ion-exchange phases can be used equally well in HPLC and SFC modes, however, the latter has the advantage of utilizing salt-free mobile phases due to the in situ generation of transient ionic species, for instance methylcarbonic acid from CO_2 and methanol [41]. The concentration of the acid and salt components formed is controlled by varying the amount of polar organic modifier, the concentration of basic additives, and system pressure and temperature.

Strong cation exchange-type chiral phases, based on syringic acid, have been used in subcritical fluid chromatography for the separation of amines [43]. Systematic variation of the concentration of the amine additive in the modifier resulted in an ion-exchange retention mechanism, conforming to the stoichiometric displacement model. The separation of diverse chiral amines on a strong cation exchange-type chiral phase by HPLC and subcritical fluid chromatography was the basis of an additional study [44].

12.4.6 Molecularly Imprinted Polymers

Molecularly imprinted polymer (MIP) chiral phases can be synthesized to favor the retention of one enantiomer over another. These phases were initially synthesized for the isolation of specific enantiomers from complex matrices for sample preparation purposes [45].

Ellwanger et al. [46] utilized a MIP stationary phase based on racemic propranolol and the L-enantiomer of phenylalanine anilide for the separation of enatiomers by SFC. Broad peaks and sample size–dependent retention dependency were indicated as limitations.

Another study investigated the use of MIPs for the resolution of ephedrine by SFC [47]. Peak asymmetry and low efficiency were identified as the main problem for the general development of MIP stationary phases.

12.4.7 Chiral Phases with Sub-2-μ Particles

A promising strategy for fast and highly efficient enantiomer separations for high-throughput screening methods is to immobilize chiral selectors on small totally porous or core-shell particles as opposed to the conventional 3 or 5 μm particles [48,49] The availability of 1.9- μm silica particles facilitate subminute chiral separations with reasonable efficiency. Teicoplanin and teicoplanin aglycone phases on 1.9 μm particles have been used for the separation of chiral heterocyclic compounds, such as oxazolindones and hydantoins with plate numbers of up to 190,000 [38]. A sub-2-μm Whelk-O 1 CSP was used to screen over 120 pharmaceutical racemates with diverse physico-chemical properties [49]. Partial or baseline resolution of 63% of the racemates was obtained with a 7-minute gradient. Of the neutral and acidic racemates, 85% were resolved. Basic compounds also showed partial separation even though it is well known that this type of compound are difficult to resolve on Pirkle-type phases [49].

12.5 CHROMATOGRAPHIC PARAMETERS IN CHIRAL SFC

12.5.1 Mobile Phase

Carbon dioxide is typically used as a mobile phase for SFC. However, since the polarity of CO_2 is comparable to that of hexane, the addition of polar modifiers is necessary to elute polar compounds [50]. The use of modifiers increases the critical parameters, and therefore working conditions often become subcritical. Fortunately the advantages of a supercritical fluid are not lost when working in subcritical conditions [51,52]. Methanol is the most frequently used modifier, with other solvents, such as ethanol, 2-propanol, and acetonitrile used as alternatives. The modifier influences the retention of analytes, due to a change in polarity, as well as a change in the density and solubilizing properties of the mobile phase. Additionally, the modifier can also adsorb onto the CSP and alter the three-dimensional structure of the selector, and therefore its enantioselectivity [17,51,53]. Dichloromethane or tetrahydrofuran did not lead to lower average selective in the separation of racemates. In some cases, however, drastic changes in selectivity with nonalcoholic modifiers and immobilized and coated polysaccharide chiral phases were observed compared to methanol [54]. Byrne et al. [55] used 2,2,2-trifluoroethanol as a modifier for the separation of alcohol-sensitive compounds on a variety of coated polysaccharide and Pirkle-type chiral phases, such as Chiralcel-OD-H, Chiralcel-OJ-H, Sepapak-3, and Whelk-O1. Compounds containing alcohol-sensitive groups may undergo side-reactions such as ester exchange, nucleophilic cleavage, or substitution reactions in the presence of alcohol modifiers.

12.5.2 Additives

The addition of a polar modifier to the mobile phase might not be sufficient to achieve the desired separation. Compounds containing basic (amine) functional groups may interact strongly with the silica backbone of the stationary phase and either elute with distorted peak shapes or fail to elute at all. This problem can be overcome by the addition of small amounts, typically between 0.1% and 2.0%, of polar additives to the modifier, which serve to suppress ionization of the analytes in the mobile phase or to suppress ionization of the silanol groups on the stationary phase [56]. Acidic additives are commonly used for acidic compounds and basic additives for basic compounds, although acidic compounds may not require additives because of the acidic nature of carbon dioxide in the presence of methanol [57]. Trifluoroacetic acid and formic acid are commonly used acidic additives, while isopropylamine and triethylamine are used for the separation of basic compounds. Acidic and basic additives have also been used simultaneously. For example, trifluoroacetic acid in combination with isopropylamine was used for the separation of acidic, basic, neutral, and amphoteric compounds on polysaccharide-based CSPs [58]. However, the removal of these additives from the stationary phase or from analyte-containing fractions are not straightforward. For this reason, volatile additives, such as nonaqueous ammonia [59] or ammonium hydroxide [60], are preferred since they can be more easily removed from the mobile phase, simplifying postpurification for preparative-scale applications. Ammonium salts or other volatile additives are also compatibile with mass spectrometric detection [2].

The addition of small percentages of water to the modifier has been shown to facilitate the elution of polar compounds [61]. Even though water is poorly soluble in supercritical CO_2, in ternary mixtures with an organic solvent it enhances the mobile phase's solvating power, which simplifies preparative-scale purifications and is compatibile with mass spectrometry [2]. The effects of the use of water as an additive have not been widely investigated, however, the results seem promising.

Additives can be used with other chiral phases, and not only those containing polysaccharide selectors. For instance, acetic acid was used for the analysis of flavanones, thiazides, and amino acid derivatives on cyclodextrin phases [32]; Trimethylamine was used for the separation of α-aryl ketones enantiomers on cyclofructan chiral phases [35]; and isopropylamine and acetic acid were used for the separation of ephedrine enantiomers on (−)-ephedrine-MIPs [47].

12.5.3 Column Temperature and Back-Pressure

In the early days of SFC, relatively low temperatures were employed (20–25°C), but more recent studies take place at 30–40°C [17]. It should be noted that the temperature for separations need not be higher than the critical temperature of carbon dioxide, since separations in the subcritical state are equally as good as those for the supercritical state [17]. Temperatures used for chiral separations often reflect the maximum allowable temperature of the CSPs recommended by the manufacturers. Subambient temperature can be used for particularly difficult separations that cannot be accomplished at ambient temperatures and for components that are stereolabile at room temperature [37]. West et al. [62] investigated the effect of temperature on the separation of fluoro-oxoindole-type enantiomers on chlorinated polysaccharide-based CSPs. The investigated temperature range (0–50°C) led to small variations in retention and separation factors and large variations in efficiency. However, the effects of temperature were strongly dependent on the stationary phase, the mobile phase, and the analytes. A design of experiments (DoE) approach was used to investigate the effect of temperature, pressure, and cosolvent on the analytical and preparative-scale separation of *trans*-stilbene oxide and 1,1′-bi-2-naphthol [63]. It was concluded that the retention factors were most influenced by the amount of cosolvent and pressure, while the selectivity was most influenced by the amount of cosolvent and temperature.

A number of studies related to the effect of the column back-pressure on chiral separations have been performed [63–65]. It is generally understood that increasing back-pressure results in a decrease in the retention factor. However, this really depends on the other experimental conditions, such as temperature and modifier concentration [37]. Tarafder et al. [66] used isopycnic (constant density plots) to gain a better visual understanding of how retention is influenced by pressure for different operating conditions. Increasing density at 20°C decreases retention factors observed for higher column inlet pressures. However, the retention factor increase is much larger for the same inlet pressure increase at a higher temperature. This is because the density of CO_2

increases by 5.2% when increasing the pressure from 2000 to 3000 psi at 20°C; but, it increases by 19% at 50°C [66].

12.5.4 Flow Rate

Because of the low viscosity of CO_2, flow rates of 2–5 mL/min can be employed with conventional instruments. Higher flow rates result in shorter separation times while preserving column efficiency, since the slope of the Van Deemter curve in SFC is less steep than is typical for HPLC [67]. This is only possible because of the higher diffusion coefficients in supercritical fluids compared with liquids. Flow rate changes in SFC lead to changes in the mobile phase density and therefore its elution strength, leading to changes in retention and selectivity [17].

12.6 COMPARISON TO OTHER TECHNIQUES

Chomatographic techniques remain the most convenient and cost-effective approaches for stereoselective separations. In addition to SFC, HPLC and gas chromatography (GC) are commonly used for this purpose. In addition, counter-current chromatography (CCC), thin layer chromatography (TLC), capillary electrophoresis (CE), and capillary electrochromatography (CEC) can also be used [68,69]. In the following subsections we will compare SFC to the main chromatographic separation techniques of LC and GC.

12.6.1 Liquid Chromatography

LC remains the mainstay technique for chiral separations in a number of areas, including drug development, pesticide analysis, food additive evaluation, natural product research, agrochemicals, and pollution analysis. This is mainly because of its high speed, sensitivity, and reproducibility [27]. Polysaccharide- and cyclodextrin-based CSPs dominate HPLC separations. Protein, ligand- and ion-exchange, and macrocyclic antibiotic-based CSPs are also widely used [70]. The use of SFC for the separation of enantiomers has been successful due to the lower viscosity, higher diffusion coefficients and favorable mass-transfer properties of supercritical fluid carbon dioxide–containing mobile phases compared with the solvents used in HPLC [71]. Like SFC, HPLC allows the effective scale-up to amounts of material needed for the different phases of drug discovery and development by increasing the column dimensions and by using other throughput-improving techniques [36]. The use of sub-2-μm particles in LC is on the rise leading to ultra-high-performance liquid chromatography (UHPLC). However, the lack of understanding of mass-transfer mechanisms in CSPs represents a hurdle to the development of sub-2-μm packing materials for chiral separations [71].

A number of studies have compared LC and SFC with different chiral phases and different test solutes. For instance, the enantioseparation of antiulcer drugs was investigated by HPLC and SFC using a Chiralpak AD (amylose tris(3,5-dimethylphenylcarbamate)) phase [72]. Two of four drugs were separated using normal-phase mobile phases, while all four were separated in a shorter time using SFC. In another study, Chiralpak AD-H was used to separate chiral proline derivatives by SFC and HPLC [73]. SFC afforded superior resolution and a shorter separation time while HPLC a lower limit of quantification.

Due to the nonpolar nature of CO_2, SFC has too often been described as a technique offering normal-phase like separation [19,61,74]. It has been shown that CO_2—modifier mobile phases are weaker eluents than liquid mobile phases, such as heptane with ethanol or isopropanol. This leads to different retention mechanisms for SFC and NPLC and differences in enantioseparations. The mobile phase components for NPLC and SFC adsorb on the CSPs in a different manner, and therefore also modify the polarity of the CSPs and three-dimensional structure in different ways. SFC yielded longer retention times and higher plate numbers when compared with NPLC. However, separation factors can be either completely different or similar [75].

The two techniques can also be coupled in a comprehensive approach (LC-SFC) to analyze complex mixtures. Venkatramani et al. [76] developed a new LC-SFC interface for the simultaneous achiral–chiral separation of pharmaceutical compounds. In the first dimension, reversed-phase LC provided the achiral separation using an Acquity HSS T3 (silica-based trifunctional C18 alkyl phase) column while in the second dimension SFC provided the chiral separation with a Chiralcel OD3 (cellulose tris(3,5-dimethylphenylcarbamate)) phase. The success of this 2-dimensional system was due to a series of low volume trapping columns, which allowed the transfer of multiple fractions from the first RP column to the second SFC chiral phase (Fig. 12.5).

12.6.2 Gas Chromatography

GC is a powerful technique for the separation of volatile chiral compounds mostly because of the development of many types of CSPs and its intrinsic high efficiency [77–80]. It is useful for the separation of nonaromatic compounds that are not easily separated in LC. However, GC comes with one main disadvantage: it is only suitable for analytes that are thermally stable and volatile, either in their native state or after derivatization. This is not a requirement for SFC. For example in GC, α-amino acids have to be transformed into volatile derivatives, such as,

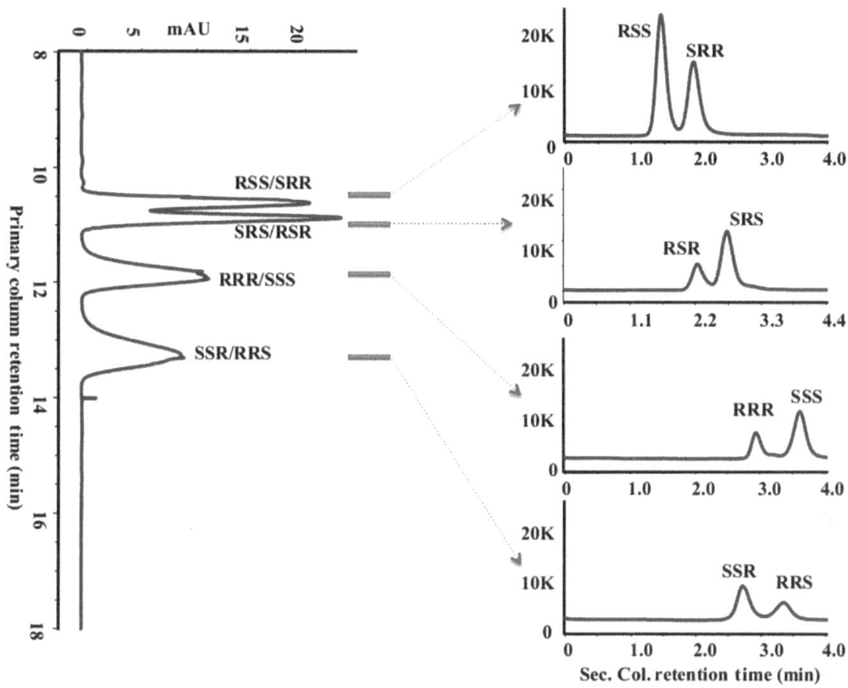

FIGURE 12.5 Efficient separation of two stereoisomers using a two-dimensional LC-SFC interface. The primary RPLC dimension resolved the four diastereomeric pairs, while the secondary SFC dimension resolved the corresponding enantiomeric pairs [76].

N-perfluoroacyl-*O*-alkyl esters of *N*-alkoxycarbonyl-*O*-alkyl esters [81]. Moreover, it is difficult to scale-up capillary GC [36].

SFC offers lower efficiency compared to GC because supercritical fluids have a higher viscosity than gases, resulting in lower mass-transfer rates. However, the elution strength of supercritical mobile phases can be modified by variation of a number of system parameters, leading to more possibilities to optimize separations [17]. The use of lower temperatures in SFC than in GC may also lead to improved resolutions with some efficiency loss, since the increase in fluid viscosity results in lower mass-transfer rates.

12.7 RECENT DEVELOPMENTS IN STEREOSELECTIVE SFC

12.7.1 Column Coupling

The low viscosity and high diffusivity of supercritical fluid mobile phase facilitates the serial coupling of columns, thereby increasing the plate number without the issue of increased column pressures [67]. The use of tandem columns in chiral SFC can be seen in two ways: (1) the coupling of an achiral and chiral column and (2) the coupling of two or more chiral phases.

A combination of achiral and chiral phases can be used to improve enantioselectivity and increase plate numbers [82] Alexander and Staab [83] separated the intermediates from a three-step stereoselective synthesis of cinnamonitrile and hydrocinnamonitrile in a single separation using a coupled achiral/chiral SFC/MS method. The achiral phase was a silica phase and the chiral phase was Chiralcel OD-H. With the tandem achiral and chiral columns, both enantiomers and diastereomers were separated without a significant increase in the separation time. Welch et al. [84] developed a SFC tandem-column screening tool in which 10 different individual chiral phases and 25 different tandem-column combinations could be exploited for the separation of complex mixtures of stereoisomers. The SFC instrument could call upon any of the different tandem-column arrangements by using simple software commands.

12.7.2 Preparative-Scale SFC

Preparative-scale SFC offers significantly reduced manufacturing costs and pollution compared to conventional preparative-scale LC. In addition, the loadability of compounds and SFC productivity can be up to 5–10 times higher than for LC. Recent application of SFC takes advantage of its high speed, low solvent costs, and ease of removal of the residual mobile phase. Small-scale purifications are usually carried out using stacked sample injections, while simulated moving-bed SFC (SMB-SFC) has comparable potential to SMB-LC when large amounts of drugs of high value are produced [82].

12.7.2.1 Stacked Sample Injections

Stacked sample injections are established as a methodology for fast new drug development, especially for chiral separations. With this approach, small sample volumes are repetitively loaded on an analytical SFC column allowing a considerable amount of purified enantiomers to be accumulated in a short time [82]. Stacked sample injections were used for the purification of four stereoisomers of derivatized β-methylphenylalanine [85]. The method was capable of purifying more than 10 g with greater than 90% total recovery. The derivatized amino acid was then available in pure enantiomeric forms (>99.9% purity) for use in reactions to synthesize new chemical entities. In another study, a residue containing tengeretin and 3,5,6,7,8,3′,3′-heptamethoxyflavone was separated using 25 cycles of stacked injections on a

Chiralpak AD column [86]. These polymethylflavones together with nobiletin and 5,6,7,4′-tetramethoxyflavone were isolated on a large scale from orange peel extract. Large-scale extraction of these compounds is particularly interesting since they possess a broad spectrum of biological activity.

12.7.2.2 Supercritical Fluid Simulated Moving–Bed Chromatography

SMB is a multicolumn configuration used for large-scale production (>20 kg scale) [87]. The purpose of SMB is to have a configuration in which the stationary and mobile phases move in counter-current directions as opposed to a fixed-bed operation encountered in single column chromatography [88,89]. Since it is not possible to physically move the stationary phase, movement is simulated by switching the feed and product ports between the fixed solid beds. Fig. 12.6 shows the setup of a four-section, two columns per section SMB [88]. Each section of the SMB is divided into several subsections so as to simulate the counter-current movement of the stationary phase. The process operates in such a way that the light component (B) moves in the direction of the mobile phase and is collected at the raffinate port. The heavy component (A), moves in the direction of the stationary phase and is collected at the extract port.

Depta et al. [90] presented the details for a plant-scale SMB-SFC for the first time in 1999. The system was tested using the separation of phytol *cis-trans* isomers and resulted in fraction purities of 99%. The effect of the modifier, modifier content, and column setup on the

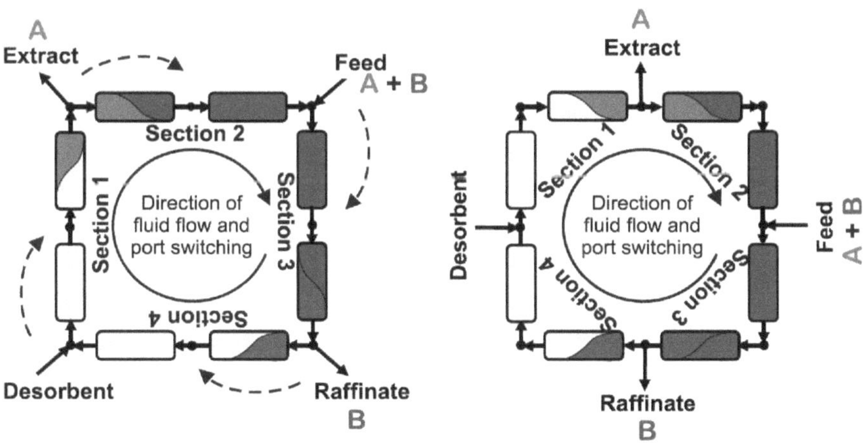

FIGURE 12.6 Scheme of a four-section, two columns per section, simulated-moving-bed (SMB) process [88].

productivity of the SMB-SFC process was later studied by the same group [91]. Initially, low concentrations of phytol isomers were used to test the feasibility of the process. Then the feed concentration was increased to 54 g/L stationary phase per hour. In another application of SMB-SFC phenyl-1-propanol enantiomers were purified with low feed concentrations, obtaining an extract purity of 99.5% [92].

Opposed to SMB-HPLC, SMB-SFC uses the pressure gradient of the mobile phase as a significant factor to modify selectivity. Pressure, temperature, and the fraction of organic modifier influence the density of the supercritical mobile phase. A small change in mobile phase density has a substantial impact on the solvation capability of the mobile phase as well as resolution [93].

12.8 OVERVIEW OF SFC STEREOISOMERIC APPLICATIONS

Table 12.2 provides an overview of enantiomer applications in SFC. Table 12.2 is divided into several subcategories reflecting applications in different fields, mainly pharmaceutical, food, and agriculture. The list of applications is not exhaustive, and is intended to illustrate possibilities for further applications, focusing on studies executed in the last 10 years.

12.9 SEPARATION OF DIASTEREOMERS

Separation of enantiomers is well-documented, however, only limited literature is available for the separation of diastereomers. A number of compounds exist as diastereomers, and a synthetic route for the synthesis of one pure diastereomer is not always possible. In addition, incomplete selectivity in the synthesis of a complex product leads to the formation of several stereoisomeric impurities together with the desired compound. In such cases efficient separation of the diastereomers is required. The separation of diastereomers using SFC has been investigated using both chiral [94–97], and achiral phases [98]. An overview of these applications by SFC is given in Table 12.3. Extra care should be taken when using chiral phases for the separation of complex mixtures comprising racemic or enantiomeric compounds, since the enantiomers in the mixture may or may not be resolved on the chiral phase. Such a situation can be overcome by using a racemic version of the chiral phase. Achiral components, diastereomers, and constitutional isomers can then be resolved but the phase is not capable of separating enantiomers [99].

TABLE 12.2 Overview of Enantiomeric Applications in Various Fields

Analytes	Stationary phases	Analysis conditions	References
PHARMACEUTICAL COMPOUNDS			
Diuretics: Bendroflumethiazide, trichlormethiazide, althiazide, indapamide, chlorthalidone	In-house bonded cationic 6A-(3-vinylimidazolium)-6-deoxyperphenylcarbamate-β-cyclodextrin chloride (VIMPCCD) CSP	CO_2 with 10% MeOH; flow rate 1 mL/min; temperature 40°C; 150 bar back-pressure	[32]
Nonsteroidal antiinflammatory drugs: Flurbiprofen, ibuprofen, naproxen	Chiralpak QN-AX- and Chiralpak QD-AX CSP (150 × 4 mm i.d.)	CO_2 with 25% MeOH, 200 mM HOAc, 100 mM NH_3; flow rate 4.0 mL/min; temperature 40°C, 150 bar back-pressure	[41]
Psychoactive substances: Mephedrone, 4-methylethcathinone, buphedrone, pentedrone	In-house naphthalene-based chiral ion-exchange-type CSP	MeOH/ACN (1/9), 50 mM FA, 25 mM DEA; flow 1 mL/min; temperature 20°C	[100]
Carbamazepine and six analogs	Chiralpak IB (150 × 4.6 mm, 3 μm)	CO_2 with ethanol (25 mM IBA); Gradient 1–40% in 5 min; flow 3 mL/min; temperature 40°C	[101]
5-Methyl-5-phenylhydantoin	Chirobiotic T and Chirobiotic TAG (50 × 4.6 mm i.d., 1.9 μm)	For Chirobiotic T: 60:40:0.1:0.1 CO_2:MeOH:TFA:TEA; Flow 7.0 mL/min; For Chirobiotic TAG: 60:40 CO_2:MeOH; Flow 7.0 mL/min; room temperature; 80 bar back-pressure for both CSPs	[38]
Alcohol-sensitive compounds: Isradipine, felodipine	Chiralpak IA and Chiralpak IC and Whelk-O1 (S,S) (250 × 4.6 mm i.d.)	CO_2 with 20% TFE; flow 3 mL/min; temperature 40°C; 100 bar back-pressure	[55]

(*Continued*)

TABLE 12.2 (Continued)

Analytes	Stationary phases	Analysis conditions	References
Peptides: Nine proline derivatives	Chiralpak AD-H (250 × 4.6 mm i.d., 5 μm)	CO_2 with 5% EtOH (0.1% TFA); flow 2.5 mL/min; temperature 35°C; 100 bar back-pressure	[73]
Antiulcer drug: Lanzoprazole	Chiralpak AD (250 × 4.6 mm i.d., 10 μm)	CO_2 with 20% MeOH; flow 2 mL/min; temperature 35°C; 200 bar back-pressure	[102]
Anti-malarial: Primaquine	Chiralpak AD-H (250 × 4.6 mm i.d., 5 μm)	CO_2 with 20% MeOH (0.4% DEA); flow 4 mL/min; temperature 35°C	[103]
In-house acidic drug AZY in an aqueous formulation	Chiralpak AD-H (250 × 4.6 mm i.d., 5 μm)	CO_2 with 30% EtOH (0.3% DMEA); flow 3 mL/min; temperature 40°C; 100 bar back-pressure	[104]
β-blocking agents: Atenolol, metoprolol, propranolol	Chiralpak IB (250 × 4.6 mm i.d., 5 μm)	CO_2 with 18% MeOH (0.5% TFA and ammonia (2:1)); flow 1.5 mL/min; temperature 40°C; 172 bar back-pressure	[105]
AGRICULTURAL COMPOUNDS			
Pesticides: Diclofop-methyl, benalaxy, acetofenate, myclobutanil, difenoconazole	Chiralpak IB-H (250 × 4.6 mm i.d., 5 μm)	CO_2 with 10% 2-PrOH; flow 2 mL/min; temperature 35°C; 150 bar back-pressure	[106]
Three neonicotinoid insecticides	Chiralpak AD and Chiralcel IB (250 × 4.6 mm i.d., 5 μm)	CO_2 with 20% or 30% EtOH; flow 2 mL/min; temperature 35 °C;	[107]

(Continued)

TABLE 12.2 (Continued)

Analytes	Stationary phases	Analysis conditions	References
		150 bar back-pressure	
Fungicides: Miconazole, econazole, sulconazole	Chiralpak AD (250 × 4.6 mm i.d., 10 μm)	CO_2 with 10% MeOH; flow 2 mL/min; temperature 35°C; 200 bar back-pressure	[108]
triazole fungicide: Tebuconazole	Chiralpak IA-3 (150 × 4.6 mm i.d., 3 μm)	CO_2 with 13% MeOH; flow 2 mL/min; temperature 30°C; 152 bar back-pressure	[109]
triazole fungicide: Flutriafol	Chiralpak IA-3 (150 × 4.6 mm i.d., 3 μm)	CO_2 with 12% MeOH; flow 2.2 mL/min; temperature 30°C; 152 bar back-pressure	[110]
FOOD ANALYSIS			
Vitamin K1	RegisPack (250 × 4.6 mm i. d., 5 μm)	CO_2 with 5% MeOH; flow 2 mL/min; temperature 30 °C; 150 bar back-pressure	[111]
Aromatic amino acids in food supplements: DL-Phenylalanine and DL-tryptophan	Chirobiotic T2 (250 × 2.1 mm i.d., 5 μm)	CO_2 with 40% modifier (90% MeOH/10% water); flow 2 mL/min; temperature 35°C; 100 bar back-pressure	[112]
γ-Lactone flavors	Chiralpak AD (250 × 4.6 mm i.d., 10 μm)	CO_2 with 1–4% 2-PrOH; flow 1.5 mL/min; temperature 30°C; 120 bar back-pressure	[113]

(*Continued*)

TABLE 12.2 (Continued)

Analytes	Stationary phases	Analysis conditions	References
Spirocyclic terpenoid flavors	Chiralpak IA and IF (250 × 4.6 mm i.d., 5 μm)	CO_2 with different modifiers and gradients; flow 2.5 mL/min; temperature 30°C; 138 bar back-pressure	[114]
δ-Lactone flavors	Chiralcel OB (250 × 4.6 mm i.d., 10 μm)	CO_2 with 1–3% 2-PrOH; flow 1.0–1.5 mL/min; temperature 30–33°C; variable back-pressure	[115]
Sotolon flavor	Chiralpak AD-H (250 × 4.6 mm i.d., 5 μm)	CO_2 with 3% MeOH; flow 2 mL/min; temperature 30°C; 90 bar back-pressure	[116]
OTHERS			
Provitamin B5 in cosmetic formulations	Chiralpak IA (250 × 4.6 mm i.d., 3 μm)	CO_2 with 11% MeOH; flow 2.3 mL/min; temperature 25°C; 150 bar back-pressure	[117]
Drugs of abuse: Methamphetamine	Trefoil AMY1 (150 × 2.1 mm i.d., 2.5 μm)	CO_2 with 8–30% EtOH (1% cyclohexylamine) in 6 min; flow 2.5 mL/min; temperature 40°C; 138 bar back-pressure	[118]

2-PrOH, 2-propanol; *AmmAc*, ammonium acetate; *DEA*, diethylamine; *DMEA*, dimethylethylamine; *EtOH*, ethanol; *HOAc*, acetic acid; *FA*, formic acid; *IBA*, isobutylamine; *MeOH*, methanol; *NSAIDs*, nonsteroidal antiinflammatory drugs; *TFA*, trifluoroacetic acid; *TFE*, 2,2,2-trifluoroethanol.

Regalado and Welch demonstrated that achiral UHPLC and SFC screening of several stationary and mobile phase combinations showed essentially no separation of a mixture of four tricycle-furan isomers, with only partial separation achieved on a Luna CN phase with SFC

TABLE 12.3 Overview of Diastereomeric Applications in Various Fields

Analyte	Stationary phase	Mobile phase	References
Isomeric cinnamonitrile/ hydrocinnamonitrile products	Chiralcel OD-H column (150 × 2.0 mm i.d., 5 μm) coupled to a silica column (150 × 4.6 mm i.d., 5 μm)	CO_2 with 5–15% MeOH; flow 1 mL/ min; temperature 30°C; 120 bar back-pressure	[83]
258 Synthetic research samples containing diastereomeric pairs from ongoing drug discovery	XBridge HILIC, Cosmosil PYE (150 × 4.6 mm, 5 μm)	CO_2 with MeOH (10 mM AmmAc); Gradient 5–60% in 6 min; flow 5 mL/ min; temperature 35°C; 100 bar back-pressure	[98]
4-Hydroxynonenal adducts of deoxyguanosine	Lux Cellulose-2 (250 × 4.6 mm)	CO_2 with 10% PrOH (0.1% DEA); flow 3.5 mL/min; temperature 40°C; 100 bar back-pressure	[95]
Narangin diastereomers	Chiralpak IC (250 × 30 mm, 5 μm)	CO_2 with 40% MeOH (5% water); flow 70 mL/min; temperature 35°C; 100 bar back-pressure	[119]
Fulvestrant diastereomers	Chiralpak AD-H (150 × 4.6 mm, 5 μm)	CO_2 with 25% MeOH: ACN (95:5); flow 2.5 mL/min; temperature 55°C	[96]

ACN, acetonitrile; *AmmAc*, ammonium acetate; *MeOH*, methanol; *PrOH*, propanol.

[94]. However, chiral SFC resulted in baseline separation of all isomers (Fig. 12.7) on a Chiralcel OJ phase, which allowed the preparative-scale purification of 100 mg of the mixture.

Rao et al. [96] used SFC for the separation of fulvestrant diastereomers. They employed a Chiralpak AD-H polysaccharide phase and 25% MeOH/ACN (95/5) cosolvent for the separation of fulvestrant sulfoxide A and B in less than 8 minutes. The method was shown to be specific, precise, accurate, linear, and robust and suitable for routine quality control of production batches. In a study by Ebinger and Weller [98], the success rate for the separation of over 205 synthetic diastereomeric mixtures using achiral SFC and achiral HPLC was determined. Bare silica and a 2-pyrenyl-ethyl phase with a mobile phase gradient, separated twice the number of compounds as RPLC. This study was based on a very limited set of stationary and mobile phases. Therefore

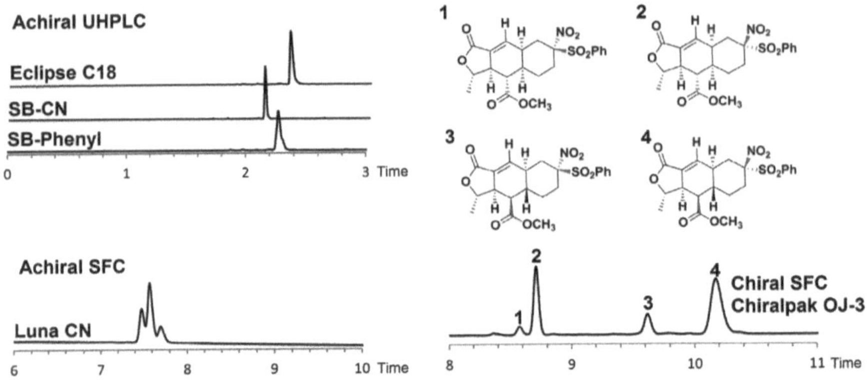

FIGURE 12.7 RP achiral UHPLC-DAD and chiral SFC-DAD chromatographic profiles of a mixture of tricycle-furan isomers. UHPLC experiments were carried out using a mobile phase of 0.1% phosphoric acid in water and acetonitrile with a gradient of 5–95% acetonitrile; SFC experiments were carried out using CO_2 with 1–40% propanol (25 mM isobutylamine) over 5 minutes [94]. IUPAC names: (1) Methyl (3S,3aR,4S,4aR,7S,8aS)-3-methyl-7-nitro-1-oxo-7-(phenoxysulfinyl)-1H,3H,3aH,4H,4aH,5H,6H,7H,8H,8aH-naphtho [2,3-c]furan-4-carboxylate (2) Methyl (3S,3aR,4S,4aR,7R,8aS)-3-methyl-7-nitro-1-oxo-7-(phenoxysulfinyl)-1H,3H,3aH,4H,4aH,5H,6H,7H,8H,8aH-naphtho[2,3-c]furan-4-carboxylate (3) Methyl (3S,3aR,4R,4aR,7S,8aS)-3-methyl-7-nitro-1-oxo-7-(phenoxysulfinyl)-1H,3H,3aH,4H, 4aH,5H,6H,7H,8H,8aH-naphtho[2,3-c]furan-4-carboxylate (4) Methyl (3S,3aR,4R,4aR,7R, 8aS)-3-methyl-7-nitro-1-oxo-7-(phenoxysulfinyl)-1H,3H,3aH,4H,4aH,5H,6H,7H,8H,8aH-naphtho [2,3-c]furan-4-carboxylate.

additional studies using more diverse experimental conditions would help to confirm the findings from this study.

12.10 CONCLUSION

With a renewed interest in supercritical fluids as eluents for chromatographic separations, an increase in research for chiral compounds has been observed. This is mainly due to instrumental improvements, and the availability of a wide range of CSPs. Although the majority of chiral separations in SFC are accomplished on polysaccharide-based phases, other CSPs, for instance, Pirkle-type and antibiotic-based phases are also used. The advantages of SFC over other techniques commonly used for chiral separations include reduced solvent consumption and costs, as well as decreased waste. In addition, faster method development and separation than for HPLC have increased interest in the technique. Unlike GC, SFC is not limited to the analysis of thermally stable or volatile compounds. To conclude, SFC has acquired a place in routine chiral analysis for a number of applications. The use of SFC for chiral separations is on a steady increase and due to the introduction of new CSPs and new technologies such as the use of sub-2-μm particles.

References

[1] Yamada T, Okada T, Sakaguchi K, Ohfune Y, Ueki H, Soloshonok VA. Efficient asymmetric synthesis of novel 4-substituted and configurationally stable analogues of thalidomide. Org Lett 2006;8:5625–8.

[2] Nováková L, Grand-Guillaume Perrenoud A, Francois I, West C, Lesellier E, Guillarme D. Modern analytical supercritical fluid chromatography using columns packed with sub-2 μm particles: a tutorial. Anal Chim Acta 2014;824:18–35.

[3] Taylor LT. Supercritical fluid chromatography. Anal Chem 2010;82:4925–35.

[4] Yan TQ, Orihuela C. Rapid and high throughput separation technologies—steady state recycling and supercritical fluid chromatography for chiral resolution of pharmaceutical intermediates. J Chromatogr A 2007;1156:220–7.

[5] McNaught AD, Wilkinson A. 2nd ed IUPAC: compendium of chemical terminology (the "Gold Book"), vol. 2193. Oxford: Blackwell Scientific Publications; 1997.

[6] Smith MB, March J. Stereochemistry. March's advanced organic chemistry. 6th ed. Hoboken, New Jersey: Wiley-Interscience; 2007. p. 136–233.

[7] Carey FA, Sundberg RJ. Stereochemistry, conformation, and stereoselectivity. advanced organic chemistry. Part A: structure and mechanisms. 5th ed Charlottesville, Virginia: Springer Science and Business Media; 2007. p. 119–240.

[8] Shekunov BY, York P. Crystallization processes in pharmaceutical technology and drug delivery design. J Cryst Growth 2000;211:122–36.

[9] Rekoske JE. Chiral separations. Am Inst Chem Eng J 2001;47:2–5.

[10] Keith JM, Larrow JF, Jacobsen EN. Practical considerations in kinetic resolution reactions. Adv Synth Catal 2001;343:5–26.

[11] Koter I. Separation of enantiomers by chirally modified membranes. In: Koter I, editor. Proceedings of the 23rd international symposium on physico-chemical methods of separation-ARS Separatoria. Poland: Nicolaus Copernicus University in Torun; 2008. p. 384.

[12] Afonso CAM, Crespo JG. Recent advances in chiral resolution through membrane-based approaches. Angew Chem Int Ed Engl 2004;43:5293–5.

[13] Gorog S, Gazdag M. Enantiomeric derivatization for biomedical chromatography. J Chromatogr B 1994;659:51–84.

[14] Ilisz I, Berkecz R, Péter A. Application of chiral derivatizing agents in the high-performance liquid chromatographic separation of amino acid enantiomers: a review. J Pharm Biomed Anal 2008;47:1–15.

[15] Gyllenhaal O, Karlsson A. Enantiomeric separations of amino alcohols by packed-column SFC on Hypercarb with L-(+)-tartaric acid as chiral selector. J Biochem Biophys Methods 2002;54:169–85.

[16] Sun Q, Olesik SV. Chiral separation by simultaneous use of vancomycin as stationary phase chiral selector and chiral mobile phase additive. J Chromatogr B 2000;745:159–66.

[17] West C. Enantioselective separations with supercritical fluids—review. Curr Anal Chem 2014;10:99–120.

[18] Phinney KW. SFC of drug enantiomers. Anal Chem 2000;75:204–11.

[19] Taylor LT. Supercritical fluid chromatography for the 21st century. J Supercrit Fluids 2009;47:566–73.

[20] Terfloth G. Enantioseparations in super- and subcritical fluid chromatography. J Chromatogr A 2001;906:301–7.

[21] Chankvetadze B. Chloromethylphenylcarbamate derivatives of cellulose as chiral stationary phases for high-performance liquid chromatography. J Chromatogr A 1994;670:39–49.

[22] Chankvetadze B, Yashima E, Okamoto Y. Dimethyl-, dichloro- and chloromethylphenylcarbamates of amylose as chiral stationary phases for high-performance liquid chromatography. J Chromatogr A 1995;694:101–9.

[23] West C, Zhang Y, Morin-Allory L. Insights into chiral recognition mechanisms in supercritical fluid chromatography. I. Non-enantiospecific interactions contributing to the retention on tris-(3,5-dimethylphenylcarbamate) amylose and cellulose stationary phases. J Chromatogr A 2011;1218:2019–32.
[24] West C, Guenegou G, Zhang Y, Morin-Allory L. Insights into chiral recognition mechanisms in supercritical fluid chromatography. II. Factors contributing to enantiomer separation on tris-(3,5-dimethylphenylcarbamate) of amylose and cellulose stationary phases. J Chromatogr A 2011;1218:2033–57.
[25] Thunberg L, Hashemi J, Andersson S. Comparative study of coated and immobilized polysaccharide-based chiral stationary phases and their applicability in the resolution of enantiomers. J Chromatogr B 2008;875:72–80.
[26] Zhang T, Kientzy C, Franco P, Ohnishi A, Kagamihara Y, Kurosawa H. Solvent versatility of immobilized 3,5-dimethylphenylcarbamate of amylose in enantiomeric separations by HPLC. J Chromatogr A 2005;1075:65–75.
[27] Cavazzini A, Pasti L, Massi A, Marchetti N, Dondi F. Recent applications in chiral high performance liquid chromatography: a review. Anal Chim Acta 2011;706: 205–22.
[28] De Klerck K, Mangelings D, Vander Heyden Y. Supercritical fluid chromatography for the enantioseparation of pharmaceuticals. J Pharm Biomed Anal 2012;69:77–92.
[29] Schneiderman E, Stalcup AM. Cyclodextrins: a versatile tool in separation science. J Chromatogr B 2000;745:83–102.
[30] Ong T-T, Wang R-Q, Muderawan IW, Ng S-C. Synthesis and application of mono-6-(3-methylimidazolium)-6-deoxyperphenylcarbamoyl-beta-cyclodextrin chloride as chiral stationary phases for high-performance liquid chromatography and supercritical fluid chromatography. J Chromatogr A 2008;1182:136–40.
[31] Wang R-Q, Ong T-T, Ng S-C. Synthesis of cationic beta-cyclodextrin derivatives and their applications as chiral stationary phases for high-performance liquid chromatography and supercritical fluid chromatography. J Chromatogr A 2008;1203:185–92.
[32] Wang R-Q, Ong T-T, Ng S-C. Chemically bonded cationic β-cyclodextrin derivatives and their applications in supercritical fluid chromatography. J Chromatogr A 2012; 1224:97–103.
[33] Janeckova L, Kalikova K, Vozka J, Bosakova Z, Tesarova E. Characterization of cyclofructan-based chiral stationary phases by linear free energy relationship. J Sep Sci 2011;34:2639–44.
[34] Vozka J, Kalikova K, Roussel C, Armstrong DW, Tesařová E. An insight into the use of dimethylphenyl carbamate cyclofructan 7 chiral stationary phase in supercritical fluid chromatography: The basic comparison with HPLC. J Sep Sci 2013;36:1711–19.
[35] Breitbach AS, Lim Y, Xu Q, Kürti L, Armstrong DW, Breitbach ZS. Enantiomeric separations of α-aryl ketones with cyclofructan chiral stationary phases via high performance liquid chromatography and supercritical fluid chromatography. J Chromatogr A 2016;1427:45–54.
[36] Zhang Y, Wu DR, Wang-Iverson DB, Tymiak AA. Enantioselective chromatography in drug discovery. Drug Discov Today 2005;10:571–7.
[37] Płotka JM, Biziuk M, Morrison C, Namieśnik J. Pharmaceutical and forensic drug applications of chiral supercritical fluid chromatography. Trends Anal Chem 2014;56:74–89.
[38] Barhate CL, Wahab MF, Breitbach ZS, Bell DS, Armstrong DW. High efficiency, narrow particle size distribution, sub-2 μm based macrocyclic glycopeptide chiral stationary phases in HPLC and SFC. Anal Chim Acta 2015;898:128–37.
[39] Mourier PA, Eliot E, Caude MH, Rosset RH, Cedex P, Tambute AG, et al. Supercritical and subcritical fluid chromatography on a chiral stationary phase for the resolution of phosphine oxide enantiomers. Anal Chem 1985;57:2819–23.

[40] Kraml CM, Zhou D, Byrne N, McConnell O. Enhanced chromatographic resolution of amine enantiomers as carbobenzyloxy derivatives in high-performance liquid chromatography and supercritical fluid chromatography. J Chromatogr A 2005;1100:108–15.
[41] Pell R, Lindner W. Potential of chiral anion-exchangers operated in various subcritical fluid chromatography modes for resolution of chiral acids. J Chromatogr A 2012;1245:175–82.
[42] Lämmerhofer M, Pell R, Mahut M, Richter M, Schiesel S, Zettl H, et al. Enantiomer separation and indirect chromatographic absolute configuration prediction of chiral pirinixic acid derivatives: limitations of polysaccharide-type chiral stationary phases in comparison to chiral anion-exchangers. J Chromatogr A 2010;1217:1033–40.
[43] Wolrab D, Kohout M, Boras M, Lindner W. Strong cation exchange-type chiral stationary phase for enantioseparation of chiral amines in subcritical fluid chromatography. J Chromatogr A 2013;1289:94–104.
[44] Wolrab D, Macíková P, Boras M, Kohout M, Lindner W. Strong cation exchange chiral stationary phase—a comparative study in high-performance liquid chromatography and subcritical fluid chromatography. J Chromatogr A 2013;1317:59–66.
[45] Maier NM, Lindner W. Chiral recognition applications of molecularly imprinted polymers: a critical review. Anal Bioanal Chem 2007;389:377–97.
[46] Ellwanger A, Owens PK, Karlsson L, Bayoudh S, Cormack P. Application of molecularly imprinted polymers in supercritical fluid chromatography. J Chromatogr A 2000;897:317–27.
[47] Ansell RJ, Kuah JKL, Wang D, Jackson CE, Bartle KD, Clifford AA. Imprinted polymers for chiral resolution of (±)-ephedrine, 4: packed column supercritical fluid chromatography using molecularly imprinted chiral stationary phases. J Chromatogr A 2012;1264:117–23.
[48] Regalado EL, Welch CJ. Pushing the speed limit in enantioselective supercritical fluid chromatography. J Sep Sci 2015;38:2826–32.
[49] Sciascera L, Ismail O, Ciogli A, Kotoni D, Cavazzini A, Botta L, et al. Expanding the potential of chiral chromatography for high-throughput screening of large compound libraries by means of sub-2 μm Whelk-O 1 stationary phase in supercritical fluid conditions. J Chromatogr A 2015;1383:160–8.
[50] Majewski W, Valery E, Ludemann Hombourger O. Principle and applications of supercritical fluid chromatography. J Liq Chromatogr Relat Technol 2005;28: 1233–52.
[51] Lesellier E, West C. The many faces of packed column supercritical fluid chromatography—a critical review. J Chromatogr A 2015;1382:2–46.
[52] Berger TA. Packed column SFC. Delaware, USA: The Royal Society of Chemistry; 1995.
[53] Maftouh M, Granier-Loyaux C, Chavana E, Marini J, Pradines A, Vander Heyden Y, et al. Screening approach for chiral separation of pharmaceuticals. Part III. Supercritical fluid chromatography for analysis and purification in drug discovery. J Chromatogr A 2005;1088:67–81.
[54] Miller L. Evaluation of non-traditional modifiers for analytical and preparative enantioseparations using supercritical fluid chromatography. J Chromatogr A 2012; 1256:261–6.
[55] Byrne N, Hayes-Larson E, Liao W-W, Kraml CM. Analysis and purification of alcohol-sensitive chiral compounds using 2,2,2-trifluoroethanol as a modifier in supercritical fluid chromatography. J Chromatogr B 2008;875:237–42.
[56] Phinney KW, Sander LC. Additive concentration effects on enantioselective separations in supercritical fluid chromatography. Chirality 2003;15:287–94.

[57] Phinney KW. Enantioselective separations by packed column subcritical and supercritical fluid chromatography. Anal Bioanal Chem 2005;382:639–45.

[58] De Klerck K, Mangelings D, Clicq D, De Boever F, Vander Heyden Y. Combined use of isopropylamine and trifluoroacetic acid in methanol-containing mobile phases for chiral supercritical fluid chromatography. J Chromatogr A 2012;1234:72–9.

[59] Ventura M, Murphy B, Goetzinger W. Ammonia as a preferred additive in chiral and achiral applications of supercritical fluid chromatography for small, drug-like molecules. J Chromatogr A 2012;1220:147–55.

[60] Hamman C, Schmidt DE, Wong M, Hayes M. The use of ammonium hydroxide as an additive in supercritical fluid chromatography for achiral and chiral separations and purifications of small, basic medicinal molecules. J Chromatogr A 2011; 1218:7886–94.

[61] Tarafder A. Metamorphosis of supercritical fluid chromatography to SFC: an overview. Trends Anal Chem 2016;81:3–10.

[62] West C, Bouet A, Routier S, Lesellier E. Effects of mobile phase composition and temperature on the supercritical fluid chromatography enantioseparation of chiral fluoro-oxoindole-type compounds with chlorinated polysaccharide stationary phases. J Chromatogr A 2012;1269:325–35.

[63] Åsberg D, Enmark M, Samuelsson J, Fornstedt T. Evaluation of co-solvent fraction, pressure and temperature effects in analytical and preparative supercritical fluid chromatography. J Chromatogr A 2014;1374:254–60.

[64] Wang C, Zhang Y. Effects of column back pressure on supercritical fluid chromatography separations of enantiomers using binary mobile phases on 10 chiral stationary phases. J Chromatogr A 2013;1281:127–34.

[65] Enmark M, Åsberg D, Samuelsson J, Fornstedt T. The effect of temperature, pressure and co-solvent on a chiral supercritical fluid chromatography separation. Chromatogr Today 2014;7:14–17.

[66] Tarafder A, Hill JF, Iraneta PC, Fountain KJ. Use of isopycnic plots to understand the role of density in SFC - I. Effect of pressure variation on retention factors. J Chromatogr A 2015;1406:316–23.

[67] Lesellier E. Retention mechanisms in super/subcritical fluid chromatography on packed columns. J Chromatogr A 2009;1216:1881–90.

[68] Declerck S, Vander Heyden Y, Mangelings D. Enantioseparations of pharmaceuticals with capillary electrochromatography: a review. J Pharm Biomed Anal 2016;130:81–99.

[69] Albals D, Vander Heyden Y, Schmid MG, Chankvetadze B, Mangelings D. Chiral separations of cathinone and amphetamine-derivatives: comparative study between capillary electrochromatography, supercritical fluid chromatography and three liquid chromatographic modes. J Pharm Biomed Anal 2016;121:232–43.

[70] Ward T, Ward K. Chiral separations: a review of current topics and trends. Anal Chem 2011;84:626–35.

[71] Cavazzini A, Marchetti N, Guzzinati R, Pierini M, Ciogli A, Kotoni D, et al. Enantioseparation by ultra-high-performance liquid chromatography. Trends Anal Chem 2014;63:95–103.

[72] Toribio L, del Nozal MJ, Bernal JL, Alonso C, Jiménez JJ. Comparative study of the enantioselective separation of several antiulcer drugs by high-performance liquid chromatography and supercritical fluid chromatography. J Chromatogr A 2005;1091:118–23.

[73] Zhao Y, Pritts WA, Zhang S. Chiral separation of selected proline derivatives using a polysaccharide-type stationary phase by supercritical fluid chromatography and comparison with high-performance liquid chromatography. J Chromatogr A 2008;1189:245–53.

[74] Periat A, Grand-Guillaume Perrenoud A, Guillarme D. Evaluation of various chromatographic approaches for the retention of hydrophilic compounds and MS compatibility. J Sep Sci 2013;36:3141–51.
[75] Khater S, Lozac'h M, Adam I, Francotte E, West C. Comparison of liquid and supercritical fluid chromatography mobile phases for enantioselectove separations on polysaccharide stationary phases. J Chromatogr A 2016;1467:463–72.
[76] Venkatramani CJ, Al-Sayah M, Li G, Goel M, Girotti J, Zang L, et al. Simultaneous achiral-chiral analysis of pharmaceutical compounds using two-dimensional reversed phase liquid chromatography-supercritical fluid chromatography. Talanta 2016;148:548–55.
[77] Xie S-M, Zhang Z-J, Wang Z-Y, Yuan L-M. Chiral metal–Organic frameworks for high-resolution gas chromatographic separations. J Am Chem Soc 2011;133:11892–5.
[78] Xie S-M, Zhang X-H, Zhang Z-J, Zhang M, Jia J, Yuan L-M. A 3-D open-framework material with intrinsic chiral topology used as a stationary phase in gas chromatography. Anal Bioanal Chem 2013;405:3407–12.
[79] Huang K, Zhang X, Armstrong DW. Ionic cyclodextrins in ionic liquid matrices as chiral stationary phases for gas chromatography. J Chromatogr A 2010;1217:5261–73.
[80] Yao C, Anderson JL. Retention characteristics of organic compounds on molten salt and ionic liquid-based gas chromatography stationary phases. J Chromatogr A 2009;1216:1658–712.
[81] Schurig V. Gas chromatographic enantioseparation of derivatized α-amino acids on chiral stationary phases—past and present. J Chromatogr B 2011;879:3122–40.
[82] Ren-Qi W, Teng-Teng O, Weihua T, Siu-Choon N. Recent advances in pharmaceutical separations with supercritical fluid chromatography using chiral stationary phases. Trends Anal Chem 2012;37:83–100.
[83] Alexander AJ, Staab A. Use of achiral/chiral SFC/MS for the profiling of isomeric cinnamonitrile/hydrocinnamonitrile products in chiral drug synthesis. Anal Chem 2006;78:3835–8.
[84] Welch CJ, Biba M, Gouker JR, Kath G, Augustine P, Hosek P. Solving multicomponent chiral separation challenges using a new SFC tandem column screening tool. Chirality 2007;189:184–9.
[85] Nogle LM, Mann CW, Watts WL, Zhang Y. Preparative separation and identification of derivatized β-methylphenylalanine enantiomers by chiral SFC, HPLC and NMR for development of new peptide ligand mimetics in drug discovery. J Pharm Biomed Anal 2006;40:901–9.
[86] Li S, Lambros T, Wang Z, Goodnow R, Ho C-T. Efficient and scalable method in isolation of polymethoxyflavones from orange peel extract by supercritical fluid chromatography. J Chromatogr B 2007;846:291–7.
[87] Yan TQ, Orihuela C, Swanson D. The application of preparative batch HPLC, supercritical fluid chromatography, steady-state recycling, and simulated moving bed for the resolution of a racemic pharmaceutical intermediate. Chirality 2008;146:139–46.
[88] Rajendran A. Design of preparative-supercritical fluid chromatography. J Chromatogr A 2012;1250:227–49.
[89] Kalíková K, Slechtová T, Vozka J, Tesařová E. Supercritical fluid chromatography as a tool for enantioselective separation: a review. Anal Chim Acta 2014;821:1–33.
[90] Depta A, Giese T, Johannsen M, Brunner G. Separation of stereoisomers in a simulated moving bed-supercritical fluid chromatography plant. J Chromatogr A 1999;865:175–86.
[91] Johannsen M, Peper S, Depta A. Simulated moving bed chromatography with supercritical fluids for the resolution of bi-naphthol enantiomers and phytol isomers. J Biochem Biophys Methods 2002;54:85–102.

[92] Rajendran A, Peper S, Johannsen M, Mazzotti M, Morbidelli M, Brunner G. Enantioseparation of 1-phenyl-1-propanol by supercritical fluid-simulated moving bed chromatography. J Chromatogr A 2005;1092:55–64.
[93] Juza M, Mazzotti M, Morbidelli M. Simulated moving-bed chromatography and its application to chirotechnology. Tibtech 2000;18:108–18.
[94] Regalado EL, Welch CJ. Separation of achiral analytes using supercritical fluid chromatography with chiral stationary phases. Trends Anal Chem 2015;67:74–81.
[95] Christov PP, Hawkins EK, Kett NR, Rizzo CJ. Simplified synthesis of individual stereoisomers of the 4-hydroxynonenal adducts of deoxyguanosine. Tetrahedron Lett 2013;54:4289–91.
[96] Venkata Narasimha Rao G, Gnanadev G, Ravi B, Dhananjaya D, Manoj P, Indu B, et al. Supercritical fluid (carbon dioxide) based ultra performance convergence chromatography for the separation and determination of fulvestrant diastereomers. Anal Methods 2013;5:4832–7.
[97] Zhao Y, McCauley J, Pang X, Kang L, Yu H, Zhang J, et al. Analytical and semipreparative separation of 25 (R/S)-spirostanol saponin diastereomers using supercritical fluid chromatography. J Sep Sci 2013;36:3270–6.
[98] Ebinger K, Weller HN. Comparison of chromatographic techniques for diastereomer separation of a diverse set of drug-like compounds. J Chromatogr A 2013; 1272:150–4.
[99] Pirkle WH, Dappen R, Reno DS. Applications for racemic versions of chiral stationary phases. J Chromatogr 1987;407:211–16.
[100] Wolrab D, Frühauf P, Moulisová A, Kuchař M, Gerner C, Lindner W, et al. Chiral separation of new designer drugs (cathinones) on chiral ion-exchange type stationary phases. J Pharm Biomed Anal 2016;120:306–15.
[101] Regalado EL, Helmy R, Green MD, Welch CJ. Chromatographic resolution of closely related species: drug metabolites and analogs. J Sep Sci 2014;37:1094–102.
[102] Toribio L, Del Nozal MJ, Bernal YL, Alonso C, Jiménez JJ. Semipreparative chiral supercritical fluid chromatography in the fractionation of lansoprazole and two related antiulcer drugs enantiomers. J Sep Sci 2008;31:1307–13.
[103] Brondz I, Ekeberg D, Bell DS, Annino AR, Hustad JA, Svendsen R, et al. Nature of the main contaminant in the drug primaquine diphosphate: SFC and SFC-MS methods of analysis. J Pharm Biomed Anal 2007;43:937–44.
[104] Mukherjee PS. Validation of direct assay of an aqueous formulation of a drug compound AZY by chiral supercritical fluid chromatography (SFC). J Pharm Biomed Anal 2007;43:464–70.
[105] Svan A, Hedeland M, Arvidsson T, Jasper JT, Sedlak DL, Pettersson CE. Rapid chiral separation of atenolol, metoprolol, propranolol and the zwitterionic metoprolol acid using supercritical fluid chromatography—tandem mass spectrometry—application to wetland microcosms. J Chromatogr A 2015;1409:251–8.
[106] Jin L, Gao W, Yang H, Lin C, Liu W. Enantiomeric resolution of five chiral pesticides on a Chiralpak IB-H column by SFC. J Chromatogr Sci 2011;49:739–43.
[107] Zhang C, Jin L, Zhou S, Zhang Y, Feng S, Zhou Q. Chiral separation of neonicotinoid insecticides by polysaccharide-type stationary phases using high-performance liquid chromatography and supercritical fluid chromatography. Chirality 2011;221:215–21.
[108] Toribio L, Del Nozal MJ, Bernal JL, Alonso C, Jiménez JJ. Enantiomeric separation of several antimycotic azole drugs using supercritical fluid chromatography. J Chromatogr A 2007;1144:255–61.
[109] Liu N, Dong F, Xu J, Liu X, Chen Z, Tao Y, et al. Stereoselective determination of tebuconazole in water and zebrafish by supercritical fluid chromatography tandem mass spectrometry. J Agric Food Chem 2015;63:6297–303.

[110] Tao Y, Dong F, Xu J, Liu X, Cheng Y, Liu N, et al. Green and sensitive supercritical fluid chromatographic—tandem mass spectrometric method for the separation and determination of flutriafol enantiomers in vegetables, fruits, and soil. J Agric Food Chem 2014;62:11457–64.

[111] Berger TA, Berger BK. Chromatographic resolution of 7 of 8 stereoisomers of vitamin K1 on an amylose stationary phase using supercritical fluid chromatography. Chromatographia 2013;76:549–52.

[112] Sánchez-Hernández L, Bernal JL, Del Nozal MJ, Toribio L. Chiral analysis of aromatic amino acids in food supplements using subcritical fluid chromatography and Chirobiotic T2 column. J Supercrit Fluids 2016;107:519–25.

[113] Xie J, Cheng J, Han H, Sun B, Yanik GW. Resolution of racemic γ-lactone flavors on Chiralpak AD by packed column supercritical fluid chromatography. Food Chem 2011;124:1107–12.

[114] Schaffrath M, Weidmann V, Maison W. Enantioselective high performance liquid chromatography and supercritical fluid chromatography separation of spirocyclic terpenoid flavor compounds. J Chromatogr A 2014;1363:270–7.

[115] Xie J-C, Cheng J, Sun B-G. Resolution of racemic δ-lactone flavors on Chiralcel OB by packed column super critical fluid chromatography. Eur Food Res Technol 2009;230:521–6.

[116] Xie J, Han H, Sun L, Zeng H, Sun B. Resolution of racemic 3-hydroxy-4,5-dimethyl-2(5H)-furanone (sotolon) by packed column supercritical fluid chromatography. Flavour Fragr J 2012;27:244–9.

[117] Khater S, West C. Development and validation of a supercritical fluid chromatography method for the direct determination of enantiomeric purity of provitamin B5 in cosmetic formulations with mass spectrometric detection. J Pharm Biomed Anal 2015;102:321–5.

[118] Li L. Enantiomeric determination of methamphetamine, amphetamine, ephedrine, and pseudoephedrine using chiral supercritical fluid chromatography with mass spectrometric detection. Microgram J 2015;12:1–4.

[119] Liu J, Regalado EL, Mergelsberg I, Welch CJ. Extending the range of supercritical fluid chromatography by use of water-rich modifiers. Org Biomol Chem 2013; 11:4925–9.

CHAPTER

13

Supercritical Fluid Chromatography of Petroleum Products

D. Thiébaut

CNRS, UMR8231 CBI CNRS–ESPCI, PSL Research University, Paris, France

OUTLINE

Supercritical Fluid Chromatography.
DOI: http://dx.doi.org/10.1016/B978-0-12-809207-1.00013-6

13.1 INTRODUCTION

Application of supercritical fluid chromatography (SFC) to the separation of petroleum compounds really started in the '80s. It came from the fact that SFC, using neat carbon dioxide as the mobile phase, features advantages of gas chromatography (GC) and liquid chromatography (LC) for the separation of hydrocarbons and related compounds in the field of petroleum compounds [1]: efficiency is similar or better than LC [2], it enables the use of LC and GC detectors and the dense mobile phase of supercritical fluids (SFs) enables LC-like solvating power that can be tuned as in LC; thus it appeared SFC could be a bridge between GC and LC [3].

Nowadays, most SFC applications are performed using polar modifiers added to carbon dioxide in order to enlarge the polarity range of the mobile phase [4]. It can be seen in the number of applications published in chromatographic journals, some sentences in reviews such as "Pure CO_2 is almost never used today, CO_2-modified with a solvent is the norm..." [5] and in the design of the commercial SFC systems that look like or are LC-like systems modified to perform SFC. The recent "rebirth"[1] of SFC was partly related to the commercial availability of new systems. However, the bad news for neat CO_2 users are some of these systems are of no interest when no modifier is added in the mobile phase because they do not enable pressure/density programing during the separation. As will be shown, density programing is mandatory in most cases when neat CO_2 is used for petroleum analysis, whatever the compounds to be addressed. It is mandatory for the separation of a wide range of molecular weights, such as in simulated distillation (Simdis).

As a result, SFC remains a niche for petroleum applications, also because it combines a LC-like mobile phase with GC detectors that are not available in commercial SFC systems; it will be shown high molecular-weight compounds that could not elute from a GC column can be eluted in SFC, while the universal and sensitive GC detector, the flame-ionization detector (FID) enables easy hydrocarbon group-type quantification. The main SFC separations of petroleum-related compounds [1,6–10] are either GC-derived separations, e.g., simulated distillation (Simdis), or LC-like separations, e.g., hydrocarbon group separations. Both packed LC columns and open-tubular GC capillary columns are implemented in the SFC of petroleum compounds, both having their advantages and drawbacks. However, in SFC, packed columns are much easier to handle than open-tubular capillary columns,

[1]How many births did SFC have? Hopefully, after a time, it should become a mature technology.

they provide faster separations than capillary columns [3] and they provide the same efficiency (but "faster") in LC and in SFC [2]; further to the use of modifiers, the use of packed columns is the norm in SFC mainly in pharmaceutical analysis, especially chiral separations, using the same packed columns as in LC. (It must be reminded here that replacing an LC by a SFC method is also worth considering in order to reduce solvent consumption and costs, or for environmental reasons. Most of normal-phase LC separations could easily be replaced by SFC with minimum optimization [4].)

In the field of petroleum analysis, Simdis is a routine GC application [11]; however, SFC is attractive for the elution of heavy hydrocarbons, i.e., hydrocarbons having more than 80–100 carbon atoms that are difficult to elute by high-temperature GC (HTGC) without cracking. Their elution can be obtained by SFC on packed or open-tubular columns at much lower temperatures than in HTGC. As SFC can elute hydrocarbons starting from C_{20} to more than C_{200}, it is more than an alternative to GC for the Simdis of heavy fractions.

One of the most studied LC applications in the petroleum industry is hydrocarbon group-type analysis [1,10]. The best feature of SFC for this application is the detection capabilities of FID and the properties of carbon dioxide as a mobile phase: FID provides easy quantification of hydrocarbon groups because of the similarity of response factors of most hydrocarbons. Thus SFC has long been registered by ASTM for this application.

The main critical point in Group-type applications of SFC is the selectivity between hydrocarbon families; it has been studied and improved via the implementation of different mobile phases and stationary phases, including the use of columns in series or in the multidimensional mode. Selectivity can also be improved via the implementation of multidetection for group-type analysis by the combination of ultraviolet (UV) detection and FID. The use of a "hypernated" system including atomic emission detection (AED), Fourier transform infrared spectrometry (FTIR), mass spectrometry (MS), UV, and FID has been described as an efficient way to improve the analysis of lubricants [1].

Another interesting feature of carbon dioxide used as the mobile phase lies in the hyphenation with GC: very early in SFC development, the hyphenation of SFC to GC was reported [12,13]. Recently, SFC has also been coupled successfully to comprehensive two-dimensional GC (GC × GC) [11]. The coupling is obvious because, after decompression, the supercritical carbon dioxide used as the mobile phase is a gas. The features of this strategy for the analysis of hydrocarbon groups in heavy gas oils and vacuum distillates will be presented. A separation according to polarity (number of double bounds, presence of polar functional groups...) can be performed by SFC prior to the on-line transfer to

GC × GC; owing to GC × GC separation performance, it is possible to discriminate each hydrocarbon group type for (1) easier quantification of group types and (2) for plotting simulated distillation curves per hydrocarbon groups.

Last but not least, it seems quite reasonable that an SFC × SFC system could be built to offer performances in the same range as in GC × GC: long columns could be used in the first dimension while a quite fast separation can be performed in the second dimension after modulation of the first dimension effluent as in GC × GC.

Thus we will give an overview of the most attractive SFC trends and applications in the petroleum industry: Simdis, group-type analysis, and related applications including the implementation of multidetection featuring high-resolution mass spectrometry, as well as the hyphenation to GC × GC for improved group-type separation, SFC × GC and the promising SFC × SFC results. Some specific technical information related to the use of open-tubular capillary columns or conventional packed columns in combination with FID (or detectors that require decompression and in some instances splitting of the mobile phase prior detection) will also be given to assist newcomers in the field.

13.2 TECHNICAL PART: THE COUPLING AND USE OF SUPERCRITICAL FLUID CHROMATOGRAPHY-FLAME-IONIZATION DETECTOR

The use of SFC with FID was described in the early days of SFC development. However, it deserves a few explanations because it is no longer commercially available, despite the fact that some of SFC manufacturers are also GC manufacturers and could propose a suitable adaptation of their instruments for this application.

Moreover, as Simdis can be performed efficiently in open-tubular capillary SFC—in these conditions, SFC is a miniaturized (nano) chromatographic technique performed under both high-pressure and high-temperature conditions (!); thus similar to nano-LC techniques, it requires some precautions although the SFC system to be used is quite simple (Fig. 13.1). Of course, the apparatus presented for Simdis can also be used for all applications of SFC-FID on capillary or micropacked columns.

When an FID is to be used, as in Simdis and many applications devoted to the separation of saturated hydrocarbons, a "fixed" restrictor (integral or frit type [10]) must be connected to the column outlet to transfer and decompress the SF in the chimney of the FID; to limit dead volumes in the detector, the outlet of the restrictor must be placed a few millimeters below the top of the chimney in order to transfer

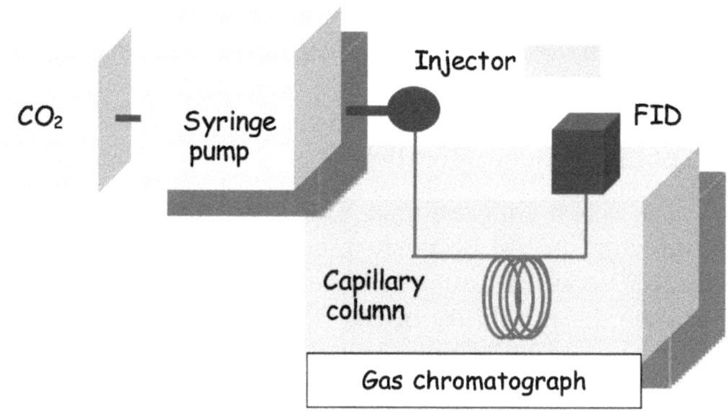

FIGURE 13.1 Diagram of the apparatus used in open-tubular capillary or micro-packed column SFC for simulated distillation. Source: *Reprinted from Dulaurent A, Dahan L, Thiébaut D, Bertoncini F, Espinat D. Extended simulated distillation by capillary supercritical fluid chromatography. Oil Gas Sci Technol 2007;62:33–42. Copyright 2007 IFP Energies Nouvelles.*

column effluent into the flame of the detector [14]. For neat CO_2 use, only one pump is necessary to deliver carbon dioxide. For Simdis, the maximum operating pressure of the pump should be as high as possible, higher than 60 MPa, and pressure programing is mandatory. This is possible using syringe pumps [15].

1. SFC-FID using low ID columns (micro chromatography):
   a. high-pressure syringe pumps that enable the accurate delivery of very low flow rate (less than 0.1 μL/min) can be used without splitting the pump flow before the column. An important point is that the pressure in the system will depend on both the dimensions of the fixed restrictor and the flow rate of the mobile phase: pressure and flow rate in the column are linked; pressure/density gradients can simply be performed by increasing the flow rate of the pump during the separation. The pressure or density programing can be available directly from the pump software [15].
   b. HPLC-like reciprocating pumps available in most of commercial SFC systems can also be implemented using a flow splitter prior to the column (like in nano-LC systems); the software-controlled automatic pressure-regulator is fed by the diverted flow and enables pressure/density programing [16,17]. However, the pressure regulator and the conventional reciprocating pumps have a quite low-pressure resistance, 40–60 MPa (attempts to use modified ultra-high-performance liquid chromatography (UHPLC) reciprocating pumps have been reported to exceed 50 MPa but the lifetime of some parts in the pump head was questionable [18]). For Simdis applications the SFC pump should

enable the highest pressure as possible. As a consequence, the pressure regulator should also withstand similar pressures as the pump; it must be pointed out that, generally, the pressure regulator must be fed by a flow rate higher than ca. 0.25–0.5 mL/min to obtain proper operation.

In both cases a test separation or measurement of the CO_2 gas flow rate in the detector should be performed daily; the accuracy of the flow rate in a small ID column must be checked either when syringe pumps are used because the reliability of the flow meter is questionable at very low flow rates. Using HPLC-like piston pumps the microcolumn flow rate can be negligible versus the diverted flow; a leak or plugging of the column cannot affect significantly the flow rate measured by the diverted flow.

2. SFC-FID using regular size HPLC column (2.1–4.6 mm ID): a conventional SFC system or a modified UHPLC can be used [18]. For FID implementation, a splitter should be placed at the column outlet so ca. 10–50 mL of CO_2 per minute will be sent into the flame of the detector (the flow rate is expressed as gas flow rate measured at the restrictor outlet, in the chimney of the detector, as for GC operation of the detector); air and hydrogen flow rates must be optimized in order (1) to obtain a stable flame *in the whole range of pressure used for the separation* and (2) the maximum response of the detector, as in GC [14]. It should also be emphasized that the same set up can be used to hyphenate any detector requiring decompression prior to detection such as AED, sulfur chemiluminescence detector (SCD), etc. With modern columns (packed or open-tubular capillary), fixed restrictors can deliver a constant flow rate for months [15,19,20].

Columns, injectors, and connection tubings must stand the maximum operating pressures for the separation. Owing to the high-temperature injection conditions used in Simdis, the lifetime of the rotor seal—and, somehow, of the stator—of the injector is shorter; it must be replaced more frequently in comparison with its use in standard operating conditions. Various types of GC or LC connectors can be used. PEEK tubings should not be used due to the combination of high-pressure/temperature conditions in Simdis. GC ferules can be used if more firmly tightened than for their use in GC.

- Samples/solvents: In order to reduce the size of the solvent peak, carbon disulfide (CS_2) is generally used in applications where FID is implemented because its response is very low (like that for CO_2). Thus, for quantitative purposes, depending on the concentration of the samples to be injected, the CS_2 response can be neglected. For Simdis, Polywax calibration standards can be dissolved in xylene at

100°C; to avoid precipitation the injection valve should be at high temperature too. Heavy samples to be analyzed in Simdis such as vacuum residues (boiling point >700°C), can be easily dissolved using such conditions despite that a high concentration is required for detection [15].

- Any commercial gas chromatograph equipped with FID can be used. A GC from Varian was used successfully for many years without modification [14,15]; vintage and very robust SFC systems from Hewlett-Packard/Berger Instruments were provided with a FID having a second heating zone in order to compensate for the cooling effect of the CO_2 decompression.

13.3 SIMULATED DISTILLATION

Simulated distillation is a standardized GC technique used for petroleum analysis. It is described in ASTM methods [11]. It is widely used for the evaluation of many types of samples, feeds, and cuts treated in refining and conversion processes. It is based on the comparison of the sample chromatogram and a calibration curve between the boiling point of normal paraffins and their elution temperature or retention time; the calibration curve is constructed using a chromatogram of normal paraffins obtained in the same experimental conditions as for the analysis of the sample. The hydrocarbon distribution of the sample (in weight percent) is given versus the boiling range of the fraction (expressed in atmospheric equivalent boiling point, AEBP). Using proper column and experimental conditions, Simdis results are in good agreement with the true boiling point (TBP) distillation curve described in ASTM D2892 [21]. The final boiling point that can be reached using Simdis in GC is above 538°C (ASTM D2887 [22]) to be compared with 400°C for ASTM D2892 and 565°C for ASTM D5236 [23] involving lower pressure (ca. 10 Pa). The use of high-temperature GC (ASTM D6352 [24], 7169 [25], 7500 [26]), at oven temperatures up to 450°C allow the determination in routine analysis of higher AEBP, around 700°C, using nonpolar dedicated stationary phases; however, for the operating conditions employed, the resistance of high molecular-weight hydrocarbons (HMHs) to cracking reactions is questionable [27], although it would likely have only a minor effect on the results [28]; of course, the use of high-stability GC columns is mandatory [11].

SFC using neat carbon dioxide as the mobile phase allows the solvent strength of the mobile phase to be varied and, from this point of view, offers unsurpassed advantages over GC techniques. Depending on the operating conditions the polarity of carbon dioxide varies between that of pentane and, at least, of toluene. This is the reason why CO_2 is a good solvent for hydrocarbons and SFC is a powerful technique for the elution of HMHs at much lower temperatures than GC. In SFC-Simdis, both open-tubular capillary GC-like columns and LC packed columns can be used. Whatever column type is employed, Simdis requires a pressure/density programing during the separation in order to elute compounds having a wide range of boiling points. It should also be emphasized that compounds with the same boiling point will coelute regardless of their structure for a truly accurate simulated distillation; this can be referred as "Simdis selectivity." The undesired "Simdis selectivity" can be expressed in retention time when considering compounds having the same boiling points or in degrees when considering compounds having the same retention time but different boiling points.

13.3.1 Packed Columns

The implementation of micro or narrow bore packed-column SFC (pSFC) for Simdis application is quite straightforward because the packed columns provide a high loadability. Thus no split injection is required. The first report of packed column SFC Simdis was in 1988 by Schwartz [29]: it described the use of a 1-mm ID column packed with a polysiloxane material and the elution of an n-C_{108} alkane from polyethylene PE740 was reported. Further developments of SFC Simdis were carried out using alkyl-bonded silica stationary phases packed in small ID columns (microbore columns).

The main results are summarized in Table 13.1 together with some comments.

Using alkyl-bonded phases, SFC performs as nonaqueous reverse phase chromatography; the longer the alkyl chain, the stronger the retention of hydrocarbons; no minimum is observed [31,32]. From this point of view a stationary phase bonded with a short chain such as butyl was preferred by Huynh et al. [32] to favor the elution of heavy hydrocarbons: it was operated at quite a high temperature (170°C) for SFC so the elution of a C_{136} hydrocarbon could be obtained instead of C_{106} at 130°C. However, the higher the temperature of the mobile phase, the higher the operating pressure required in order to maintain a high CO_2 density in the column during the separation; thus the pressure limits of the whole system (pump, injector, column) were reached. However, Shariff et al. [31] stressed on the fact that the lowest difference between the retention of different compounds having the same boiling point ("Simdis selectivity") was obtained using phases bonded with long alkyl chains, C_8 and higher. As the retention increased with the length of the alkyl bonded chain, a compromise had to be found

TABLE 13.1 Main Published Results in Packed Column SFC-Simulated Distillation

Ref.	Column oven temperature	Eluted carbon # sample	Comments	Final pressure (MPa)
[16]	Packed column	C100	FID/UV (flow splitting)	36
	C18-bonded silica	Polywax 655		
[30]	C11-bonded silica	C132	Injection T: 130°C	>50
	Packed capillary 180°C	Polywax 1000		
[31]	Various chain length	C130	Better RSD on C6	41.5
	Packed capillary 120°C	Polywax	Capillary packed column	
[32]	C_4-bonded silica	C_{136}	Routine between C80 and C120	48
			Better selectivity with alkyl bonded chains >C4	
	Custom packed capillary 160–170°C	Polywax 1000	Increase of retention vs alkyl bonded chain length	

Adapted from Thiébaut D. Separation of petroleum products involving supercritical fluid chromatography. J Chromatogr A 2012;1252:177–88; and Thiébaut D. Gas chromatography and 2D gas chromatography for petroleum industry. Paris: Technip; 2013.

[32]: to favor a lower retention of the compounds, a deactivated C4-bonded silica packing was used to keep the final operating pressure around 50 MPa (this value is now compatible with the pressure limit of some SFC reciprocating pumps; however, some syringe pumps can handle higher pressures); the "Simdis selectivity" was less than 10°C. No precipitation of heavy aromatic compounds was reported; nevertheless, a blank injection following oil or standard samples is highly recommended. SFC results were consistent with GC results [32].

Concluding remarks and expectations

- Elution of heavy compounds is obtained at much lower temperatures in packed column SFC than in GC
- Nonpolar alkyl bonded silica LC stationary phases enable elution of hydrocarbons with minimum "Simdis selectivity"; using neat CO_2, they can be used at elevated temperatures (more than 150°C) without evidence of degradation of their chromatographic performances because it is a nonaqueous mobile phase [32]; thus the columns can be used for routine analysis.

- However, ideally routine SFC-Simdis apparatus and columns should be designed for operation at elevated pressures: at least *they should withstand 60 MPa*, including the pressure regulator, so that working at about 50 MPa would not be a problem. Moreover, the injection system should allow sample introduction at high temperature (as in size exclusion chromatography of synthetic polymers). UHPLC injectors and pumps are desirable for this application. Recently, a very fast separation of a standard alkane mixture from C16 to C80 was reported using a sub-2 μm packed column [18]. The separation time was less than 4 min using a C18 column and 2.5 min with a C4-bonded phase; a density gradient from 80 bars to 300 bars and a temperature of 100°C were applied. To our knowledge, this is the fastest separation of this type of standard mixture because it takes 4 min in GC while the last eluting alkane is only C60 [33]. This is the reason why SFC Simdis on sub-2 μm packed columns deserves further investigation in order to extend the mass range of eluting compounds by increasing both the temperature and the pressure of the mobile phase, for these conditions a high-pressure system allowing pressure gradient programing is required.

13.3.2 Capillary Columns

Generally speaking, open-tubular capillary SFC (cSFC) is not popular in the SFC community: for more than 25 years, the main applications of SFC have been carried out using LC-like packed columns and polar modifiers, generally for the separation of pharmaceutical compounds. Mostly, capillary columns are not discussed in the recent SFC literature [4]. However, Simdis seems to be the last niche for open-tubular capillary SFC applications. The main results obtained by cSFC Simdis are summarized in Table 13.2.

In all the cases, nonpolar stationary-phases were used, including octyl phases, in similar operating conditions to those described above for packed column SFC Simdis: A very high pressure is required to elute HMHs above C_{100} at high temperature (>100°C) and the density of the mobile phase must be increased during the separation via pressure programing.

Comparing the TBP and SFC Simdis BP of aromatic hydrocarbons from naphthalene to chrysene, octyl-bonded phases were reported to provide a lower deviation than polydimethyl-siloxane and phenylmethylpolysiloxane stationary-phases [31,34,36].

The heaviest paraffin eluted by cSFC Simdis is C_{126} [15] using a 5% phenylpolydimethylsiloxane stationary phase and a pressure programing of the CO_2 starting at 10 min from 100 to 550 bar at 13.3 bar/min

TABLE 13.2 Main Published Results in Open-Tubular Capillary Column SFC-Simulated Distillation

Authors (year) [ref.]	Stationary phase	Column temperature (°C)	Maximum pressure (MPa)	Heaviest paraffin eluted (BP)
Raynie (1991) [34]	*n*-Octylpolysiloxane	150	32	C_{100} (719)
Shariff (1994) [31]	*n*-Octylpolysiloxane	NA	NA	C_{90} (700)
Bouigeon (1996) Unpublished results	SB-octyl	180	50	C_{96} (712)
	5% phenyl-methylpolysiloxane			C_{92} (704)
	Polydimethyl-siloxane			C_{108} (732)
Dahan (2002) [35]	5% phenyl-methylpolysiloxane	160	55	C_{120} (750)
Dulaurent (2005) [15]	5% phenyl-methylpolysiloxane	160	55	C_{126} (759)

Adapted from Thiébaut D. Separation of petroleum products involving supercritical fluid chromatography. J Chromatogr A 2012;1252:177–88; and Thiébaut D. Gas chromatography and 2D gas chromatography for petroleum industry. Paris: Technip; 2013.

(Fig. 13.2A). The operating temperature was high, 160°C. Identification of paraffins having ca. 120 carbon atoms was possible; heavier paraffins were eluted but unresolved so they could not be identified.

More impressive and interesting were the chromatograms of real samples obtained with the same operating conditions: the chromatogram of a vacuum distillation (Fig. 13.2B) demonstrated that elution of very HMHs was obtained; in the last third of the chromatogram, it was estimated that compounds having more than 200 carbon atoms (BP > 900°C) were eluted; for this application their BP were out of the calibration constructed using a standard chromatogram such as presented in Fig. 13.2A. Thus the calibration curve had to be extrapolated via a logarithmic regression [15]. The whole global calibration curve was the combination of a first part, where standards were available (<C100) and identified in the chromatogram of standards, with a second part, for which the retention times of standards had to be extrapolated. The points between C100 and C120 were used to check the accuracy of the model. It should be emphasized that most of the TBP of standard compounds are not available in this range of high molecular weight and had to be estimated using a correlation curve [32]. The method developed was used to compare the composition of vacuum distillation residues from different sources via the comparison of their Simdis curves [15].

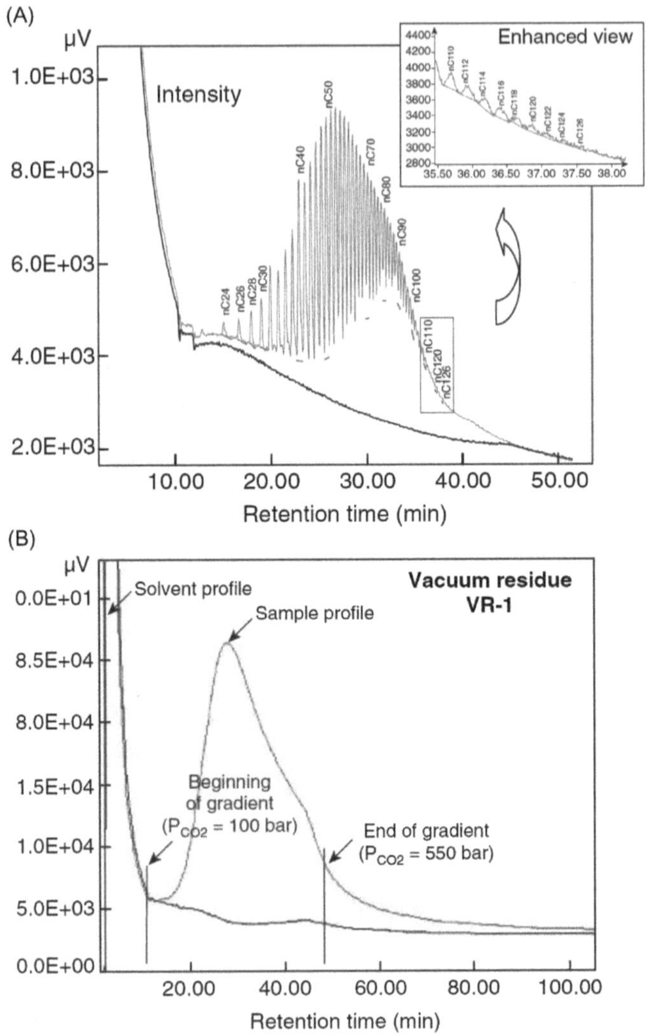

FIGURE 13.2 Capillary SFC Simdis of heavy samples: (A) Example of calibration chromatogram of paraffin standard. (B) Open-tubular capillary SFC chromatogram of a vacuum distillation residue. Conditions: column DB55m × 0.05 mm, 0.2 μm using CO_2 at 160°C, pressure programing starting at 10 min from 100 to 550 bar at 13.3 bar/min. Source: *Reprinted from Dulaurent A, Dahan L, Thiébaut D, Bertoncini F, Espinat D. Extended simulated distillation by capillary supercritical fluid chromatography. Oil Gas Sci Technol 2007;62:33–42. Copyright 2007 IFP Energies Nouvelles.*

Despite the inherent uncertainties to the lack of TBP and the poor resolution of reference HMHs, samples with a significant amount of low mass compounds presented another difficulty related to the elution of the lighter compounds in the tail of the solvent peak. To solve this problem, the initial part of the curve (no more than 2 points) was corrected using the GC Simdis curve of the same sample for which there is no elution interference of light compounds and the solvent peak. Thus the SFC and GC curves could be superimposed. No information could be obtained from GC beyond C_{120} [15].

Simdis of sulfur-containing compounds was also reported in similar conditions: SCD was hyphenated to open-tubular capillary SFC [11] to monitor the sulfur trace of three vacuum residues. As in classical simulated distillation, the cumulated sulfur content of samples could be plotted versus the BP (or carbon number) of reference compounds (alkanes). The total sulfur content of the investigated residues determined by SFC-SCD was in good agreement with the information provided with the sample (between 2.5% and 5.6%). As the ratio of the response of sulfur versus that of nonsulfur containing species (S/C) was better than 10^5, there was no significant signal of the solvent peak (xylene) and, therefore no interference with the peaks of low BP paraffin standards. Thus it was not necessary to use GC information to correct SFC information. Owing to the high level of sulfur in the samples, classical and sulfur distillation curves were very similar and almost coincided. Thus it could be assessed that almost all the detected molecules contained a sulfur atom.

However, the signal to noise ratio was not sufficient to use this method to monitor samples having a lower sulfur content.

13.4 GROUP-TYPE ANALYSIS IN SUPERCRITICAL FLUID CHROMATOGRAPHY AND RELATED APPLICATIONS

13.4.1 Principle

Group-type analysis refers to the separation and quantification of the hydrocarbon groups, i.e., saturates, olefins, aromatic hydrocarbons, and "polar (or resins)" compounds. For the main hydrocarbons types, subgroups can be further separated: paraffins, branched paraffins, and cycloalkanes for saturates ("naphthenes"), and mono-, di-, tri- and polycyclic aromatic hydrocarbons for aromatic hydrocarbons. Thus the corresponding acronyms used for group-type separations are SAR, SARA, or SOARA for Saturates, Olefins, Aromatics, Resins, and Asphaltenes. SFC is very efficient for group-type separations because, as in GC, flame-ionization detection can be used and does not require correction of response versus the hydrocarbon type. In LC the use of refractometric detection requires a calibration that can be inaccurate [10,36].

SFC allows group-type separations of middle distillates and heavy fractions. The main ASTM method, D5186, was released in 1991 [37] for the separation of hydrocarbon groups in diesel fuels. The goal was the determination of total aromatic and nonaromatic content. Updates of the method were released regularly and the application range was extended to aviation turbine fuels and blend stocks; the goal of the method was also extended to the determination of nonaromatic, mono and polyaromatic hydrocarbon groups.

Methods D6550 [38] and D7347 [39] are devoted to the SFC determination of the olefin content in gasoline and of denatured ethanol, respectively; they require an improvement of the separation to isolate the olefins from the nonaromatic fraction.

A large number of methods were reported by Robert [10]. This chapter will focus on SFC applications described in ASTM methods, applied to light and to heavy samples, to olefins, and some recent trends (1) to obtain improved or faster separations, (2) to extend the application to heavier fractions, or (3) to enhance separation capabilities by hyphenating SFC to GC × GC or developing SFC × GC or SFC × SFC.

13.4.2 Practical Supercritical Fluid Chromatography Group-Type Separations

13.4.2.1 Light Fractions (Gasoline, Gas Oil, Jet Fuel, etc.)

13.4.2.1.1 ASTM D5186 Method

In the D5186 ASTM method [37] a successful separation of nonaromatic, mono, and polyaromatic hydrocarbons requires simple criteria. The evaluation of the separation performance is based on the resolution between selected test compounds used as markers for the groups to be separated. In ASTM D5186-03 the resolution must be greater than four between the nonaromatics (hexadecane) and monoaromatics (toluene) and greater than two between the monoaromatics (tetraline) and polynuclear aromatics (naphthalene). As a consequence, this also means that all the compounds eluted before the retention time of hexadecane should be saturated hydrocarbons because they will be assigned to the nonaromatic hydrocarbons. The compounds eluting after hexadecane but before naphthalene will be assigned to monoaromatics; thus the area of the peaks eluted from the retention time of the naphthalene peak to the final return to baseline is assigned to the polyaromatic hydrocarbons.

The resolution values given in the ASTM D5186 method can be reached easily using adsorption chromatography on high-specific area bare silica and carbon dioxide at 30°C as the mobile phase [10,40,41]. Some suitable stationary phases are indicated in the method. However, for the analysis of gas oils, using a single hydrocarbon-group separation column (250 mm × 4.6 mm) only a partial resolution between the nonaromatic fraction and monoaromatic hydrocarbons could be obtained [42]. Thus some attempts to modify the eluent strength of carbon dioxide to improve the selectivity between hydrocarbon groups were reported: the mobile phase was a mixture of carbon dioxide and sulfur hexafluoride [42]; it allowed the extension of the ASTM D5186 method to the determination of all aromatic groups except olefins that could only be partly separated from alkanes using neat SF_6 [43]. However, the use of a complex mixture involving corrosive SF_6 is not recommended.

13.4.2.1.2 Improvement of the Separations: Columns in Series, Stationary Phases and Detection

The resolution between the groups or the separation of more fractions/sub groups in diesel fuels was also achieved by the use of two different columns connected in series.

The selectivity between aromatic groups was improved by the use of a cyanopropyl-bonded silica column connected in series to a bare silica column [44,45].

Owing to the low-pressure drop in the columns, the use of longer columns than in LC is also possible in SFC [4,46]. This can enhance the overall resolution but the separation duration is increased and the quantification can be difficult due to the broadening of zones for the hydrocarbon groups.

Another way to improve the discrimination between aromatic and saturated hydrocarbons is the use of the double detection mode, UV and FID. This was proposed almost 25 years ago and reported in application notes from Hewlett Packard [47], Analytical Control [48], and a US patent [49]. UV detection is also reported to allow the detection of conjugated dienes eluting in the nonaromatic fraction using two silica columns in series [50]; it was also implemented by Paproski et al. [51,52].

Paprosky et al. [51] also investigated the use of stationary phases as an alternative or in combination with silica: titania, zirconia, and their combination with silica were studied for the separation of hydrocarbon groups in diesel fuels according to the requirements of the ASTM D 5186 method. The combination of a titania column coupled to a silica column gave the best results as it improved the resolution between families in the heaviest diesel fuel investigated in this work.

The same authors [52] investigated the use of monolithic silica (Chromolith) or of columns packed with smaller silica particles in order to speed up the separation. As expected, a 13-fold reduction of the separation time was obtained with the Chromolith column; however, it was not considered as an interesting alternative to the silica column because the resolution between families was lower owing to the lower surface area of the silica monolith. The titania-silica combination of stationary phases in shorter columns than in Paprosky et al. [51] gave resolution and sensitivity values that compared favorably with the results obtained with longer columns [52].

13.4.2.1.3 Hyphenation of Supercritical Fluid Chromatography With High-Resolution Mass Spectrometry

In addition to UV detection and FID, the hyphenation of SFC with field ionization time-of-flight high-resolution mass spectrometry (FI-ToF HRMS) was reported for the quantitative analysis of petroleum middle distillates [53]. As expected, SFC provided the separation of petroleum middle distillates into saturates and 1- to 3-ring aromatics while soft ionization conditions of FI-ToF HRMS could generate molecular ions for hydrocarbon species eluted from the SFC. The combination of high-chromatographic resolution to the exact mass measurements by ToF mass spectrometry provided elemental composition fraction components. However, SFC-FID-UV was used for the quantification of saturates and aromatic subgroup types using a carbon number calibration.

13.4.2.1.4 Hyphenation of Supercritical Fluid Chromatography to Comprehensive Two-Dimensional Gas Chromatography (Supercritical Fluid Chromatography−Gas Chromatography × Gas Chromatography)

In order to improve the analysis of saturates and aromatics groups and subgroups of gas oils by GC × GC, experiments were carried out in IFPen to perform a separation between saturated and unsaturated hydrocarbons prior to their separation by GC × GC. After the demonstration that GC could be used for a preseparation of fractions having hydrocarbons with a carbon number below C15 only [54] SFC was investigated [55]. For this purpose, SFC was online coupled to GC × GC following the pioneering work of Levy et al. [12,13] for SFC−GC coupling. Using neat carbon dioxide as the mobile phase in SFC, the on-line hyphenation of SFC to GC or GC × GC only required decompression of the dense SFC mobile phase to obtain a gas phase in the injector of the GC × GC chromatograph; no dedicated interface was necessary. To perform the transfer of solutes from SFC to GC × GC, an integral restrictor was inserted in the GC × GC injector; the cryogenic capabilities of the GC × GC gas chromatograph oven permitted the implementation of a low temperature in the GC × GC oven for the cryogenic focusing of the compounds eluted from the SFC column. Band broadening during the transfer of fractions from the SFC column to the GC × GC was so low that it was very difficult to differentiate a GC × GC chromatogram obtained after coupling on-line to SFC from a direct injection [55,56]. For this application to gas oil samples, group-type analysis by SFC was restricted to saturated and unsaturated hydrocarbons separation. Thus only two GC × GC separations per sample were needed: the saturated fraction was transferred in a first GC × GC system of columns while the unsaturated fraction was backflushed from the SFC column to a second GC × GC system of columns placed in the same oven. After trapping the separation of the two SFC fractions transferred in the two GC × GC injectors connected to the two GC × GC column systems in a so-called twin GC × GC apparatus, could be performed in only one run; unsurpassed information according to both the carbon number and group type was reported because it enabled the determination of families that would coelute without the SFC separation; indeed naphthenic families would elute with aromatic hydrocarbons and olefins would elute with alkanes and branched alkanes shown in Fig. 13.3 for the GC × GC chromatograms of the saturated and unsaturated fractions of a light cycle gas-oil obtained by SFC [55].

Recently, SFC–GC × GC was utilized by Sasol for analysis of Kerosene [57]. To improve the trapping of compounds coming from the SFC fractionation in the first dimension column, a programed temperature vaporizing (PTV) injector was aptly used. Compared with the cooling of the whole GC oven described in [55] better trapping of volatile compounds was obtained using the PTV and the consumption of cryogenic fluid was reduced. The authors also reported they obtained more reliable results using SFC fractionation than for HPLC fractionation prior to GC × GC separation because interferences by oxygenates eluting in the unsaturated hydrocarbons fraction were suppressed.

13.4.2.2 Determination of Olefins

Olefins are not found in substantial amount in unprocessed crude oil but are generated during the cracking processes [58]. Their amount in petroleum products must be known in order to assess fuel quality and to adopt the adequate processes. The detailed analysis of olefins is still quite challenging because they are difficult to fully separate from saturated hydrocarbons, especially in a wide range of boiling point, and they have the same molecular weight as naphthens (cycloalkanes). Thus they deserve for some discussion in this chapter.

In SFC, Norris and Rawdon reported in 1984 the specific retention of olefins and aromatic hydrocarbons from gasoline [59] on silver-loaded silica. The retention of olefins relies on the formation of adducts between Ag ions with the π-donor olefins and aromatic hydrocarbons. Based upon this work approaches involving multidimensional methods [60] carbon dioxide/SF_6 mixtures [61] were published and the ASTM method D6550 [38] was released for their determination in gasoline. The ASTM D6550 method involves two columns connected via switching valves: a bare silica column having a high-surface area is used to retain aromatic and oxygenate compounds while a Ag^+ strong cation

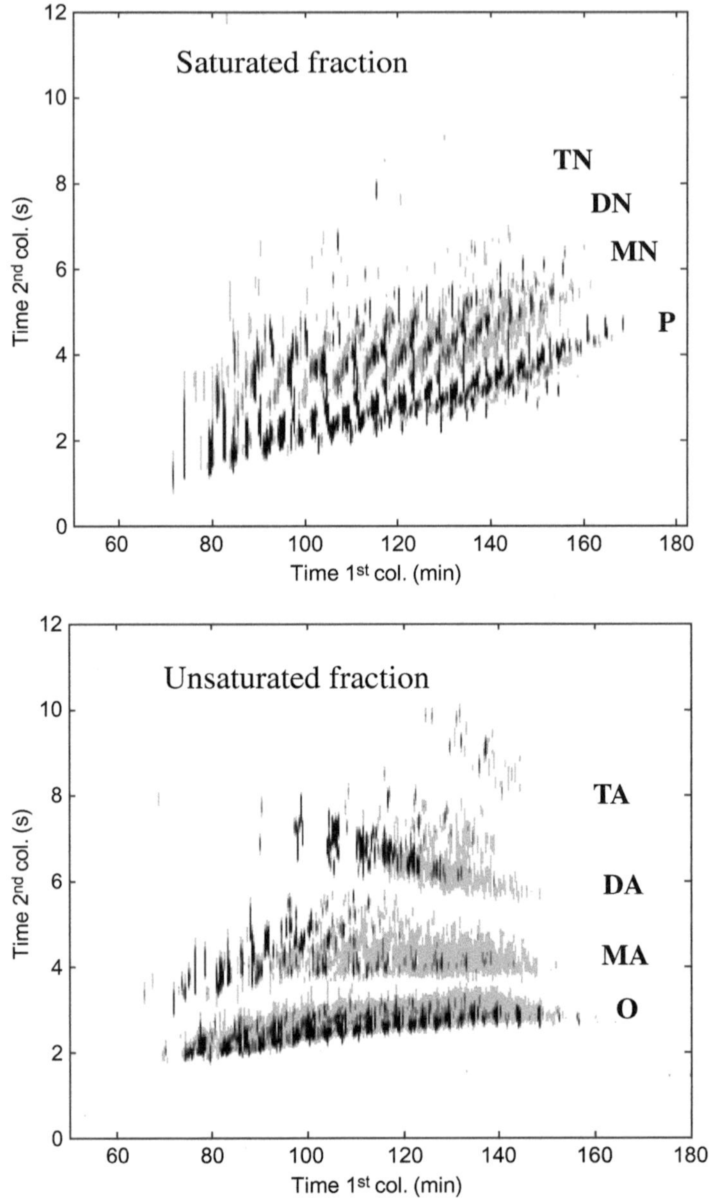

FIGURE 13.3 Two-dimensional chromatograms of saturated and unsaturated fractions obtained by SFC–twin-GC × GC for light cycle oil diesel. Source: *Reprinted with permission from Adam F, Bertoncini F, Thiébaut D, Espinat D, Hennion MC. Supercritical fluid chromatography hyphenated with twin comprehensive two-dimensional gas chromatography for ultimate analysis of middle distillates. J Chromatogr A 2010;1217:1386–1394. Copyright 2010 Elsevier B.V.*

exchange column or silver-loaded bare silica column is used to trap olefins. The columns are connected in series via a valve. Using a proper setting for valve actuation, adequate setting of column switching can be performed so saturates elute first from the two columns; then aromatics and oxygenates are backflushed from the silica column to the detector; finally, olefins are backflushed to the detector from the silver-loaded column. This complex separation must be well defined prior to its implementation via the separation of the so-called performance mixture prepared with selected compounds from the hydrocarbon types to be separated.

A modification of the previous method was released as ASTM method D7346 [39] in order to deal with the separation of olefins in denatured ethanol. In this method a third column packed with polyvinyl alcohol-bonded silica is added to the previous system upstream of the silica and silver-loaded columns in order to trap the alcohol while hydrocarbons go to the silica column in a forward mode. Then, alcohol can be backflushed to the detector while the hydrocarbons are in the silica column. Then, the silica column is flushed to proceed as described in the ASTM D6550 method for the separation of saturates, aromatics, and olefins. The method set up is carried out using a performance mixture of hydrocarbons (no more than 2%) dissolved in ethanol.

Recently [58], a combination of the procedure derived from the ASTM D6550 method in combination with FI-ToF MS was reported for the detailed analysis of olefins and other hydrocarbon groups in total petroleum liquid products from catalytic cracking experiments. With a silver-modified silica column associated with four silica columns connected in series with a dinitroanilinopropyl-silica for multidimensional operation (Fig. 13.4), SFC was able to separate the petroleum samples into saturates, aromatics, and olefins. The authors reported that ambiguities in mass spectral assignments between olefins and naphthenes were overcome by the SFC separation. SFC-FID provided accurate quantitation of the three major hydrocarbon group types, as shown in Fig. 13.5; the values reported on the chromatogram indicate the amount of each hydrocarbon group determined using FID and ToF MS peak area; the discrepancy between the values obtained from the two detectors showed that ToF MS cannot be used for quantitative analysis without using proper response factors and confirmed the value of FID for quantitative analysis of hydrocarbon groups.

13.4.2.3 *Heavy Samples*

Because the complexity of oil samples increases with the range of carbon atom numbers, group-type separation of heavy samples is more difficult. Thus multidimensional separations are used for group-type

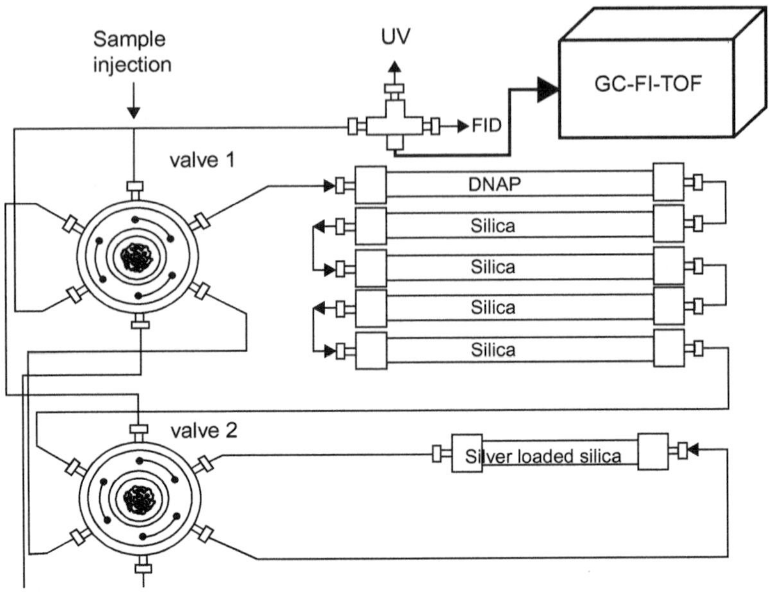

FIGURE 13.4 Multidimensional SFC for olefin separation. Source: *Reprinted with permission from Qian K, DiSanzo FP. Detailed analysis of olefins in processed petroleum streams by combined multi-dimensional supercritical fluid chromatography and field ionization time-of-flight mass spectrometry. Energy Fuels 2016;30:98–103. Copyright 2016 American Chemical Society.*

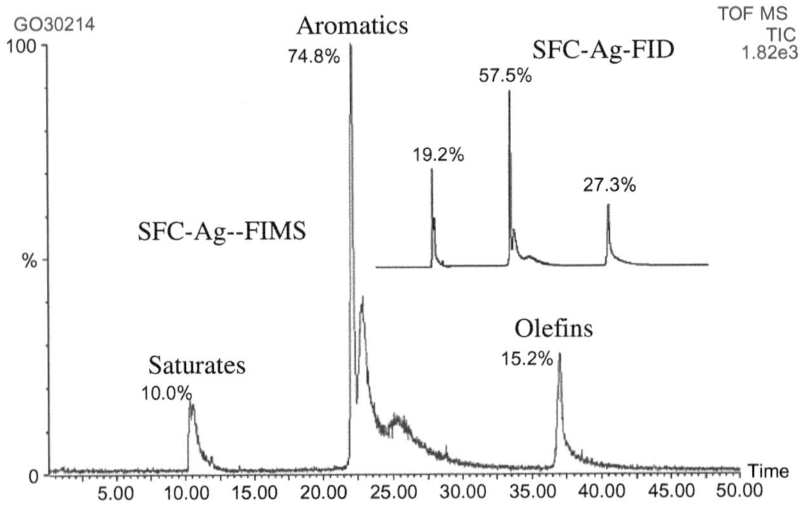

FIGURE 13.5 Total ion chromatogram and FID trace of a total petroleum liquid product obtained using the multidimensional SFC depicted in Fig. 13.4. Source: *Reprinted with permission from Qian K, DiSanzo FP. Detailed analysis of olefins in processed petroleum streams by combined multi-dimensional supercritical fluid chromatography and field ionization time-of-flight mass spectrometry. Energy Fuels 2016;30:98–103. Copyright 2016 American Chemical Society.*

separation of heavy samples too utilizing the same principles as described for the separations of lighter samples and of olefins. The main difference lies in the replacement of the silica column to reduce the retention of polar compounds as shown in the approach developed by Skaar et al. [62] for group-type separation of crude oil. At least, a normal-phase column such as cyanopropyl-bonded silica is preferred to the silica column. This approach was recently adapted by Dutriez et al. [63–66] for the analysis of vacuum distillates using on-line multidimensional SFC coupled to GC × GC similar to the system developed by Adam et al. for gas oils.

Two column or three column systems are used for this approach. The simpler approach utilized a cyanopropyl-bonded and a silver-loaded silica connected by two 6-port valves. In a first step, polar compounds were trapped on the cyanopropyl stationary phase while hydrocarbons were eluted and transferred to the silver-loaded silica. Once the aliphatic fraction had eluted from the second column, the six-port valves were switched in order to backflush (1) the unsaturated compounds initially adsorbed on the silver column and (2) the polar compounds retained on the cyanopropyl column. Using this method a baseline separation between aliphatic, unsaturated, and polar group types was achieved as illustrated in Fig. 13.6 for the analysis of vacuum distillates; the SFC conditions used for this heavy sample were quite different (250 bars, 65°C) compared to the usual conditions applied for group-type separation of lighter samples (150 bars, 30°C) [41]. Again, owing to the high separation power of GC × GC applied to each of the transferred SFC fractions, a very detailed separation of compounds from the different hydrocarbon groups was obtained and provided more information for the distribution of the hydrocarbon families. In the more sophisticated approach, another valve was added to the multidimensional SFC for the connection of an aminopropyl-bonded silica column to perform a partial subgroup types of the aromatic fraction according to the number of aromatic rings after the aromatic fraction had been backflushed from the silver-loaded column [11].

In order to perform GC × GC analysis of the SFC fractions obtained from this system, a dedicated interface was designed [66] which allowed the intermediate collection of SFC fractions. The interface enabled (1) the temporary storage of each SFC fraction in a loop and (2) their subsequent and consecutive introduction into the GC × GC system. Thus the new interface enabled the GC × GC analysis of one SFC fraction while the others fractions were stored in the collection loops. The transfer to the GC × GC columns was similar to the procedure described for the analysis of saturated and unsaturated fractions of gas oils by SFC–GC × GC.

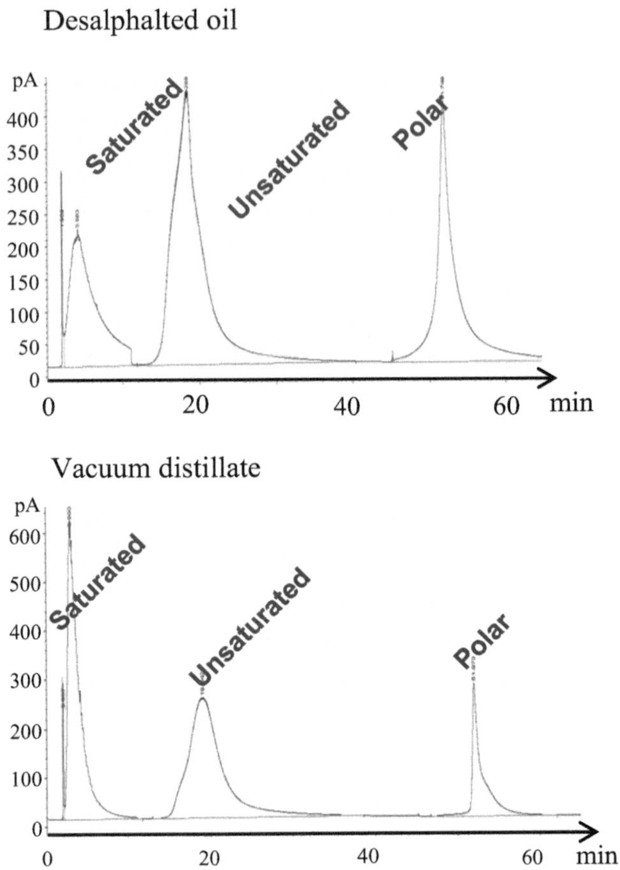

FIGURE 13.6 Group-type separation of heavy fraction from vacuum distillate using two-dimensional SFC prior to on-line GC × GC. Separation is performed at 250 bars and 65°C. Source: *Adapted from Omais B, Dutriez T, Courtiade M, Charon N, Ponthus J, Dulot H, et al. SFC–GC × GC to analyse matrices from petroleum and coal. LC GC Europe 2011;24 (7):352–65.*

This setup has been applied to the analysis of heavy vacuum distillates owing to the recent advances in the separation of heavy boiling point compounds by GC × GC [67–69]. However, such a detailed analysis would not be possible without the sophisticated and selective SFC group-type separation performed prior to the GC × GC analyses. The price to pay for this analysis is the rather quite complex instrumentation and the duration of the analysis as soon as more than one GC × GC analysis is needed per sample. When the number of fractions collected increased, the method moved towards comprehensive (SFC)–(GC × GC) or SFC × GC × GC!

13.4.2.4 *Separation of Unusual Families*

13.4.2.4.1 Oxygenates in Commercial Gasoline

A comprehensive SFC × GC instrument was developed for the detailed separation of polar compounds in a commercial gasoline [70]. The first dimension utilized a porous layer open-tubular silica gel PLOT column (30-m, 0.32-mm-i.d.) with subcritical CO_2 as the mobile phase to perform the group separation of polar oxygenate compounds. In the second dimension a resistively heated nonpolar GC column was used to perform a fast GC separation according to the volatility of the compounds. The stop flow mode was used for transferring 5-s fractions from the SFC column to the GC column. As an example, the chromatogram for a commercial gasoline is shown in Fig. 13.7: ethers were fully separated from aliphatic and aromatic compounds that eluted together. It was also reported that ethers could be separated from compounds containing aldehydes or alcohols while elution of carboxylic acids from the silica column was also possible. This work is a nice illustration of a valuable and orthogonal combination of columns and techniques.

13.4.2.4.2 Analysis of Lipids in Fuels

SFC is a widely used technique for the analysis of lipids. If polar modifiers are required for the elution of polar lipids (an impressive separation was recently reported by Holcapek [71]), fatty methyl esters (FAMEs) used in biofuels can be eluted without modifier:

- FAMEs in jet fuel.

An impressive 3 min separation was reported recently by Langley and coworkers [72] for the analysis of rapeseed methyl esters (RME) and coconut methyl esters in jet fuel. The current international reference method is GC–MS (IP585/10) for the determination of (RME) in jet fuel (US ASTM 1655 and DEF STAN 91–91). Owing to the complex fuel matrix a slow temperature gradient method (50 min) is required and the GC method is restricted to the detection and quantification of low-carbon-number fatty acid methyl esters (C8 – C14). On the contrary the UHPSFC – MS method is approximately 20 times faster than the reference method. The authors also reported SFC–MS afforded a comparable linear dynamic range for the detection of total FAME content up to 100 ppm with an acceptable linearity.

- Biodiesel fuel (B100).

Ultra-high performance supercritical fluid chromatography (UHPSFC) in combination with sub-2 μm particles and either diode array UV, evaporative light scattering (ELSD), or high-resolution mass spectrometric (MS) detection was evaluated for the determination of

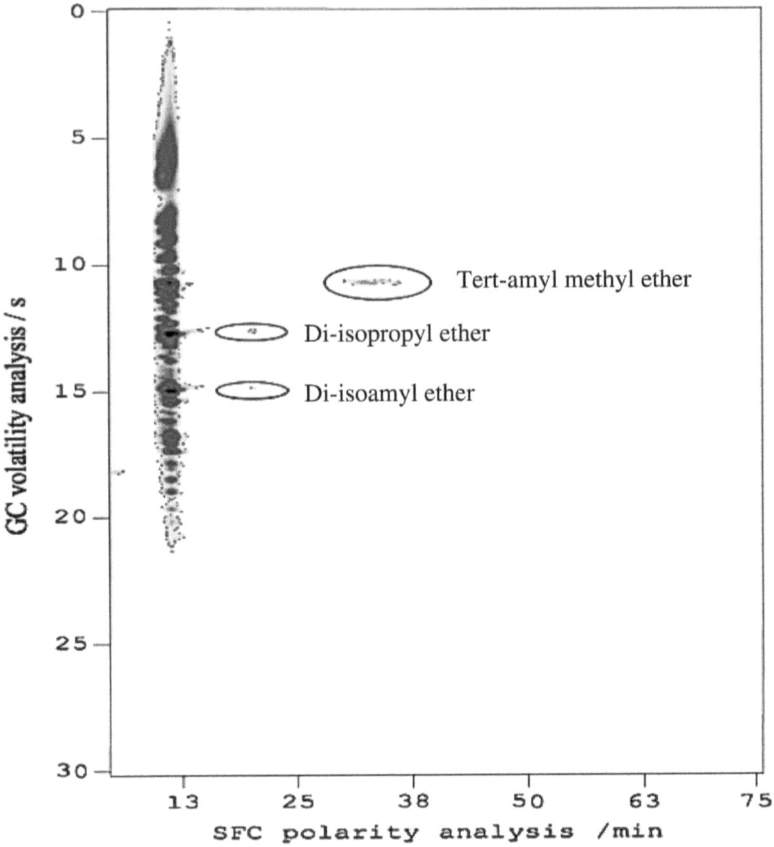

FIGURE 13.7 SFC × GC analysis of a commercial lead-free gasoline sample showing the presence of Tert-amyl methyl ester, diisopropyl ether, and diisoamyl ether. Conditions: SFC analysis, CO_2 at a pressure of 150 atm and a temperature of 28°C. The flow through the PLOT column was collected for intervals of 5 s. The GC was repeatedly temperature programed from −50 to +250°C at 450°C/min; hydrogen was used as carrier gas. Source: *Reprinted with permission from Venter A, Makgwane PR, Rohwer ER. Group-type analysis of oxygenated compounds with a silica gel, gel porous layer open tubular column and comprehensive two-dimensional supercritical fluid and gas chromatography. Anal Chem 2006;78:2051–2054. Copyright 2006 American Chemical Society.*

acylglycerols in soybean, corn, sesame, and tobacco seed oils [73]. As the authors used LC detectors rather than the FID because group-type separation was not required, CO_2 could be modified by acetonitrile using a modifier gradient from 2% to 20%. It is reported that excellent resolution of target compounds was obtained on an unendcapped single C18 column (3.0 mm × 150 mm) in less than 10 min; a partly normal-phase separation is assumed because the retention of the acyl glycerols increased with the number of free hydroxyl groups. No derivatization step was required and residual tri-, di-, and mono-acylglycerols alongside glycerol could be quantified at 0.02% (w/w) and 0.05% (w/w) respectively using ELSD. Thus the method should be compatible with the requirements of ASTM D6751 method for the monitoring of biodiesel B100 impurities.

13.4.2.4.3 Coal Tar Liquefaction Distillates

Coal liquefaction products belong to the new generation of fuel substitutes. However, their composition is different to petroleum fuel specifications: they are mainly composed of naphthenes, aromatics, polycondensed naphthenic and aromatic structures, and heteroatomic compounds (bearing nitrogen or oxygen atoms). Their paraffin content is very low. Oxygen is found in alkylated phenolic and furanic structures. Owing to the properties of this kind of fuel, the identification and quantification of oxygen-containing species is of major importance prior to the understanding of their behavior for their use, such as new fuels. As they belong to petroleum-related samples, the oil industry uses the same nomenclature based on distillation as for regular oil samples to describe the different fractions obtained from processing.

- SFC–GC × GC of middle distillates

For this purpose, SFC was investigated recently by Omais et al. [65]. They studied the performances of three polar stationary phases for the SFC separation of the unusual families of compounds supposed to be present in these samples. Thus unusual stationary phases for SFC group-type separation were also investigated including a pyridine-bonded silica. Fig. 13.8A provides a schematic representation of the retention features of the investigated columns. On the pyridine column, saturated hydrocarbons eluted rapidly since there are no strong interactions with such a phase; silica, as expected, could provide a partial separation of the test compounds owing to its strong affinity for polar compounds and the cyano-bonded phase was not retentive enough to provide a sufficient selectivity between the different groups of compounds (this stationary phase of intermediate polarity is used in normal-phase and reversed-phase LC). As expected, nitrogen-containing compounds were strongly retained by all the phases, especially bare silica. Interestingly, the pyridine-bonded phase enabled a full separation of phenols owing to strong O − H···N hydrogen bonds than O − H···O for silica (29 kJ/mol against 21 kJ/mol). As a result, the implementation of the pyridine column in SFC prior to GC × GC allowed the full elution of phenolic

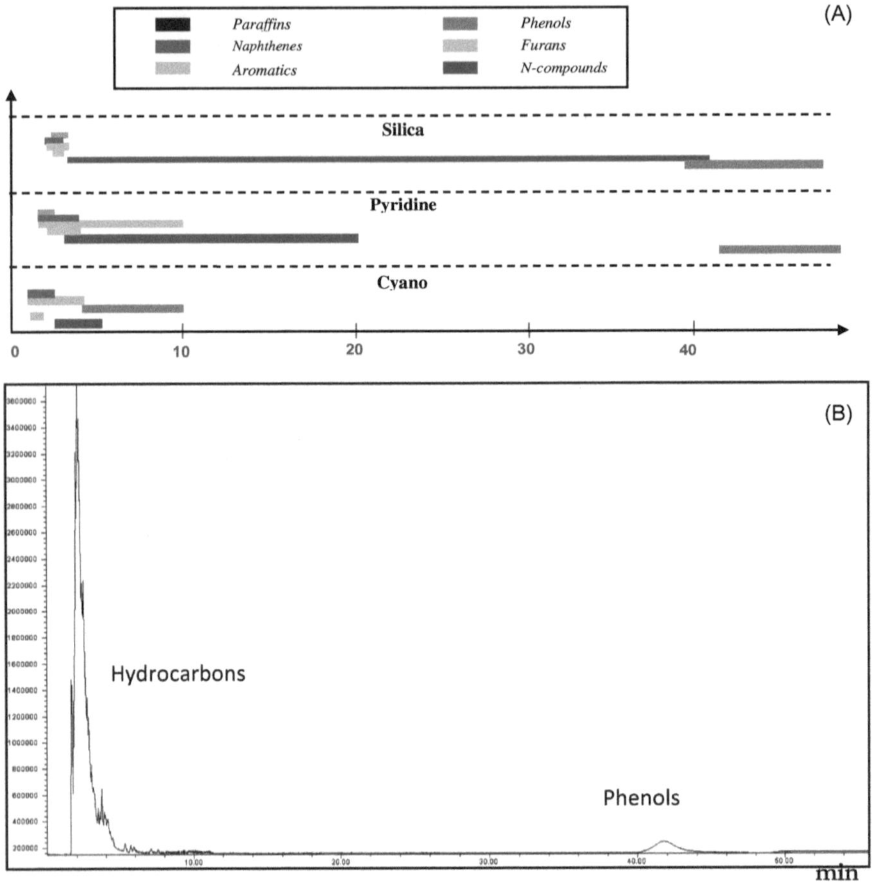

FIGURE 13.8 (A) Schematic representation of elution zones of different chemical families representative of a coal-derived middle distillate; (B) SFC-FID chromatogram of a coal-derived middle distillate using the pyridine bonded phase showing the huge resolution of phenols from hydrocarbons. Source: *Reprinted from Omais B, Dutriez T, Courtiade M, Charon N, Ponthus J, Dulot H, et al. SFC–GC × GC to analyse matrices from petroleum and coal. LC GC Europe 2011;24(7):352–65.*

compounds and their resolution from hydrocarbons in a small section of the chromatogram (from 40 to 46 min); they could be collected without any interference from other compounds (Fig. 13.8B) as established by the GC × GC separation of this fraction (nothing other than phenols was visible in the chromatogram) [65]: in case a detailed separation of phenolic compounds is required, GC × GC could be optimized in order to provide a full space occupancy for a better identification of the species.

• SFC × SFC of vacuum distillate

In order to analyze this heavier fraction, two conventional SFCs were utilized to build a SFC × SFC system without decompression between the two separations (unlike the pioneering SFC × SFC work presented by Hirata et al. [74]). The instrument was constructed on the basis of on-line comprehensive 2D LC and used a LC × LC-like interface [75]; it consisted of a two-loop switching valve allowing the collection of the first dimension column effluent and the subsequent transfer to the second dimension; as in GC × GC, separation of a fraction in the second dimension had to be performed during the time allowed for the collection of the next fraction of the first dimension eluent. Owing to the conventional diameter of the column used in the first dimension, the first dimension column flow rate had to be split prior to modulation. In both dimensions, separations could be monitored by UV detection; in the second dimension, the main flow was diverted to implement a FID for the detection of hydrocarbons. The FID trace was used to obtain the color plots for the 2D separations. The interface was modified to perform solute focusing between the two dimensions (active modulation) as reported in [76]. Compared to the first SFC × SFC-FID chromatogram of a vacuum distillate reported by Guibal et al. [75], Fig. 13.9 shows a real improvement of the quality of the separation owing to the increase in efficiency evidenced by the reduction of the size of the spots. The chromatogram b is very similar to a GC × GC chromatogram. Although this preliminary work is very promising, the system is too complex for routine use.

13.5 SEPARATION OF BASE STOCKS AND LUBRICANT ADDITIVES

Lubricant such as those used in car engines are complex mixtures; indeed, they are composed of both a base stock and a large variety of additives added to the base as mixtures ("packages"). Thus their analysis is a difficult task because the base stock can be derived from a complex oil fraction and the additives can represent around 25% of the lubricant composition.

Direct methods such as UV or infrared spectrometry [77] and NMR [78], on carbon or hetero-elements, can provide fast but general information on the lubricant and they generally suffer from low specificity. Chromatographic methods, GC [79–81], and LC [82–84] can be used for the separation and the identification of additive (most often polymers additives) but they cannot provide the total determination of the composition of a lubricant [85,86].

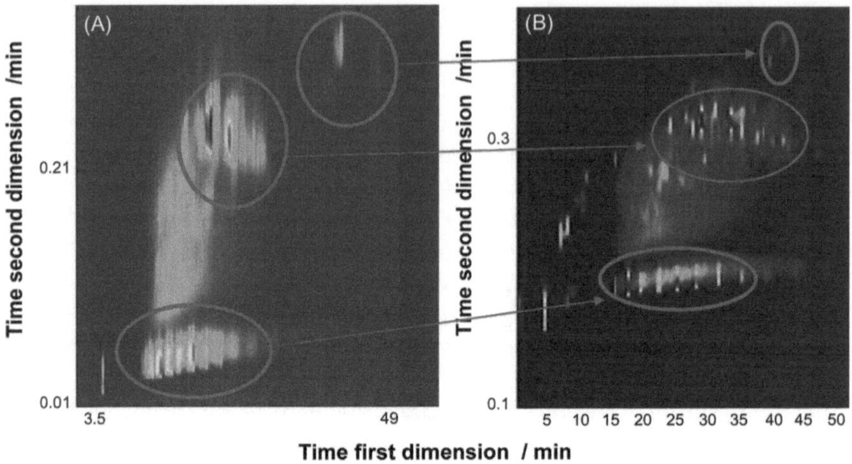

FIGURE 13.9 Comprehensive 2D SFC chromatogram of a coal tar vacuum distillate. (A) Without active modulation; (B) with active modulation. Mobile phase: carbon dioxide pressure programed, 60°C; first dimension, three Ascentis Express fused core columns (each 15 cm × 4.6 mm, 2.7 μm); second dimension column Syncronis Silica, 5 cm × 2.1 mm, 1.8 μm; CO_2 flow rate: 1.6 mL/min. Modulation period: 21 s.

SFC is a good choice for this application because it is well established for group-type separations that fit with part of the analytical challenge. Another important advantage is the compatibility with most of the detectors used in GC and LC. Again, a two-dimensional column switching system was implemented to (1) trap polar additives on a polar stationary phase (bare silica) while the base stock was partially separated and (2) to backflush the additives and transfer them to an apolar stationary phase (an encapsulated silica C18 column) [20]. Elution was carried out without adding polar modifiers: it enabled implementing specific detectors to improve the overall analysis because the complexity of the sample is so high that coelution could not be avoided despite using a 2D system. Pure CO_2 allowed the elution of quite polar compounds [20] using a very deactivated C18 stationary phase unlike the column used in [73] for the separation of acyl esters where the residual silanols were found to be partly responsible for the separation mechanism.

13.5.1 Multihyphenated System

In addition to FID and UV detections that are mandatory for quantitative analysis of hydrocarbons from oil samples, SFC was hyphenated to AED, MS, and FTIR. For FTIR the high-pressure cell was modified in order to tolerate the pressure used in SFC [10].

The AED used a microwave-induced plasma; its hyphenation to SFC is reported in [19]. For an operator familiar with SFC-FID the adaptation of AED to SFC requires the same parameters to be optimized, i.e., the position of the restrictor in the detector, the flow rates (effluent, plasma gases (He and Ar), and the usual reactant gases for AED, O_2, H_2), and the wavelength; other conditions were set as for GC-AED operation. The weakpoints were (1) signal depletion induced by the diminution of plasma energy owing to the introduction of CO_2 in the plasma; (2) continuous and specific emissions from the CO_2 that interfere with that of some elements. It mainly occurs in the range 170–200 nm. Thus the detection of nitrogen was performed at 388 nm (emission of CN formed in the plasma) instead of 174 nm [19]. It was found the addition of 10–15% argon in He could reduce the CO_2 interference [19,87]. The implementation of atmospheric pressure chemical ionization source for the hyphenation of an ion trap MS to SFC describes full optimization of experimental conditions [88]. The use of a make-up fluid [89] was investigated: Interestingly, when no make-up fluid was added to the mobile phase prior entering the source, ionization could still occur because traces of water in the CO_2 were able to form $(M+H)^+$ adducts or charge transfer with the CO_2 could lead to $(M)^{\bullet +}$ ions [90–92]. Using a special device, automatic tuning of the MS operating parameters in the SF could be carried out in the same solvent conditions as during the SFC separation (in the presence of supercritical CO_2) rather than in the liquid phase used in LC conditions [88,89,92]. When the tuning in SFC conditions was applied, the detection sensitivity was improved by a factor 1.5–7.7 depending on the target compound [88,92].

The hypernated system FID-UV-FTIR-AED-MS-SFC is presented in [1,88,92,93].

13.5.2 Base Stocks

For base stocks obtained from mineral sources the idea was to start from the ASTM Method D 5186 [37], described above. As it lacked sufficient resolution for group-type separation of mineral base stock [87,88,92] improvements were required such as the use of silver-loaded silica column and a multidimensional separation.

For synthetic base stocks, a partial resolution of PolyAlphaOlefins and aliphatics was achieved on bare silica by pressure programing providing an approximate estimate of their concentration [87,88,92].

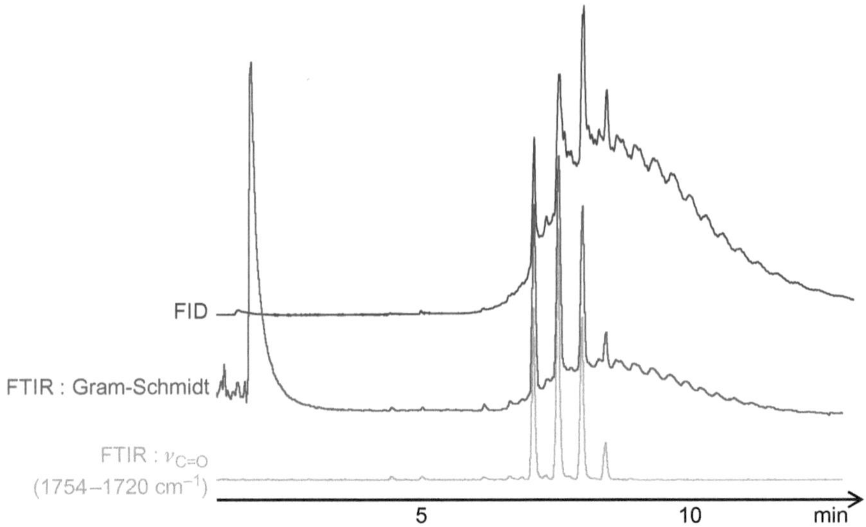

FIGURE 13.10 Separation of esters in a hemi-synthetic base stock by hypernated SFC Column Capcell Pak C_{18} (250 × 4.6 mm), 80°C, pressure program from 100 to 300 bars at 20 bars/min, CO_2 flow rate 2 mL/min. Source: *Reprinted from Lavison-Bompard G, Bertoncini F, Thiébaut D, Beziau JF, Carrazé B, Valette P, et al. Hypernated supercritical fluid chromatography: potential application for car lubricant analysis. J Chromatogr A 2012;1270:318–23. Copyright 2010 Elsevier B.V.*

Base stocks containing bioesters could be monitored very easily via their intense and specific FTIR trace utilizing the carbonyl band stretching vibration at 1736 cm^{-1} [92,93] (Fig. 13.10).

13.5.3 Additives

Additives are added to the base stock in order to trap particles, improve viscosity at high and low temperatures, improve resistance to oxidation, limit engine wear, etc. For car manufacturers, their monitoring can help (1) to better understand their behavior, (2) to study their properties, and (3) to study their ageing in the engine.

In the proposed SFC hypernated system (see Fig. 13.12), most of the detectors involved played a significant role for monitoring the different classes of investigated additives. AED capabilities were useful for the monitoring of Zn, S, P, and N (via CN) [19,92–94] in order to define specific elution zones in order to simplify FTIR and/or MS spectra interpretations.

Further to the specific data obtained from sophisticated detectors the information gained through UV detection should not be neglected. For example, chromatograms based on the UV trace at 254 nm of used and

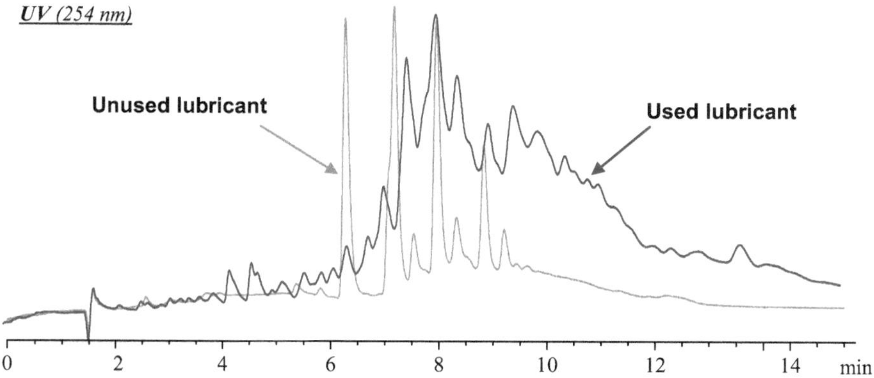

FIGURE 13.11 Comparison of SFC/UV analysis of unused (lower signal in blue) and used (upper signal in red) lubricant. Same conditions as in Fig. 13.10. Source: *Reprinted with permission from Lavison-Bompard G, Bertoncini F, Thiébaut D, Beziau JF, Carrazé B, Valette P, et al. Hypernated supercritical fluid chromatography: potential application for car lubricant analysis. J Chromatogr A 2012;1270:318–23. Copyright 2010 Elsevier B.V.*

unused lubricant (Fig. 13.11) exhibited strong differences; in the used lubricant, Irganox L57, and Zn-dithiophosphates could not be detected; the presence of more retained compounds indicated that new species were generated during ageing standardized tests. Some could be attributed to nitrogen-containing compounds.

The information gained from MS was critical for structural identification of compounds as shown in Fig. 13.12 for Irganox L57 additive, and in [92,93] for Lowinox DLTDP additive for example.

13.6 CONCLUSION—PERSPECTIVES

SFC is a complementary technique to LC and GC; however, for the separation of heavy hydrocarbons, SFC is better adapted than GC because supercritical CO_2 is a dense fluid having real solvating properties. Thus SFC enables simulated distillation to be performed in a range of boiling points exceeding 900°C and efficient group-type separations of heavy fractions while, in both cases, FID can be used as in GC. For the separation of hydrocarbons, it is an essential advantage over LC and mass spectrometry, particularly for quantitative analysis. This is the reason why ASTM methods involving SFC have been released and SFC is used in routine analysis for this purpose.

Unfortunately, the widest detection capability of SFC is mainly obtained with neat CO_2 while most applications are now performed using modifiers. As a result, most commercial systems do not provide

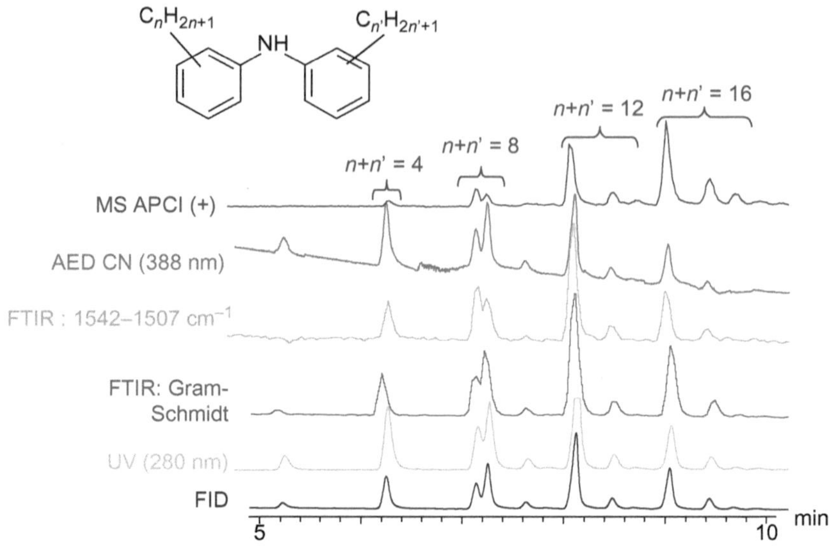

FIGURE 13.12 Hypernated SFC/FID/UV/FTIR/AED/MS analysis of Irganox L57 reference additive. Same conditions as in Fig. 13.10. See Lavison-Bompard et al. [93] for detailed conditions. Source: *Reprinted with permission from Lavison-Bompard G, Bertoncini F, Thiébaut D, Beziau JF, Carrazé B, Valette P, et al. Hypernated supercritical fluid chromatography: potential application for car lubricant analysis. J Chromatogr A 2012;1270:318–23. Copyright 2010 Elsevier B.V.*

these detection capabilities and, hyphenated systems are restricted to a few experienced teams, mostly in research laboratories.

SFC has also an important role to play in hyphenated techniques for the analysis of very complex mixtures such as petroleum fractions: for such samples, efficiency is not the only important criterion (on that, SFC also has a decisive advantage over LC because longer columns can be implemented): searching for selectivity via the coupling of conjugated different separation mechanisms has proven its capabilities in GC × GC [11] and LC × LC [95]; SFC also, as shown in this chapter for SFC–GC × GC, and SFC × LC, and LC × SFC [4]. Despite its high complexity, the feasibility and capability of SFC × SFC has also been demonstrated.

Acknowledgments

D.T. thanks everyone who participated in the experiments reported in this chapter. He is particularly grateful to the students and postdoctoral fellows (at the time) who worked with him in SFC: Frédérick Adam, Thomas Dutriez, Badaoui Omais, and Laure Mahé at IFP Energies Nouvelles; Rachid M'Hamdi, Vy-Khanh Huynh, Fabrice Bertoncini, Gwenaelle Lavison, Laure Dahan, Alina Dulaurent, Jiri Urban, Cédric Sarazin, Pierre

Guibal, and Orjen Petkovic at ESPCI; and their supervisors at IFP Energies Nouvelles, Fabrice Bertoncini, Marion Courtiade, Hughes Dulot, Cyril Dartiguelongue, Jeremie Ponthus, Eric Robert, and Vincent Souchon, and, last but not least, his colleagues at ESPCI, Patrick Sassiat and Jérôme Vial.

References

[1] Thiébaut D. Separation of petroleum products involving supercritical fluid chromatography. J Chromatogr A 2012;1252:177–88.
[2] Lambert N, Felinger A. Performance of the same column in supercritical fluid chromatography and in liquid chromatography. J Chromatogr A 2015;1409:234–40.
[3] Taylor LT. Supercritical fluid chromatography for the 21st century. J Supercrit Fluids 2009;47:566–73.
[4] Lesellier E, West C. The many faces of packed column supercritical fluid chromatography—a critical review. J Chromatogr A 2015;1382:2–46.
[5] Berger TA. Minimizing ultraviolet noise due to mismatches between detector flow cell and post column mobile phase temperatures in supercritical fluid chromatography: effect of flow cell design. J Chromatogr A 2014;1364:249–60.
[6] Smith RD, Wright BW, Yonker CR. Supercritical fluid chromatography: current status and prognosis. Anal Chem 1988;60:1323A–36A.
[7] Chester TL, Pinkston JD, Raynie DE. Supercritical fluid chromatography and extraction. Anal Chem 1998;70:301–20.
[8] Peaden PA. Simulated distillation of petroleum and its products by gas and supercritical fluid chromatography: a review. J High Resolut Chromatogr 1994;17:203–11.
[9] Roberts I. Chromatography in the petroleum industry. Amsterdam: Elsevier; 1995.
[10] Robert E. Practical supercritical-fluid chromatography and extraction. Reading: Harwood Academic Publishers; 1999.
[11] Thiébaut D. Gas chromatography and 2D gas chromatography for petroleum industry. Paris: Technip; 2013.
[12] Guzowski JP, Huhak WE. On-line multidimensional supercritical fluid chromatography capillary gas chromatography. J High Resolut Chromatogr 1987;10:337–41.
[13] Levy JM, Cavalier RA, Bosch TN, Rynaski AM, Huhak WE. Multidimensional supercritical fluid chromatography and supercritical fluid extraction. J Chromatogr Sci 1989;27:341–6.
[14] Thiébaut D, Caude M, Rosset R. Couplage de la chromatographie en phase dioxyde de carbone supercritique avec la détection par ionisation de flamme: application à l'analyse des produits pétroliers. Analusis 1987;15:528–39.
[15] Dulaurent A, Dahan L, Thiébaut D, Bertoncini F, Espinat D. Extended simulated distillation by capillary supercritical fluid chromatography. Oil Gas Sci Technol 2007;62:33–42.
[16] Hewlett Packard Application notes 228-169 SFC of polywaxes and heavy petroleum fractions and 228-176, separation of Polywax 655 by capillary SFC; 1992.
[17] Kelemidou K, Severin D. Simulated distillation by chromatography with supercritical fluid on high boiling petroleum fractions. Erdöl Erdgas Kohle 1996;112:25–7.
[18] Sarazin C, Thiébaut D, Sassiat P, Vial J. Feasibility of ultra-high performance supercritical carbon dioxide chromatography at conventional pressures. J Sep Sci 2011;34:1773–8.
[19] Bertoncini F, Thiébaut D, Caude M, Gagean M, Carraze B, Beurdouche P, et al. On-line packed column supercritical fluid chromatography-microwave-induced plasma atomic emission. J Chromatogr A 2001;910:127–35.
[20] Lavison G, Bertoncini F, Thiébaut D, Beziau JF, Carrazé B, Valette P, et al. Supercritical fluid chromatography and two-dimensional supercritical fluid

chromatography of polar car lubricant additives with neat CO_2 as mobile phase. J Chromatogr A 2007;1161:300–7.
[21] ASTM D2892. Standard test method for distillation of crude petroleum; 2005.
[22] ASTM D2887. Standard test method for boiling range distribution of petroleum fractions by gas chromatography; 2008.
[23] ASTM D5236. Standard test method for distillation of heavy hydrocarbon mixtures; 2007.
[24] ASTM 6352. Standard test method for boiling range distribution of petroleum distillates in boiling range from 174 to 700°c by gas chromatography; 2009.
[25] ASTM 7160. Boiling point distribution of samples with residues such as crude oils and atmospheric and vacuum residues by high temperature gas chromatography; 2005.
[26] ASTM 7500. Standard test method for determination of boiling range distribution of distillates and lubricating base oils—in boiling range from 100 to 735°C by gas chromatography; 2008.
[27] Schwartz MM, Brownlee HE, Boduszynski RG, Su F. Simulated distillation of high boiling petroleum fractions by capillary supercritical fluid chromatography and vacuum thermal gravimetric analysis. Anal Chem 1987;59:1393–401.
[28] Carbognani L, Lubkowitz J, Gonzalez MF, Perreira-Almaro P. High temperature simulated distillation of athabasca vacuum residue fractions. Bimodal distributions and evidence for secondary "on-column" cracking of heavy hydrocarbons. Energy Fuels 2007;21:2831–9.
[29] Schwartz HE. Simulated distillation by packed column supercritical fluid chromatography. J Chromatogr Sci 1988;26:275–9.
[30] Sotty Ph, Rocca J.L., Grand C. 15th Riva del Garda; 1993.
[31] Shariff SM, Tong D, Bartle KD. Simulated distillation by supercritical fluid chromatography on packed capillary columns. J Chromatogr Sci 1994;32:541–6.
[32] Huynh V.K. PhD Dissertation, Université Pierre et Marie Curie. Paris; 1998.
[33] DiSanzo F., Nicholas M., Cadoppi A., Munari F. Oral communication, 32nd Int. Symp. on Capillary Chromatography. Riva Del Garda; 2008.
[34] Raynie DE, Markides KE, Lee ML. Boiling range distribution of petroleum and coal-derived heavy ends by supercritical fluid chromatography. J Microcol Sep 1991;3:423–33.
[35] Dahan L. PhD Dissertation, Université Pierre et Marie Curie. Paris; 2004.
[36] Thiébaut DRP, Robert EC. Group-type separation and simulated distillation: a niche for SFC. Analusis 1999;27:681–90.
[37] ASTM 5186. Standard test method for determination of the aromatic content and polynuclear aromatic content of diesel fuels and aviation turbine fuels by supercritical fluid chromatography; 2009.
[38] ASTM 6550. Standard test method for determination of olefin content of gasolines by supercritical-fluid chromatography; 2005.
[39] ASTM 7346. Standard test method for determination of olefin content in denatured ethanol by supercritical fluid chromatography; 2008.
[40] Di Sanzo FP, Yoder RE. Determination of aromatics in jet and diesel fuels by supercritical fluid chromatography with flame ionization detection (SFC—FID): a quantitative study. J Chromatogr Sci 1991;29:4–7.
[41] M'Hamdi R, Thiébaut D, Caude M. Packed column SFC of gas-oils. (I) Hydrocarbon group separation using pure carbon dioxide. J High Resolut Chromatogr 1997;20:545–54.
[42] M'Hamdi R, Thiébaut D, Caude, Robert E, Grand C. Packed column SFC of gas oils. Part II: Experimental design optimization of the group-type analysis using CO_2-SF_6 as the mobile phase. J High Resolut Chromatogr 1998;21:94–102.

[43] Schwartz HE, Brownlee RG. Hydrocarbon group analysis of gasolines with microbore supercritical fluid chromatography and flame ionization detection. J Chromatogr 1986;353:77–93.
[44] Chen EN, Cusatis PD, Popiel EJ. Validation of the aromatic ring distribution in diesel fuel refinery streams by supercritical fluid chromatography and mass spectrometry. J Chromatogr 1993;637:181–6.
[45] M'Hamdi R. PhD Dissertation. Université Pierre et Marie Curie, Paris; 1996.
[46] Bouigeon C, Thiébaut D, Caude M. Long packed column supercritical fluid chromatography: influence of pressure drop on efficiency. Anal Chem 1996;68:3622–30.
[47] Klee M., Wang M.Z. Optimisation of group seprartions for determination of the aromatics content in diesel fuels. Hewlett Packard Application note 228-167; 1992.
[48] Analytical Control Application notes 9317-9318; 1993.
[49] Shultz W.W., Genowitz M.W. Method for quantitatively determining the amount of saturates, olefins, and aromatics in a mixture thereof. US Patent 5,190,882; 1993.
[50] Albuquerque FC. Determination of conjugated dienes in petroleum products by supercritical fluid chromatography and ultraviolet detection. J Chromatogr Sci 2003;26:1403–6.
[51] Paproski RE, Cooley J, Lucy CA. Comparison of titania, zirconia, and silica stationary phases for separating diesel fuels according to hydrocarbon group-type by supercritical fluid chromatography. J Chromatogr A 2005;1095:156–63.
[52] Paproski RE, Cooley J, Lucy CA. Fast supercritical fluid chromatography hydrocarbon group-type separations of diesel fuels using packed and monolithic columns. Analyst 2006;131:422–8.
[53] Qian K, Diehl JW, Dechert GJ, DiSanzo FP. The coupling of supercritical fluid chromatography and field ionization time-of-flight high-resolution mass spectrometry for rapid and quantitative analysis of petroleum middle distillates. Eur J Mass Spectrom 2004;10:187–96.
[54] Vendeuvre C, Bertoncini F, Espinat D, Thiébaut D, Hennion MC. Multidimensional gas chromatography for the detailed PIONA analysis of heavy naphtha: hyphenation of an olefin trap to comprehensive two-dimensional gas chromatography. J Chromatogr A 2005;1090:116–25.
[55] Adam F, Bertoncini F, Thiébaut D, Espinat D, Hennion MC. Supercritical fluid chromatography hyphenated with twin comprehensive two-dimensional gas chromatography for ultimate analysis of middle distillates. J Chromatogr A 2010;1217:1386–94.
[56] Adam F., PhD Dissertation, Université Pierre et Marie Curie; 2008.
[57] Potgieter H, van der Westhuizen R, Rohwer E, Malan D. Hyphenation of supercritical fluid chromatography and two-dimensional gas chromatography–mass spectrometry for group type separations. J Chromatogr A 2013;1294:137–44.
[58] Qian K, DiSanzo FP. Detailed analysis of olefins in processed petroleum streams by combined multi-dimensional supercritical fluid chromatography and field ionization time-of-flight mass spectrometry. Energy Fuels 2016;30:98–103.
[59] Norris TA, Rawdon MG. Determination of hydrocarbon types in petroleum liquids by supercritical fluid chromatography with flame ionization detection. Anal Chem 1984;56:1767–9.
[60] Anderson PE, Demirbücker M, Blomberg LG. Characterisation of fuels by multidimensional supercritical fluid chromatography and supercritical fluid–mass spectrometry. J Chromatogr 1993;641:347–55.
[61] Fuhr BJ, Klein LL, Reichert C, Lee SW. Determination of aromatic types in middle distillates by supercritical fluid chromatography. LC-GC 1990;8:800–4.
[62] Skaar H, Norli HR, Lundanes E, Greibrook T. Group separation of crude oil by supercritical fluid chromatography using packed narrow bore columns, column switching, and backflushing. J Microcol Sep 1990;2:222–8.

[63] Dutriez T, Thiébaut D, Courtiade M, Dulot H, Bertoncini F, Hennion MC. Supercritical fluid chromatography hyphenated to bidimensional gas chromatography in comprehensive and heart-cutting mode: design of the instrumentation. J Chromatogr A 2012;1255:153–62.

[64] Dutriez T, Thiébaut D, Courtiade M, Dulot H, Bertoncini F, Hennion MC. Application to SFC–GC×GC to heavy petroleum fractions analysis. Fuel 2013;104:583–92.

[65] Omais B, Dutriez T, Courtiade M, Charon N, Ponthus J, Dulot H, et al. SFC–GC×GC to analyse matrices from petroleum and coal. LC GC Europe 2011;24 (7):352–65.

[66] Dutriez T., Courtiade M., Thiébaut D., Dulot H., Bertoncini F. Dispositif de couplage modulable entre des chromatographies en phase supercritique et en phase gazeuse bidimensionnelle, méthode pour réaliser une analyse quantitative en 3D. French Patent FR1002248; 2010.

[67] Dutriez T, Courtiade M, Thiébaut D, Dulot H, Bertoncini F, Vial J, et al. High temperature two dimensional gas chromatography of hydrocarbons up to nC60 for analysis of vacuum gas oils. J Chromatogr A 2009;1216:2905–12.

[68] Mahé L, Courtiade M, Dartiguelongue C, Ponthus J, Souchon V, Thiébaut D. Overcoming the high-temperature two-dimensional gas chromatography limits to elute heavy compounds. J Chromatogr A 2012;1229:298–301.

[69] Boursier L, Souchon V, Dartiguelongue C, Ponthus J, Courtiade M, Thiébaut D. Complete elution of vacuum gas oil resins by comprehensive high-temperature two-dimensional gas chromatography. J Chromatogr A 2013;1280:98–103.

[70] Venter A, Makgwane PR, Rohwer ER. Group type analysis of oxygenated compounds with a silica gel, gel porous layer open tubular column and comprehensive two-dimensional supercritical fluid and gas chromatography. Anal Chem 2006;78:2051–4.

[71] Lísa M, Holčapek M. High-throughput and comprehensive lipidomic analysis using ultrahigh-performance supercritical fluid chromatography–mass spectrometry. Anal. Chem. 2015;87:7187–95.

[72] Ratsameepakai W, Herniman JM, Jenkins TJ, Langley GJ. Evaluation of ultrahigh-performance supercritical fluid chromatography–mass spectrometry as an alternative approach for the analysis of fatty acid methyl esters in aviation turbine fuel. Energy Fuels 2015;29:2485–92.

[73] Ashraf-Khorassani M, Yang J, Rainville P, Jones MD, Fountain KJ, Isaac G, et al. Ultrahigh performance supercritical fluid chromatography of lipophilic compounds with application to synthetic and commercial biodiesel. J Chromatogr B 2015;983-984:94–100.

[74] Hirata Y, Hashiguchi T, Kawata E. Development of comprehensive two-dimensional packed column supercritical fluid chromatography. J Sep Sci 2003;26:531–5.

[75] Guibal P, Thiébaut D, Sassiat P, Vial J. Feasability of neat carbon dioxide packed column comprehensive two dimensional supercritical fluid chromatography. J Chromatogr A 2012;1255:252–8.

[76] Petkovic O., Guibal P., Sassiat P., Vial J., Thiébaut D. Active modulation in neat carbon dioxide packed column comprehensive two dimensional supercritical fluid chromatography. To be submitted.

[77] Bansal V, Chopra A, Sastry MIS, Kagdiyal V, Sarpal AS. Detailed characterisation of sulphur and nitrogen components in a multifunctional, ashless lubricant additive by spectroscopic and chromatographic techniques. Tribotest 2003;9:317–32.

[78] Sarpal AS, Kapur GS, Chopra A, Jain KS, Srivastava SP, Bhatnagar AK. Hydrocarbon characterization of hydrocracked base stocks by one- and twodimensional N.M.R. spectroscopy. Fuel 1996;75:483–90.

[79] Wolf M, Ries M, Heitmann D, Schreiner M, Thoma H, Vierle O, et al. Application of a purge and trap TDS-GC/MS procedure for the determination of emissions from flame retarded polymers. Chemosphere 2000;41:693–9.
[80] Thiébaut B, Lattuati-Derieux A, Hocevar M, Vilmont L. Application of headspace SPME-GC–MS in characterisation of odorous volatile organic compounds emitted from magnetic tape coatings based on poly(urethane-ester) after natural and artificial ageing. Polym Test 2007;26:243–56.
[81] Lehrle RS, Duncan R, Liu Y, Parsons IW, Rollinson M, Lamb G, et al. Mass spectrometric methods for assessing the thermal stability of liquid polymers and oils: study of some liquid polyisobutylenes used in the production of crankcase oil additives. J Anal Appl Pyrolisis 2002;64:207–27.
[82] Carbognani L. Preparative isolation and characterization of zincdialkyldithiophosphates from commercial antiwearadditives. J Sep Sci 2003;26:1575–81.
[83] Schlummer M, Brandl F, Mäurer A, Eldik RJ. Analysis of flame retardant additives in polymer fractions of waste of electric and electronic equipment (WEEE) by means of HPLC–UV/MS and GPC–HPLC–UV. J Chromatogr A 2005;1064:39–51.
[84] Block C, Wynants L, Kelchtermans M, De Boer R, Compernolle F. Identification of polymer additives by liquid chromatography–mass spectrometry. Pol Deg Stabil 2006;91:3163–73.
[85] Permanyer A, Douifi L, Lahcini A, Lamontagne J, Kister J. FTIR and SUVF spectroscopy applied to reservoir compartmentalization: a comparative study with gas chromatography fingerprints results. Fuel 2002;81:861–6.
[86] Lambropoulos N, Cardwell TJ, Caridi D, Marriott PJ. Separation of zinc dialkyldithiophosphates in lubricating oil additives by normal-phase high-performance liquid chromatography. J Chromatogr 1996;749:87–94.
[87] Bertoncini F. PhD dissertation. Université Pierre et Marie Curie, Paris; 2002.
[88] Lavison-Bompard G, Thiébaut D, Beziau JF, Carrazé B, Valette P, Duteurtre X, et al. Hyphenation of APCI-MS to supercritical fluid chromatography for polar car lubricant additives with neat CO_2 as mobile phase. J Chromatogr A 2009;1216:837–44.
[89] Morgan DG, Harbol KL, Kitrinos Jr. NP. Optimization of a supercritical fluid chromatograph–atmospheric pressure chemical ionization mass spectrometer interface using an ion trap and two quadrupole mass spectrometers. J Chromatogr A 1998;800:39–49.
[90] Sadoun F, Virelizier H, Arpino PJ. Packed-column supercritical fluid chromatography coupled with electrospray ionization mass spectrometry. J Chromatogr A 1993;647:351–9.
[91] Sjoberg PJR, Markides KE. Capillary column supercritical fluid chromatography–atmospheric pressure ionisation mass spectrometry Interface performance of atmospheric pressure chemical ionisation and electrospray ionisation. J Chromatogr A 1999;855:317–27.
[92] Lavison G. PhD dissertation. Université Pierre et Marie Curie, Paris; 2004.
[93] Lavison-Bompard G, Bertoncini F, Thiébaut D, Beziau JF, Carrazé B, Valette P, et al. Hypernated supercritical fluid chromatography: potential application for car lubricant analysis. J Chromatogr A 2012;1270:318–23.
[94] Bertoncini F, Thiébaut D, Gagean M, Carrazé B, Valette P, Duteurtre X. Easy hyphenation of supercritical fluid chromatography to atomic emission detection for analysing lubricant additives. Chromatographia 2001;53:427–33.
[95] Mondello L. Comprehensive chromatography in combination with mass spectrometry. Hoboken, NJ: Wiley; 2011.

CHAPTER

14

Separation of Lipids

T. Yamada[1], *K. Taguchi*[1] *and T. Bamba*[1,2]

[1]Osaka University, Suita, Osaka, Japan
[2]Kyushu University, Fukuoka, Japan

OUTLINE

Supercritical Fluid Chromatography.
DOI: http://dx.doi.org/10.1016/B978-0-12-809207-1.00014-8

14.1 INTRODUCTION

14.1.1 Diversity and Complexity of Lipids

Lipids are classified into eight categories defined by the Lipid Metabolites and Pathways Strategy (LIPID MAPS) consortium [1]. Fatty acyls, glycerolipids, glycerophospholipids, sphingolipids, saccharolipids, and polyketides have a ketoacyl subunit (called a "building block"), while sterol lipids and prenol lipids have an isoprene building block. Each category is further divided into classes, subclasses, and 4th-level classes, increasing the diversity of lipid molecular species.

Lipids play crucial roles in biological systems and their functions differ according to their categories, classes, or subclasses. Fatty acyls including fatty acids and their functional variants, affect the regulation of various metabolic processes. It is well known that the functions of fatty acids depend on the carbon chain length (number of carbon atoms), degree of unsaturation (number of double bonds), and double bond configurations (position and *cis/trans*). Glycerolipids, composed of several classes of neutral lipids, serve as a source of energy through the β-oxidation of fatty acids esterified onto the glycerol backbone. Glycerophospholipids, sphingolipids, and sterol lipids are the major components of cell membranes and occasionally function as signal transducers. Lipids, therefore, are one of the most important target molecules in metabolic analysis. In recent years, there has been rapid progress in the field of lipidomics, which seeks to comprehensively profile the individual lipid molecular species occurring in biological systems.

Lipids also include structural analogs and isomers of common substructures, which often have vastly different, sometimes even opposite, biological functions than their counterparts. Therefore separation of individual molecular species including structural analogs in a complex mixture is a crucial task in the field of lipid biochemistry.

14.1.2 Mass Spectrometry-Based Techniques for Lipid Profiling

A UV detector, evaporative light scattering detector and corona charged aerosol detector have been used for lipid analysis in the past. For each technique, retention time is the key piece of information used to identify individual lipid molecular species. However, retention time alone is not enough to identify each molecular species when wide-targeted or comprehensive lipid analyses are performed. As mentioned earlier, the structures of lipids are diverse, which makes it difficult to prepare all of the targeted standards in order to confirm the retention times for each molecular species. In such cases, another piece of information is required to assist in molecular identification. Mass spectrometry (MS) can be used to determine the molecular weight and substructures of each component, which is helpful in identifying individual molecular species and for increasing the selectivity and specificity of the detection method.

MS with atmospheric pressure chemical ionization (APCI) has been used for the analysis of neutral lipids, whereas MS with electrospray ionization (ESI) enables ionization of not only neutral lipids but also polar lipids. Therefore if diverse lipids are targeted, ESI-MS should be the first candidate to perform the analysis.

In general lipid profiling by MS utilizes the information of precursor ions and product ions to identify individual molecular species. Triple quadrupole mass spectrometer (TQ MS), time-of-flight mass spectrometer (TOF MS) and Orbitrap Fourier transform mass spectrometer (Orbitrap-FT MS) have been used for lipid profiling.

TQ MS consists of two quadrupole mass analyzers, which enables the selection of precursor ions with a particular *m/z* value in the first quadrupole and product ions fragmented from the precursor ions in the third quadrupole. As a result, TQ MS can detect individual molecular species with high selectivity, and TQ MS is capable of selecting numerous precursor/product ion pairs in a single run. As such, it is a powerful tool for widely-targeted lipid profiling. However, in TQ MS, ions pass through a quadrupole with about a 1 Da window, thereby making it difficult to discriminate between isomeric molecular species.

TOF MS and Orbitrap-FT MS enable accurate mass analysis in lipid profiling, and are useful for comprehensive lipid profiling by utilizing a combination of full scan and product ion scan. These different scan modes allow for higher mass accuracy and resolution, which can narrow down the compositional formula and identify a specific substructure of molecular species. As such, these techniques provide informative results assisting in the identification of lipid molecular species. However, these MS instruments are relatively expensive and, due to their lower scan speeds produce problematic narrow peaks as compared with TQ MS.

MS is a powerful tool for lipid profiling. Because there are several MS options, it is important to choose a suitable technique based on the objectives of each lipid separation. Coupling the correct instrument with a robust separation technique further enhances the utility of MS in lipid analysis.

Liquid chromatography-mass spectrometry (LC-MS) has been utilized frequently for lipid profiling. Separation of individual molecular species by chromatography provides additional specificity and reduces the potential matrix effects in MS. However, many lipid molecular species are difficult to separate using LC, which often requires undesirable lengthy run times.

Supercritical fluid chromatography (SFC) is a promising technique for the separation of diverse lipid molecular species in a crude mixture. The high diffusivity and low viscosity of supercritical carbon dioxide ($SCCO_2$), which is used as an SFC mobile phase, enables the separation of lipid molecular species with high speed and resolution. Using gradient elution with carbon dioxide and methanol modifier is useful for separating lipids with a wide range of polarities. This chapter focuses on the applications of SFC in the field of lipid profiling.

14.2 SFC-MS METHODS FOR VARIOUS LIPID CLASSES

14.2.1 Fatty Acids

Fatty acids are involved in the regulation of many metabolic processes. Their functions differ depending on the length of carbon chain, the degree of unsaturation, and the position and configuration of double bonds. Therefore the separation of individual molecular species of fatty acids is crucial for metabolic analyses.

SFC facilitates the separation of free fatty acids based on their carbon chain length, degree of unsaturation, and double bond configurations using nonend capped octadecylsilyl (ODS) columns with methanol modifier containing 0.1% (v/v) formic acid [2]. Notably, isomeric species with different double bond positions, which cannot be separated using reversed-phase liquid chromatography (RPLC), are separated successfully using the SFC method (Fig. 14.1). For example, the ω-6 family is proinflammatory, while the ω-3 family is antiinflammatory. Therefore chromatographic separation of such isomers with opposite biological functions allows for their detailed metabolic analyses. Gas chromatography is also capable of separating isomers of some fatty acids; however, the analytes must be derivatized prior to analysis. However, this time-consuming and complicated derivatization process is not required for SFC-MS analysis.

14.2.2 Acylglycerols

Acylglycerols are the major component of edible and plant oil, and also play important roles in energy storage through the β-oxidation of fatty acids. Therefore determination of the composition and distribution of fatty acids esterified to the glycerol backbone is a crucial task. Due to the hydrophobicity of CO_2, SFC is suitable for the high-throughput separation of triacylglycerols (TAGs), diacylglycerols (DAGs),

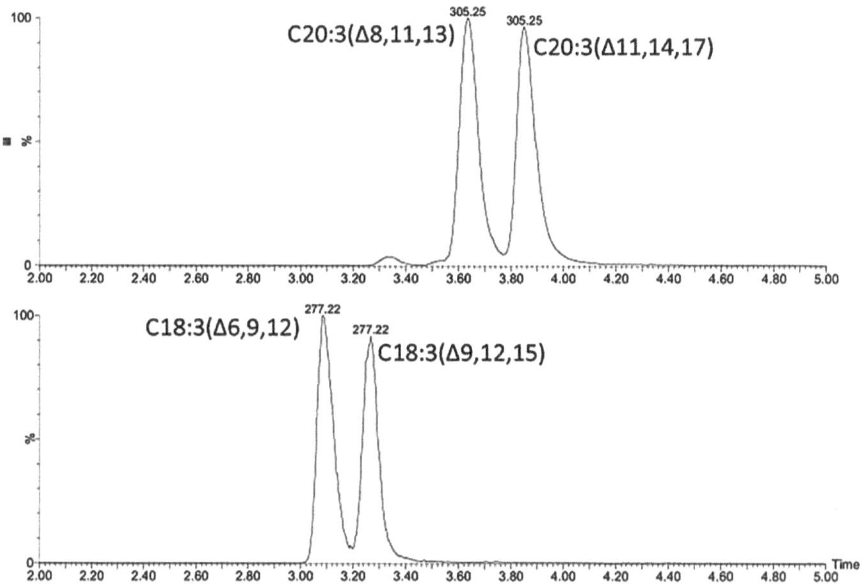

FIGURE 14.1 Extracted ion chromatogram showing the separation of isomeric free fatty acid molecular species based on the chain position of the double bonds. *Source: Reproduced from Ref. [2] with permission.*

monoacylglycerols (MAGs), monogalactosyldiacylglycerols (MGDGs), and digalactosyldiacylglycerols (DGDGs) [3,4]. These molecular species are separated based on their carbon chain lengths and degrees of unsaturation using ODS or phenyl columns (Fig. 14.2) [3,5]. Notably, C30 columns enable the high-resolution separation of regioisomeric TAG molecular species with the same fatty acid moiety in different positions along the glycerol backbone (Fig. 14.3) [6]. The analysis time is also shorter than LC.

Normal-phase columns are also applicable for the separation of TAGs. TAGs are poorly separated in normal-phase LC because they are highly hydrophobic and are poorly retained by the columns. They are, however, well resolved using a 2-etylpyridine (2-EP) column in SFC (Fig. 14.4) [7]. In this method, TAGs with longer carbon chains on their fatty acyl moieties exhibit longer retention times (RTs), as they do on reversed-phase columns. In contrast, as the degree of unsaturation in the fatty acyl moieties increases, RTs become longer, and the elution order is opposite that of reversed-phase columns. The degree of unsaturation often increases with the increased number of carbon chain length on fatty acyl moieties: therefore, many molecular species coelute when a reversed-phase column is used. As such, 2-EP columns provide favorable selectivity for resolving TAGs from crude mixtures in biological samples.

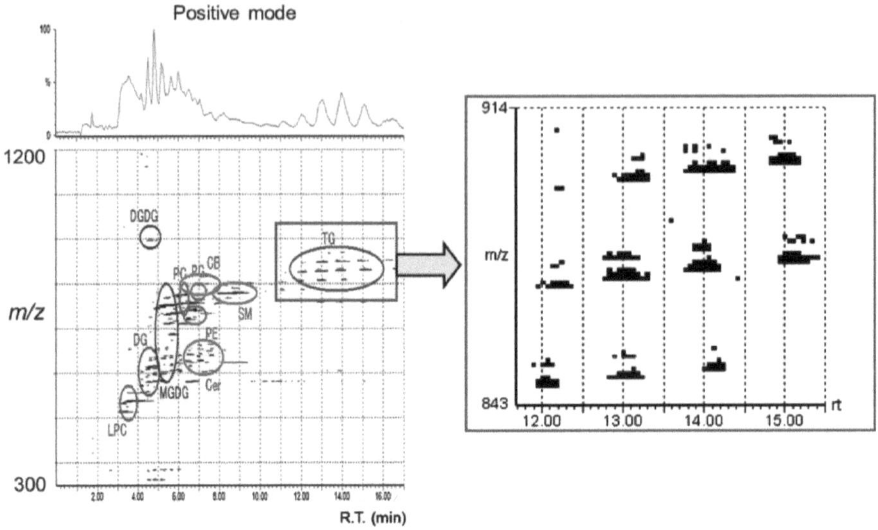

FIGURE 14.2 Base peak ion (BPI) chromatograms and two-dimensional (2D) displays obtained from SFC-MS analysis of a standard mixture using an ODS column.

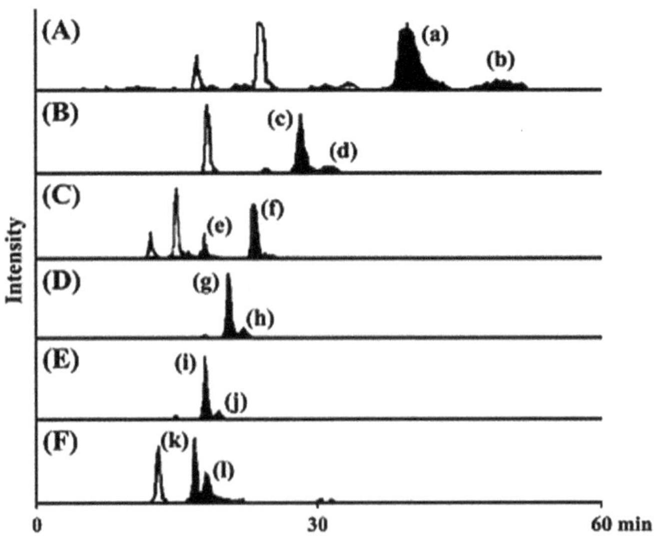

FIGURE 14.3 MRM chromatograms of six regioisomeric TAG pairs ((A) TAG 54:1, (B) TAG 52:1, (C) TAG 52:3, (D) TAG 50:1, (E) TAG 50:2, (F) TAG 50:3) in palm and canola oils. (a) TAG 18:0/18:1/18:0, (b) TAG 18:0/18:0/18:1, (c) TAG 18:0/18:1/16:0, (d) TAG 18:0/16:0/18:1, (e) TAG 18:0/18:3/16:0, (f) TAG 18:0/16:0/18:3, (g) TAG 16:0/18:1/16:0, (h) TAG 16:0/16:0/18:1, (i) TAG 16:0/18:2/16:0, (j) TAG 16:0/16:0/18:2, (k) TAG 16:0/18:3/16:0, and (l) TAG 16:0/16:0/18:3. *Source: Reproduced from Ref. [6] with permission.*

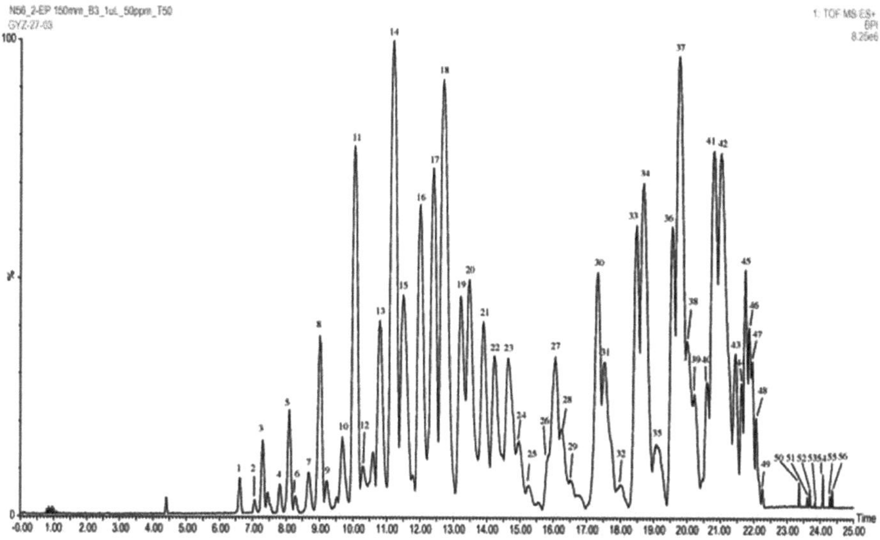

FIGURE 14.4 Base peak ion (BPI) chromatogram of triacylglycerols and diacylglycerols in a lipid extract of cow milk fat. *Source: Reproduced from Ref. [7] with permission.*

SFC can also be used to identify chloropropanol fatty acid esters, which are structural analogs of acylglycerols that exist in edible oils [8]. SFC facilitates the rapid and simultaneous separation of mono-, di-, and tri-esters, which are difficult to separate in a single LC run.

14.2.3 Phospholipids and Sphingolipids

Phospholipids and sphingolipids play crucial roles in cellular signaling and as the major components of cell membranes. The combination of substructures including polar head groups and fatty acyl moieties enhances structural diversity and complexity of molecular species, resulting in several classes: glycerophospholipids, sphingoid bases, ceramides, phosphosphingolipids, and glycosphingolipids. In addition, oxidized phospholipids are produced by oxidative stress. There are a variety of molecular species with different functional groups such as hydroxides, epoxides, and hydroperoxides with a large number of structural and positional isomers. Therefore analytical methods that facilitate the separation of diverse molecular species with a wide range of polarities are required for profiling phospholipids and sphingolipids. SFC is ideal for polar lipid profiling because the polarity of the mobile phase can be adjusted by changing the ratio of CO_2 and modifier.

Two separation modes are applicable to profiling of phospholipids and sphingolipids. These polar lipid molecular species are separated

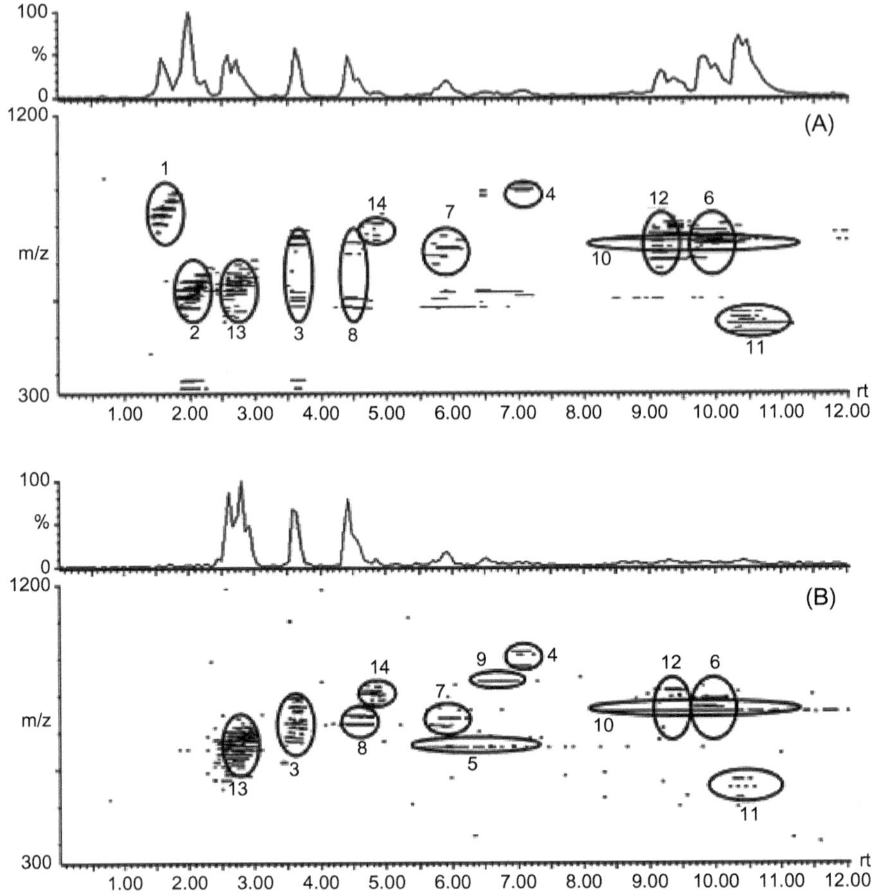

FIGURE 14.5 Base peak ion (BPI) chromatograms and 2D displays obtained from SFC-MS analysis of a standard mixture using a cyano column. 1, Triacylglycerol (TAG); 2, diacylglycerol (DAG); (3) monogalactosyldiacylglycerol (MGDG); 4, digalactosyldiacylglycerol (DGDG); 5, phosphatidic acid (PA); 6, phosphatidylcholine (PC); 7, phosphatidylethanolamine (PE); 8, phosphatidylglycerol (PG); 9, phosphatidylinositol (PI); 10, phosphatidylserine (PS); 11, lysophosphatidylcholine (LPC); 12, sphingomyelin (SM); 13, ceramide (Cer); and 14, cerebroside (CB). *Source: Reproduced from Ref. [3] with permission.*

according to their polar head groups using a normal-phase column (Figs. 14.2 and 14.5), while they are separated according to their fatty acyl moieties using a reversed-phase column [3]. With a polar-embedded ODS column, however, polar lipid molecular species are resolved based on both their polar head groups and their fatty acyl moieties (Fig. 14.6) [9]. Thus SFC enables high-resolution separation of polar lipid molecular species in a short time using only a single column, while longer time is required for two-dimensional LC, which employs two columns.

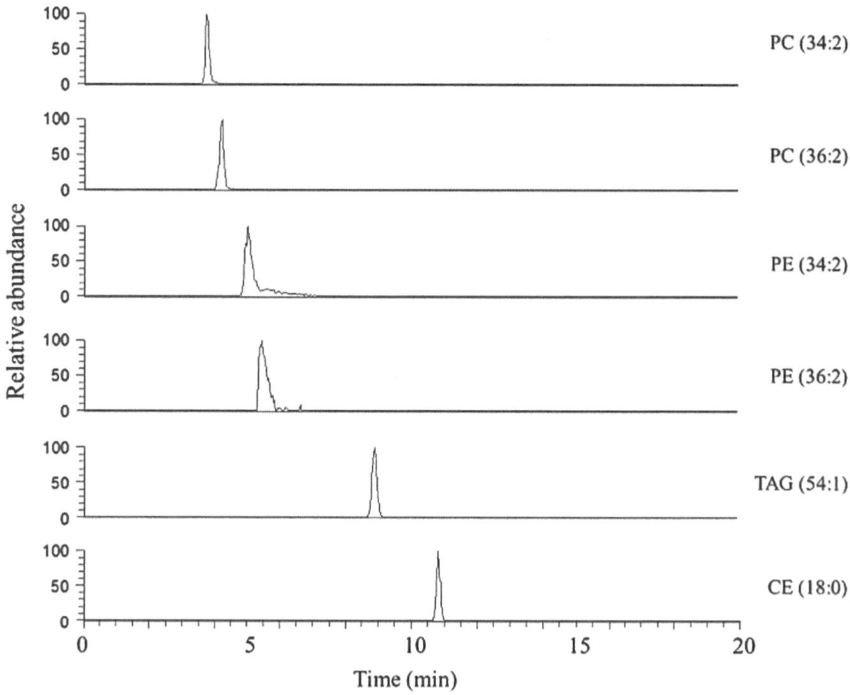

FIGURE 14.6 Extracted ion chromatograms obtained from SFC-MS analysis of a lipid standard mixture using a polar-embedded ODS column. *PC*, phosphatidylcholine; *PE*, phosphatidylethanolamine; *TAG*, triacylglycerol; *CE*, cholesteryl ester. *Source: Reproduced from Ref. [9] with permission.*

SFC is also eminently suited for the separation of isomeric oxidized phospholipids. Positional isomers of epoxides, which are difficult to separate using LC, have been separated successfully using a 2-EP column in SFC (Fig. 14.7) [10]. It is presumed that the interactions between the positively-charged column stationary phase and the negatively-charged oxidized phospholipids contributed to the good separation. Thus SFC again enables the precise identification of individual isomeric molecular species.

Several classes of phospholipids and sphingolipids including phosphatidylserine (PS), phosphatidic acid (PA), and ceramide-1-phosphate (C1P), often exhibit tailing peaks because of the interactions between the analytes and the column stationary phase are too strong. This issue can be resolved by silylation or methylation of the polar functional groups (Figs. 14.8 and 14.9), resulting in sharper peaks and an improved separation of individual species and easier detection [11,12].

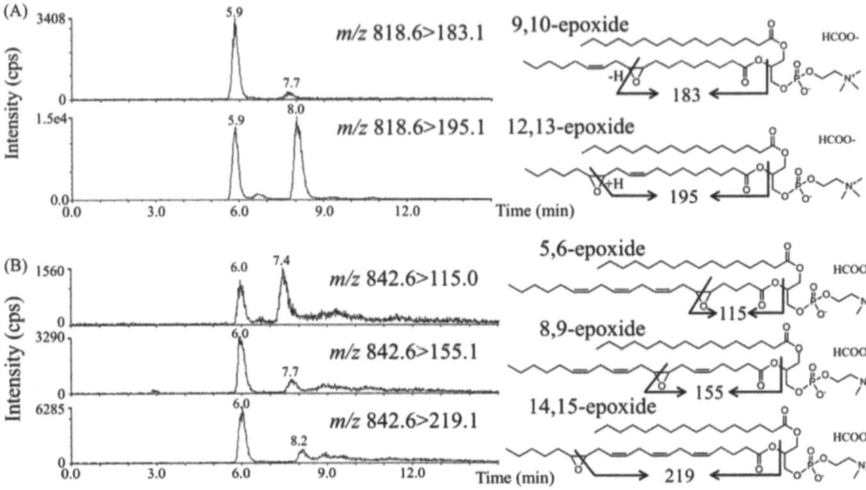

FIGURE 14.7 MRM chromatograms of positional isomers of (A) palmitoyl-linoleoyl- and (B) palmitoyl-arachidonyl-phosphatidylcholine (PC) epoxides. *Source: Reproduced from Ref. [10] with permission.*

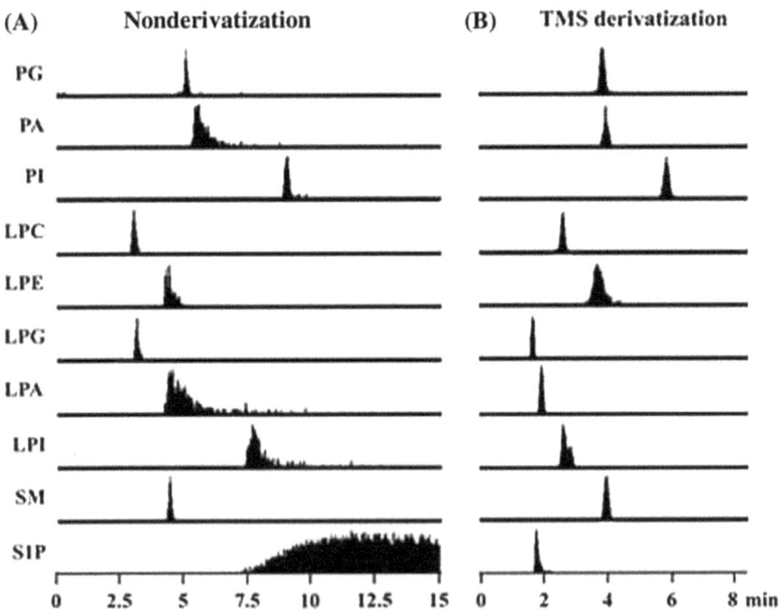

FIGURE 14.8 MRM chromatograms of polar lipids in a mixture of standard samples (B) with and (A) without trimethylsilyl (TMS) derivatization. *Source: Reproduced from Ref. [11] with permission.*

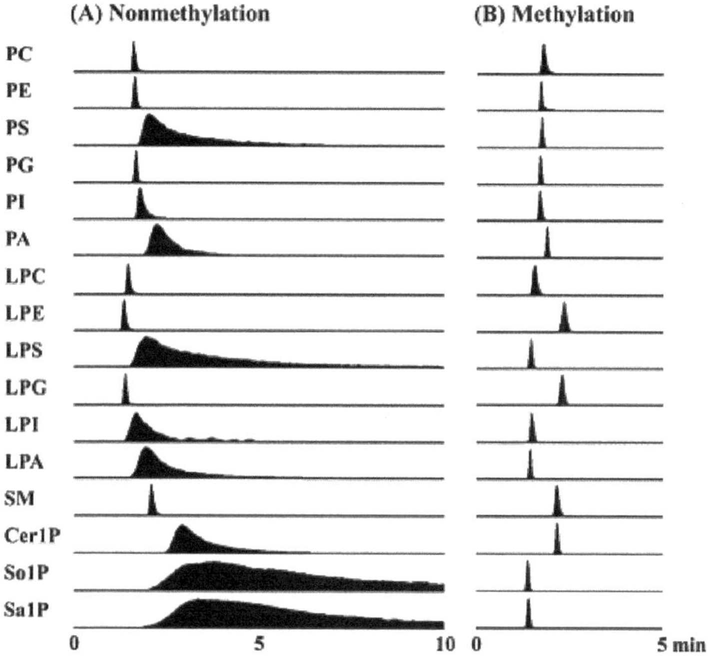

FIGURE 14.9 MRM chromatograms of polar lipids in a mixture of standard samples (B) with and (A) without methylation. *PC*, phosphatidylcholine; *PE*, phosphatidylethanolamine; *PS*, phosphatidylserine; *PG*, phosphatidylglycerol; *PI*, phosphatidylinositol; *PA*, phosphatidic acid; *LPC*, lysophosphatidylcholine; *LPE*, lysophosphatidylethanolamine; *LPS*, lysophosphatidylserine; *LPG*, lysophosphatidylglycerol; *LPI*, lysophosphatidylinositol; *LPA*, lysophosphatidic acid; *SM*, sphingomyelin; *Cer1P, ceramide-1-phosphate; So1P, sphingosine-1-phosphate; Sa1P, sphinganine-1-phosphate. Source: Reproduced from Ref. [12] with permission.*

14.2.4 Sterol Lipids and Prenol Lipids

Sterol lipids, such as cholesterol and its derivatives contain a four-membered hydrocarbon core called as cyclopentenoperhydrophenanthrene, and have different biological roles as signaling molecules and are important components of membrane lipids.

Prenol lipids are synthesized from the 5-carbon precursors; isopentenyl diphosphate and dimethylallyl diphosphate. Two pathways for these precursors are known but they are mainly produced through the mevalonic acid pathway. Carotenoids functioning as antioxidants, and Vitamins E, K, and the ubiquinones are examples of this class.

SFC has been used for the analysis of sterol and prenol lipids. Steroids have a broad spectrum of polarity but they do have some structural similarities. This is one of the reasons that it is difficult to separate various steroids simultaneously. Since SFC is capable simultaneously separating compounds of a wide range of polarity, it has also proven suitable for steroid analysis.

Scalia et al. demonstrated the separation of bile acids by SFC in the 1990s and reported the separation of five unconjugated bile acids using a cyanopropylsiloxane-bonded silica column or a phenyl column. The simultaneous separation of estrone, estradiol, estriol, and their metabolites is difficult because of the large number of estrogens with a similar structure. However, Xu et al. were able to separate 15 estrogen metabolites using SFC with a cyanopropylsiloxane-bonded silica column connected in series with a diol column, both 2.1 mm i.d. × 150 mm long, packed with 5 μm particles [13]. Despite having two columns in series, the flow rate was 2 mL/min resulting in a separation time less than 10 min. Because the mobile phase is CO_2, which is endowed with low viscosity and high diffusivity, SFC allows for the use of longer columns, or multiple columns in tandem, with little flow rate restriction, thereby promoting throughput and separation. In the case of the estrogen analysis, the throughput was further improved using sub-2-μm particles and a charged surface hybrid column with pentafluorophenyl bonded ligands, providing ultra-fast separation of nine estrogens within 1.6 minutes [14].

The previously mentioned analysis targeted only unconjugated estrogens. Other studies have demonstrated that SFC is suitable for the separation of conjugated steroids. Conjugation is a crucial metabolic reaction and regulatory mechanism controlling the concentration of steroids with strong physiological activity. Therefore it is important to profile conjugated steroids. Scalia et al. reported the separation of eight conjugated bile acids by an ODS column in the 1990s. In the 2010s Doue et al. developed SFC methods for the analysis of two conjugated steroids, glucuronide, and sulfate esters of androstane and estrane including their isomeric forms. Ten sulfate-conjugated forms were separated on a 2-EP column, while, eight glucuronide-conjugated forms were separated on a silica column.

These studies clearly indicated that SFC is suitable for steroid analysis, but none of these studies achieved simultaneous analysis for both unconjugated and conjugated forms. However, a further study revealed that SFC has the potential for the simultaneous analysis of steroids including both unconjugated and conjugated forms [15].

Another study was the first SFC-ESI-MS/MS method used for the simultaneous profiling of 25 bile acids by SFC as shown in Fig. 14.10, which analyzed both conjugated and unconjugated bile acids in a single run. This method was also applied to rat serum analysis as shown in Fig. 14.11.

SFC has long been considered a suitable analysis technique for hydrophobic compounds due to its use of a low-polarity mobile phase. This study showed that the simultaneous analysis of bile acids,

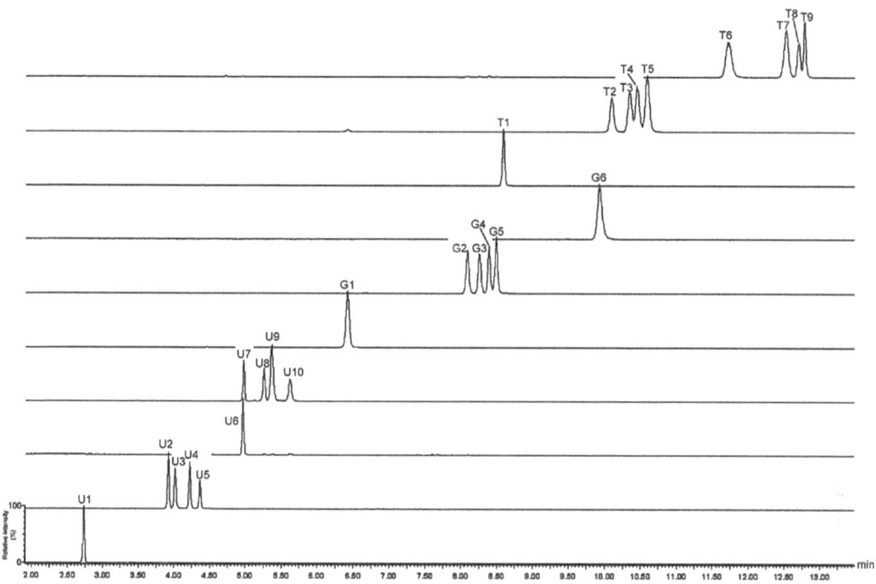

FIGURE 14.10 MRM chromatogram of a 25 bile acid standard mixture. U1, LCA; U2, DCA; U3, CDCA; U4, UDCA; U5, HDCA; U6, 7-oxo-DCA; U7, βMCA; U8, ωMCA; U9, CA; U10, αMCA; G1, GLCA; G2, GDCA; G3, GCDCA; G4, GUDCA; G5, GHDCA; G6, GCA; T1, TLCA; T2, TDCA; T3, TCDCA; T4, TUDCA; T5, THDCA; T6, TβMCA; T7, TωMCA; T8, TCA; T9, TαMCA. Gradient condition: 5–25% (4.5 min), 25% (1.5 min), 25–37.5% (2.5 min), 37.5% (1 min), 37.5–40% (1 min), 40% (1.5 min), 40–50% (0.5 min), 50% (1 min), 50–5% (0.5 min), 5% (1 min). Flow rate: 2.0 mL/min, Column Temperature: 70°C, BPR: 13.8 MPa. *Source: Reproduced from Ref. [15] with permission.*

including conjugates, was possible when method conditions such as column, modifier, and additives were optimized. SFC is not only beneficial for the analysis of sterol lipids, but it is also applicable to the simultaneous analysis across a wide range of polarity. Recently SFC was used to analyze fat-soluble vitamins (FSV) including prenol lipids as well as water-soluble vitamins (WSV).

Vitamins are essential dietary nutrients required for the normal growth, self-maintenance, and functioning of humans. Vitamins are classified into two groups, FSV and WSV, based on their solubility. The two groups have clearly different water solubility, in that their log *P* values range from −2 to 10 as shown in Table 14.1. This vast polarity difference makes it difficult to analyze FSV and WSV simultaneously.

However, SFC was able to analyze both FSV and WSV in a single run [16]. In this approach, the gradient started at almost 100% CO_2 and reached and maintained at 100% methanol at the end, thereby allowing even further extension of the compound polarity range in a single run.

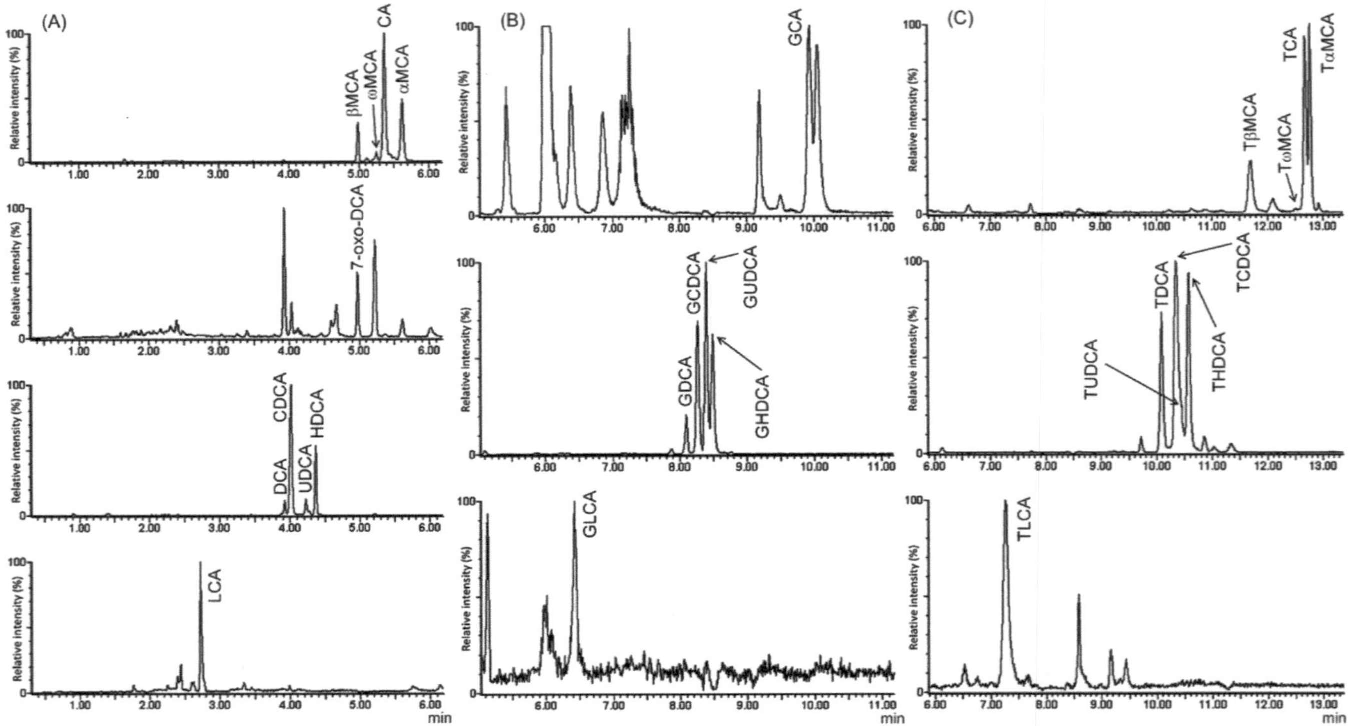

FIGURE 14.11 MRM chromatograms of bile acids in rat serum. MRM data shown peaks of (A) unconjugated bile acids, (B) glycine conjugates, and (C) taurine conjugates. *Source: Reproduced from Ref. [15] with permission.*

TABLE 14.1 Classification of Vitamins

	Vitamin	Class	Log *P*
1	Retinol	FSV	6.38
2	Retinyl acetate (A acetate)	FSV	6.56
3	Retinyl palmitate (A palmitate)	FSV	10.12
4	α-Tocopherol	FSV	7.59
5	Ergocalciferol (D2)	FSV	8.84
6	K1	FSV	8.48
7	K2	FSV	9.01
8	β-Carotene	FSV	9.72
9	Tiamine	WSV	− 2.11
10	Riboflavin	WSV	− 1.05
11	Nicotinic acid	WSV	0.29
12	Niacinamide	WSV	− 0.45
13	D-Pantothenic acid	WSV	− 1.12
14	Pyridoxine	WSV	− 0.57
15	Biotin	WSV	0.17
16	Cyanocobalamin (B12)	WSV	0.67
17	Ascorbic acid (VC)	WSV	− 1.58

Reproduced from Ref. [16] with permission.

As shown in Fig. 14.12, FSV and WSV were simultaneously analyzed on a C18 column without end capping. An octadecylsiloxane-bonded silica column was employed for hydrophobic interactions, and FSV was retained as well. The column is also able to retain hydrophilic compounds with residual silanol groups on a nonend capped stationary phase.

Prenol lipid separations are well studied in SFC. Matsubara et al. developed profiling methods for carotenoids and their epoxidized products, which are potential oxidative stress makers, and applied the method to serum samples [17,18]. A fast tocopherol separation method that can also separate isomeric forms was reported and used for human serum analysis [19]. A method for the determination of prenol lipids in food was also reported [20–22].

14.2.5 Polyketides

Citrus polymethoxyflavones are plant-occurring flavonoids, and characterized by the presence of several methoxy groups. Isolation

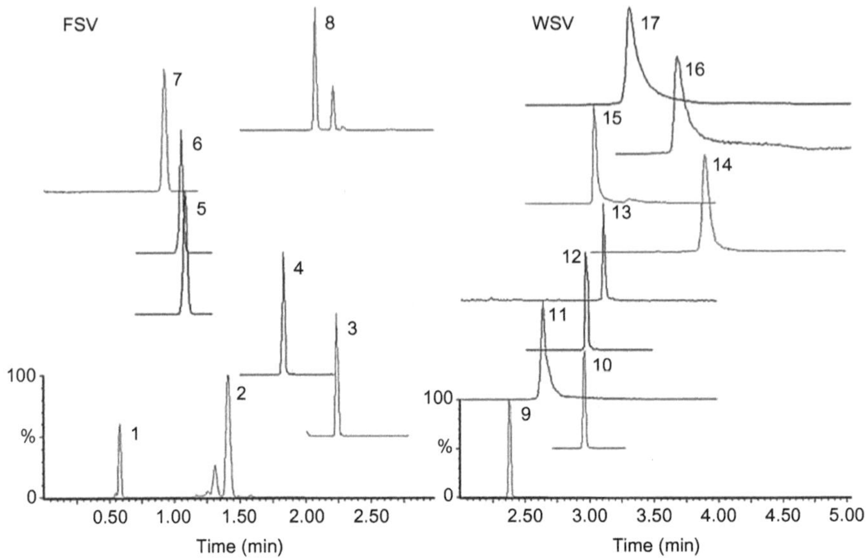

FIGURE 14.12 MRM chromatograms of 17 vitamins. 1, A acetate; 2, A palmitate; 3, D2; 4, α-Tocopherol; 5, K2; 6, K1; 7, α-Tocopherol acetate; 8, β-Carotene; 9, Nicotinamide; 10, Nicotinic acid; 11, Pyridoxine; 12, D-Pantothenic acid; 13, Biotin; 14, Thiamine; 15, Riboflavin; 16, B12; 17, VC. Method conditions as follows; modifier: methanol/water (95/5, v/v) with 0.2% ammonium formate; gradient condition: 2% (0.5 min), 2–30% (2.0 min), 30–85% (0.8 min), 85% (2.7 min), 85–100% (0.2 min), 100% (1.3 min), 100–2% (1 min), 2% (1.5 min); flow rate: 1.2 mL/min at a column temperature of 40°C; back pressure: 15.2 MPa (6.0 min), 15.2–10.3 MPa (0.2 min), 10.3 MPa (1.6 min), 10.3–15.2 MPa (0.5 min), 15.2 MPa (1.7 min). *FSV*, fat-soluble vitamin; *WSV*, water-soluble vitamin *Source: Reproduced from Ref. [16] with permission.*

of these molecules is essential because they have positive effects on human cells, including prevention of inflammation and mutagenicity. SFC can be utilized for the analysis of the polymethoxyflavones with CO_2-MeOH mobile phases or pure CO_2 at high column temperature (150°C) [23]. It was also demonstrated that these molecules can be extracted using CO_2 and ethanol at high temperature (80°C) and pressure (30 MPa).

14.3 ONLINE SFE-SFC-MS METHODS FOR LIPID PROFILING

14.3.1 Online Coupling of SFE and SFC

Supercritical fluid extraction (SFE) is suitable for rapid and efficient recovery of hydrophobic compounds with high selectivity due to the low polarity and high diffusivity of CO_2. In addition, the consumption of organic solvents can be decreased as compared to conventional liquid–liquid extraction methods. SFE has been employed for the extraction of carotenoids and fatty acids [24,25]. It can also be utilized for hydrophilic metabolites by using a higher ratio of modifier during the extraction [26].

Online SFE-SFC-MS analysis facilitates continuous procedures for extraction, separation, and detection in a short time without photolytic degradation and oxidation. This means that online SFE-SFC-MS is eminently suitable for the high-throughput analysis of thermally labile and oxidizable hydrophobic compounds. In this section, applications of online SFE-SFC-MS methods in lipid analysis are reviewed.

14.3.2 Highly Accurate Redox Status Analysis of Coenzyme Q_{10}

Coenzyme Q_{10} (CoQ_{10}) plays a crucial role in the mitochondrial respiratory chain as an enzyme cofactor. It exists in a reduced form (ubiquinol-10) and an oxidized form (ubiquinone-10); the redox status (i.e., ratio of these two forms) of CoQ_{10} is correlated with several diseases. Therefore accurate analysis of the reduced form of CoQ_{10} without oxidation is required in clinical studies.

In SFE the analytes are extracted in a closed system without oxygen and light, allowing online SFE-SFC-MS to facilitate more efficient recovery of the reduced form without degradation by oxidation as compared with offline SFE-SFC-MS (Fig. 14.13) [27]. Thus online SFE-SFC-MS is suitable for redox status analysis of CoQ_{10}.

14.3.3 High-Throughput Analysis of Phospholipids From Dried Plasma Spots

Dried plasma spots (DPS) are used for biomarker screening and diagnosis as an inexpensive blood sampling method. Phospholipid profiling is a typical application using online SFE-SFC = MS and DPS testing. Using this method, more than 130 molecular species were detected from DPS within 20 minutes, and the recovery by SFE was significantly higher than that for liquid–liquid extraction [28].

The major difficulty in the development of this method was the optimization of the extraction conditions and the separation conditions to obtain high extraction efficiency and good peak shapes. A HILIC column was selected to retain the targeted phospholipids and separate isobaric molecular species. However, the optimum extraction conditions with 15% (v/v) methanol modifier produced broad peaks because the extracted phospholipids were not retained in the column during the extraction (Fig. 14.14). Therefore extraction conditions with a lower ratio of modifier were adopted to obtain sharp peaks.

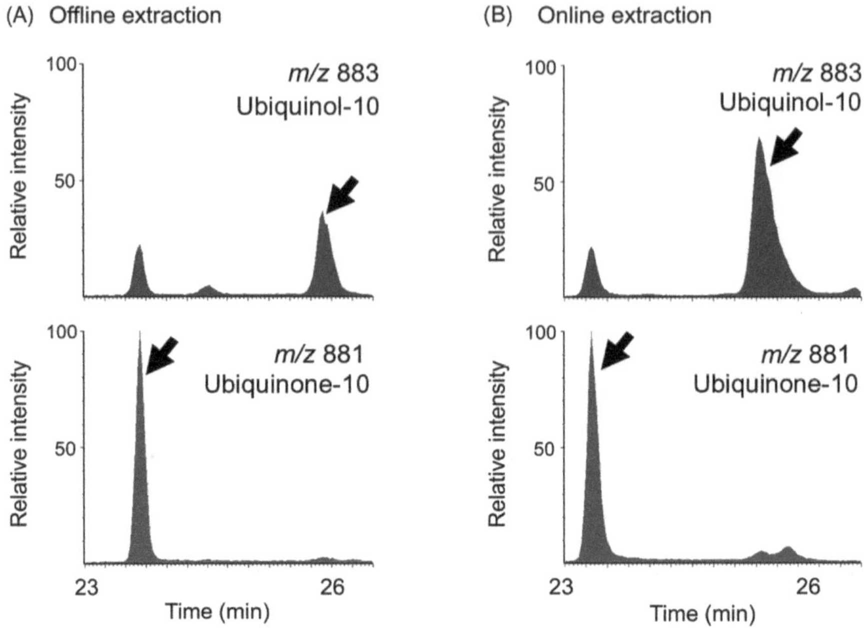

FIGURE 14.13 Extracted ion chromatograms of ubiquinone-10 and ubiquinol-10 in the photosynthesis bacterium, *Rhodobium marinum* A501 obtained by (A) offline extraction (using hexane) or (B) online extraction. *Source: Reproduced from Ref. [27] with permission.*

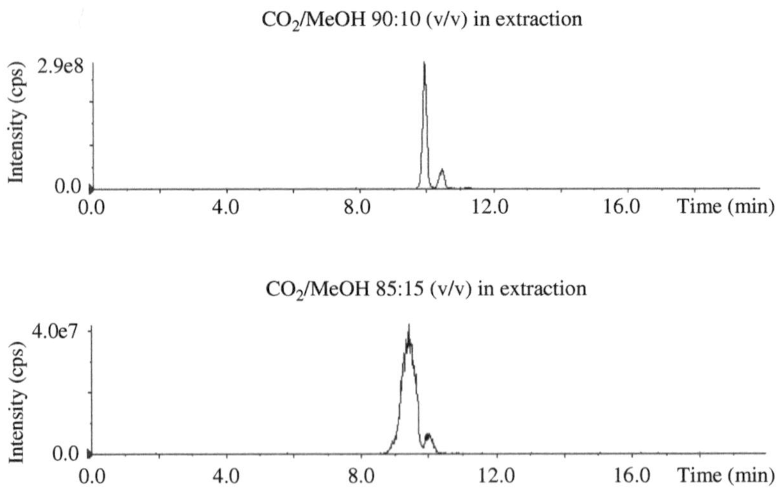

FIGURE 14.14 Extracted ion chromatograms of phospholipid molecular species with different ratios of CO_2 and methanol in extraction.

14.4 APPLICATION OF SFC-MS METHODS IN LIFE SCIENCE

Lipidome analysis using SFC-MS is applied to pathological and physiological research. SFC-MS analysis of plasma and its lipoproteins collected from normal and myocardial infarction-prone (Watanabe heritable hyperlipidemic: WHHLMI) rabbits revealed specific alterations in lipid profiles in each lipoprotein fraction [29]. Elevated levels of bioactive lipids, including alkyl-phosphatidylcholines, phospholipids possessing ω-6 fatty acids, and plasmalogens, were observed in WHHLMI rabbit plasma. In addition, the levels of phosphatidylethanolamine plasmalogens were higher in low-density lipoprotein (LDL) than in very-low-density lipoprotein (VLDL).

SFC-MS-based lipidomic profiling was also performed for plant leaves, soybeans, blood plasma, liver, and brain [3–5,9]. From the lipid extract of each sample, several hundred molecular species were identified using the combination of MS and software tools. SFC-MS is expected to be a powerful tool for high-throughput analysis of multiple samples in various research fields.

14.5 CONCLUSION AND FUTURE PERSPECTIVES

In this chapter, the applications of SFC/MS methods in lipid analyses were reviewed. SFC has an advantage in simultaneous analysis of lipids with wide-ranging polarities, high-resolution separation of isomeric molecular species, and high-throughput analysis of complex lipid mixtures. Online coupling of SFE and SFC also enables faster analyses and more accurate quantification of thermally labile and oxidizable metabolites.

Elucidation of detailed mechanisms of separation in SFC would be useful for method development for other targets, and separation of isomers and structural analogs would be enhanced by combining several columns with different stationary phases.

References

[1] Fahy E, Cotter D, Sud M, Subramaniam S. Biochim Biophys Acta 2011;1811:637.
[2] Ashraf-Khorassani M, Isaac G, Rainville P, Fountain K, Taylor LT. J Chromatogr B 2015;997:45.
[3] Bamba T, Shimonishi N, Matsubara A, Hirata K, Nakazawa Y, Kobayashi A, et al. J Biosci Bioeng 2008;105:460.
[4] Lísa M, Holcapek M. Anal Chem 2015;87:7187.
[5] Lee JW, Uchikata T, Matsubara A, Nakamura T, Fukusaki E, Bamba T. J Biosci Bioeng 2012;113:262.

[6] Lee JW, Nagai T, Gotoh N, Fukusaki E, Bamba T. J Chromatogr B Analyt Technol Biomed Life Sci 2014;966:193.
[7] Zhou Q, Gao B, Zhang X, Xu Y, Shi H, Yu LL. Food Chem 2014;143:199.
[8] Hori K, Matsubara A, Uchikata T, Tsumura K, Fukusaki E, Bamba T. J Chromatogr A 2012;1250:99.
[9] Yamada T, Uchikata T, Sakamoto S, Yokoi Y, Nishiumi S, Yoshida M, et al. J Chromatogr A 2013;1301:237.
[10] Uchikata T, Matsubara A, Nishiumi S, Yoshida M, Fukusaki E, Bamba T. J Chromatogr A 2012;1250:205.
[11] Lee JW, Yamamoto T, Uchikata T, Matsubara A, Fukusaki E, Bamba T. J Sep Sci 2011;34:3553.
[12] Lee JW, Nishiumi S, Yoshida M, Fukusaki E, Bamba T. J Chromatogr A 2013;1279:98.
[13] Xu X, Roman JM, Veenstra TD, Van Anda J, Ziegler RG, Issaq HJ. Anal Chem 2006;78:1553.
[14] Nováková L, Chocholouš P, Solich P. Talanta 2014;121:178.
[15] Taguchi K, Fukusaki E, Bamba T. J Chromatogr A 2013;1299:103.
[16] Taguchi K, Fukusaki E, Bamba T. J Chromatogr A 2014;1362:270.
[17] Matsubara A, Bamba T, Ishida H, Fukusaki E, Hirata K. J Sep Sci 2009;32:1459.
[18] Matsubara A, Uchikata T, Shinohara M, Nishiumi S, Yoshida M, Fukusaki E, et al. J Biosci Bioeng 2012;113:782.
[19] Pilařováa V, Gottvalda T, Svobodaa P, Novákb O, Benešovác K, Bělákovác S. Anal Chim Acta 2016;934:252.
[20] Geea PT, Liewb CY, Thongc MC, Gayd MCL. Food Chem 2016;196:367.
[21] Li B, Zhao H, Liu J, Liu W, Fan S, Wu G, et al. J Chromatogr A 2015;1425:287.
[22] Jumaah F, Plaza M, Abrahamsson V, Turner C, Sandahl M. Anal Bioanal Chem 2016;408:5883.
[23] Hadj-Mahammed M, Badjah-Hadj-Ahmed Y, Meklati BY. Phytochem Anal 1993;4:275.
[24] Cao X, Ito Y. J Chromatogr A 2003;1021:117.
[25] Careri M, Furlattini L, Mangia A, Musci M, Anklam E, Theobald A, et al. J Chromatogr A 2001;912:61.
[26] Matsubara A, Izumi Y, Nishiumi S, Suzuki M, Azuma T, Fukusaki E, et al. J Chromatogr B 2014;969:199.
[27] Matsubara A, Harada K, Hirata K, Fukusaki E, Bamba T. J Chromatogr A 2012;1250:76.
[28] Uchikata T, Matsubara A, Fukusaki E, Bamba T. J Chromatogr A 2012;1250:69.
[29] Takeda H, Koike T, Izumi Y, Yamada T, Yoshida M, Shiomi M, et al. J Biosci Bioeng 2015;120:476.

CHAPTER

15

Separation of Natural Products

M. Ganzera and A. Murauer
University of Innsbruck, Innsbruck, Austria

OUTLINE

15.1 INTRODUCTION

Without doubt natural products are indispensable for healthcare and pharmaceutical research because they either represent bioactive constituents themselves or serve as a starting point for further product development. In this respect, it was reported that the majority (about 80%) of commercial drugs are derived from natural products [1]. Lachance et al. even termed them "privileged structures." Their biosynthesis proceeds with sequential binding of the biosynthetic intermediates to different proteins, and therefore natural products more likely have the potential

Supercritical Fluid Chromatography.
DOI: http://dx.doi.org/10.1016/B978-0-12-809207-1.00015-X

to alter their function, rendering them as promising leads for drug discovery [2]. Natural products are derived from primary or secondary metabolism, and their number and structural diversity is almost endless. The latter varies from rather simple monoterpenes in essential oils to complex structures like ginkgolides in *Ginkgo biloba* leaves. Additionally, herbal extracts usually contain a multitude of constituents including bioactive compounds, which are often present at low concentration only. A natural variation even within the same species, depending on factors like collecting season, soil or climatic conditions, further complicates their analysis. All these facts indicate that any technique suitable for natural product analysis has to meet high demands in respect to selectivity, versatility, and efficiency. The most widely used technique in this area of research is high-performance liquid chromatography (HPLC), often in combination with mass spectrometric (MS) detection [3]. It fulfills all requirements with impressive sensitivity and speed, especially if sub-2-μm particles are used as the stationary phase. Other analytical techniques commonly used include gas chromatography (GC) and capillary electrophoresis (CE); however, even if they are sometimes the methods of choice their application is limited to a large extent by the properties of the analytes (is the compound volatile, ionizable, etc.) [4]. Among the more exotic options for natural products analysis are capillary electrochromatography (CEC) and supercritical fluid chromatography (SFC). The first is a hybrid technique of chromatography and electrophoresis; but its practical relevance is limited because the required stationary phases (capillaries) are virtually not commercially available [5]. On the other hand, even if it is still on a manageable level, the use of SFC has steadily increased over the last few years [6]. This trend is a clear indication of recent technical advances and the general benefits of this separation technique. To confirm these observations, this chapter summarizes the most relevant and interesting SFC applications for natural products analysis, particularly focusing on compounds from (medicinal) plants. The separation of lipids is not covered, because this topic is addressed in another chapter.

15.2 SELECTED APPLICATIONS

15.2.1 Alkaloids

One of the first reports on the separation of alkaloids by SFC was published in 1987 [7]. In this study Holzer and coworkers used a typical GC column (a 10 m × 50 μm fused silica capillary coated with a 0.25-μm film of cross-linked polysiloxane or SB-biphenyl 30). Pure CO_2 served as a mobile phase (130°C, pressure gradient from 100 to 280 bar at 3 bar/min, and mobile phase velocity 1.84 cm/s) in combination with a flame ionization detector (FID). This setup resembles the OTC (open tubular column) variant of SFC, which was popular in the earlier days of SFC but is rarely used today. The separation time of 80 minutes was long, yet up to eight pyrrolizidine alkaloids in *Senecio anonymus* were separated and the method had good sensitivity (LOD values from 1 to 3 ng on-column). Only a year later packed columns were used for the separation of opium alkaloids (narcotine, papaverine, thebaine, codeine, cryptopine, and morphine), including their identification in poppy straw extracts [8]. Aminopropylsiloxane-bonded silica with a particle size of 10 μm in combination with a mobile phase of CO_2, methanol, methylamine, and water (82.95/16.25/0.50/0.30) enabled their separation in less than 3 minutes, but the authors preferred bare silica material (LiChrosorb Si 60, 5 μm) due to better resolution. The separation time increased to 11 minutes, but this was still half the time required for HPLC procedures then in use.

When turning to more recent literature the number of relevant publications is rather small. Even so, interesting and innovative technical options are reported. Three studies emphasized the use of SFC-MS, one describing the determination of sesquiterpene pyridine alkaloids in the root bark of *Tripterygium wilfordii* [9]. This plant, also known as thunder god vine, plays an important role in traditional Chinese medicine to treat fever, edema, and carbuncles. By SFC-DAD-MS/MS it was possible to identify more than 70 alkaloids, belonging to several subtypes (wilfordate/evoninate, *iso*-wilfordate/evoninate, or hydroxyl-wilfordate/evoninate). For separation on an ethylene-bridged hybrid material (Aquity UPC^2 BEH-2EP) a CO_2-methanol gradient was employed, and MS detection was performed in the positive electrospray ionization (ESI) mode, to monitor a mass range from 200 to 500 Da. Zhao et al. reported the quantitative analysis of five toxic alkaloids in *Aconitum pendulum* by SFC-MS [10]. Stationary and mobile phases were the same as mentioned earlier, only with 10 mM ammonium acetate added to the methanol to improve peak shape and ionization. With a separation time of 4 minutes and LOD values ranging from 0.03 to 0.07 ng/mL this approach is an excellent alternative to established HPLC methods. Especially since all validation criteria were excellently met and highly reproducible quantitative results were obtained. Aconitine was identified as the dominant alkaloid in the plant with a concentration up to 154 μg/g. The third SFC-MS publication focused on the determination of atropine in *Atropa belladonna* extracts [11]. Optimum resolution was obtained using a mixture of supercritical CO_2 with 15% methanol, 0.5% trifluoroacetic acid, and 0.5% diethylamine on a 5 μm cyanopropylsiloxane-bonded silica sorbent (Spherisorb) as the stationary phase. The preferred detection was by APCI (atmospheric pressure chemical ionization) in the positive mode, resulting in an LOD of 700 pg for atropine (10 μL injected). This study was published more than 15 years ago, and it is possible that it no longer reflects the state-of-the-art concerning instrumental design or column technology.

Other reports describe the use of an imprinted polymer for the separation of ephedrine isomers [12] or the hyphenation of SFC and UPLC [13]. The latter is probably of more practical relevance, because it was used for the qualitative analysis of amine alkaloids in pepper (*Piper longum*). Due to their structural similarity high column orthogonality was an objective of this study. A combination of an XAmide column (150 × 4.6 mm, 5 μm, Acchrom, first dimension: SFC) with an HSS T3 column (100 × 2.1 mm, 1.8 μm, Waters, second dimension: UPLC) was selected. Both were hyphenated off-line, meaning the SFC effluent was fractionated manually in 0.5 minute intervals, the dried fractions re-dissolved and then analyzed by liquid chromatography. Although elaborate, the authors considered several aspects of their method advantageous: it has high resolving capabilities not achievable with one-dimensional separations, it is upgradable to the preparative scale, and it permits the detection of low-abundance compounds.

15.2.2 Carotenoids

Due to their nonpolar character carotenoids are ideal targets for SFC analysis. They can be divided in two major groups, xanthophylls (molecules containing oxygen) and carotenes (structures only based on carbon and hydrogen), and are found as natural fat-soluble pigments in algae, plants, and microorganisms. They cannot be synthetized by humans, thus their sufficient supply through food (carrots, oranges, tomatoes, etc.) is relevant for several reasons. First, derivatives like α- and β-carotene, or β-cryptoxanthin are vitamin A precursors, and second, they possess a number of desired pharmacological effects. They are antioxidants and therefore beneficial for the prevention of cancer, sclerosis, Alzheimer's disease, or skin aging. Their immense ecological relevance for photosynthesis and their role in the coloration of birds and reptiles should not be forgotten as well.

Early SFC studies utilized carotenoids mainly for methodological investigations, covering topics like the impact of temperature, pressure, or modifier on the separation of *cis/trans*-isomers [14] or the effects of different stationary phases on selectivity and retention [15,16]. They describe the use of regular HPLC sorbents (RP18, 3 or 5 μm particle size) from different sources with the following conclusions: The dominant separation mechanism in SFC is partition, with adsorption only being of minor importance. Selectivity is not changed by pressure or temperature, but by the modifiers used. To resolve carotenoid isomers a ternary mixture of acetonitrile, methanol, and carbon dioxide was the most promising.

Abrahamson and colleagues combined a lab-built supercritical fluid extraction (SFE) unit and a commercial SFC instrument (Thar Investigator) for the analysis of eight carotenoids in the microalgae *Scenedesmus* sp. [17]. The compounds were initially extracted from the freeze-dried material using CO_2 with 10% ethanol. The target analytes (astaxanthin, β-carotene, canthaxanthin, echinenone, lutein, neoxanthin, violaxanthin, and zeaxanthin) were then separated by SFC using two columns (SunFire C18 and Viridis SFC silica 2-ethylpyridine, both from Waters) in tandem. Even though both columns were of 25 cm length, it was possible to set a flow rate of 5 mL/min (CO_2/methanol gradient), so that all compounds were separated in less than 15 minutes (Fig. 15.1). Column temperature was found to be most important for obtaining satisfactory results, in particular for β-carotene. At lower temperatures (28°C) it coeluted with neoxanthin, whereas at 40°C it could not be separated from lutein anymore. The LOD values, ranging from 0.02 to 0.05 mg/L, were comparable to conventional RP-HPLC, and recovery rates (99–103%) and repeatability ($\sigma_{rel} \leq 2.4\%$) were acceptable. Lutein (436.1 μg/g) and neoxanthin (670.8 μg/g) were identified as the dominant carotenoids in the investigated species. The same or related compounds were determined by SFC in dietary supplements [18]. At first the authors screened eight stationary phases

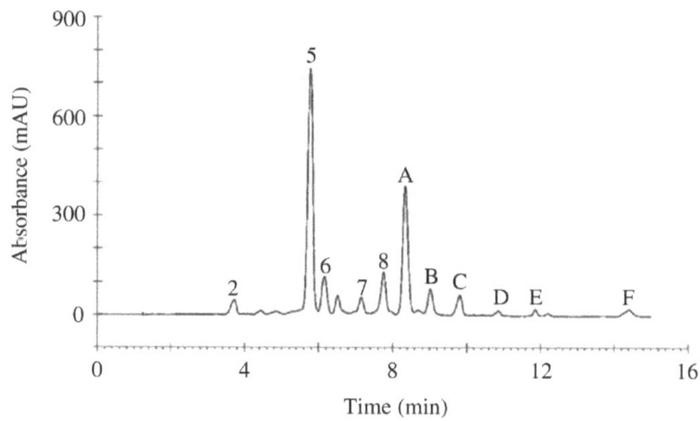

FIGURE 15.1 Determination of carotenoids (astaxanthin (2), lutein (5), zeaxanthin (6), β-carotene (7), and neoxanthin (8)) in the microalgae *Scenedesmus* sp. by SFC; compounds A–F are unidentified chlorophylls. *Source: Reproduced with permission of Elsevier Abrahamsson V, Rodriguez-Meizoso I, Turner C. Determination of carotenoids in microalgae using supercritical fluid extraction and chromatography. J Chromatogr A 2012;1250:63–68 [17].*

(including Torus columns from Waters, all of 1.7 or 1.8 μm particle size), and optimized cosolvent, pressure, flow rate, and temperature for the separation. The preferred procedure (column: Acquity UPC^2 HSS C18 SB, 150 × 3.0 mm, 1.8 μm; binary mobile phase comprising CO_2 and methanol/ethanol (1:2); flow rate: 1.0 mL/min; sample volume: 1 μL; ABPR: 152 bar) is typical for SFC. For quantification the products (tablets or capsules) were powdered and dissolved in a mixture of DMSO and water (3:1). This solution was then partitioned between dichloromethane/ethanol (1:2), and aliquots of the upper layer used for analysis. All samples contained 0.1% of the antioxidant BHT (butylated hydroxytoluene), and β-apo-8′-carotenal was added as an internal standard to account for possible losses during the extraction procedure. A simple handling and short separation time (10 minutes for 9 carotenoids) were considered the main advantages, especially as the separation of isomers like α- and β-carotene, or lutein and zeaxanthin is difficult even on C-30 phases, which usually are recommended for this separation by HPLC.

In two publications from Matsubara et al. the analysis of carotenoids by SFC-MS is described. They first focused on the rapid determination of seven common carotenoids in the acetone extract of the green algae *Chlamydomonas reinhardtii* [19]. The use of a monolithic stationary phase (Chromolith Performance RP-18e, 100 × 4.6 mm; Merck) was advantageous to improve throughput and resolution. Quadrupole MS afforded sufficient selectivity and sensitivity (LOD below 50 pg) for the analysis. The mobile phase consisted of carbon dioxide and methanol with 0.1% ammonium formate; the latter was also used as make-up liquid and infused into the ESI source at a flow rate of 0.1 mL/min. Due to the generally low back pressure of monolithic columns, it was possible to increase the column flow rate up to 9 mL/min or use three columns in connection, without exceeding the pressure limit of the instrument. Quantitative results were not presented, but the combination of an innovative stationary phase and SFC-MS seems to have a lot of potential. In a study from 2012, carotenoids and their epoxy derivatives (degradation products and oxidative stress markers) were determined in human serum [20]. Analysis of the latter is tricky by conventional methods because of their low concentration and poor separation from hydroxycarotenoids. An ESI source in the positive mode was preferred for MS detection, because by APCI some of the structure-specific product ions were not observed. This step was followed by evaluating the optimum chromatographic conditions. A C-30 (YMC Carotenoid) as well as an ODS (Merck Puroshere RP-18e) column provided sufficient selectivity for the separation. The latter column was preferred because of a shorter separation time (20 minutes). The optimum mobile phase composition was the same as in the earlier study, however for serum samples the precise detection of some carotenoids was disturbed by the occurrence of several isomers. This problem could be resolved by monitoring specific product ions by MRM (multiple reaction monitoring). As a result, five major carotenoids (α-carotene, β-carotene, β-cryptoxanthin, lutein, and zeaxanthin) and six epoxy carotenoids were successfully detected and quantified in human serum. The procedure affords low LOD values in the sub-fmol range, and a 0.1-mL sample size was sufficient for analysis.

15.2.3 Essential Oils

Most essential oils are composed of a complex mixture of volatile compounds, usually belonging to the class of terpenes (monoterpenes or sesquiterpenes), coumarins, or phenylpropanoids. Their occurrence in plants is variable from total absence or minute amounts (e.g., 0.05% in *Melissa officinalis*) to more than 15% (e.g., in nutmeg, *Myristica fragrans*). They possess a characteristic smell, and are used in perfumes, as spices, or to treat minor health complaints (stomach ache, mouth or throat infections, etc.). Because of their lipophilic character they can be extracted from plant material by supercritical fluids (SFE), an option that has been described many times [21], also considering underlying theoretical aspects [22]. Two reports from the mid-1990s describe the hyphenation of SFC and GC for the analysis of volatiles in citrus [23] or cloud berry (*Rubus chamaemorus*) oil [24]. The setup was similar; e.g., in the latter capillary SFC (SB-Cyanopropyl-25, 10 m × 50 μm, Dionex) was used for preseparation of the oils. The volatiles eluting first, were trapped on a short (2 m) and cooled (−2°C) SE-54 column, then thermally desorbed for analysis by GC. More than 70 compounds were identified by GC-MS based on database matches. In 1997 Blum et al. reported an improved SFE-MS interface in combination with an OTC type column for the separation of essential oils [25]. Incorporation of a retention gap in their apparatus resulted in a nearly 10-fold increase in sensitivity. To confirm practical applicability a thyme (*Thymus vulgaris*) extract was analyzed; however, no method validation or quantification was attempted. In another fundamental study the main constituents of citrus essential oil, limonene, and linalool, were used to investigate SFC adsorption mechanisms on silica gel in the presence of supercritical CO_2 [26]. A lab-built, two-column system (pre- and adsorption column) was utilized to study the effects of temperature and pressure on adsorption equilibria and mass transport properties, as well as intraparticle and molecular diffusivity.

The unusual combination of SFC with fourier-transform infrared spectroscopy detection (SFC-FTIR) was employed for characterizing the essential oil from hops (*Humulus lupulus*; [27]). The samples were first separated by OTC SFC on a SB-Phenyl-5 column (10 m × 50 μm with 0.25 μm film), at a constant oven temperature of 80°C and a pressure gradient (80 bar for 10 minutes, then up to 350 bar at 3 bar/min) with supercritical CO_2 without modifier as mobile phase. Online IR spectra were recorded using a flow-cell interface. The spectra were compared with library spectra of the major oil constituents recorded as films deposited on AgCl disks. Interestingly most of the bands shifted to approximately 8–10 cm^{-1} higher wave numbers in the SFC-FTIR spectra; an explanation for this deviation was not provided. The method was suitable to differentiate the oil of different hops cultivars and permitted the identification of one major constituent, α-humulene; while several other compounds could only be partially elucidated.

The most recent study of essential oil constituents focused on the determination of furocoumarins [28]. Lemon oil residue, the remains of an Italian lemon oil distillate, was used as sample, and different approaches for analyzing this complex matrix were developed and compared. They ranged from isocratic elution or a two-step gradient, to a two-dimensional design with orthogonal stationary phases. Concerning the last option, a first, manual fractionation of compounds with different polarity was achieved on an ethylpyridine phase (2-Ethylpyridine (EP), Princeton Chromatography), then each fraction was separated in a second step on a pentafluorophenyl phase (Discovery HS F5) from Supelco. The mobile phase was a 91:9 mixture of CO_2 and ethanol. However, the first column (250 × 4.6 mm, 5 μm) alone already enabled the identification of 16 furocoumarins in less than 10 minutes (Fig. 15.2). A variety of stationary phases were tested, and the impact of modifier concentration, temperature, outlet pressure, and flow rate on the retention factors (log k) was investigated. These authors point to the particular importance of selecting the optimum stationary phase in SFC. Firstly, the absence of water in the mobile phase results in more varied interactions than in HPLC, and secondly, even if materials are nominally identical they can result in significantly different separations.

15.2.4 Miscellaneous

Anthraquinones represent an important class of natural products, because they are the active ingredients in popular laxatives like senna (*Cassia angustifolia, Cassia acutifolia*), aloe (*Aloe ferrox, Aloe barbadensis*) and rhubarb (*Rheum officinale, Rheum palmatum*). The major aglyca (chrysophanol, physcion, emodin, aloeemodin, and rhein) in rhubarb could be resolved in less than 5 minutes on a SFC-specific stationary phase

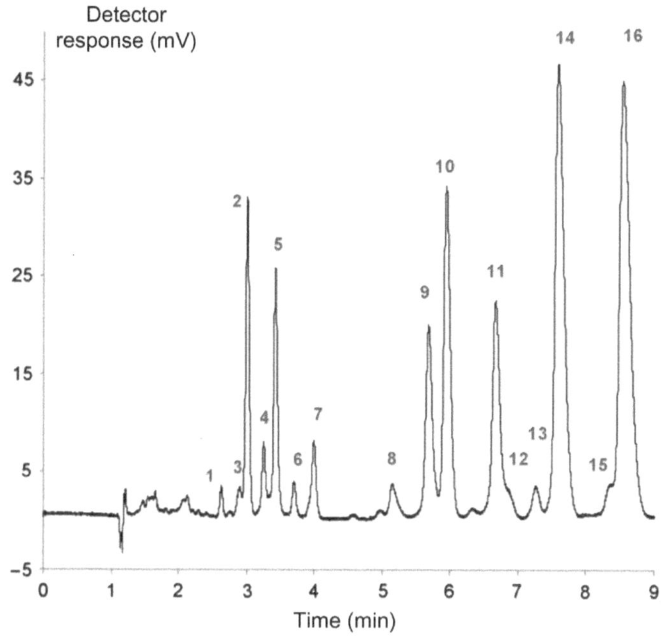

FIGURE 15.2 SFC analysis of furocoumarins in lemon oil residue; assignment of the major signals: 8-geranyloxypsoralen (2), byakangelicol (5), oxypeucedanin hydrate (9), citropten (10), oxypeucedanin (11), 5-geranyloxy-7-methoxycoumarin (14), and bergamottin (16). *Source: Reproduced with permission of Elsevier Desmortreux C, Rothaupt M, West C, Lesellier E. Improved separation of furocoumarins of essential oils by supercritical fluid chromatography. J Chromatogr A 2009;1216:7088–95 [28].*

(Acquity UPC^2 HSS C18 SB, Waters) with a small particle size (1.8 μm), in combination with a mobile phase containing CO_2, methanol, and 0.05% diethylamine [29]. Gradient elution at a flow rate of 2 mL/min was used for the separation. The method was selective, precise, and accurate (recovery rates from 95.4% to 103.1%), with an LOD below 0.5 ng on-column, and was suitable for the analysis of plant material without any specific sample clean-up. Quantitative results were reproducible ($\sigma_{rel} \leq 2\%$) and in agreement with published data (total content in samples before hydrolysis 0.32–0.73%).

Destruxins are cyclic hexadepsipeptides with insecticidal and phytotoxic activities. They are produced as secondary metabolites by the entomopathogenic fungus *Metarhizium brunneum,* which is used as biological pest control agent against cockchafers and garden-chafers. To facilitate the analysis of destruxins SFC was evaluated and this approach permitted the identification of 13 derivatives in purified *M. brunneum* culture broth [30]. Five compounds (destruxins A, B, D, and E, as well as destruxin E-diol) were assigned using reference compounds, the tentative identification of other substances was possible by SFC-DAD-MS/MS. For the latter an Acquity UPC^2 was coupled to a Xevo TQD triple quadrupole mass spectrometer operated in the positive ESI mode; methanol was added as make-up solvent. The mobile phase was the same as for stand-alone SFE experiments (CO_2 and an 8/2 mixture of methanol and acetonitrile), only for SFE-MS 0.02% formic acid was added to the modifier instead of 0.02% TFA. Method validation and quantification of the compounds was achieved using a diode-array detector. Compared to the established UPLC method the authors noted many advantages (shorter analysis time, rapid equilibration, higher throughput), but also a slightly lower sensitivity (LOD by UPLC: 0.1–1.4 μg/mL; LOD by SFC: 4.4–6.5 μg/mL). They explained this observation by the low injection volume of 1 μL and the diminished sensitivity of SFC-DAD.

The fact that chirality may determine pharmacological activity is relevant for natural products as well. One out of many examples is *Isatis indigotica*, a plant also known as woad. A blue dye can be produced from its leaves and the roots ("Ban Lan Gen" in Chinese) are an important remedy in Asian traditional medicine to treat fever. They contain two sulfur-containing oxazolidine enantiomers, R- and S-goitrin, as active ingredients, which differ significantly in their biological activity. The R-variant (epigoitrin) is antiviral, whereas S-goitrin is goitrogen (i.e. it increases thyroid tissue due to disrupted production of thyroid hormones). For quality control purposes, it is important to quantify both enantiomers. SFC with a chiral stationary phase, a (S,S)-Whelk-O1 column (4.6 × 50 mm, 10 μm) and a simple CO_2-methanol gradient enabled their separation in less than 6 minutes [31]. Detection was performed at 244 nm (DAD) and also by a single quadrupole MS with APCI source (positive mode). Not surprisingly, the sensitivity of the latter was 50-times higher, but other validation parameters (correlation coefficients, repeatability, precision, etc.) were comparable. Three commercial formulations all labeled as "Ban Lan Gen" powder were quantitatively analyzed. One of them contained none of the goitrins, in the remaining R-goitrin was dominant. However, with a maximum content of 0.03% its concentration was rather low.

15.2.5 Phenols including Flavonoids

Flavonoids are widely distributed secondary metabolites in plants, but they are never synthesized by bacteria and fungi. They exhibit a broad spectrum of pharmacological activities that can be recapped under the term "antioxidant"; accordingly, with an estimated intake of about 1 g per day they are also important factors of a healthy diet.

A rich source of flavonoids are sweet oranges (*Citrus sinensis*), with nobiletin being one of the dominant compounds. The SFC separation of its major metabolites, which are formed during biotransformation (demethylation) by hepatic P450 enzymes, in mouse urine was described. Chiral and achiral columns were evaluated for the chromatographic separation of 3′-demethylnobiletin and 4′-demethylnobiletin by SFC and HPLC. Only a chiral phase (Chiralpak AD-H column, 250×4.6 mm, 5 μm, Chiral Technologies) in combination with SFC provided adequate separation. Supercritical CO_2 with 20% methanol was used as eluent, with temperature, flow rate, and pressure set to 30°C, 2.0 mL/min, and 100 bar. Under these conditions the two regio-isomers could be resolved with a retention time difference of 10 min; with a concentration of 28.9 μg/mL 4′-demethylnobiletin was found to be the major metabolite in the samples [32,33].

Ganzera studied the separation of isoflavones by SFC [34]. These compounds have a phenyl residue in position C-3 of the chromone scaffold instead of C-2, and because of their hormone-like activities they are assigned to the group of phytoestrogens. They are found in medicinal plants which are used to treat menopausal symptoms such as hot flushes, bone loss, or mood swings. Nine isoflavones (aglyca and glycosides) could be baseline resolved in less than 9 minutes on an Acquity UPC^2 BEH column (100×3.0 mm, 1.7 μm), using a solvent gradient comprising CO_2 and methanol with 0.05% phosphoric acid. The column temperature was maintained at 50°C, and 254 nm selected for detection. After validation (e.g., LOD $\leq$ 0.2 ng on-column) the practical applicability of the method was confirmed by analyzing dietary supplements containing red glover (*Trifolium pratense*), soy (*Glycine max*), and kudzu (*Pueraria lobata*) extracts (Fig. 15.3).

Polyphenols in *Mangifera indica* (mango) leaves, which contain the xanthoid mangiferin, quercetin glycosides, and other phenolic compounds (gallic acid, methyl gallate, etc.) were fractionated by RP-SFC at pilot plant scale [35]. A 25×3 cm column containing Synergi Hydro-RP C18 silica, 5 μm particle size, was used. Except that the separation time seems extremely long with 450 minutes other parameters are typical of conventional SFC. For example, elution was performed in the gradient mode (5–50% modifier) using CO_2 and methanol with 0.5% formic acid as mobile phase. At the beginning of a run the crude extract (30 mg) was dried with the packing material and then this trituration applied at ambient pressure and temperature at the top of the column. Fractions were collected in 5 minute intervals and analyzed by HPLC-MS for their composition. Due to the high polarity of the studied phenolic compounds, their low solubility in the mobile phase and the large-scale used, only poor separations were achieved. Yet the first semiindustrial application of SFC was described, which enabled the fractionation of a

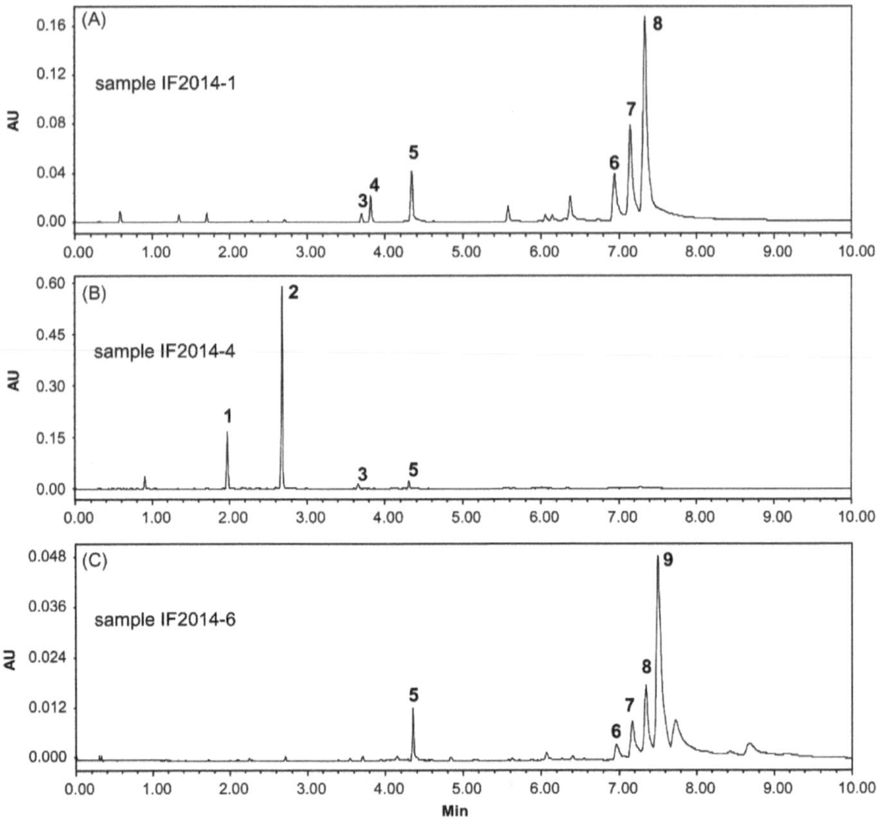

FIGURE 15.3 Determination of isoflavones (biochanin A (1), formononetin (2), genistein (3), glycitein (4), daidzein (5), genistin (6), glycitin (7), daidzin (8), and puerarin (9)) in dietary supplements containing extracts of soy (**A**), red glover (**B**), and kudzu (**C**) by SFC. *Source: Reproduced with permission of Elsevier Ganzera M. Supercritical fluid chromatography for the separation of isoflavones. J Pharm Biomed Anal 2015;107:364–69 [34].*

mango leaf extract into different compound classes (phenolic acids, benzophenones, and xanthones).

Two SFC studies that are not directly related to natural products should briefly be mentioned as well. They focus on the determination of toxic phenolic pollutants like pentachlorphenol (PCP) or 4-nitrophenol (4-NP) in water, which is the result of industrial effluents or use of pesticides. Bernal et al. described the online coupling of solid-phase extraction (SPE) and SFC-DAD through two six-port rotary valves, and used this setup for quantifying eleven phenolic compounds in water designated by the Environmental Protection Agency as priority contaminants. Water samples were percolated through 20 μm PLRP-S material and then the adsorbed phenols were adequately separated on two interconnected LiChrospher Diol columns (250 × 4.6 mm, 5 μm, flow rate: 2.4 mL/min, temperature: 40°C). The observed detection limits ranged from 0.5 to 5 mg/L [36]. On the other hand, Ramsey and coworkers utilized a similar preconcentration strategy to determine 2,4-dibromophenol, 2,4,5-trichlorophenol, and 4-nitrophenol by SFE-SFC-MS. Two Appex amino RP columns (150 × 4.6 mm, 5 μm) were required, with one of them serving as the trap column. Low ppb levels (40 ppb) of the pollutants could be detected in the negative APCI mode [37].

15.2.6 Nonvolatile Terpenes and Steroids

The interest in triterpene acids, especially in betulinic and platanic acid, has significantly increased since they were identified as potent and selective HIV-1 antivirals. In 2001 a capillary SFC method for the analysis of several acids (betulinic, oleanolic, ursolic, and polpunonic acid) was published. This study focused on the optimization of separation conditions utilizing standards, and applications to plant material are missing. However, with carbon dioxide as mobile phase all triterpenes eluted within 13 mintes from an LM-5 capillary-column (20 m × 0.10 mm) coated with cross-linked 5% phenyl-95% methylpolysiloxane. The column temperature was kept isothermal at 80°C and the detector (FID) at 310°C. The latter was required because the analytes possess no chromophore [38]. Another detector for monitoring such compounds is evaporative light lcattering detection (ELSD), an option that was described by Lesellier et al. for the separation of triterpenoids in apple pomace extracts [39]. Apple peel is especially rich in these compounds and it is available in large quantities as a byproduct of apple juice production. In all seven columns were screened, as well as separation conditions refined (mobile phase: 3% methanol in CO_2, temperature: 20°C, flow rate 3 mL/min, back pressure: 120 bar) before choosing a Synergi Polar-RP column (250 × 4.6 mm, 5 μm) for the analysis. It enabled the SFC-ELSD separation of a standard mixture of oleanolic acid, erythrodiol, β-amyrin, ursolic acid, uvaol, betulinic acid, botulin, and lupeol as well as other triterpenes in less than 20 minutes. The tentative identification of additional compounds was possible by flow injection analysis (FIA) using an AB Sciex API 3000 triple quadrupole mass spectrometer. A qualitative comparison of different apple varieties (Royal Gala, Golden, Granny Smith, and Pink Lady) showed no major differences in the composition of triterpenes. Other related SFC applications include the analysis of bacoside A3 and bacopaside II in *Bacopa monnieri* [40], of ginkgolides and their hydrolyzed metabolites by SFC-MS [41], the chiral separation of egostane triterpenes in the mushroom *Antrodia camphorata* [42], and the determination of triterpene saponins in diverse TCM plants [43]. This last example is of particular interest because a mobile phase containing water was used. It was the authors aim to develop a method suitable for the analysis of saponins isolated from *Illex latifolia* (kudinosides), *Panax ginseng*, and *Panax quinquefolius* (ginsenosides). During method development they observed that a small amount of water was beneficial to improve chromatographic performance and MS sensitivity. For example, nine kudinosides, glycosides with three to four sugar residues, could be resolved on a Zorbax RX-SIL column (150 × 4.6 mm, 5 μm) with a mobile phase containing CO_2 and methanol with 10% water and 0.05% formic acid. Within 10 minutes, the modifier content was increased from 30% to 50%, with column temperature, flow rate, and back pressure adjusted to 20°C, 3 mL/min, and 160 bar, respectively. The addition of acid significantly increased the MS response, but at the same time the resolution of some compounds like kudinosides C/G and latifolosides Q/H decreased; using 0.05% of formic acid was a good compromise. Compared with an established reversed-phase LC-MS method, only half the time was required for analysis by this SFC-MS method (Fig. 15.4).

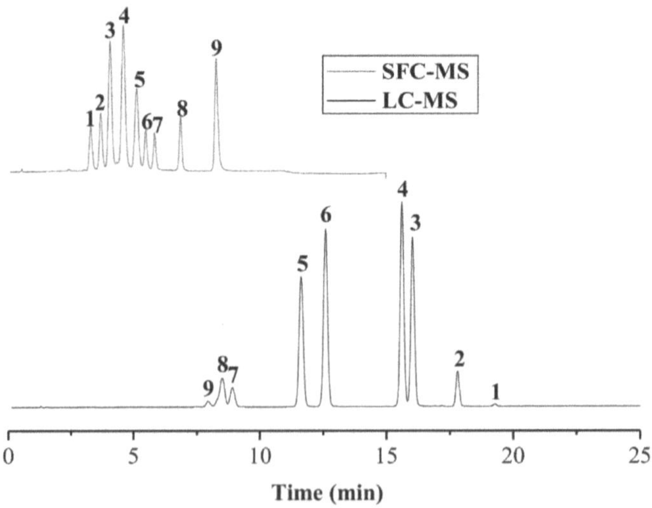

FIGURE 15.4 Comparison of SFC-MS and LC-MS for the separation of nine kudinosides from *Ilex latifolia* leaves; peak assignment: kudinoside F (1), kudinoside A (2), ilekudinoside G (3), kudinoside E (4), kudinoside C (5), kudinoside G (6), latifoloside Q (7), latifoloside H (8), and kudinoside O (9). *Source: Reproduced with permission of Elsevier Huang Y, Zhang TT, Zhou HB, Feng Y, Fan CL, Chen WJ, et al. Fast separation of triterpenoid saponins using supercritical fluid chromatography coupled with single quadrupole mass spectrometry. J Pharm Biomed Anal 2016;121:22–29 [43].*

The use of SFC is not limited to small molecules, since comparatively large structures such as polyprenols have also been studied. Commonly referred to as "rubber" they represent linear polymers of 1,4-polyprenyl alcohols, and interestingly, the type of polyprenol and degree of polymerization varies with the plant species (*Hevea brasiliensis, Palaquium gutta, Eucommia ulmoides*, etc.). Bamba et al. have published several papers on this topic, including one investigating the occurrence of polyprenol isomers in *E. ulmoides* [44]. A phenylalkylsiloxane-bonded silica sorbent (Inertsil Ph-3) permitted the separation of poly-*trans* and -*cis* prenols up to a degree of polymerization of 30. On RP-phases their separation was not possible. For structural elucidation preparative SFC on the same stationary phase was used together with NMR analysis. For the first time the presence of long chain poly-*trans* prenols was confirmed in plants. In a follow-up study the method was extended to even lager polymers containing over 100 monomers [45]. Chromatographic conditions were similar, however, tetrahydrofuran was used as a modifier instead of ethanol. Average molecular masses were calculated by MALDI-TOF-MS.

Spirostanol saponins, which are allocated to steroids, are found as bioactive compounds in many medicinal plants of the Dioscoreaceae, Agavaceae, Alliaceae, and Liliaceae families. A major analytical challenge is the separation of the 25 R/S diastereomers. Their analysis in *Trigonella foenum-graecum* (fenugreek) seeds by SFC is possible by using two chiral columns (Chiralpak IC, 250 × 4.5 mm, 5 μm) in tandem. Six pairs of diastereomers in prepurified mixtures containing "only" a single racemate were investigated [46]. Detection was achieved by ELSD, and the system was suitable for analytical as well as semipreparative use without major modification; only the sample volume was adjusted. Zhu et al. compared UHPLC and UHPSFC for the separation of the same saponins [47]. They found that UHPSFC was sensitive to the number and position of hydroxyl groups in the aglycone. UHPLC was complementary because the results on a BEH C18 column were mainly influenced by the number of double bonds in the analytes. Concerning the choice of SFC stationary phase, a HSS C18 SB column was able to resolve individual sapogenins, while spirostanol saponins with up to four sugars could be separated on a Torus Diol column. Again an ELSD was used for detection and other settings were typical for SFC; no application to plant extracts was presented. In older publications (>15 years) pharmaceuticals with steroidal structure were used as standards. They utilized capillary SFC in combination with electron-capture [48] and phosphorus-selective detectors [49], or studied the retention behavior of packed columns [50]. More recently the metabolites of anabolic steroids were determined in human urine by SFC-MS, using a fully validated assay with high sensitivity (LOD for glucuronides: 0.5 ng/mL, for sulfates: 0.1 ng/mL) and speed (separation time: 15 minutes). The make-up solvent, which was infused right after the back pressure regulator, was a water–methanol mixture (1:1) containing 0.1% formic acid [51].

15.2.7 Vitamins

The separation of the fat-soluble vitamins A, D, E, and K is a challenging task due to their similar structures (e.g., only differing in the number or configuration of double bonds) and limited stability. Their analysis is commonly achieved by normal- or reversed-phase liquid chromatography using UV–vis, fluorescence, or MS detection. GC methods have also been reported, but generally require derivatization of the thermosensitive vitamins prior to analysis. A few aspects render SFC to an optimum technique for their determination. Probably the most relevant are the exclusion of oxygen and minimization of heat during their analysis. Shorter extraction times and higher selectivity are also advantageous. Accordingly, vitamins with widely different molecular masses are accessible with this technique without sample pretreatment or derivatization [52].

In 1997 the separation of five *cis/trans*-isomers of vitamin A acetate by SFC-^{1}H-NMR was reported. This uncommon hyphenation was possible by using a specially designed pressure-proof continuous-flow NMR probe head, from which a signal line width of 1.5 Hz was obtained in the supercritical state [53]. To assure supercritical conditions in the flow cell (5 × 3 mm sapphire tube with a volume of 120 μL) its temperature was maintained at 37°C. The chromatographic system was connected to

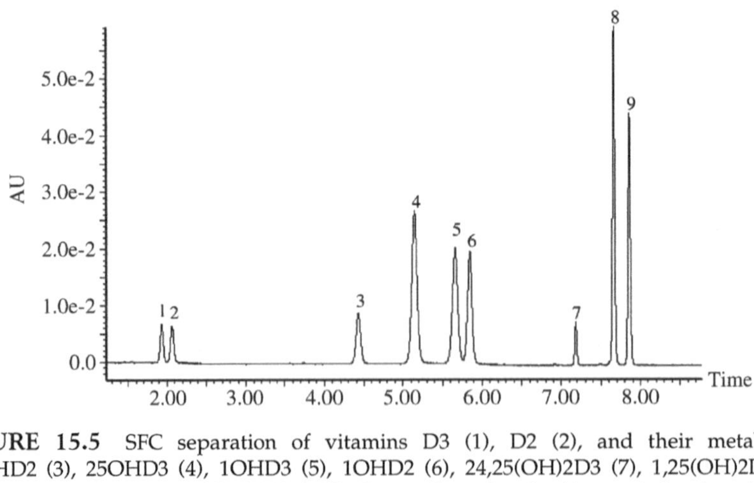

FIGURE 15.5 SFC separation of vitamins D3 (1), D2 (2), and their metabolites (25OHD2 (3), 25OHD3 (4), 1OHD3 (5), 1OHD2 (6), 24,25(OH)2D3 (7), 1,25(OH)2D2 (8), 1,25(OH)2D3 (9), and 1,24(OH)2D3 (10)). *Source: Reproduced with permission from Elsevier Jumaah F, Larsson S, Essen S, Cunico LP, Holm C, Turner C, et al. A rapid method for the separation of vitamin D and its metabolites by ultra-high-performance supercritical fluid chromatography-mass spectrometry. J Chromatogr A 2016;1440:191–200 [54].*

the probe by a 3 m PEEK tubing; this was required to avoid any disturbance of the magnetic field. For SFC separations a Vydac peptide and protein C_{18} column (250 × 4.6 mm) was utilized with CO_2 at 130 bar, a temperature of 60°C, and a flow rate of 0.65 mL/min. The analyzed sample was thermally isomerized vitamin A. Within 25 minutes the major isomers were eluted and NMR-spectra obtained with a quality equivalent to those recorded in liquids. The main advantage of carbon dioxide as eluent is that the whole NMR spectral range can be observed without the need for solvent suppression.

Just this year Jumaah and coworkers published the first SFC-MS method for the separation of vitamin D metabolites [54]. This is important because the hydroxylated metabolites of vitamins D2 and D3 are the actual biologically active form. Most existing methods only permit the determination of total vitamin D metabolites without differentiating between the D2 and D3 forms. Additionally, even the gold standard method HPLC-MS requires time consuming derivatization steps or the use of isotopic standards. By SFC 9 vitamin D metabolites were resolved ($R_s > 1.5$) in less than 8 minutes without derivatization (Fig. 15.5). The separation conditions were carefully investigated and a Torus 1-aminoanthracene (1-AA) column was selected for the separation. It was speculated that $\pi-\pi$ interactions between the unsaturated bonds of the vitamin metabolites and the aromatic rings of anthracene in the stationary phase contributed significantly to the separation mechanism. The outlet pressure (200 bar), column temperature (35°C), and flow rate (2 mL/min) had an substantial impact on retention and selectivity of early eluting compounds. The sensitivity of SFC-APCI was shown to be sixfold lower than SFC-ESI (both in positive ion mode), a fact that was explained by the mass flow dependency of APCI. Using this approach, vitamin D3 (6.6 ± 3.0 ng/mL), 24,25-dihydroxyvitamin D3 (23.8 ± 9.2 ng/mL), and 1-hydroxyvitamin D2 (5.4 ± 2.7 ng/mL) could be determined in human plasma samples.

In two publications the SFC analysis of vitamin K isomers was addressed. One focused on the separation of *cis*- and *trans*-isomers of vitamin K1 on an RX-Sil column with a 1.8 μm particle size [55]. With a separation time of only 2 minutes, this method was at least three times faster than previously reported HPLC procedures. In a follow-up study the different stereoisomers of K1 were investigated [56]. The side chain of this vitamin contains two chiral centers, and by SFC it was possible to partially resolve at least seven of the eight stereoisomers. A commercial vitamin K1 standard from Sigma was the sample, and a single enantiomer, probably *trans*-4R,11R-philloquinone accounted for 58% of the total peak area. The ratio of *cis*/*trans*-isomers was approximately 11/89. This information is useful because until now it is not clear if only the before mentioned enantiomer is biologically active as a vitamin.

The term vitamin E does not describe a single compound but includes, among others, tocopherols and tocotrienols. Both are based on a chromanol ring system, whereby tocopherols possess a saturated side-chain and tocotrienols have a side-chain with three double bonds. For each group four representatives (α, β, γ, and δ) are known, depending on the substitution pattern of the chromanol substructure. They are important antioxidants and are found in many food products (e.g., wheat germ or almond oil). Yarita et al. used SFC for the determination of tocopherols in vegetable oils [57]. They utilized a custom made apparatus consisting of two pumps (for CO_2 and methanol), a mixer, a sample loop (20 μL), a column oven and UV-detector, as well as a restrictor (200 mm × 50 μm capillary tube) to control the flow rate. With this simple setup they were able to separate four tocopherols in 15 minutes on a 5 μm ODS sorbent, with quantitative results almost identical to normal-phase HPLC. The simultaneous analysis of tocopherols and tocotrienols in soybean oil by SFC-MS was reported much later [58]. The chromatographic conditions are typical for SFC with the exception that an aminopropylsiloxane-bonded sorbent column (Luna NH_2, 150 × 2 mm, 3 μm) was used together with MS detection in the APPI (atmospheric pressure photoionization) mode. The authors evaluated the efficiency of ESI, APCI, and APPI sources and found the latter much more robust and sensitive. The LOD values (e.g., 10 μg/L for β-tocopherol) determined by SFC-APPI-MS were in the same order of magnitude as reported for standard LC-MS methods. Additionally, all validation criteria were fulfilled and eight compounds could be separated and quantified in real samples in less than 5 minutes. Last but not least, in a study published by Ibanez different stationary phases were evaluated for their potential to resolve tocopherol analogs [59]. The stationary phases were prepared by coating 10- μm porous silica particles (6 Å pore size) with different amounts of polyethylene glycol (Carbowax 20 M); the latter is a typical stationary phase for GC. Using a mixture of four tocopherols it was observed that increasing the percentage of stationary phase coating from 3% to 10% resulted in enhanced resolution; more than that impaired the results. This adjustable polarity and selectivity was considered as the main advantage, especially as no separation of vitamin E isomers was possible with conventional RP-18 sorbents like Spherisorb ODS2. Application to a sample of plant origin was not demonstrated.

CONCLUSION

When summarizing the relevance of SFC for natural products analysis based on the currently available literature it is obvious that it offers much more than acting as a substitute for normal-phase liquid chromatography only. Even if the majority of successfully analyzed analytes are of nonpolar type (carotenoids, fat-soluble vitamins, etc.) more polar compounds, such as saponins or alkaloids are also accessible. Additionally, there is an enormous variability in adaptable stationary phases, detectors or hyphenation options, so that in many cases SFC can be considered as an acceptable alternative to state-of-the-art techniques like UPLC or UPLC-MS. Some features, such as its green character, scalability for preparative applications, and the potential to resolve chiral compounds distinguish SFC from other established approaches. Consequently it only is a matter of time until this powerful and user-friendly technique will play a greater role in research and routine analysis.

References

[1] Bauer A, Broenstrup M. Industrial natural product chemistry for drug discovery and development. Nat Prod Rep 2014;31:35–60.

[2] Lachance H, Wetzel S, Kumar K, Waldmann H. Charting, navigating, and populating natural product chemical space for drug discovery. J Med Chem 2012;55:5989–6001.

[3] Steinmann D, Ganzera A. Recent advances on HPLC/MS in medicinal plants analysis. J Pharm Biomed Anal 2011;55:744–57.

[4] Ganzera A. Quality control of herbal medicines by capillary electrophoresis: potential, requirements and applications. Electrophoresis 2008;29:3489–503.

[5] Ganzera M, Nischang I. The use of capillary electrochromatography for natural product analysis—theoretical background and recent applications. Curr Org Chem 2010;14:1769–80.

[6] Hartmann A, Ganzera M. Supercritical fluid chromatography—theoretical background and applications on natural products. Planta Med 2015;81:1570–81.

[7] Holzer G, Zalkow LH, Asibal CF. Capillary supercritical fluid chromatography of pyrrolizidine alkaloids. J Chromatogr 1987;400:317–22.

[8] Lanicot JL, Caude M, Rosset R. Separation of opium alkaloids by carbon dioxide sub- and supercritical fluid chromatography with packed columns. J Chromatogr 1988;437:351–64.

[9] Fu Q, Li ZY, Sun CC, Xin HX, Ke YX, Jin Y, et al. Rapid and simultaneous analysis of sesquiterpene pyridine alkaloids from *Tripterygium wilfordii* Hook. f. using supercritical fluid chromatography-diode array detector-tandem mass spectrometry. J Supercrit Fluid 2015;104:85–93.

[10] Zhao TJ, Qi HY, Chen J, Shi YP. Quantitative analysis of five toxic alkaloids in *Aconitum pendulum* using ultra-performance convergence chromatography (UPC^2) coupled with mass spectrometry. RSC Adv 2015;5:103869–75.

[11] Dost K, Davidson G. Development of a packed-column supercritical fluid chromatography/atmospheric pressure chemical-ionisation mass spectrometric technique for the analysis of atropine. J Biochem Bioph Meth 2000;43:125–34.

[12] Ansell RJ, Kuah JKL, Wang DW, Jackson CE, Bartle KD, Clifford AA. Imprinted polymers for chiral resolution of (+/-)-ephedrine, 4: packed column supercritical fluid chromatography using molecularly imprinted chiral stationary phases. J Chromatogr A 2012;1264:117–23.

[13] Li KY, Fu Q, Xin HX, Ke YX, Jin Y, Liang XM. Alkaloids analysis using off-line two-dimensional supercritical fluid chromatography x ultra-high performance liquid chromatography. Analyst 2014;139:3577–87.
[14] Aubert MC, Lee CR, Krstulocic AM, Lesellier E, Pechard MR, Tchapla A. Separation of trans/cis- alpha- and beta-carotenes by supercritical fluid chromatography - 1. Effects of temperature, pressure and organic modifiers on retention of carotenes. J Chromatogr 1991;557:47–58.
[15] Lesellier E, Tchapla A, Pechard MR, Lee CR, Krstulocic AM. Separation of trans/cis alpha- and beta-carotenes by supercritical fluid chromatography; II. Effect of type of octadecyl-bonded stationary phase on retention and selectivity of carotenes. J Chromatogr 1991;557:59–67.
[16] Lesellier E, Krstulocic AM, Tchapla A. Influence of the modifiers on the nature of the stationary phase and the separation of carotenes in sub-critical fluid chromatography. Chromatographia 1993;36:275–82.
[17] Abrahamsson V, Rodriguez-Meizoso I, Turner C. Determination of carotenoids in microalgae using supercritical fluid extraction and chromatography. J Chromatogr A 2012;1250:63–8.
[18] Li B, Zhao HY, Liu J, Liu W, Fan S, Wu GH, et al. Application of ultra-high performance supercritical fluid chromatography for the determination of carotenoids in dietary supplements. J Chromatogr A 2015;1425:287–92.
[19] Matsubara A, Bamba T, Ishida H, Fukusaki E, Hirata K. Highly sensitive and accurate profiling of carotenoids by supercritical fluid chromatography coupled with mass spectrometry. J Sep Sci 2009;32:1459–64.
[20] Matsubara A, Uchikata T, Shinohara M, Nishiumi S, Yoshida M, Fukusaki E, et al. Highly sensitive and rapid profiling method for carotenoids and their epoxidized products using supercritical fluid chromatography coupled with electrospray ionzation-triple quadrupole mass spectrometry. J Biosci Bioeng 2012;113:782–7.
[21] Fornari T, Vicente G, Vazquez E, Garcia-Risco MR, Reglero G. Isolation of essential oil from different plants and herbs by supercritical fluid extraction. J Chromatogr A 2012;1250:34–48.
[22] Sovova H. Modeling the supercritical fluid extraction of essential oils from plant materials. J Chromatogr A 2012;1250:27–33.
[23] Yarita T, Nomura A, Horimoto Y. Type analysis of citrus essential oils by multidimensional supercritical fluid chromatography/gas chromatography. Anal Sci 1994;10:25–9.
[24] Manninen P, Kallio H. Supercritical fluid chromatography-gas chromatography of volatiles in cloud berry (*Rubus chamaemorus*) oil extracted with supercritical carbon dioxide. J Chromatogr A 1997;787:276–82.
[25] Blum C, Kubeczka KH, Becker K. Supercritical fluid chromatography-mass spectrometry of thyme extracts (*Thymus vulgaris* L.). J Chromatogr A 1997;773:377–80.
[26] Sato M, Goto M, Kodama A, Hirose T. Chromatographic analysis of limonene and linalool on silica gel in supercritical carbon dioxide. Sep Sci Technol 1998;33:1283–301.
[27] Auerbach RH, Dost K, Davidson G. Characterization of varietat differences in essential oil components of Hops (*Humulus lupulus*) by SFC-FTIR spectroscopy. J AOAC Int 2000;83:621–6.
[28] Desmortreux C, Rothaupt M, West C, Lesellier E. Improved separation of furocoumarins of essential oils by supercritical fluid chromatography. J Chromatogr A 2009;1216:7088–95.
[29] Aichner D, Ganzera M. Analysis of anthraquinones in rhubarb (*Rheum palmatum* and *Rheum officinale*) by supercritical fluid chromatography. Talanta 2015;144:1239–44.

[30] Taibon J, Sturm S, Seger C, Werth M, Strasser H, Stuppner H. Supercritical fluid chromatography as an alternative tool for the qualitative and quantitative analysis of *Metarhizium brunneum* metabolites from culture broth. Planta Med 2015;81:1736–43.

[31] Wang R, Runco J, Yang L, Yu K, Li YM, Chen R, et al. Qualitative and quantitative analyses of goitrin-epigoitrin in *Isatis indigotica* using supercritical fluid chromatography-photodiode array detector-mass spectrometry. RSC Adv 2014;4: 49257–63.

[32] Wang Z, Li S, Jonca M, Lambros T, Fergusson S, Goodnow R, et al. Comparison of supercritical fluid chromatography and liquid chromatography for the separation of urinary metabolites of nobiletin with chiral and non-chiral stationary phases. Biomed Chromatogr 2006;20:1206–15.

[33] Li S, Wang Z, Sang S, Huang MT, Ho CT. Identification of nobiletin metabolites in mouse urine. Mol Nutr Food Res 2006;50:291–9.

[34] Ganzera M. Supercritical fluid chromatography for the separation of isoflavones. J Pharm Biomed Anal 2015;107:364–9.

[35] Fernandez-Ponce MT, Casas L, Mantell C, de la Ossa EM. Fractionation of *Mangifera indica* Linn polyphenols by reverse phase supercritical fluid chromatography (RP-SFC) at pilot plant scale. J Supercrit Fluids 2014;95:444–56.

[36] Bernal JL, Nozal MJ, Toribio L, Serna ML, Borrull F, Marce RM, et al. Determination of phenolic compounds in water samples by online solid-phase extraction—supercritical-fluid chromatography with diode-array detection. Chromatographia 1997;46:295–300.

[37] Ramsey ED, Minty B, Mccullagh MA, Games DE, Rees AT. Analysis of phenols in water at the ppb level using direct supercritical fluid extraction of aqueous samples combined online with supercritical fluid chromatography-mass spectrometry. Anal Commun 1997;34:3–6.

[38] Tavares MCH, Yariwake Vilegas JH, Lancas FM. Separation of underivatized triterpene acids by capillary supercritical fluid chromatography. Phytochem Anal 2001;12:134–7.

[39] Lesellier E, Destandau E, Grigoras C, Fougere L, Elfakir C. Fast separation of triterpenoids by supercritical fluid chromatography/evaporative light scattering detector. J Chromatogr A 2012;1268:157–65.

[40] Agrawal H, Kaul N, Paradkar AR, Mahadik KR. Separation of bacoside A3 and bacopaside II, major triterpenoid saponins in *Bacopa monnieri*, by HPTLC and SFC. Application of SFC in implementation of uniform design for herbal drug standardization, with thermodynamic study. Acta Chrom 2006;17:125–50.

[41] Liu XG, Qi LW, Fan ZY, Dong X, Guo RZ, Lou FC, et al. Accurate analysis of ginkgolides and their hydrolyzed metabolites by analytical supercritical fluid chromatography hybrid tandem mass spectrometry. J Chromatogr A 2015;1388:251–8.

[42] Qiao X, An R, Hunag Y, Ji S, Li L, Tzeng YM, et al. Separation of 25R/S-ergostane triterpenoids in the medicinal mushroom *Antrodia camphorata* using analytical supercritical-fluid chromatography. J Chromatogr A 2014;1358:252–60.

[43] Huang Y, Zhang TT, Zhou HB, Feng Y, Fan CL, Chen WJ, et al. Fast separation of triterpenoid saponins using supercritical fluid chromatography coupled with single quadrupole mass spectrometry. J Pharm Biomed Anal 2016;121:22–9.

[44] Bamba T, Fukusaki E, Kajiyama S, Ute K, Kitayama T, Kobayashi A. The occurrence of geometric polyprenol isomers in the rubber-producing plant, *Eucommia ulmoides* Oliver. Lipids 2001;7:727–32.

[45] Bamba T, Fukusaki E, Nakazawa Y, Sato H, Ute K, Kitayama T, et al. Analysis of long-chain polyprenols using supercritical fluid chromatography and matrix-assisted laser desorption ionization time-of-flight mass spectrometry. J Chromatogr A 2003;995:203–7.

[46] Zhao Y, McCauley J, Pang X, Kang LP, Yu HS, Zhang J, et al. Analytical and semi-preparative separation of 25 (R/S)-spirostanolsaponin diastereomers using supercritical fluid chromatography. J Sep Sci 2013;36:3270–6.
[47] Zhu LL, Zhao Y, Xu YW, Sun QL, Sun XG, Kang LP, et al. Comparison of ultra-high performance supercritical fluid chromatography and ultra-high performance liquid chromatography for the separation of spirostanol saponins. J Pharm Biomed Anal 2016;120:72–8.
[48] Baiocchi C, Giacosa D, Roggero MA, Marengo E. Analysis of steroids by capillary supercritical fluid chromatography with flame-ionization and electron-capture detectors. J Chrom Sci 1996;34:399–404.
[49] David PA, Novotny M. Analysis of steroids by capillary supercritical fluid chromatography with phosphorus-selective detection. J Chromatogr 1989;461:111–20.
[50] Hanson M. Aspects of retention behaviour of steroids in packed column supercritical fluid chromatography. Chromatographia 1995;40:58–68.
[51] Doue M, Dervilly-Pinel G, Pouponneau K, Monteau F, Le Bizec B. Analysis of glucuronide and sulfate steroids in urine by ultra-high-performance supercritical-fluid chromatography hyphenated tandem mass spectrometry. Anal Bioanal Chem 2015;407:4473–84.
[52] Turner C, King JW, Mathiasson L. Supercritical fluid extraction and chromatography for fat-soluble vitamin analysis. J Chromatogr A 2001;936:215–17.
[53] Braumann U, Händel H, Strohschein S, Spraul M, Krack G, Ecker R, et al. Separation and identification of vitamin A acetate isomers by supercritical fluid chromatography—^{1}H NMR coupling. J Chromatogr A 1997;761:336–40.
[54] Jumaah F, Larsson S, Essen S, Cunico LP, Holm C, Turner C, et al. A rapid method for the separation of vitamin D and its metabolites by ultra-high performance supercritical fluid chromatography-mass spectrometry. J Chromatogr A 2016;1440:191–200.
[55] Berger T, Berger B. Two minute separation of the cis- and trans-isomers of vitamin K1 without chlorinated solvents or acetonitrile. Chromatographia 2013;76:109–15.
[56] Berger TA, Berger BK. Chromatographic resolution of 7 of 8 stereoisomers of vitamin K1 on an amylose stationary phase using supercritical fluid chromatography. Chromatographia 2013;76:549–52.
[57] Yarita T, Nomura A, Abe K, Takeshita Y. Supercritical fluid chromatographic determination of tocopherols on an ODS-silica gel column. J Chromatogr A 1994;679:329–34.
[58] Mejean M, Brunelle A, Touboul D. Quantification of tocopherols and tocotrienols in soybean oil by supercritical-fluid chromatography coupled to high-resolution mass spectrometry. Anal Bioanal Chem 2015;407:5133–42.
[59] Ibanez E, Palacios J, Reglero G. Analysis of tocopherols by on-line coupling supercritical fluid extraction—supercritical fluid chromatography. J Microcolumn Sep 1999;11:605–11.

CHAPTER

16

Pharmaceutical Applications

L. Nováková and K. Plachká

Charles University, Hradec Králové, Czech Republic

OUTLINE

Supercritical Fluid Chromatography.

DOI: http://dx.doi.org/10.1016/B978-0-12-809207-1.00016-1

16.1 INTRODUCTION

16.1.1 Drug Development

Drug development is the process of bringing a new drug to the market and to patients. First, a lead compound is identified within the drug discovery stage. Subsequently, formulation into a dosage form, the preclinical research, and clinical trials follow, as shown in Fig. 16.1. After a successful completion of these stages and preparation of appropriate documentation (new drug application, NDA), the drug is then approved, by the Drug regulatory authority and brought to the market [1].

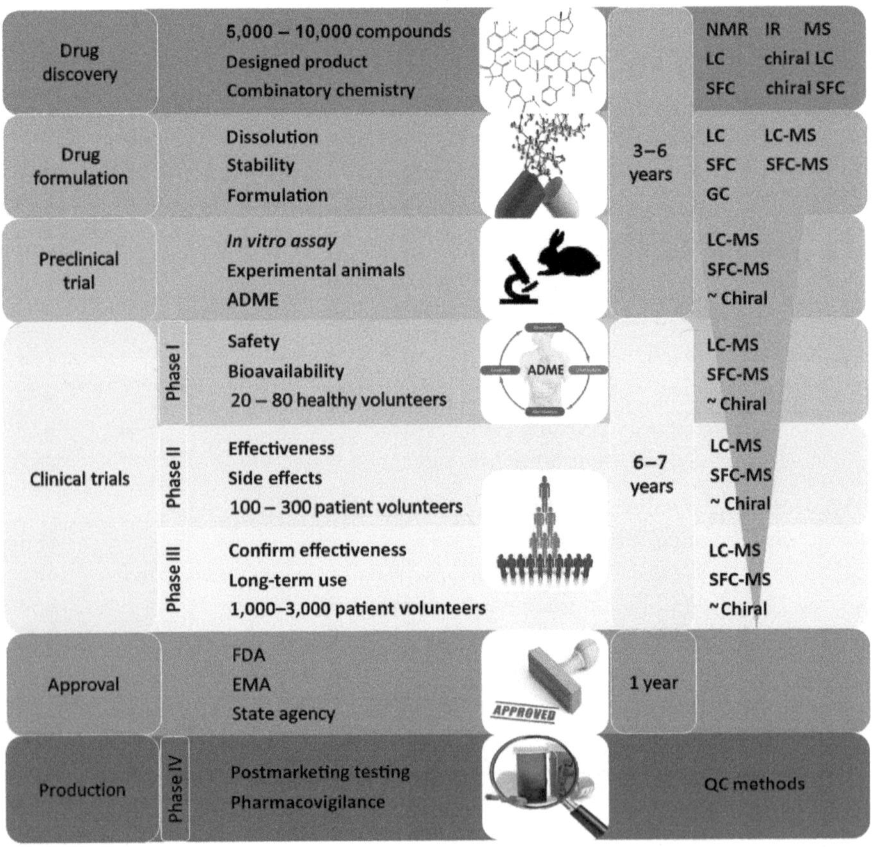

FIGURE 16.1 The individual phases of drug research and development. Analytical methods the most widely used in different stages of the drug research and development are shown in the last column.

The main goal of the first stage of drug development is to discover an active chemical entity (lead compound). There are many sources for lead compounds: (1) random screening (plants, animals, microbial, synthesis), (2) observed side effects of known drugs or their metabolites, (3) natural products with known pharmacological properties, (4) structure of natural body substances, (5) in-silico approach, and (6) ligand- or receptor-based designs. In the next step, the exploratory series of analogs of the lead compound are synthesized and tested [2]. A few effective candidates are selected by screening using in vitro bioassays and/or pharmacological tests [3]. The promising candidate compounds are then formulated into the drug product. In this process the lead compound with various excipients are combined to produce a drug formulation [4]. All of these substances must be compatible and the formulated drug product must be stable and acceptable to the patient. Besides that, formulation studies are focused on other factors, such as particle size, pH, and solubility which can affect bioavailability of the drug. This formulation ensures the proper drug delivery parameters ahead of clinical trials, but it is subsequently modified based on findings obtained during preclinical and clinical trials [5]. Several types of stability studies are necessary at this stage including short-term stability, long-term drug substances and drug product stability studies, photostability and temperature cycle studies. Forced degradation studies are used to develop sensitive and specific analytical methods for identification and quantification of impurities and degradation products [6,7].

The formulated drug candidates are submitted to preclinical testing in order to obtain preliminary information about absorption, distribution, metabolism, and excretion (ADME) of a drug candidate, to evaluate acute and short-term toxicity and to determine lethal doses. In vitro (test tube or cell culture) as well as in vivo (animal) experiments using a wide range of doses are performed at this stage. In the next stage, the drug is tested on humans (clinical trials Phase I-III). In Phase I, the drug is administered to healthy volunteers in order to determine bioavailability, dosage, and side effects associated with increasing doses. In Phase II (100–300 patient volunteers), effectiveness in treating a particular disease is evaluated and side effects are determined. All the side effects as well as adverse reactions for the long-term use are confirmed on thousands of patient volunteers during Phase III. The testing during preclinical and clinical trials requires monitoring of drug concentrations in biological fluids such as plasma, serum, tissue extracts, and urine from experimental animals or human subjects [1,2]. Therefore sample preparation techniques and chromatographic methods are essential at these stages of drug development. Before bringing to market, the new drug has to be approved by the Drug regulatory authority.

The evidence of drug safety and efficiency as well as the description of manufacturing protocol and quality control methods ensuring drug identity, content, quality, and purity are required in the registration application. To ensure the drug quality also in the production phase, the development of analytical methods is necessary for the confirmation of the cleanliness of the pharmaceutical manufacturing equipment and for the in-process control of the progress in the manufacturing of a drug. All this information is part of the application documentation (NDA). Even after approval, the drug is subjected to additional post-marketing testing known as clinical trial Phase IV [1].

16.1.2 Regulation in Drug Development

The goal of the regulation in drug development process is to ensure that drugs approved for the market will be safe, effective, and high quality. The main Drug regulatory authorities are the Food and Drug Administration (FDA) [1] for the United States and European Medicines Agency (EMA) [8] for Europe, with most other countries having their own Drug regulatory authorities. To achieve greater harmonization worldwide and to ensure that safe, effective, and high quality medicines are developed and registered in the most resource-efficient manner, the *International Council for Harmonization of Technical Requirements for Pharmaceuticals for Human Use (ICH)* [9] was established in 1990. ICH brings together the regulatory authorities and pharmaceutical industry to discuss scientific and technical aspects of drug registration. The main goal of ICH is to make recommendations toward achieving greater harmonization in the interpretation and application of technical guidelines and requirements for pharmaceutical product registration, thereby reducing or obviating duplication of testing carried out during the research and development of new human medicines. In line with the main goal of the regulation process, these guidelines are focused on quality, efficacy, and safety and are thus classified as Q (quality), S (safety), E (efficacy), and M (multidisciplinary) [9]. Quality guidelines (Q1–Q12) are focused on quality issues, such as stability studies and impurities, method validation, good manufacturing practices, specifications, and others. Safety guidelines (S1-S11) deal with carcinogenicity, genotoxicity, and other toxicity and safety studies. Efficacy guidelines (E1–E18) deal with the clinical trials. Multidisciplinary guidelines (M1–M9) cover diverse topics that do not fit uniquely into one of the previous groups. These include, e.g., genotoxic impurities, gene therapy or electronic standards, and others [9]. Many of these guidelines were adopted by both EMA and FDA. The most important guidelines for the development of analytical methods in Pharmacy include Q1A and Q1C–Stability [6,7], Q2–Analytical validation [10], Q3A, Q3B, Q3C and Q3D–Impurities [11,12], M7–Genotoxic impurities [13], and Q6A and Q6B–Specifications [14].

16.1.3 Analytical Methods in Drug Development

The quality of drug substances and drug products as well as their concentration in biological fluids are evaluated with the help of different analytical methods. The most common analytical procedures in drug development based on Q2 ICH guideline include: (1) identification tests, (2) quantitative tests for impurity content, (3) limit tests for the control of impurities and (4) quantitative tests (assay) for the active component (active pharmaceutical ingredient, API) in samples of drug substances or drug products. Eventually, another selected component, such as excipient, may be determined as well. All these methods need to be properly validated including the determination of typical parameters, such as accuracy, precision, specificity, detection (LOD) and quantification limit (LOQ), linearity, and range (Table 16.1). However, the extent of method validation depends on the particular method [10]. These analytical procedures are combined to provide information about the identity, purity, content, and stability of raw materials, excipients and APIs. During the synthetic process and storage, impurities may arise and have to be monitored simultaneously with the API. Any impurity at a level greater than the reporting threshold and total impurities should be reported. The reporting threshold is calculated based on the maximum daily dose with 0.05% for the maximum daily dose ≤ 2 g and 0.03% for the maximum daily dose > 2 g. The impurity must be also identified once its content exceeds 0.05–0.1% (identification threshold) and quantified once its content exceeds 0.05–0.15% (qualification threshold) again, based on the daily dose of the drug substance [11]. As the drug products are more complex, consisting of the API/APIs and at least one pharmaceutical excipient, different analytical methods are often required for the analysis of APIs and drug products taking into consideration also various forms of the drug products, such as solutions, tablets, capsules, pills, infusions, drops, ointments, creams, suppositories, and others. At this stage, also degradation products have to be monitored besides API. Reporting threshold is again dependent on the daily dose with 0.1% for the maximum daily dose ≤ 1 g and 0.05% for the maximum daily doses > 1 g, while identification and qualification thresholds are accordingly higher [12]. Similar rules apply also for chiral impurities.

Where a new drug substance is predominantly one enantiomer, the opposite enantiomer is excluded from the qualification [14] and the identification threshold is given in the ICH Q3 guidelines for both drug substance and drug product [11,12]. A specific group of impurities are potential genotoxic impurities, which have received greater interest recently. Potential genotoxic impurities are compounds that may cause cancer in humans due to their potential to induce genetic mutations, chromosomal breaks, and/or chromosomal rearrangements. Because of their significant toxicity, reporting, identification, and qualification limits are very low. A limit for the qualification threshold, when considering a long-term use of the drug, corresponds to an intake level for the genotoxic impurity of 1.5 μg/day [13].

Apparently, the separation of mixtures of API, impurities, and/or other excipients can be demanding, especially when considering the related requirements for speed and high-throughput. Different challenges are presented when the variability of monitored analytes in physicochemical properties (log P, pK_a, MW) is large (acidic, basic, neutral structures of different molecular weight in one mixture) and in the case of negligible structural differences (various structural isomers and closely related compounds). The impurity profiling of API or pharmaceutical products requires a complete separation of all potential impurities and/or degradation products from the API as well as adequate method sensitivity to obtain the reporting threshold. Moreover, the resolution of 1.5 may be insufficient for the API and a neighboring impurity, although it is generally adopted as an acceptance criterion for adequate separation. Enantioselective approaches are needed for the separation of racemic impurities due to the same physicochemical properties of both isomers providing thus exactly the same behavior in conventional achiral separation systems [15,16].

Until recently, HPLC/UHPLC coupled to various detection techniques, mostly UV (ultra-violet) or MS (mass spectrometry), was considered the gold standard. While UV detection is sufficient for most of applications, MS detection is important when high sensitivity or additional selectivity is needed, as in case of the determination of genotoxic impurities, for example. The flexibility in method development due to an extensive variety of available modes including RP (reversed phase), NP (normal phase), and HILIC (hydrophilic interaction chromatography), and stationary phases together with the relatively wide range of parameters to optimize (i.e., mobile phase composition, pH, gradient elution profile, and temperature) enables separation of wide range of compounds differing in physicochemical properties. However, despite this flexibility provided by the different possible modes for HPLC, it is not always possible to separate all components required for a pharmaceutical product or clinical sample. Orthogonal selectivity by different chromatographic modes, particularly different separation methods, are important. Today, supercritical fluid chromatography (SFC)/ultra-high performance supercritical fluid chromatography (UHPSFC) has become a viable option in quality control of drug substances and drug products, see Section 16.3 for details [15,16].

While the Q2 ICH guideline is more suitable for drug substances and drug products, determination of API and its metabolites in biological materials is better addressed in EMA [17] and FDA guidelines [18]. Such bioanalytical methods are needed for the determination of pharmacokinetic profiles, metabolic profiles, and for the separation of enantiomers that may possess different pharmacological activity and/or are metabolized at different rates. As defined by EMA, this data may be required to support applications for new active substances and generics as well as variations to authorized drug products. The results of animal toxicokinetic studies and of clinical trials, including bioequivalence studies are used to make critical decisions supporting the safety and efficacy of a drug substance or product [17]. Bioanalytical methods are thus needed starting from preclinical evaluation using animal models and later on in all clinical trial phases. Biological samples are complex and usually contain the analytes of interest at low concentrations (low ng/mL or less). The presence of abundant interfering compounds frequently requires sample preparation steps for matrix simplification and possibly analyte preconcentration. High sensitivity and selectivity are additional features required for many methods. Therefore LC-MS is currently the method of choice for most cases. Triple quadrupole analyzers using selective SRM (selected reaction monitoring) or high-resolution mass spectrometry (HRMS) with narrow extraction window, such as 10 mDa, are the two main approaches for quantitative bioanalysis. The main bottleneck of LC-MS methods is the presence of matrix effects. Their assessment has become an integrated part of method validation (Table 16.1). Matrix effects can be a source of serious error in quanfication, due to the presence of coeluting compounds from the matrix (or other sources) that can alter the signal response causing either ion suppression or enhancement [17,18]. Most of the important validation parameters including accuracy, linearity, inter-/intraday precision, and sensitivity might be negatively influenced when matrix effects are present. In this regard, since SFC provides an orthogonal selectivity to RP-HPLC, it has become a suitable alternative for reduction/elimination of matrix effects, as will be discussed in Section 16.4.

TABLE 16.1 A Comparison of Method Validation Requirements by Various Authorities

Parameter	ICH	FDA	FDA limits	EMA	EMA limits
Accuracy (%)	3×3	3×5	±15%, LOQ±20%	4×5	±15%, LOQ±20%
Precision (% RSD)	3×3	3×5	<15%, LOQ<20%	4×5	<15%, LOQ<20%
Recovery (%)	x	3	Precise and consistent	x	
Selectivity	√	6	No interference	6	No interference
Calibration curve linearity	5×3	6–8	Accuracy±15%, LOQ 20%	6×3	Accuracy±15%, LOQ 20%
LOQ (limit of quantitation)	√	√	Accuracy±20% and precision<20%	√	Accuracy±20% and precision<20%
Matrix effects (%)	x	√	To ensure lack of matrix effects	2×6	<15%
Carry-over	x	x	x	√	No influence to accuracy/precision
Dilution integrity	x	√	To ensure method linearity	5	Accuracy±15% and precision<15%
Stability	x	√	*	√	*

The numbers describe number of concentration levels times number of requested replicates. √, the parameter is required; x, the parameter is not required; *, very detailed stability study is required including stability of stock and working solution of standards, the stability of the analyte in matrix, long-term storage stability, stability in the autosampler, freeze-thaw cycle stability, etc.
Adapted and updated according to reference Nováková L. Challenges in the development of bioanalytical liquid chromatography-mass spectrometry method with emphasis on fast analysis. J Chromatogr A 2013;1292: 25–37 [19] and appropriate guidelines.

16.2 SFC IN DRUG DISCOVERY

The discovery stage is typical by synthesis and subsequent structure confirmation of drug candidates and their structural analogs (synthetic compound libraries). Identification of newly developed and synthetized candidates for drug substances requires especially good method specificity and identification capability. Typically, a combination of infrared spectroscopy (IR), nuclear magnetic resonance (NMR), and MS, especially HRMS, data are required. Chromatographic techniques are often used prior to MS to ensure the separation of contaminants. LC is currently the technique of choice. However, the role of SFC may also increase in the near future due to its different separation selectivity, wide range of applicability and greater potential for the analysis of low-polarity compounds too strongly retained in RP-HPLC. The drug candidates successful in in vitro bioassays are subject to further testing, and should be available in an enantiomerically pure form, since the FDA requires evaluation of the individual enantiomers for stereoisomers [20].

16.2.1 Determination of Enantiomeric Purity by SFC

Stereochemistry is an important feature for the selection of lead compounds for clinical trials. The human body is stereospecific, therefore enantiomers frequently exhibit very different biological activity because of interactions with active sites of enzymes resulting in differences in pharmacological activity and pharmacokinetic and pharmacodynamic effects. Chirality is thus a critical factor having a large impact on drug profiles, action, metabolism, and toxicity. In many cases, one enantiomer may produce the desired therapeutic effect, while the other may be inactive or even toxic. The importance of separation and testing of single-enantiomer drugs for pharmacological properties, metabolism and toxicity has become standard within the pharmaceutical industry [20–22].

Enantioseparations employ normal-phase and reversed-phase LC, gas chromatography (GC), thin layer chromatography (TLC), capillary zone electrophoresis (CZE), and SFC [22]. SFC has many benefits compared with LC due to the mobile phase containing a large fraction of carbon dioxide having much lower viscosity and much higher diffusivity compared with solvents typically used in LC. As a result, higher flow-rates may be used without losing resolution and efficiency. Faster equilibration times are another benefit leading to high throughput. Additionally, the possibility of scale-up for purification, which is impossible for CZE or GC, brings further advantages. In many pharmaceutical companies, SFC has progressively become the first choice for enantioseparation [15,23,24]. A wide range of SFC applications for the separation of enantiomers of pharmaceutical interested is reviewed in [24–28] and described in a separate chapter in this book.

In a discovery environment the number of compounds submitted for enantioseparation can be large, the variety of compounds great (including raw materials and intermediates), often only small amounts are available, and the achiral purity can be less than 90%. Achiral impurities may severely interfere with the chiral method development. To tackle these challenges efficiently, method development using rapid screening approaches is utilized. These screening strategies are not aimed at optimal separation, but to identify rapidly those conditions that can achieve an acceptable separation as a starting point for further method development [23]. The generic screening consists of a limited number of fast separations with different stationary phase and mobile phase combinations. A large number of chiral stationary phases, including polysaccharides, oligosaccharides (cyclodextrins and cyclofructans), macrocyclic glycopeptides, Pirkle-type, and protein-bonded phases are currently available [22,25]. Since the effect of the organic modifier on retention and enantioselectivity is compound and chiral stationary phase-dependent and often difficult to predict, various organic modifiers have to be tested during the initial screening of conditions. Some insight into the chiral recognition mechanism in SFC was presented by West et al. [29].

for tris-(3,5-dimethylphenylcarbamate) amylose and cellulose-based chiral stationary phases. Chiral separations in SFC will be focused in details in Chapter 12, Separation of Stereoisomers, of this book. Therefore Table 16.2 summarizes briefly typical chromatographic conditions used for separations in SFC screening methods in pharmaceutical analysis.

Polysaccharide chiral stationary phases based on amylose and cellulose (see Table 16.2 for details) are the most widely used. In some approaches separate procedures for the separation of acidic, basic, and neutral compounds are employed [23,30], while as far as possible generic approaches are more popular [31–33]. The generic methods employ additives in order to facilitate the screening step with no need for compound classification. There is a growing trend in the use of volatile additives, such as ammonia [34–35] or volatile buffers to facilitate the coupling of SFC with MS and for preparative applications. Both isocratic and gradient elution may be used with the latter providing more flexibility and faster separations. Further increase in screening effectiveness may be reached when using sub-2-μm particle chiral stationary phases [36] or parallel multiple-column screening systems [37–38].

16.2.2 Purification of Compounds Using Preparative SFC

Preparative SFC will be discussed in details in Chapter 9, Theoretical Aspects of Preparative SFC and Chapter 10, Practical Aspects and Applications of Preparative SFC. Therefore only a brief summary is presented here in order to emphasize its importance in pharmaceutical industry. Preparative chromatography is routinely used for the purification of from a few grams to the tons per day scale. For small-scale methods (typically up to tens of kilogram per day) a single column processes are used, while for larger scales, complex multicolumn processes are needed. The choice of a process or preparative-scale separation depends on productivity, solvent consumption, and complexity of the design and cost factors [39]. Preparative SFC is used for purification of both achiral and chiral compounds.

Although a large number of approaches have been used to isolate single enantiomers, enatioseparations using LC or SFC on chiral stationary phases has become the most widely used technique in the context of obtaining limited quantities of pure enantiomers quickly, particularly in drug discovery. At the discovery stage, minimizing the time required to obtain pure enantiomers is of significant importance. In addition, enantioselective chromatography is able to isolate both enantiomers in high purity, which is important for biological, pharmacological, and toxicological evaluation. Nowadays SFC has become the predominant technique for the isolation of enantiomers in the pharmaceutical industry. There are several advantages relative to LC: (1) increased flow-rates allow

TABLE 16.2 Typical Chromatographic Conditions for Chiral SFC Separation in Pharmaceutical Analysis

Stationary phase	**Chiral selector**		**Examples of commercial columns**	
	Amylose tris(3,5-dimethylphenylcarbamate) Amylose tris(3-chlorophenylcarbamate) Amylose tris(5-chloro-2-methylphenylcarbamate) Amylose tris(3,5-dichlorophenylcarbamate) Cellulose tris(3,5-dimethylphenylcarbamate) Cellulose tris(3-chloro-4-dimethylphenylcarbamate) Cellulose tris(3,5-dichlorophenylcarbamate) Cellulose tris(4-methylbenzoate)		Chiralpak AD, Chiralpak IA, Amycoat Chiralpak ID Chiralpak AY, Lux Amylose 2 Chiralpak IE Chiralcel OD, Chiralpak IB, Lux-Cellulose 1 Chiralcel OZ, Lux-Cellulose 2 Chiralpak IC Chiralcel OJ, Lux-Cellulose 3	
Organic modifier	Methanol, ethanol, isopropanol Acetonitrile Mixtures of alcohols, mixtures of alcohol and acetonitrile			
Additive 0.1–1%	**Acidic compounds**	**Basic compounds**	**Neutral compounds**	**Generic approach**
	TFA acetic acid	DEA, TEA, IPA NH_4OH Isobutylamine	Not necessary	IPA/TFA DEA/TFA TEA/TFA AmAc, AmF
Temperature	20–40°C			
BPR	100–150 bar			

TFA, trifluoracetic acid; *DEA*, diethylamine; *TEA*, trimethylamine; *IPA*, isopropylamine; *AmAc*, ammonium acetate; *AmF*, ammonium formate; *BPR*, back pressure regulator.

higher productivity, (2) typical mobile phases being based on CO_2 make the method more environmentally friendly and cost effective, and finally, (3) CO_2 is easily removed postseparation by decreasing pressure allowing a highly concentrated product in the modifier to be obtained, which is easily removed. In addition, the solvent volumes in preparative SFC are 2–10 times less than in preparative LC. The majority of preparative SFC enantioseparations utilizes polysaccharide-based or Pirkle-type stationary phases due to their high loadability allowing application of increased amounts of racemates [22,40–42].

The primary goal of high-throughput combinatorial synthesis in drug discovery is to provide drug candidates for further testing. These products are often not adequately pure for biological screening, since impurities can severely complicate the interpretation of structure-activity relationship data. To provide libraries of highly pure compounds for biological screening, purification process usually follows using either LC or SFC. Beyond already discussed benefits of preparative SFC, in achiral purification further advantages include the possibility to analyze compounds of low solubility and/or stability in aqueous mobile phases used in RP-HPLC and also compounds unretained and/or unresolved in RP-HPLC [43].

16.3 SFC IN ANALYSIS OF DRUG SUBSTANCES AND DRUG PRODUCTS

While SFC has long been the method of choice for chiral analysis and purification in support of pharmaceutical discovery and development, implementation of this technique in regulated pharmaceutical laboratories has been slower due to limitations in sensitivity, reproducibility, accuracy, and robustness. Such limitations did not allow for appropriate method validation and robustness required by regulatory authorities, as discussed in Section 16.1.2. Indeed, one significant limitations of SFC systems used to be the lower sensitivity of UV detection compared to LC-UV systems, mostly related to higher baseline noise. This noise arises from pressure fluctuations resulting in refractive index and density changes. Using new BPR designs enabling more precise pressure control this issue has been substantially improved during the last few years. This together with other improvements, such as degassing of the modifier and thermal matching of the mobile phase to the detector temperature the baseline noise of new SFC systems is much improved. New BPR designs also contributed significantly to improved system stability and repeatability of retention times. Since about 2012 the commercialization of the next generation of analytical SFC instrumentation has facilitated the use of SFC in a GMP (good manufacturing practice) environment. In addition, these SFC platforms are suitable for use with sub-2-μm particles and are designated as UHPSFC [15,44].

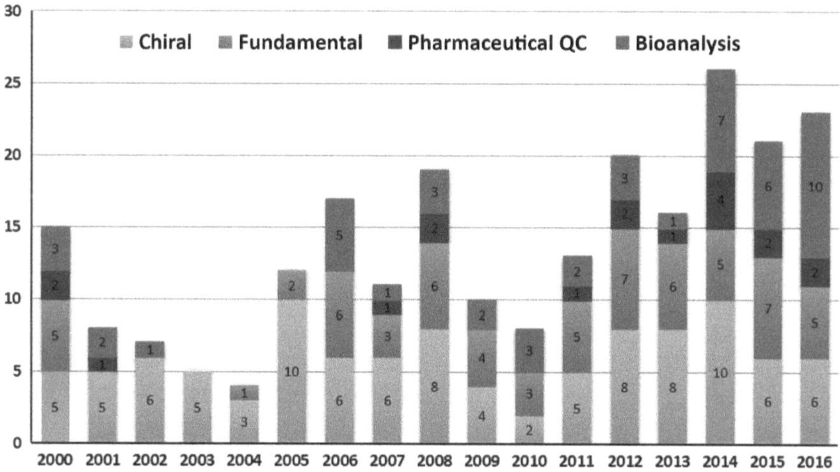

FIGURE 16.2 The number of published articles dealing with SFC in various fields of pharmaceutical analysis.

The separation potential and applicability of SFC in pharmaceutical analysis has already been shown in many fundamental studies using standard mixtures of various pharmaceuticals or QC samples [45]. These compounds were used as probes to study individual variables of the SFC system, such as stationary phase [46,47], mobile phase composition [47–49], influence of mobile phase additives [48,50], effect of pressure drop and temperature, as well as the evaluation of different detection systems (e.g., evaporative light scattering detector (ELSD) [51] and charged aerosol detector (CAD) [52]). On the other hand, studies demonstrating the potential of SFC with real samples and looking at the method performance from the rigorous point of view of method validation are still quite rare despite the obvious advantages provided by SFC.

In spite of the significant increase in the number of articles using SFC in general (30–40 articles/year before 2012 and 70–80 articles/year after 2012) it is still utilized largely for fundamental studies and enantiomer separations more so than for quantitative analysis and impurity profiling for quality control (QC) methods in pharmacy (Fig. 16.2). In our opinion, the number of reports of SFC for pharmaceutical analysis is not representative of its use in the pharmaceutical industry, since there are many pharmaceutical companies who are using SFC routinely without publishing their work [15]. A few reviews [15,45,53,54] published to date support this conclusion as well.

Various methods for the quality control of drug substances and drug products are used at different stages of drug development as described in Section 16.1.3. A principal difference is whether the method is aimed solely at quantification of an API, and further components, such as in a drug assay, stability-indicating methods and dissolution studies or whether the method is intended for impurity profiling, as in the case of quality control methods and forced degradation studies. The different approaches are discussed further.

16.3.1 Drug Substance and Drug Product Assay Using SFC

Analytical methods for the assay of drug substance and drug product must follow the requirements discussed in Section 16.1.2 and must be thoroughly validated. ICH guideline [10] requires determination of accuracy, precision, linearity, range, and selectivity. In this case, ultimate sensitivity and determination of LOQ and LOD is not needed, as very low concentrations will never be analyzed for the API. On the other hand, method selectivity is extremely important in order to ensure that coelution of the API and impurity/degradation products do not occur.

Using CO_2 as a mobile phase, SFC has essentially the same polarity profile as normal-phase chromatography. SFC can be used to separate both polar and nonpolar substances. However, to enable the elution of relatively polar compounds, an organic modifier is required. Organic solvents such as alcohols (methanol, ethanol, isopropanol) and acetonitrile are used to improve the solvent strength. Methanol is the first choice owing to its higher eluotropic strength, low viscosity, and favorable properties for MS detection. To ensure symmetrical peaks for ionizable compounds, a small amount of an additive is usually added to the mobile phase. For acidic compounds, additives such as formic acid, acetic acid, or trifluoroacetic acid are used. For basic compounds, diethylamine, isopropylamine, or triethylamine is recommended. In recent times, volatile buffers (ammonium formate, ammonium acetate) have become the most popular approach to improve peak shapes for both acidic and basic compounds and at the same time retain compatibility with MS detection. Another option for the analysis of polar compounds is the use of water as an additive. Water is not completely miscible with CO_2, but in reasonable proportion (1–5%) can be added to SFC mobile phases containing CO_2 and an organic solvent [15,44,55].

Even though additives are used mainly to ensure symmetrical and narrow peaks, they also affect the elution order of the analytes. Fig. 16.3 illustrates the influence of ammonium acetate, ammonium formate, and ammonium formate +5% water on the separation of agomelatin (API) and its six impurities on two stationary phases. While the separation improves on the BEH 2-EP stationary phase when ammonium acetate is replaced by ammonium formate and the resolution between impurities 2 and 3 is improved by the addition of 5% water to ammonium formate, on the CSH PFP stationary phase water addition leads to the coelution of another critical pair, API and impurity 5 in this case. These

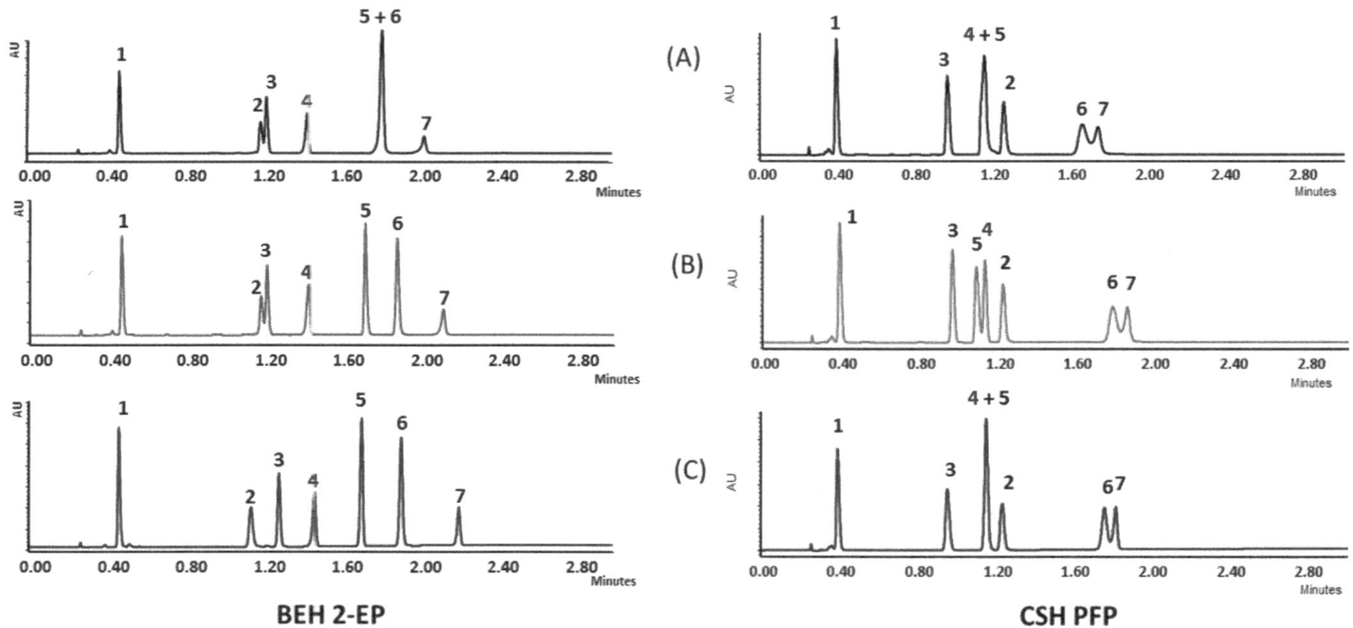

FIGURE 16.3 The influence of additive type on UHPSFC separation of agomelatine and its impurities on BEH 2-EP and CSH PFP stationary phases. Gradient elution with CO_2−methanol + additive, 5-30% in 3 minutes, 2.0 mL/min, temperature 40°C, BPR pressure 2000 psi. (A) 20 mM ammonium acetate, (B) 20 mM ammonium formate, (C) 20 mM ammonium formate with 5% of water. (1) nitrile, (2) dimer, (3) amide, (4) agomelatine (API), (5) acid, (6) amine, (7) AgoSalt. *Source: Unpublished figure created from the data obtained in Plachká K, Chrenková L, Douša M, Nováková L. Development, validation and comparison of UHPSFC and UHPLC methods for the determination of agomelatine and its impurities. J Pharm Biomed Anal 2016;125:376–84 [56].*

compounds are separated, with partial resolution, only when ammonium acetate is replaced by ammonium formate. However, different studies reach different conclusions based on analyte type and experimental conditions, as will be discussed in Section 16.3.2. Therefore it is useful to test the influence of additive experimentally.

In SFC method development, the selectivity of the separation is mainly driven by the type of stationary phase and the mobile phase composition (modifiers, and additives and their concentration). Fine tuning parameters include temperature, pressure, and gradient slope [44]. More details can be found in Chapter 5, Method Development in Supercritical Fluid Chromatography. In terms of method development and separation efficiency, SFC is a complementary option to LC. The limiting step in practice was the quantitative features and low robustness of previous SFC platforms, as discussed previously.

Only few recent reports discuss the development and validation of SFC methods for drug assay control. One reported the quantification of chlorzoxazone, paracetamol, and aceclofenac in individual and combined dosage forms [57]. These methods met the rigorous criteria of ICH guidelines in terms of accuracy, precision, and linearity. However, in this particular case method selectivity was neglected, which is an important drawback. Additionally, it is not clear, if the validation parameters of accuracy and precision were evaluated using real samples or just standard mixtures. In another study, SFC-MS/MS was used for the identification and quantification of isoniazid and pyrazinamide in dosage forms [58]. Method validation met the requirements of ICH guidelines including selectivity and robustness. Since MS/MS detection was employed, matrix effects should have been evaluated as well. Assay and uniformity content were determined for ramipril and telmisartan in their dosage forms [59]. This method was properly validated and was suitable for routine QC analysis in both combined and individual dosage forms. The complete validation using the total error approach was reported by Dispas et al. [60] for several antibiotic drugs. Two high efficiency methods, UHPSFC and UHPLC, were compared in terms of quantitative aspects, method validation, and application to the real samples. Both methods were applicable for quality control purposes with slightly better performance for UHPLC. Moreover, lower method sensitivity for UHPSFC was indicated in this study. However, it was not an issue since only a QC assay method was the object of this study.

Of course, it is more desirable to develop multiple-purpose quality control methods, i.e., assay and impurity profiling in one step. This may also partially explain the low number of SFC drug assay control methods reported so far. However, in some cases, due to time-consuming method development and due to economic reasons, two different methods are preferred anyway in routine practice.

16.3.2 Impurity Profiling Using SFC

The synthesis of APIs is usually a multistep process and cannot be undertaken economically without the acceptance of low, but controlled levels of residual impurities arising from the synthesis. Impurities can arise during synthesis, purification, or storage, and can include raw materials, by-products, catalysts, reagents, and/or intermediates from the synthesis, degradation products, and/or ligands [11]. Equally the same considerations apply for the drug product. The presence of oxygen and moisture in the environment leads to several degradation pathways, such as hydrolysis and oxidation, despite careful selection of formulation and manufacturing processes and packaging [16]. The identification and characterization of impurities is important to ensure that their presence will not cause an adverse response, either pharmacological or toxicological, when taking the medication. Furthermore, the impact of impurities on the production process must be taken into account as well. The identification and characterization should be finished before the first human clinical trials and should also fit to the production schedules. Impurity control is regulated by Q3 ICH guidelines [11,12] and the details were discussed in Section 16.1.2.

The comprehensive determination of all drug impurities is a challenging task. To minimize the possibility that an impurity remains unnoticed under the peak of the active ingredient(s) or under another impurity, several chromatographic systems are usually employed. The combination of chromatographic conditions as orthogonal as possible are recommended [16]. Impurity profiling of drug substance and drug product must also follow the rigorous requirements of the regulation authorities and must be thoroughly validated to demonstrate the method's ability to unequivocally assess each of the analytes in the presence of others. In this case, ICH guideline [10] requires determination of accuracy, precision, linearity, range, and selectivity similar to assays and determination of LOD and LOQ. Method sensitivity in impurity profiling is of great importance as the criteria required to quantitate impurity peaks are strict, as discussed in Section 16.1.2. The specificity is a critical attribute for the methods used in impurity profiling and is also one of the most challenging and time-consuming tasks. One approach to ensure method specificity is to develop two orthogonal chromatographic methods with different retention mechanisms, as it is shown in Fig. 16.4 for UHPSFC and UHPLC.

A great importance of SFC for impurity profiling is witnessed by the interest of scientific community in this research topic. Several groups have focused on the development of generic SFC screening approaches

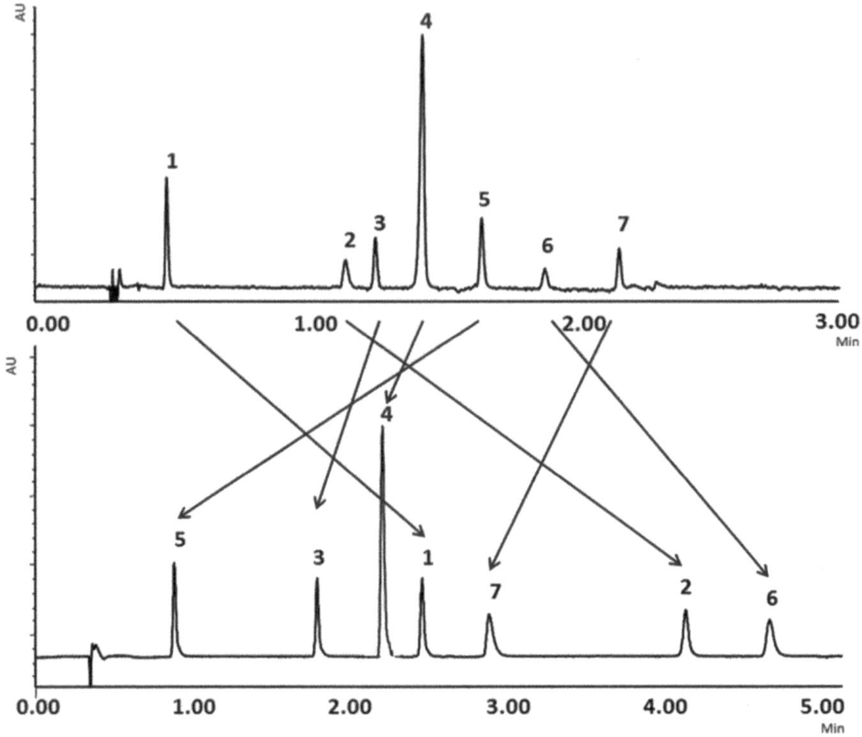

FIGURE 16.4 A comparison between UHPSFC (upper chromatogram) and UHPLC (lower chromatogram) separation selectivity. (1) nitrile, (2) dimer, (3) amide, (4) agomelatine (API), (5) acid, (6) amine, (7) AgoSalt. *Source: Unpublished figure created from the data obtained in Plachká K, Chrenková L, Douša M, Nováková L. Development, validation and comparison of UHPSFC and UHPLC methods for the determination of agomelatine and its impurities. J Pharm Biomed Anal 2016;125:376–84 [56].*

for achiral impurity profiling [46,48,61,62]. Although these studies are fundamentally based using artificial mixtures or large sets of pharmaceuticals and do not separate real API/impurity mixtures, they are very helpful in defining initial SFC screening conditions. Mobile phase composition was studied by Lemasson et al. [48]. using a Derringer desirability function with the criteria including elution capability, peak shapes, UV baseline drift, and UV and MS response based on S/N. Ammonium acetate and ammonium hydroxide were selected as the best additives, among other additives tested, and were further evaluated using larger set of stationary phases. Both additives provided comparable scores with slightly better chromatographic quality observed for ammonium acetate. Finally, a mixture of ammonium acetate with the addition of 2% (v/v) water was selected as the preferred additive. Ammonium formate was identified as the preferred additive in [61]. Also the addition of water was confirmed to have a beneficial effect on peak shapes and stationary phase stability. Detailed evaluation of more than 20 stationary phases was performed in two studies with the aim of identifying orthogonal stationary phases [46,62]. Lemasson et al. [46] proposed two stationary phases, Acquity UPC^2 HSS C18 SB and Nucleosil HILIC, while Galea et al. [62] proposed a set of six stationary phases including Luna Silica, Sunfire C18, Luna NH2, Interstil-Phenyl, Luna HILIC, and Luna CN. None of these studies addressed the quantitative aspect of impurity profiling.

Initially, SFC was rarely used for impurity profiling, in part due to the limitation of instruments available at that time. First impurity profiles by SFC were reported in 2000 and 2001. However, these methods only dealt with method development for the separation of impurities and the evaluation of impurity concentrations relative to the API. Comprehensive method validation was not undertaken. Both methods provided surprisingly good sensitivity for the determination of impurities. In the first case, for the compound SC-65872 reaching up to 0.01% relative to the API [63] and in the second case for isosorbide-5-mononitrate and related compounds in bulk substance and tablets up to 0.1% relative to the API [64]. The loading concentrations were relatively high at 12.5 mg/mL and 10 mg/mL, respectively. After the appearance of these two studies, reports of SFC for impurity profiling became rare until about to 2011. Recently, several studies have focused on all aspects of impurity profiling involving method development targeted on particular application, quantitative aspect, and method validation taking into consideration low concentrations of impurities (see Table 16.3). The separation conditions summarized in Table 16.3 demonstrate a large variation and do not always correspond to the conclusions cited in the fundamental studies earlier. This is attributed to targeted method development focused on specific groups of compounds and the physicochemical properties of these particular compounds.

Wang et al. [65] described the first comprehensive impurity profiling method using SFC in 2011 for the determination of impurities in mometasone furoate. A high sample concentration was required to monitor the trace levels of potential impurities at 0.05% of the area for the API. The method was partially validated and compared to a UHPLC method. All the impurities were separated using both UHPLC and SFC with a shorter separation time for SFC (15 vs 50 minutes). Good orthogonal selectivity for the UHPLC and SFC methods was noted ($R^2 = 0.2163$) similarly, as was shown in other studies comparing SFC and LC methods (see also Fig. 16.4).

TABLE 16.3 An Overview of Impurity Profiling Methods by SFC With Separation Conditions and Percentage Level of Impurities Relative to the API

Determined compounds	Stationary phase	Mobile phase	Percent level relative to API (loading concentration)	References
Agomelatine +6 Impurities	BEH 2-EP (100 × 3.0 mm, 1.7 μm)	CO_2/MeOH +20 mM AmF + 5% H_2O	0.5% Quantitation (0.05 mg/mL)	[56]
Mometasone furoate + 5 impurities	Silica (250 × 4.6 mm, 5 μm)	CO_2/MeOH	0.1% Quantitation (2 mg/mL)	[65]
1. API + 6 impurities **2.** Aripiprazole + 8 impurities	**1.** Diethylaminopropyl (150 × 4.6 mm, 5 μm) **2.** Aminophenyl (150 × 4.6 mm, 5 μm)	**1.** CO_2/MeOH + 1% H_2O **2.** CO_2/MeOH + 0.3% IPA + 5% ACN	0.1% LOD (≤2 mg/mL)	[66] [a]
Lamivudine + 6 impurities Festinavir Efavirenz + 6 impurities	2-EP (150 × 4.6 mm, 3 μm)	CO_2/MeOH + 10 mM AmAc + 0.1% IPA	0.02% Reporting limit 0.05–0.1% quantitation (0.3 or 0.6 mg/mL)	[67] [a]
Rifampicin +4 impurities	Torus diol (100 × 3.0 mm, 1.7 μm)	CO_2/MeOH + 0.1% AmF + 2% H_2O	0.006% LOD (10 mg/mL)	[68]

AmF, ammonium formate; *AmAc*, ammonium acetate; *IPA*, isopropylamine.
[a]*The method deals with quantitative aspect, but was not fully validated.*

Adequate sensitivity for quantifying impurities at different drug loadings was demonstrated by all methods (see Table 16.3). SFC impurity profiling methods were usually faster [56,65,66,68], while higher sensitivity was reported for UHPLC. Another advantage for SFC was the more even distribution of peaks across the chromatogram, fewer system peaks and easier method development compared to UHPLC [56,67]. Other validation parameters, such as retention time and peak area repeatability at low concentration levels of impurities, method precision and accuracy were found comparable for both methods. This has shown SFC to be a viable and complementary option for impurity profiling in pharmaceutical quality control, but further studies and more validated methods are still needed. There are no reports of the use of SFC for genotoxic impurity determinations so far.

Interestingly in ICH guidelines it is not clearly distinguished, whether the impurity is chiral or achiral. Nevertheless there is a growing trend in the pharmaceutical industry to produce drugs in a single enantiomeric form in order to improve their therapeutic profile, such as to lower dose requirements or to obtain a faster response compared with racemic mixtures [69]. Advances related to enantioselective synthesis induced an increase of newly registered drugs containing only the active enantiomer. For some drugs there are both racemic mixtures and single-enantiomer forms available. In case of new drug substances/products containing a single enantiomer, the other enantiomer is treated as a regular impurity and its identification threshold is given in ICH Q3 guidelines [11,12]. Moreover, in the case of racemic drug mixture, the FDA requires that each enantiomer is individually investigated for their toxicity, bioactivity, and pharmacokinetic properties anyway. It is clear that chiral purity control is indispensable at all stages of drug development. As already discussed in Section 16.2.1, this opens up a great space for SFC applications. More details can be found in Chapter 12, Separation of Stereoisomers, and in Section 16.2.1.

Again it is desirable to develop multipurpose quality control methods with sufficient selectivity and sensitivity for all assay and impurity profiling of chiral and achiral impurities in one step. However, this is extremely difficult and such methods are rare both in scientific literature and routine practice. Chiral–achiral in series column arrangement enabled the determination of propionyl carnitine and related impurities using a combination of CHIROBIOTIC TAG and Spherisorb S4 SCX columns in 30 minutes [70]. Similarly using three columns in tandem (diol and two Chiralcel OD), xemilofiban and its chiral and achiral impurities were determined in

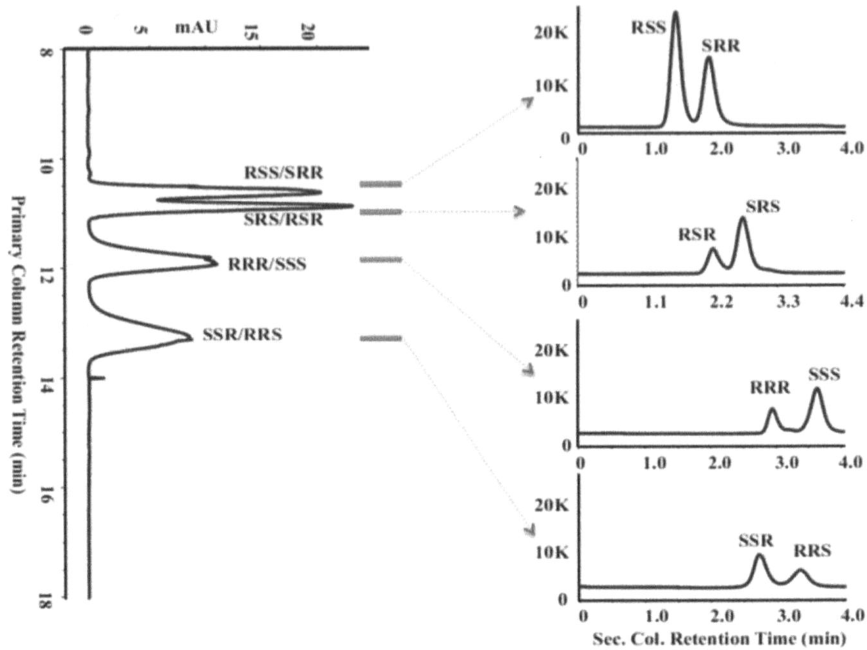

FIGURE 16.5 Separation of eight stereoisomers of drug substance using heart-cutting 2D LC-SFC. The primary RP-HPLC dimension resolves four diastereomeric pairs and the secondary SFC dimension resolves the corresponding enantiomeric pairs. *Source: Reprinted from the reference with permission Venkatramani CJ, Al-Sayah M, Li G, Goel M, Girotti J, Zang L, et al. Simultaneous achiral-chiral analysis of pharmaceutical compounds using two-dimensional reversed phase liquid chromatography-supercritical fluid chromatography. Talanta 2016;148:548–55 [72].*

30 minutes [71]. Achiral–chiral analysis using two-dimensional RP-HPLC × SFC provides an alternative approach [72]. Peaks eluting from the first RP-HPLC dimension are focused as sharp zones on a trapping column and then injected into the second dimension SFC column. The first dimension provided achiral impurity information, while the second dimension achieved the separation of the enantiomers enabling quantification of all components (Fig. 16.5).

16.3.3 Stability-Indicating Methods

A stability-indicating method is an analytical procedure used to quantitate a decrease in the amount of API due to degradation. This method should be validated to ensure its capability to monitor the stability of the drug substances and drug products over time. Stability-indicating method should measure accurately the change in API concentration without interference from degradation products, impurities, and excipients [73]. RP-HPLC with UV detection is the method of choice for this task. There are no methods reported using SFC for this task so far.

16.4 SFC IN ANALYSIS OF DRUGS AND THEIR METABOLITES IN BIOLOGICAL FLUIDS

16.4.1 Sample Preparation for SFC Analysis of Biological Fluids

Biological fluids, such as whole blood, plasma, serum, urine, or saliva are complex samples and, in most cases, unsuitable for chromatographic analysis without preliminary sample preparation. Common interfering compounds for drug analysis include salts, phospholipids, and proteins. The sample preparation step influences all further steps of the analysis with a critical impact on accuracy and precision. Since the analytes are generally present at low concentrations, while interfering compounds are abundant, method sensitivity and selectivity are critical parameters for bioanalytical procedures. Despite the high sensitivity and selectivity of detection by MS, matrix effects can seriously affect the quantitative performance and sensitivity of a method. A well-designed sample preparation technique should enable isolation, clean-up, and/or preconcentration of the analytes in the biological matrix with removal of interfering compounds [19].

The most frequently used sample preparation methods in bioanalysis include protein precipitation (PP), liquid–liquid extraction (LLE), and solid-phase extraction (SPE). When LC-MS is used for sample analysis, the evaporation and reconstitution step is required to ensure the compatibility of the RP-HPLC mobile phase with mostly organic extracts (ACN or MeOH in case of PP and SPE and nonpolar solvents, such as ethyl acetate, hexane, or heptane in case of LLE) [19]. Due to better compatibility of these solvents with SFC mobile phases the evaporation step can be omitted to decrease sample preparation time and increase sample throughput when sample preconcentration is not required. Indeed recently reported studies demonstrate this advantage for extracts from LLE using hexane/isopropanol [74] and from PP using ACN–MeOH (1:1) [75] or acetone [76]. This approach is now feasible due to the sensitivity of modern mass spectrometers in spite of the sample dilution that occurs during sample preparation. However, such an approach is impractical if sample preconcentration is necessary to achieve the desirable method sensitivity.

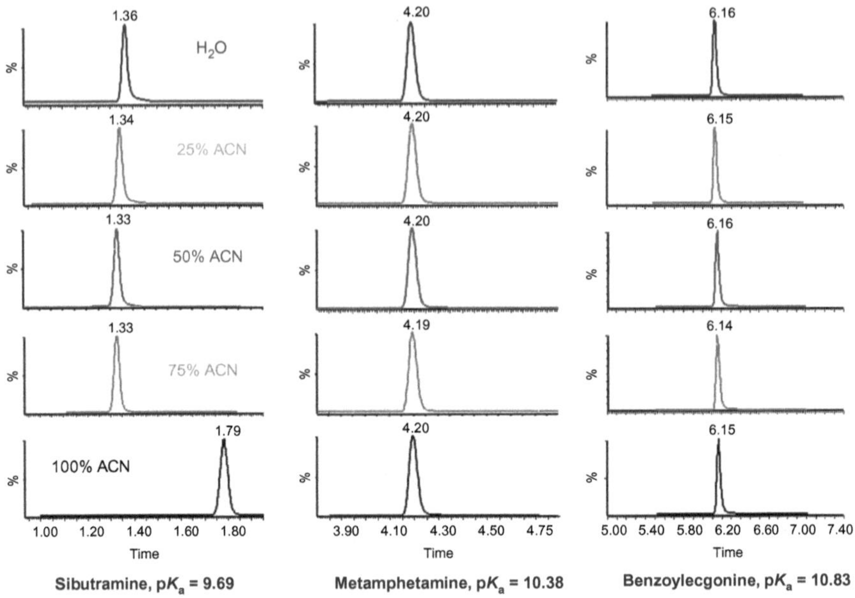

FIGURE 16.6 The influence of sample diluent on the peak shape of early eluted compound (sibutramine), moderately retained compound (metamphetamine), and lately eluted compound (benzoyleckgonine). *Source: Redrawn based on the data from the reference Nováková L, Rentsch M, Grand-Guillaume Perrenoud A, Nicoli R, Saugy M, Veuthey LJ, et al. Ultra high performance supercritical fluid chromatography coupled with tandem mass spectrometry for screening of doping agents. II: analysis of biological samples. Anal Chim Acta 2015;853:647–59 [77].*

The simple dilute and shoot approach, often used with urine samples brings an important constraint for SFC. While in LC the dilution is often made with water, peak shapes on one hand and the possibility of the precipitation of salts by organic solvents on the other hand have to be taken into account when choosing a dilution solvent, as discussed in [77]. Although water seems to be an inappropriate solvent for SFC due to its incompatibility with mobile phase properties, it was shown that injection of aqueous sample was feasible for most compounds out of 110 tested doping agents when using low injection volumes and sample concentrations (Fig. 16.6).

16.4.2 SFC-MS in Quantitative Bioanalysis

SFC-MS is still less widely utilized compared with LC-MS, although a significant increase in its use is noted recently (2014–16,

Fig. 16.2). Only three review articles addressed bioanalytical applications of SFC so far [15,53,78]. The main interest in using SFC for bioanalysis was to obtain different selectivity for the analysis of metabolites [79,80], to analyze groups of compounds that are challenging in RP-HPLC, such as antibiotics [75], lipids [81], and fat-soluble vitamins [74,82,83] and in methods for the quantification of chiral drugs and their metabolites [84].

The coupling of SFC to MS seems relatively easier compared with LC-MS due to the high proportion of volatile CO_2 contained in the mobile phase, which should enhance the evaporation step during the ionization process. Moreover, atmospheric pressure ionization sources developed initially for LC, such as ESI (electrospray ionization) and APCI (atmospheric pressure chemical ionization), are well suited for coupling with SFC [85,86]. However, the possibility of the precipitation of analytes due to mobile phase decompression in SFC must be taken into account. UHPSFC—MS coupling was studied in details by Grand-Guillaume Perrenoud et al. [87] (see chapter: Hyphenated Detectors: Mass Spectrometry, for details). The settings of individual ion source parameters and the mobile phase composition can alter ionization efficiency and thus sensitivity in SFC-MS (as is known for LC-MS). Since SFC-MS is a new approach in bioanalysis, most studies focus on the optimization of individual parameters such as ion source set-up, mobile phase composition, make-up solvent composition, temperature, and BPR pressure [74,75,79,80,82,84,87]. It is substantially more difficult to develop some generic SFC approaches in bioanalysis due to the considerable diversity of applications running across a wide range of analyte polarity and other physicochemical properties (Table 16.4).

Generally the use of volatile additives including ammonium acetate and ammonium formate is reported to be beneficial in terms of peak shapes and sample detectability. Moreover, many studies favor the addition of small amount of water (1–5%) contributing to symmetrical peaks, stationary phase stability, and MS sensitivity [79,80,83,91]. In addition, small amounts of water in the mobile phase provide improved solubility for buffers compared with organic modifiers alone.

An interesting approach for the simultaneous analysis of water- and fat-soluble vitamins employs supercritical, subcritical, and liquid conditions achieved with a mobile phase gradient starting from pure carbon dioxide and terminating with 100% organic modifier. Seventeen vitamins with a log P range from -2.11 to 10.2 were separated in 4 minutes [83], as shown in Fig. 16.7.

TABLE 16.4 An Overview of Bioanalytical SFC Methods Demonstrating the Variability of Applications

Determined compounds	Stationary phase	Mobile phase	Method validation	Matrix effects	References
Tocopherols Tocotrienols	BEH 2-EP (100 × 3.0 mm, 1.7 μm)	CO_2/MeOH + 10 mM AmF (95:5 and 98:2)	+	+	[74]
Fifteen sulfonamides and metabolites	BEH (100 × 3.0 mm, 1.7 μm)	CO_2/EtOH gradient elution	+	– PDA	[75]
Coenzyme Q10	BEH 2-EP (100 × 3.0 mm, 1.7 μm)	CO_2/MeOH (85:15)	+	+	[76]
Polar urinary metabolites	BEH, BEH 2-EP, CSH PFP, HSS C18 SB, BEH amide, BEH HILIC, BEH phenyl, HSS CN, Torus diol, Torus 2-PIC, Torus-DEA, Torus 1-AA	CO_2/MeOH, water, FA, AA, AmF, AmAc, AmOH, IPA, IBA, IPeA	–	–	[79]
Eight substrates and eight CYP-specific metabolites	HSS C18 SB (100 × 3.0 mm, 1.8 μm) CSH PFP, BEH 2-EP, HSS Cyano (100 × 3.0 mm, 1.7 μm)	CO_2/MeOH CO_2/isopropanol + 10 mM AmF ± 2% water	–	–	[80]
Vitamin D and its metabolites	1-AA (100 × 3.0 mm, 1.7 μm)	CO_2/MeOH gradient elution	+	–	[82]
Seventeen water- and fat-soluble vitamins	HSS C18 SB (100 × 3.0 mm, 1.8 μm)	CO_2/MeOH gradient elution + 0.2% AmF + 5% water	–	–	[83]
Oxcarbazepine and chiral metabolites	Trefoil Cel2 (150 × 3.0 mm, 2.5 μm)	CO_2/MeOH (60:40)	+	+	[84]
Bifendate	HSS C18 SB (100 × 3.0 mm, 1.8 μm)	CO_2/MeOH (95:5)	+	+	[88]
R-/S-warfarin	Chiralpak AD (250 × 4.6 mm, 5 μm)	CO_2/EtOH (70:30)	+	+	[89]
R-/S-propranolol R-/S-pindolol	Chiracel OD-H (250 × 4.6 mm, 5 μm)	CO_2/MeOH + 0.2% DEA	+	+	[90]

AmF, ammonium formate; *AmAc*, ammonium acetate; *AmOH*, ammonium hydroxide; *AA*, acetic acid; *FA*, formic acid; *DEA*, diethylamine; *IPA*, isopropylamine; *IBA*, isobutylamine; *IPeA*, isopentylamine.

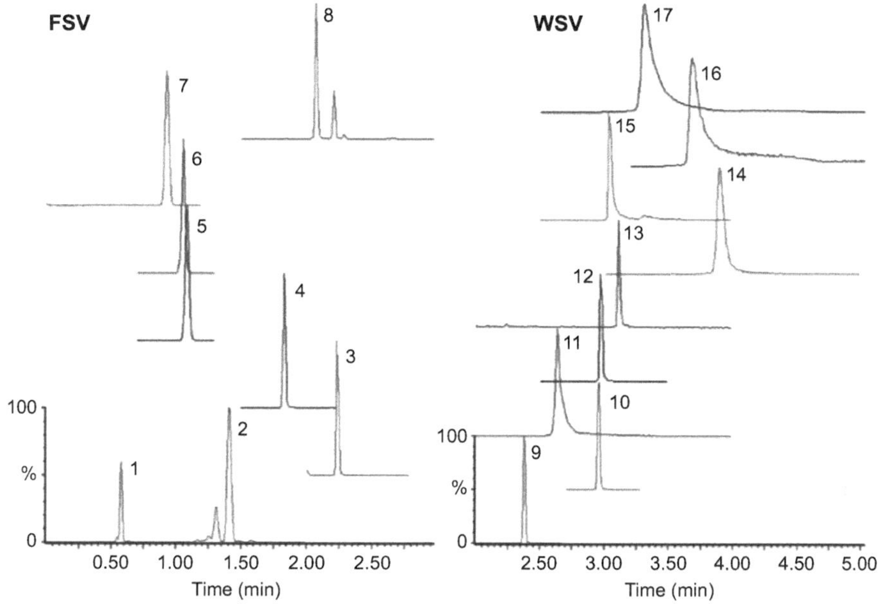

FIGURE 16.7 Separation of 17 vitamins using a gradient from 100% CO_2 to 100% organic modifier. Column: HSS C18 SB, organic modifier: methanol/water (95:5) with 0.2% ammonium formate, flow-rate 1.2 mL/min, temperature 40°C. Further details are given in Ref. [83]. *FSV*, fat-soluble vitamin; *WSV*, water-soluble vitamin. *Source: Adapted from the reference Taguchi K, Fukusaki E, Bamba T. Simultaneous analysis of water- and fat-soluble vitamins by a novel single chromatography technique unifying supercritical fluid chromatography and liquid chromatography. J Chromatogr A 2014;1362:270–77 with permission.*

16.4.3 Matrix Effects in SFC-MS

Matrix effects are an important issue for quantitative methods using MS as a detection technique. It has been widely discussed in LC-MS, while in SFC-MS it has received less attention so far (Table 16.4). Studies focusing on the fundamental aspects of matrix effects in SFC are missing. Although the determination of matrix effects is now mandatory in bioanalytical methods [17,18], it was often neglected in early SFC-MS studies and has only recently become an intrinsic part of these methods [74,76,84,88]. Some studies only focus on SFC method development itself without any quantitative aspect and full method validation. The first reports of matrix effects in SFC-MS were published in 2006. The matrix effects were evaluated using a postextraction addition method for warfarin in human plasma showing ion suppression −24% to −33% [89]. The postcolumn infusion method was used also for the evaluation of matrix effects for propranolol and pindolol in mouse blood using

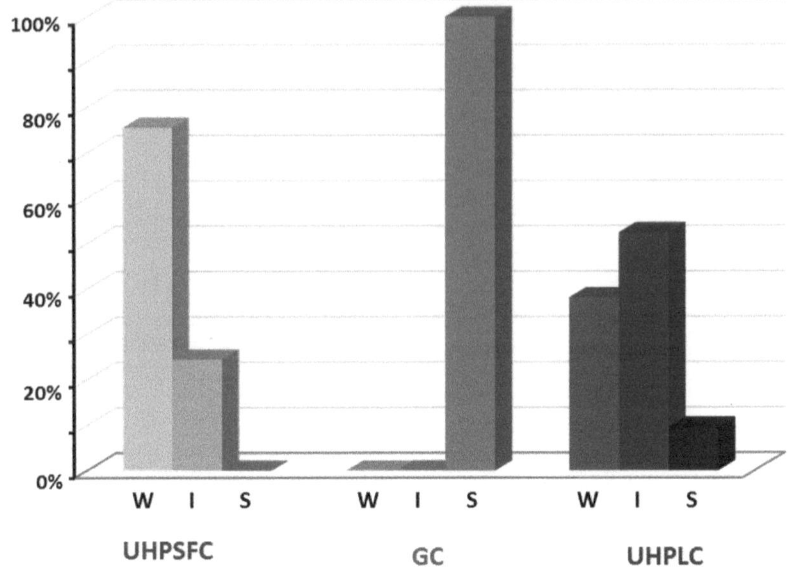

FIGURE 16.8 Graphical representation of the importance of matrix effects for UHPSFC, UHPLC, and GC. Weak matrix effects (W) correspond to $-20\% < ME_{rel} < 20$, intermediate matrix effects (I) to $20\% < |ME_{rel}| < 100\%$, and strong matrix effects (S) to ME_{rel} higher than 100% or lower than -100%. *Source: Adapted from the reference Desfontaine V, Nováková L, Ponzetto F, Nicoli R, Saugy M, Veuthey JL, et al. Liquid chromatography and supercritical fluid chromatography as alternative techniques to gas chromatography for the rapid screening of anabolic agents in urine. J Chromatogr A 2016;1451:145–55 with permission.*

APCI and established the absence of ion suppression effects [90]. Systematic evaluation of matrix effects and a comparison among UHPSFC–MS/MS, UHPLC–MS/MS, and GC-MS/MS was reported for doping agents in urine after a simple dilute and shoot sample pretreatment and supported liquid extraction, respectively [90,92]. Matrix effects were evaluated at several concentration levels in both reports. The results revealed generally a lower incidence of matrix effects for UHPSFC–MS/MS compared with UHPLC–MS/MS [90,92]. Serious matrix effects were reported for GC-MS [92] (see Fig. 16.8). Further fundamental studies and comparison with other techniques are needed.

16.5 CONCLUSION

SFC is an important technique in pharmaceutical analysis and its importance is still growing promoted by the introduction of improved SFC instrument platforms. It has been traditionally used for enantiomer separations and purification and is established as the technique of choice for these applications, alongside LC. In this way SFC is providing very good complementarity and many advantages due to favorable properties of CO2-based mobile phases. Recently also other domains of pharmaceutical analysis experience an increase of importance of SFC. While for a long time applications of SFC were limited in clinical and impurity profiling, this trend is currently reversing. In bioanalysis, especially, an increase in the number of SFC-MS methods should be noted and this trend is further expected to grow. In the future, more validated methods for both bioanalytical quantification and impurity profiling are needed. In addition, the application of SFC to stability-indicating methods and the determination of genotoxic impurities can be expected.

References

[1] http://www.fda.gov/.
[2] Levin S. High performance liquid chromatography (HPLC) in the pharmaceutical analysis. Pharmaceutical sciences Encyclopedia. Published on-line, John Willey and Sons 2010.
[3] Edger J, Sedrani R, Wiesmann CH. The discovery of first-in-class drugs: origins and evolution. Nat Rev Drug Discov 2014;13:577–87.
[4] Hassan BAR. Overview on pharmaceutical formulation and drug design. Pharmaceutica. Analytica Acta 2012;3:10–11.
[5] Hwang R, Kowalski DL. Design of experiments for formulation, development. Pharm Technol 2005;2005:1–4.
[6] International Conference on Harmonization of technical requirements for registration of pharmaceuticals for human use. ICH harmonized tripartite guideline, Q1A (R2): Stability testing of new drug substances and products; 2003.
[7] International Conference on Harmonization of technical requirements for registration of pharmaceuticals for human use. ICH harmonized tripartite guideline, Q1C: Stability testing for new dosage forms; 1996.
[8] http://www.ema.europa.eu/ema/.
[9] http://www.ich.org/home.html.
[10] International Conference on Harmonization of technical requirements for registration of pharmaceuticals for human use. ICH harmonized tripartite guideline, Q2 (R1): Validation of Analytical Procedures: Text and Methodology; 2005.
[11] International Conference on Harmonization of technical requirements for registration of pharmaceuticals for human use. ICH harmonized tripartite guideline, Q3A(R2): Impurities in New Drug Substances; 2002.
[12] International Conference on Harmonization of technical requirements for registration of pharmaceuticals for human use. ICH harmonized tripartite guideline, Q3B(R2): Impurities in New Drug Products; 2003.
[13] International Conference on Harmonization of technical requirements for registration of pharmaceuticals for human use. ICH harmonized tripartite guideline, M7: Assessment and Control of DNA Reactive (Mutagenic) Impurities in Pharmaceuticals to Limit Potential Carcinogenic Risk; 2014.
[14] International Conference on Harmonization of technical requirements for registration of pharmaceuticals for human use. ICH harmonized tripartite guideline, Q6A(R2): Test

Procedures and Acceptance Criteria for New Drug Substances and New Drug Products: Chemical Substances; 1999.
[15] Desfontaine V, Guillarme D, Francotte E, Nováková L. Supercritical fluid chromatography in pharmaceutical analysis. J Pharm Biomed Anal 2015;113:56–71.
[16] Holm R, Elder DP. Analytical advances in pharmaceutical impurity profiling.. Eur J Pharm Sci 2016;87:118–35.
[17] Committee for medicinal products for human use. Guideline on bioanalytical method validation, EMA, London, UK; 2011.
[18] U.S. Department of Health and Human Services, Food and Drug Administration. Guidance for industry, Bioanalytical method validation; 2001.
[19] Nováková L. Challenges in the development of bioanalytical liquid chromatography-mass spectrometry method with emphasis on fast analysis. J Chromatogr A 2013;1292:25–37.
[20] FDA's policy statement for the development of new stereoisomeric drugs. 1992; May 1, corrections made on 1997; January 3.
[21] Patel BK, Hutt AJ. Stereoselectivity in drug action and disposition: an overview. In: Reddy IK, Mehvar R, editors. Chirality in drug design and development. New York: Marcel Dekker Inc; 2004. p. 139–87.
[22] Zhang Y, Wu DR, Wang-Iverson DB, Tymiak AA. Enantioselective chromatography in drug discovery. Drug Discov Today 2005;10:571–7.
[23] Zhao Y, Woo G, Thomas S, Semin D, Sandra P. Rapid method development for chiral separation in drug discovery using sample pooling and supercritical fluid chromatography-mass spectrometry. J Chromatogr A 2003;1003:157–66.
[24] Płotka JM, Biziuk M, Morrison C, Namieśnik J. Pharmaceutical and forensic applications of chiral supercritical fluid chromatography. Trends Anal Chem 2014;56:74–89.
[25] West C. Enantioselective separations with supercritical fluids—review. Current Anal Chem 2014;10:99–120.
[26] Ren-Qui W, Teng-Teng O, Weihua T, Siu-Choon N. Recent advances in pharmaceutical separations with supercritical fluid chromatography using chiral stationary phases. Trends Anal Chem 2012;37:83–100.
[27] De Klerck K, Mangelings D, Vander Heyden Y. Supercritical fluid chromatography for the enantioseparation of pharmaceuticals. J Pharm Biomed Anal 2012;69:77–92.
[28] Kalíková K, Šlechtová T, Vozka J, Tesařová E. Supercritical fluid chromatography as a tool for enantioselective separation: a review. Anal Chim Acta 2014;821:1–33.
[29] West C, Guenegou G, Zhang Y, Morin-Allory L. Insights into chiral recognition mechanism in supercritical fluid chromatography. II Factors contributing to enantiomer separation on tris-(3,5-dimethylphenylcarbamate) of amylose and cellulose stationary phase. J Chromatogr A 2011;1218:2033–57.
[30] Maftouh M, Granier-Loyaux C, Chavana E, Marini J, Pradines A, Vander Heyden Y, et al. Screening approach for chiral separation of pharmaceuticals: Part III. Supercritical fluid chromatography for analysis and purification in drug discovery. J Chromatogr A 2005;1088:67–81.
[31] De Klerck K, Mangelings D, Clicq D, De Boever F, Vander Heyden Y. Combined use of isopropylamine and trifluoroacetic acid in methanol-containing mobile phases for chiral supercritical fluid chromatography. J Chromatogr A 2012;1234:72–9.
[32] De Klerck K, Tistaert CH, Mangelings D, Vander Heyden Y. Updating a generic screening approach in sub- or supercritical fluid chromatography for the enantioresolution of pharmaceuticals. J Supercrit Fluids 2013;80:50–9.
[33] Klerck De, Vander Heyden Y, Mangelings D. Generic chiral method development in supercritical fluid chromatography and ultra-performance supercritical fluid chromatography. J Chromatogr A 2014;1363:311–22.

[34] Ventura M, Murphy B, Goetzinger W. Ammonia as a preferred additive in chiral and achiral applications of supercritical fluid chromatography for small, drug-like molecules. J Chromatogr A 2012;1220:147–55.
[35] Hamman C, Schmidt Jr DE, Wong M, Hayes M. The use of ammonium hydroxide as an additive in supercritical fluid chromatography for achiral and chiral separations and purifications of small basic medicinal molecules. J Chromatogr A 2011;1218: 7886–94.
[36] Hamman C, Wong M, Aliagas I, Ortwine DF, Pease J, Schmidt DE, et al. The evaluation of 25 chiral stationary phases and the utilization of sub-2.0 μm coated polysaccharide chiral stationary phases via supercritical fluid chromatography. J Chromatogr A 2013;1305:310–19.
[37] Zeng L, Xu R, Laskar DB, Kassel DB. Parallel supercritical fluid chromatography/ mass spectrometry system for high-throughput enantioselective optimization and separation. J Chromatogr A 2007;1169:193–204.
[38] Zhang Y, Watts W, Nogle L, McConnell O. Rapid method development for chiral separation in drug discovery using multi-column parallel screening and circular dichroism signal pooling. J Chromatogr A 2004;1049:75–84.
[39] Rajendran A. Design of preparative-supercritical fluid chromatography. J Chromatogr A 2012;1250:227–49.
[40] Miller L. Preparative enantioseparations using supercritical fluid chromatography. J Chromatogr A 2012;1250:250–5.
[41] White C. Integration of supercritical fluid chromatography into drug discovery as a routine support tool Part I. Fast chiral screening and purification. J Chromatogr A 2005;1074:163–73.
[42] Speybrouck D, Lipka E. Preparative supercritical fluid chromatography: a powerful tool for chiral separations. J Chromatogr A 2016; in press, http://dx.doi.org/ 10.1016/j.chroma.2016.07.050.
[43] Ventura M, Farrell W, Aurigemma C, Tivel K, Greig M, Wheatley J, et al. High-throughput preparative process utilizing three complementary chromatographic purification technologies. J Chromatogr A 2004;1036:7–13.
[44] Nováková L, Grand-Guillaume Perrenoud A, Francois I, West C, Lesellier E, Guillarme D. Modern analytical supercritical fluid chromatography using columns packed with sub-2 μm particles: a tutorial. Anal Chim Acta 2014;824:18–35.
[45] Salvador A, Jaime MA, Becerra G, De La Guardina M. Supercritical fluid chromatography in drug analysis: a literature survey. Fresenius J Anal Chem 1996;356: 109–22.
[46] Lemasson E, Bertin S, Henning P, Boiteux H, Lesellier E, West C. Development of an achiral supercritical fluid chromatography method with ultraviolet absorbance and mass spectrometric detection for impurity profiling of drug candidates: Part II: selection of an orthogonal set of stationary phases. J Chromatogr A 2015;1408:227–35.
[47] Berger TA, Deye J. Effects of column and mobile phase polarity using steroids as probes in packed-column supercritical fluid chromatography. J Chromatogr Sci 1991;29:280–6.
[48] Lemasson E, Bertin S, Henning P, Boiteux H, Lesellier E, West C. Development of an achiral supercritical fluid chromatography method with ultraviolet absorbance and mass spectrometric detection for impurity profiling of drug candidates: Part I: optimization of mobile phase composition. J Chromatogr A 2015;1408:217–26.
[49] Brunelli C, Yining Z, Brown MH, Sandra P. Pharmaceutical analysis by supercritical fluid chromatography: optimization of the mobile phase composition on a 2-ethylpyridine column. J Sep Sci 2008;31:1299–306.

[50] Cazenave-Gassiot A, Boughtflower R, Caldwell J, Hitzel L, Holyoak C, Lane S, et al. Effect of increasing concentration of ammonium acetate as an additive in supercritical fluid chromatography using CO_2-methanol mobile phases. J Chromatogr A 2009;1216:6441–50.
[51] Lesellier E, Valarché A, West C, Dreux M. Effects of selected parameters on the response of the evaporative light scattering detector in supercritical fluid chromatography. J Chromatogr A 2012;1250:220–6.
[52] Brunelli C, Górecki T, Zhao Y, Sandra P. Corona-charged aerosol detection in supercritical fluid chromatography for pharmaceutical analysis. Anal Chem 2007;79: 2472–82.
[53] Abott E, Veenstra TD, Issaq HJ. Clinical and pharmaceutical applications of packed-column supercritical fluid chromatography. J Sep Sci 2008;31:1223–30.
[54] Yaku K, Morishita F. Separation of drugs by packed-column supercritical fluid chromatography. J Biophys Methods 2000;43:59–76.
[55] Lesellier E, West C. The many faces of packed column supercritical fluid chromatography—a critical review. J Chromatogr A 2015;1382:2–46.
[56] Plachká K, Chrenková L, Douša M, Nováková L. Development, validation and comparison of UHPSFC and UHPLC methods for the determination of agomelatine and its impurities. J Pharm Biomed Anal 2016;125:376–84.
[57] Desai PP, Patel NR, Sherikar OD, Metha PJ. Development and validation of packed column supercritical fluid chromatographic technique for quantitation of chlorzoxazone, paracetamol and aceclofenac in their individual dosage forms. J Chromatogr Sci 2012;50:769–74.
[58] Prajapati P, Agrawal YK. SFC-MS/MS for identification and simultaneous estimation of the isoniazid and pyrazinamide in its dosage form. J Supercrit Fluids 2014;95:597–602.
[59] Mehta S, Singh, Chikhalia K, Mehta P, Dadhania T. Determination of assay and uniformity of content of ramipril and telmisartan in their multiple dosage forms by a developed and validated supercritical fluid chromatographic technique. Anal Methods 2014;6:7068–74.
[60] Dispas A, Lebrun P, Ziemons E, Marini R, Rozet E, Hubert P. Evaluation of the quantitative performance of supercritical fluid chromatography: from method development to validation. J Chromatogr A 2014;1353:78–88.
[61] Dispas A, Lebrun P, Sacré PY, Hubert P. Screening study of SFC critical method parameters for the determination of pharmaceutical compounds. J Pharm Biomed Anal 2016;125:339–54.
[62] Galea Ch, Mangelings D, Vander Heyden Y. Method development for impurity profiling in SFC: the selection of a dissimilar set of stationary phases. J Pharm Biomed Anal 2015;111:333–43.
[63] Roston DA, Ahmed S, Williams D, Catalano T. Comparison of drug substance impurity profiles generated with extended length columns during packed-column SFC. J Pharm Biomed Anal 2001;26:339–55.
[64] Gyllenhaal O, Karlsson A. Packed-column supercritical fluid chromatography for the analysis of isosorbide-5-mononitrate and related compounds in bulk substance and tablets. J Biochem Biophys Methods 2000;43:135–46.
[65] Wang Z, Zhang H, Liu O, Donovan B. Development of an orthogonal method for mometasone furoate impurity analysis using supercritical fluid chromatography. J Chromatogr A 2011;1218:2311–19.
[66] Alexander AJ, Hooker TF, Tomasella FP. Evaluation of mobile phase gradient supercritical fluid chromatography for impurity profiling of pharmaceutical compounds. J Pharm Biomed Anal 2012;70:77–86.

[67] Alexander AJ, Zhang L, Hooker TF, Tomasella FP. Comparison of supercritical fluid chromatography and reverse phase liquid chromatography for the impurity profiling of the antiretroviral drugs lamivudine /BMS-986001/efavirenz in a combination tablet. J Pharm Biomed Anal 2013;78–79:243–51.
[68] Li W, Wang J, Yan ZY. Development of sensitive and rapid method for rifampicin impurity analysis using supercritical fluid chromatography. J Pharm Biomed Anal 2015;114:341–7.
[69] Sekhon BS. Exploiting the power of stereochemistry in drugs: an overview of racemic and enantiopure drugs. J Mod Med Chem 2013;1:10–36.
[70] D'Acquarica I, Gasparrini F, Giannoli B, Badaloni E, Galletti B, Giorgi F, et al. Enantio- and chemo-selective HPLC separations by chiral-achiral tandem-columns approach: the combination of CHIROBIOTIC TAGTM and SCX columns for the analysis of propionyl carnitine and related impurities. J Chromatogr A 2004;1061: 167–73.
[71] Ashraf-Khorassani M, Taylor LT, Williams DG, Roston DA, Catalano TT. Demonstrative validation study employing a packed column pressurized chromatography method that provides assay, achiral impurities, chiral impurity, and IR identity testing for a drug substance. J Pharm Biomed Anal 2001;26:725–38.
[72] Venkatramani CJ, Al-Sayah M, Li G, Goel M, Girotti J, Zang L, et al. Simultaneous achiral-chiral analysis of pharmaceutical compounds using two-dimensional reversed phase liquid chromatography-supercritical fluid chromatography. Talanta 2016;148:548–55.
[73] Blessy M, Patel RD, Prajapati PN, Agrawal YK. Development of forced degradation and stability indicating studies of drugs—a review. J Pharm Analysis 2014;4:159–65.
[74] Pilařová V, Gottvald T, Svoboda P, Novák O, Benešová K, Běláková S, et al. Development and optimization of ultra-high performance supercritical fluid chromatography mass spectrometry method for high-throughput determination of tocopherols and tocotrienols in human serum. Anal Chim Acta 2016;934:252–65.
[75] Yuan Z, Wei-E Z, Shao-Hui L, Zhi-Qin R, Wei-Qing L, Yu Z, et al. A simple, accurate, time-saving and green method for the determination of 15 sulfonamides and metabolites in serum samples by ultra-high performance supercritical fluid chromatography. J Chromatogr A 2016;1432:132–9.
[76] Yang R, Li Y, Liu C, Xu Y, Zhao L, Zhang T. An improvement of separation and response applying post-column compensation and one-step acetone protein precipitation for the determination of coenzyme Q10 in rat plasma by SFC-MS/MS. J Chromatogr B 2016;1031:221–6.
[77] Nováková L, Rentsch M, Grand-Guillaume Perrenoud A, Nicoli R, Saugy M, Veuthey LJ, et al. Ultra high performance supercritical fluid chromatography coupled with tandem mass spectrometry for screening of doping agents. II: analysis of biological samples. Anal Chim Acta 2015;853:647–59.
[78] Ríos A, Zougagh M, de Andrés R. Bioanalytical applications using supercritical fluid techniques. Bioanalysis 2010;2:9–25.
[79] Sen A, Knappy C, Lewis MR, Plumb RS, Wilson ID, Nicholson JK, et al. Analysis of polar urinary metabolites for metabolic phenotyping using supercritical fluid chromatography and mass spectrometry. J Chromatogr A 2016;1449:141–55.
[80] Spaggiari D, Mehl F, Desfontaine V, Grand-Guillaume Perrenoud A, Fekete S, Rudaz S, et al. Comparison of liquid chromatography and supercritical fluid chromatography coupled to single quadrupole mass spectrometer for targeted *in vitro* metabolism assay. J Chromatogr A 2014;1371:244–56.
[81] Bamba T, Lee JW, Matsubara A, Fukusaki E. Metabolic profiling of lipids by supercritical fluid chromatography/mass spectrometry. J Chromatogr A 2012;1250:212–19.

[82] Jumaah F, Larsson S, Essén S, Cunico LP, Holm C, Turner CH, et al. A rapid method for the separation of vitamin D and its metabolites by ultra-high performance supercritical fluid chromatography-mass spectrometry. J Chromatogr A 2016;1440:191–200.
[83] Taguchi K, Fukusaki E, Bamba T. Simultaneous analysis of water- and fat-soluble vitamins by a novel single chromatography technique unifying supercritical fluid chromatography and liquid chromatography. J Chromatogr A 2014;1362:270–7.
[84] Yang Z, Xu X, Sun L, Zhao X, Wang H, Fawcett JP, et al. Development and validation of enantioselective SFC-MS/MS method for simultaneous separation and quantification of oxcarbazepine and its chiral metabolites in beagle dog plasma. J Chromatogr B 2016;1020:36–42.
[85] Chen R, A brief review of interfacing supercritical fluid chromatography with mass spectrometry. Chromatogr Today 2009;2:11–15.
[86] Pinkston JD. Advantages and drawbacks of popular supercritical fluid chromatography interfacing approaches—a user's perspective. Eur J Mass Spectrom 2005;11:189–97.
[87] Grand-Guillaume Perrenoud A, Veuthey JL, Guillarme D. Coupling state-of-the-art supercritical fluid chromatography and mass spectrometry: from hyphenation interface optimization to high-sensitivity analysis of pharmaceutical compounds. J Chromatogr A 2014;1339:174–84.
[88] Liu M, Zhao L, Yang D, Ma J, Wang X, Zhang T. Preclinical pharmacokinetic evaluation of a new formulation of a bifendate solid dispersion using a supercritical fluid chromatography-tandem mass spectrometry method. J Pharm Biomed Anal 2014;100:387–92.
[89] Coe RA, Rathe JO, Lee JW. Supercritical fluid chromatography-tandem mass spectrometry for fast bioanalysis of R/S warfarin in human plasma. J Pharm Biomed Anal 2006;42:573–80.
[90] Chen J, Hsieh Y, Cook J, Morrison R, Korfmacher WA. Supercritical fluid chromatography-tandem mass spectrometry for the enantioselective determination of propranolol and pindolol in mouse blood by serial sampling. Anal Chem 2006;78:1212–17.
[91] Nováková L, Grand-Guillaume Perrenoud A, Nicoli R, Saugy M, Veuthey JL, Guillarme D. Ultra high performance supercritical fluid chromatography coupled with tandem mass spectrometry for screening of doping agents. I: investigation of mobile phase and MS conditions. Anal Chim Acta 2015;853:637–46.
[92] Desfontaine V, Nováková L, Ponzetto F, Nicoli R, Saugy M, Veuthey JL, et al. Liquid chromatography and supercritical fluid chromatography as alternative techniques to gas chromatography for the rapid screening of anabolic agents in urine. J Chromatogr A 2016;1451:145–55.

CHAPTER

17

Applications to Food Analysis

J. Bernal, A.M. Ares and L. Toribio

University of Valladolid, Valladolid, Spain

OUTLINE

17.1 INTRODUCTION

In recent years, the use of supercritical fluid chromatography (SFC) has attracted increasing interest form researchers in the food sector. This type of chromatography provides an alternative to liquid (LC) and gas chromatography (GC) separations for some compounds. In addition to the inherent advantages of using supercritical fluids such as carbon dioxide (CO_2), which is the most commonly used fluid because its properties (nontoxic, nonexplosive, relatively inexpensive, easily available experimental conditions) [1], SFC allows the use of high flow rates without the expected column pressure drop that happens in LC, and subsequently, it is possible to obtain similar or better efficiency in shorter separation times without the need for large amounts of organic solvents. Moreover, with the development of packed-column SFC, with its components

Supercritical Fluid Chromatography.
DOI: http://dx.doi.org/10.1016/B978-0-12-809207-1.00017-3

similar to those of LC instruments (injectors, packed columns, UV detectors, electronically controlled back pressure), and the use of organic modifiers to increase the solvating power and to decrease the retention time of polar solutes, it is currently widely accepted that SFC is a green alternative to normal-phase LC without most of its associated problems [2]. In addition, SFC does not employ the high temperatures required for GC, reducing the thermal or chemical degradation of analytes [3]. In addition, moderately polar compounds can be separated by adding organic modifiers, such as methanol, isopropanol, and acetonitrile, into the SFC mobile phase [3]. It should also be commented that recent advances in detection strategies, especially coupling with mass spectrometry (MS), and the commercialization of ultra-high performance supercritical fluid chromatography (UHPSFC) systems, which integrate SFC and ultra-high performance liquid chromatography (UHPLC) technologies [4], has also contributed to the revival of SFC in the food analysis field. The increase in applications of SFC to food analysis is demonstrated in Fig. 17.1, where the number of publications related to food matrices during the last 5 years is shown. The high number of works published in 2015 and 2016 in relation with previous years is remarkable.

These studies, can be grouped according to the nature of the food matrix: animal origin; plant origin (vegetables, fruits, and related

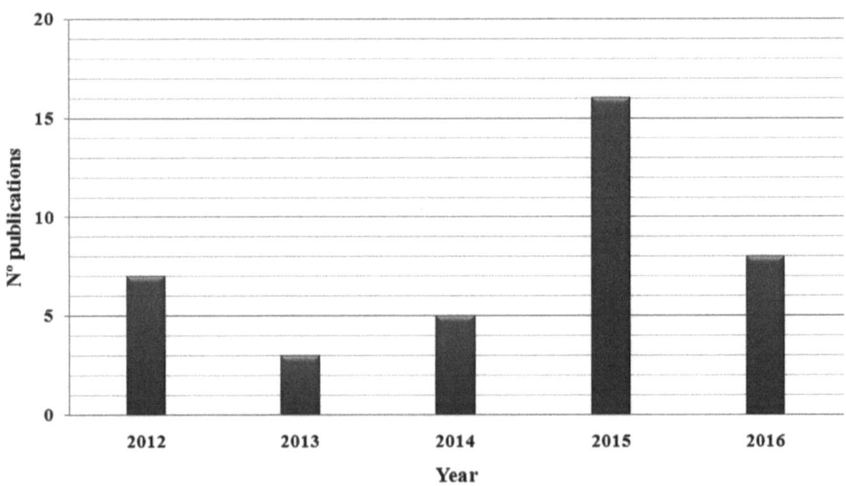

FIGURE 17.1 Evolution of the published works in the last 5 years (2012–16) in food analysis using SFC (data up to end of June 2016). The sources of information were the databases ISI-Web of Knowledge, Scirus, and Scopus. The search has been done using as keywords. ((Supercritical Fluid Chromatography) or (SFC)) and ((Food) or (Beverages) or (Drinks) or (Vegetables) or (Fruits)).

products); and other origins, such as dietary supplements or beverages. Moreover, it must be pointed out that 60% of these works are related to plant origin matrices (see Fig. 17.2).

A wide variety of compounds have been studied by SFC in food matrices, like nutritional compounds (lipids, polyphenols, biogenic amines...) or other compounds such as pesticides, preservatives, or pigments (see Fig. 17.3).

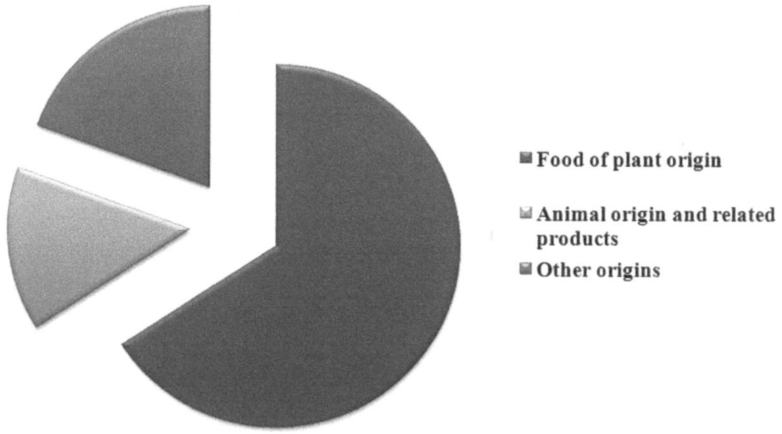

FIGURE 17.2 Summary of food matrices analyzed with SFC in the last 5 years (data up to end of June 2016).

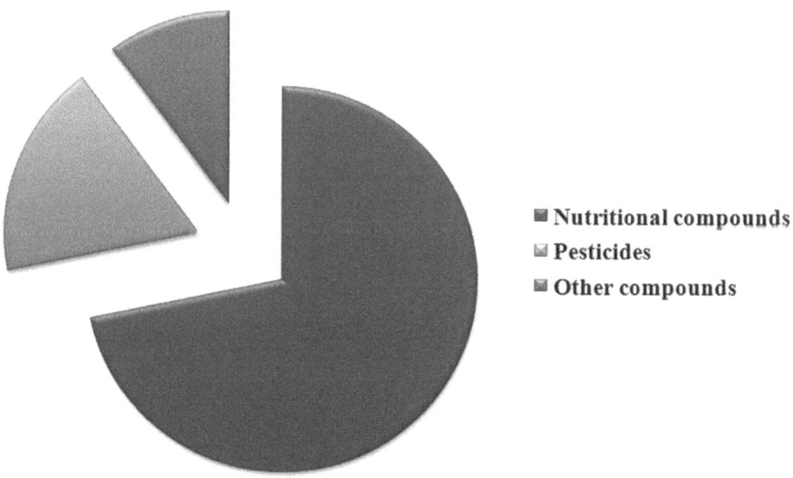

FIGURE 17.3 Summary of the compounds analyzed with SFC in the last 5 years (data up to end of June 2016).

As already stated the aim of this work is to present some of the applications of SFC to food analysis published in the last 5 years with earlier work is discussed in ref. [1]. Thus, this chapter is distributed according to the different types of food matrices, paying special attention to the compounds analyzed, and the different columns, mobile phases, and detectors employed.

17.2 FOOD OF ANIMAL ORIGIN AND RELATED PRODUCTS

Food of animal origin is an important source of nutritional compounds in the diet of humans. Animal products are important as a source of lipids, such as fatty acids, glycerolipids, sterols, or phenols which cannot be metabolically obtained. Lipids are the compounds typically analyzed in this type of foods, as can be seen in Table 17.1, where publications related to the use of SFC for the analysis of food of animal origin are summarized. It must be noted that lipidomics, which is the large-scale study of pathways and networks of cellular lipids in biological systems, is of growing interest in the last decade, since lipids possess many functions in the human organism [5]. For example, a recent work illustrated the usefulness of UHPSFC coupled to evaporative light scattering (ELSD) and MS detection compared with GC-MS for the baseline separation of free fatty acids (FFAs) in fish oil samples [6]. The method facilitated the analysis of beneficial ω-3 and ω-6 fatty acids as well as the monitoring of fish oil degradation.

Experiments were carried out using an Acquity UPC^2 HSS C_{18} SB (150 × 3.0 mm, 1.8 μm) column. The mobile phase consisted of compressed CO_2 and methanol containing formic acid applied in the gradient elution mode. Formic acid was added to both suppress ionization of the FFAs and to improve peak shapes. Analysis by UHPSFC was not only faster than GC, but it also did not require derivatization. However, it was also concluded that when UHPSFC was used with a nonspecific detector, such as ELSD it is difficult to identify components (especially partially resolved and/or coeluting components) and sample detectability is not as good as with MS detectors, which is comparable to GC-MS. UHPSFC-MS was also employed for comprehensive lipidomic analysis of a complex sample (porcine brain extract) in order to show the potential of this technique [5]. Acquity BEH UPC^2 (100 × 3 mm, 1.7 μm) column was used for the analyses. Meanwhile, the mobile phase which was applied in the gradient elution mode was a mixture of CO_2 and a methanol–water mixture containing ammonium acetate as an additive to control the peak shape and compound ionization. The UHPSFC-MS method allowed the identification of 436 lipid species belonging to

TABLE 17.1 SFC Applications in the Analysis of Foods of Animal Origin

Matrix	Compounds of interest (number)	Stationary phase	Mobile phase conditions	Instrument	References
Porcine brain	Lipids (436)	Acquity BEH UPC^2 (100 × 3 mm, 1.7 μm)	A: CO_2, B: Methanol/water (99:1, v/v) containing 30 mM of ammonium acetate Gradient elution mode T^a: 60°C; F-R: 1.9 mL/min; BP: 1800 psi	UHPSFC-MS	[5]
Fish oil	FFAs (31)	Acquity UPC^2 HSS C_{18} SB (150 × 3.0 mm, 1.8 μm)	A: CO_2, B: 0.1% Formic acid in methanol Gradient elution mode T^a: 25°C; F-R: 1.0–1.2 mL/min; BP: 1500 psi	UHPSFC-ELSD/MS	[6]
Ground beef	Lipid A	Acquity UPC^2 BEH 2-EP (150 × 3.0 mm, 1.7 μm)	A: CO_2, B: 0.2% of diethylamine in methanol Isocratic elution mode T^a: 34°C and 40°C; F-R: 2.0 mL/min; BP:2175 psi	SFC-MS/MS	[7]
Honey and bee pollen	Dinotefuran and UF metabolite	Trefoil AMY1_3.0 (150 × 3.0 mm, 2.5 μm)	A: CO_2, B: 0.2% formic acid in methanol Gradient elution mode T^a: 26°C; F-R: 1.9 mL/min; BP: 2010 psi	UHPSFC-MS/MS	[8]

A and B relate to the mobile phase components; BP, back pressure; ELSD, evaporative light scattering detection; FFAs, free fatty acids; F-R, flow-rate; MS, mass spectrometry; MS/MS, tandem mass spectrometry; SFC, supercritical fluid chromatography; Ta, temperature; UHPSFC, ultra-high performance supercritical fluid chromatography.

24 different lipid classes. The main advantage of this method is its short separation time for both nonpolar and polar lipid classes comparable to previously established MS methods. In addition, the quantitative analysis of lipids was significantly improved because the class separation completely avoids the ion suppression effects. Nevertheless, not all lipids have a positive effect for the human organism. Lipid A is a toxic component of bacteria, which can potentially contaminate the human diet, and a component of lipopolysaccharides, which are also known as endotoxins that can provoke fever and inflammation [7]. In order to analyze lipid A in ground beef and lettuce, an SFC-tandem mass spectrometry (MS/MS) method employing a cyanopropyl (30×4.6 mm, 5 μm) column was proposed, utilizing isocratic conditions and CO_2 with methanol containing 0.2% of diethylamine, which was used to improve the elution of lipid A. SFC-MS/MS was an effective tool for rapid analysis of lipid A requiring only a simple extraction procedure without additional cleanup. Other compounds with potential adverse health effects, such as pesticides, have been also analyzed in foods. The UHPSFC-MS/MS coupling was indicated as the fastest and most eco-friendliest choice for the analysis of dinetofuran, a chiral neonicotinoid insecticide and its metabolite UF, in bee products (pollen and honey) as well as environmental samples [8]. This is important as the sum of dinotefuran and UF was also recommended for the estimation of dietary intake and for compliance with the maximum residue limits, and until this study, the residue levels of the two pairs of enantiomers in those matrices were unknown. In total, 16 chiral columns were evaluated for the enantioseparation of dinotefuran and UF with the best results achieved with the Trefoil AMY1_3.0 (150×3.0 mm, 2.5 μm) column and a mobile phase of CO_2 and methanol with formic acid as additive in the gradient elution mode. A three-dimensional response surface methodology was employed to optimize the separation.

17.3 FOOD OF PLANT ORIGIN AND RELATED PRODUCTS

The consumption of foods from plant origin (vegetables, fruits, and related products) is essential for the balance of the human diet. Their contribution to the maintenance of the correct level of fibers, vitamins, and minerals in the body, which can help to prevent chronic diseases, is especially relevant [9]. The compounds analyzed in these food matrices are usually natural compounds, especially lipids, although pesticides, pigments, and preservatives have been also analyzed (see Table 17.2).

TABLE 17.2 SFC Applications in the Analysis of Foods of Plant Origin

Matrix	Compounds of interest (number)	Stationary phase	Mobile phase conditions	Instrument	References
Papaya and Avocado	Pesticides (5)	Viridis BEH 2-EP (100 × 4.6 mm, 5 μm)	A: CO_2, B: Methanol Gradient elution mode T[a]: 40°C; F-R: 1.5 mL/min; BP: 1500 psi	UHPSFC-PDA	[3]
Lettuce	Lipid A	Acquity UPC^2 BEH 2-EP (150 × 3.0 mm, 1.7 μm)	A: CO_2, B: 0.2% of diethylamine in methanol Isocratic elution mode T[a]: 34°C and 40°C; F-R: 2.0 mL/min; BP:2175 psi	SFC-MS/MS	[7]
Kniphofia uvaria seed extracts	Lipids(>50)	[a]Kinetex C_{18} (150 × 4.6 mm, 2.6 μm) [a]Accucore C_{18} core shell (150 × 4.6 mm, 2.6 μm)	A: CO_2, B: Methanol Isocratic elution mode T[a]: 9 and 17°C; F-R: 1.0-1.5 mL/min; BP: 1450 psi	SFC-UV/MS	[10]
Bilberry	Lipids(>200)	Acquity UPC^2 HSS C_{18} SB (100 × 3 mm, 1.8 μm)	A: CO_2, B: Methanol containing 2 g/L ammonium formate Gradient elution mode T[a]: 35°C; F-R: 1.0 mL/min; BP: 2900 psi	UHPSFC-MS	[11]
Soybean	TAGs (13)	[b]Chromolith Performance RP-18e (100 × 4.6 mm, NS)	A: CO_2, B: methanol containing 0.1% (w/w) ammonium formate NS T[a]: 35°C; F-R: 3 mL/min; BP: NS	SFC-MS	[12]
Vegetable oils	TAGs and DAGs (>50)	Acquity UPC^2 BEH-2-EP (100 × 3.0 mm, 1.7 μm)	A: CO_2, B:Methanol/acetonitrile/formic acid (50:50:0.1, v/v/v) Gradient elution mode T[a]: 50°C; F-R: 1.2 mL/min; BP: 1600 psi	UHPSFC-MS/MS	[13]

(Continued)

TABLE 17.2 (Continued)

Matrix	Compounds of interest (number)	Stationary phase	Mobile phase conditions	Instrument	References
Vegetable oils	FFAs (8)	Acquity UPC^2 HSS C_{18} SB (100 × 3 mm, 1.8 μm)	A: CO_2, B: Methanol/acetonitrile (50:50, v/v) containing 0.1% formic acid Gradient elution mode T^a: 40°C; F-R: 1.6 mL/min; BP:1500 psi	UHPSFC-MS/MS	[14]
Vegetable oils	3-MCPD FA esters (13)	Inertsil ODS-4 (250 × 4.6 mm, 5 μm)	A: CO_2, B: 0.1% Ammonium formate in methanol, C: Isopropanol (cleaning) Gradient elution mode T^a: 35°C; F-R: 3.0 mL/min; BP:1450 psi	SFC-MS/MS	[15]
Ginkgo biloba extracts	Terpene lactones (5) and ginkgolic acids (4)	Acquity UPC^2 BEH 2-EP (150 × 3.0 mm, 1.7 μm)	A: CO_2, B: Isopropanol/methanol (50:50, v/v) with 10 mM ammonium acetate Gradient elution mode T^a: 30°C; F-R: 1.4 mL/min; BP: 1500 psi	UHPSFC-PDA/MS	[16]
Turmeric	Curcuminoids (3)	Acquity UPC^2 BEH (100 × 3.0 mm, 1.7 μm)	A:CO_2, B:Methanol containing 10 mM oxalic acid Gradient elution mode T^a: 40°C; F-R: 0.9 mL/min; BP: 1800 psi	UHPSFC-UV	[17]
Mangifera indica L. leaves	Polyphenols (14)	Synergi Hydro-RP C18 (250 × 30 mm, 5 μm)	A: CO_2, B:0.5% Formic acid in methanol Gradient elution mode T^a: 40°C; F-R: 20 g/min; BP: 1450-5800 psi	SFC-UV	[18]
Fermented food	Biogenic amines (8)	Acquity UPC^2 HSS C_{18} SB (100 × 3 mm, 1.7 μm)	A: CO_2, B: *n*-hexane/isopropanol/ammonium hydroxide (70:30:0.15, v/v/v) Gradient elution mode T^a: 50°C; F-R: 2.0 mL/min; BP: 2000 psi	UHPSFC-DAD	[19]

Parsnip	C17-polyacetylenes (2)	Viridis silica (250 × 4.6 mm, 5 μm)	A: CO_2, B: Acetonitrile Gradient elution mode T[a]: 40°C; F-R: 5000 mL/min; BP:1740 psi	SFC-PDA/MS	[20]
Brown rice [21], onion [21], and spinach [21,22]	Pesticides (17 [21]; > 400 [22])	Inertsil ODS-EP (250 × 4.6 mm, 5 μm)	A: CO_2, B: 0.1% Ammonium formate in methanol Gradient elution mode T[a]: 35°C; F-R: 3.0 mL/min; BP:NS	SFC-MS/MS	[21,22]
Tomato, cucumber, apple, grape	Flutrafiol enantiomers (2)	Chiralpak IA-3 (150 × 4.6 mm, 3 μm)	A: CO_2, B: Methanol Isocratic elution mode T[a]: 30°C; F-R: 2.2 mL/min; BP: 2200 psi	UHPSFC-MS/MS	[23]
Wheat, corn, peanut	Isofenphos-methyl enantiomers (2)	Chiralpak IA-3 (150 × 4.6 mm, 3 μm)	A: CO_2, B: Isopropanol Isocratic elution mode T[a]: 30°C; F-R: 2.2 mL/min; BP: 2200 psi	SFC-MS/MS	[24]
Paprika, chili powder, and pepper	Paprika oleoresins (NS)	[c]SB C_{18} and [c]RX-Sil (100 × 3 mm, 1.8 μm)	A: CO_2, B: Isopropanol Gradient elution mode T[a]: 50°C; F-R: 1.95 mL/min; BP:2175 psi	UHPSFC-DAD	[25]
Chili sauces and powder	Illegal dyes (11)	Acquity UPC[2] CSH Fluoro-Phenyl (100 × 3 mm, 1.7 μm),	A: CO_2, B: Methanol/acetonitrile (1:1, v/v) containing 2.5% of formic acid Gradient elution mode T[a]: 70°C; F-R: 2.0 mL/min; BP: 2000 psi	UHPSFC-PDA	[26]

A and B relate to the mobile phase components; BP, back pressure; DAGs, diacylglycerols; FFAs, free fatty acids; F-R, flow-rate; MS, mass spectrometry; MS/MS, tandem mass spectrometry; 3-MCPD FA, 3-monochloropropane-1,2-diol fatty acid; NS, not specified; PDA, photodiode array detector; SFC, supercritical fluid chromatography; T[a], temperature; TAGs, triacylglycerols; UHPSFC, ultra-high performance supercritical fluid chromatography; UV, ultraviolet detector.

[a]*Seven (six Kinetex C_{18} and one Accucore C_{18}) and three (Kinetex C_{18}) linked columns were used for SFC-MS and SFC-UV analyses, respectively.*

[b]*Three linked columns were used.*

[c]*SB C_{18} and RX-Sil columns were used to analyze unsaponified and saponified samples, respectively.*

Lipids constitute a group of naturally occurring molecules that include fats, fatty acids, waxes, sterols, fat-soluble vitamins, terpenes, monoglycerides, diglycerides, phospholipids, carotenoids, and others. Some of the biological functions of lipids are related to energy storage, composition of cell membranes, and molecular signaling [27]. When addressing lipid analysis by SFC, there is a series of advantages with respect to other approaches, such as the absence of a derivatization step, the potential coupling to universal detectors such as ELSD or MS, and the simultaneous analysis of different lipid classes [28]. Several reviews discuss these advantages [1,28–30]. Reference [1] comprehensively covers the field of food matrices up to 2012. Reference [28] is exclusively devoted to lipidomics covering the literature to 2014, reference [29] is devoted to the analysis of natural products, including lipids, up to 2014, and reference [30] applications of packed-column SFC for the analyses and preparative separations of palm oil minor components, including lipids, such as triglycerides, diglycerides, and FFAs. In addition, several lipid classes have been simultaneously analyzed in a complex vegetable extract [10] and bilberry [11] by using SFC. For example, SFC-UV and SFC-MS systems were used for the identification of lipid derivatives in a *Kniphofia uvaria* seed extract [10]. Seven core-shell columns (six Kinetex C_{18}, 150 × 4.6 mm, 2.6 μm; one Accucore C_{18} core shell, 150 × 4.6 mm, 2.6 μm) were connected in series for the separation of the oil by SFC-MS; meanwhile, the SFC-UV analyses were performed with three Kinetex C_{18} (150 × 4.6 mm, 2.6 μm) columns. The optimization of the mobile phase composition and the temperature was performed using chemometric tools. Different modifiers were tested with CO_2, and the best separation was obtained with methanol. As a result, the main molecular families were determined. The major constituents were TAGs, diacylglycerols, and fatty acids. It was concluded that SFC coupled to UV and MS detectors is a good analytical tool to analyze vegetable oil samples. Indeed more than 50 compounds were detected and identified, considerably reducing the volume of toxic solvents used compared with HPLC. Furthermore, several lipid classes were analyzed in supercritical fluid extracts of bilberry by UHPSFC-MS and GC-MS [11]. The mobile phase consisted of CO_2 and methanol containing ammonium formate, in gradient elution mode; with an Acquity UPC^2 HSS C_{18} SB (100 × 3 mm, 1.8 μm) column. It was demonstrated that the derivatization procedure performed prior to GC-MS analysis severely underestimated the lipid content in bilberry. The UHPSFC-MS analyses of the same extracts showed that bilberry contains significant amounts of carotenoids, esters, and phospholipids that are not detected by GC-MS because they are easily hydrolyzed and methylated as glycerols. Lipid profiling is also important to the food industry in order to discriminate and evaluate different cultivars [12] or for identifying authenticity of edible oils [13,14]. For example, the content of TAGs was used for the detailed phenotyping of soybean cultivars [12]. In order to analyze the lipids in many soybean cultivars, a high-throughput and high-resolution analytical method based on SFC-MS was developed. Different columns (InertsilODS-4 column, XBridge C18 column, and three Chromolith Performance RP-18e columns in tandem) were evaluated with methanol containing ammonium formate as modifier to achieve the separation in the shortest time. The Chromolith Performance RP-18e column tandem was selected as the most effective option for TAGs profiling. The proposed method allowed a separation (<8 minutes) with sufficient resolution of the TAGs. Additionally, MS detection using a programmed cone voltage fragmentation method allowed identification of diverse TAGs. The TAGs, which are major ingredients of oils, are used as key indicators for the identification of adulteration based on their characteristic distribution in different oils.

A UHPSFC-MS/MS method without sample treatment for rapidly detecting TAGs and diacylglycerols (DAGs) in various vegetable oils was recently developed [13]. The retention mechanism on three Acquity UPC^2 columns with different polarities (HSS C_{18} SB, 100×3 mm, 1.8 μm; BEH, 100×3 mm, 1.7 μm; BEH 2-EP, 100×3 mm, 1.7 μm) was evaluated with the constraint that a baseline separation of major TAGs was necessary for the purpose of profiling and authenticating edible oils. The BEH-2-EP column was selected. The best performance was obtained with a mixture of methanol, acetonitrile, and formic acid. The results were subsequently analyzed by principal component analysis (PCA) affording a clear classification between six different kinds of vegetable oils, illustrating its potential for quality control and authenticity evaluation. The same group also proposed a UHPSFC-MS/MS method for the direct detection of FFAs in vegetable oils [14]. The FFAs content in foods is of enormous interest because it is also related to the quality and authenticity of the oils. Two columns were evaluated for the separation, Acquity UPC^2 HSS C_{18} SB, 100×3 mm, 1.8 μm and BEH 2-EP, 100×3 mm, 1.7 μm. All FFAs could not be baseline separated on the BEH 2-EP column, especially some positional isomers. Four different modifiers were studied (methanol, acetonitrile, methanol/acetonitrile, and methanol/acetonitrile with 0.1% formic acid). Only for (methanol/acetonitrile with 0.1% formic acid), α-C18:3 and γ-C18:3 could be baseline separated. Moreover, in the absence of formic acid, FFAs were eluted with strong peak tailing, poor selectivity, and long retention times. Finally, it was shown that an optimized SFC method coupled with MS, combined with PCA analysis, can be used to separate and quantify individual saturated or unsaturated FFAs from various edible oils. With this technique a range of FFAs that cannot be directly detected by other chromatographic techniques, such as GC or LC, were determined. 3-Monochloropropane-1,2-diol (3-MCPD) fatty acid esters, reported as process contaminants in refined edible oils and fats, were also analyzed in vegetable oils by SFC-MS/MS [15]. The aim was to develop a new method that did not require nonvolatile ionization reagents or time-consuming sample preparation procedures. The type of modifier and gradient conditions were examined using a cyano column (Inertsil CN-3), a diol column (Inertsil Diol), and several octadecylsiloxane-bonded silica (ODS) columns (Inertsil ODS-4, Inertsil ODS-SP, and Inertsil ODS-P). Inertsil ODS-4 (250×4.6 mm, 5 μm), was selected as the most suitable column because it demonstrated good separation of the analytes in the shortest time (<9 minutes). The selected mobile phase consisted of CO_2 and two modifiers applied in gradient elution mode. The first modifier was methanol containing ammonium formate and was used for the separation and column conditioning the second modifier, isopropanol, was employed for cleaning purposes. The effectiveness of the method was assessed by analyzing 14 fatty acid esters in various edible oils (soybean, corn, and palm).

Terpenes, a different lipid class, have been investigated in vegetables as well. Terpene lactones, flavonol glycosides, and ginkgolic acids were determined in *Ginkgo biloba* extracts using a UHPSFC-photodiode array detector (PDA)/MS method [16]. These extracts, which are derived from the medicinally used leaves of *Ginkgo* tree, are found in many herbal medicines and dietary supplements. However, little is known about the therapeutic contributions of the individual compounds to the mode of action of ginkgo extract, and for this reason there is a need for a reliable analytical method for the detection and quantification of these compounds. Their separation was achieved using a Waters Acquity UPC^2 BEH 2-EP (150×3.0 mm, 1.7 μm) column and a mobile phase consisting of CO_2 and a mixture of isopropanol and methanol with ammonium acetate. It was shown that UHPSFC can be used to determine both ginkgolic acid and terpene lactone concentrations in plant and commercial dietary supplement samples. Single quadrupole MS detection was used to quantify the analytes at low concentrations without derivatization of the ginkgolic acids, an important advantage in comparison with GC. In addition, the use of UHPSFC avoided the hydrolysis of ginkgolides that could occur during RPLC with aqueous eluents.

Other natural compounds than lipids, such as polyphenols [17,18], biogenic amines [19], and polyacetylenes [20], have been investigated as well in food from vegetable origin. Curcuminoids (curcumin, demethoxycurcumin, and bisdemethoxycurcumin), are polyphenolic compounds found in plants, with several health promoting effects. Curcumin, in particular, possesses several pharmacological effects like anticancer, antioxidant, antiinflammatory, and antidiabetic activities. A rapid and efficient UHPSFC-UV method for the isolation of curcumin, demethoxycurcumin, and bisdemethoxycurcumin directly from the methanol extract of turmeric (*Curcuma longa* L.) has been reported [17]. The isolation strategy comprises two steps. Firstly, the method was developed and optimized on an analytical scale UHPSFC instrument.

Several columns with sub-2-μm particles (Acquity UPC^2 BEH, 1.7 μm; BEH 2-EP, 1.7 μm; HSS C_{18} SB, 1.8 μm) and the same dimensions (100 × 3.0 mm) were tested. Different modifiers were also tested, including methanol, ethanol, and mixtures of methanol and acetonitrile. The results demonstrated that an Acquity UPC^2 BEH column with a mobile phase composed of CO_2 and methanol containing oxalic acid provided the best resolution. Second, the analytical method was scaled up with slight modifications for preparative scale SFC. A new column (Viridis BEH OBD 250 × 19 mm, 5 μm) was used for the separation. High purity curcuminoids were obtained from turmeric by the proposed method. A similar strategy was employed to fractionate the polyphenols of *Mangifera indica Linn* by-products, which include phenolic acids, flavonoids, xanthones, benzophenones, and gallotannins, of high value on account of their biological properties [18]. Concentrated fractions of specific families of valuable compounds used in cosmetics or nutraceutical products were isolated. The influence of some SFC parameters (pressure, temperature, and mobile phase composition) on the separation of mango polyphenols at pilot plant scale using a Synergi Hydro-RP C18 (250 × 30 mm, 5 μm) column was studied. A complex mango leaf extract was fractionated under the optimized conditions with isolation of the mango polyphenols. It was concluded that a gradient with changes in the modifier content (methanol with formic acid) was necessary for the separation of the mango polyphenols by SFC. The same conditions were used for the fractionation of mango leaf extracts. The pilot plant scale system facilitated the fractionation of large quantities of mango leaf extract for the isolation of mango polyphenols.

Biogenic amines are basic nitrogen-containing compounds found in large amounts as a consequence of microbial metabolism in a wide range of fermented foods. Biogenic amines formed during food processing partially contribute to food aroma; however, more importantly they are also potential precursors for the formation of carcinogenic *N*-nitro compounds [19]. Therefore the analysis of biogenic amines is necessary in order to evaluate their toxicological risk or the degree of food freshness. UHPSFC is a useful tool for this task [19]. The separation efficiency for the biogenic amines was evaluated with different chromatographic conditions, including selection of the column and mobile phase among others. Different columns with sub-2-μm particles including Acquity UPC^2 BEH (100 × 3.0 mm, 1.7 μm), BEH 2-EP (100 × 3.0 mm, 1.7 μm), CSH Fluoro-Phenyl (150 × 3.0 mm, 1.7 μm), and HSS C_{18} SB (150 × 3.0 mm, 1.8 μm) were evaluated. Different cosolvents, including methanol, acetonitrile, and mixtures of methanol with acetonitrile or *n*-hexane with isopropanol were investigated. The HSS C_{18} SB column and a mobile phase consisting of CO_2 and a mixture of *n*-hexane and isopropanol with a linear gradient resulted in better resolution and peak shape. In addition, it was necessary to add ammonium hydroxide to the cosolvent to improve the resolution of some compounds, as the separation of biogenic amines was pH dependent. The proposed method was used to characterize biogenic amines in Chinese fermented foods. SFC has also been used to purify two C17-polyacetelynes (falcarinol and falcarindiol) with positive health effects (anticancer and antiinflammatory effects) from parsnip (*Pastinaca sativa*) for the production of analytical standards for these compounds [20]. These standards are required for quantification purposes in processing carrot and celery waste. Different columns (10) and cosolvents (4) were evaluated to optimize the SFC purification of the C17-polyacetlynes. The best results were obtained with a Viridis silica (250 × 4.6 mm, 5 μm) column and mobile phase of CO_2 and acetonitrile employed in the gradient elution mode. This work presents a first step towards large-scale purification of C17-polyacetylenes in a sustainable manner.

The application of pesticides to agricultural products is an efficient mechanism for minimizing the loss of agricultural production. Many pesticides, however, are toxic for humans, and it is imperative to monitor their residues in agricultural products, especially in those products likely to be consumed by humans. GC or LC coupled with different detectors are typically used for the screening and/or measurement of pesticide residues in food. Several multiresidue methods using SFC have been proposed in recent years [3,21,22]. Five organophosphate, carbamate, and triazine pesticides were determined in papaya and avocado samples by UHPSFC-PDA [3]. Four columns were evaluated for the separation (two stationary phases with a polar group (Viridis BEH 2-EP and Acquity UPC^2 BEH), an octadecylsiloxane-bonded stationary phase (BEH C_{18}), and a stationary phase with both a bonded polar group and a hydrophobic chain (Shield RP_{18})). The last two columns were discarded because of poor resolution of the pesticides. The Viridis BEH 2-EP column (100 × 4.6 mm, 5 μm) with a mobile phase of CO_2 and a methanol gradient provided the best separation and peak symmetry, although the separation time was slightly longer. Seventeen pesticides (organophosphates, carbamates, bipyridinium, and pyrethroid) with a wide range of properties were determined in brown rice, onion, and spinach by SFC-MS/MS [21]. Several columns with the same dimensions (250 × 4.6 mm) were evaluated and an Inertsil ODS-EP column with a mobile phase of CO_2 and methanol containing ammonium formate selected as the most effective system for the simultaneous analysis of the pesticides. In a related study [22], a larger number of pesticides (>400) were analyzed in spinach extracts by SFC-HRMS (high-resolution mass spectrometry). The separation employed gradient elution in order to accommodate the wide range of pesticide solubility in the mobile phase. It was concluded that when a small number of pesticides are analyzed [21], a single quadrupole mass spectrometer may be sufficient for their detection. However, when higher selectivity is required, high-resolution MS techniques are needed [22], In addition, it was shown that pesticides that need to be analyzed separately by GC-MS and LC-MS can be analyzed simultaneously by SFC-MS/MS in a short time. Specific groups of pesticides, such as pyrethroids [31], triazoles [23], or organophosphorus [24] pesticides in plant matrices have also been determined by SFC-MS. For example, a UHPSFC-MS/MS method was developed to determine flutriafol enantiomers in vegetables and fruits [23]. This study provides a novel and green chiral analytical method for the determination of flutriafol enantiomers in plants. The enantiomers of isofenphos-methyl residues in wheat, corn, and peanut extracts were separated by SFC-MS/MS [24]. The separation was achieved in a relatively short time using a Chiralpak IA-3 column and an isocratic mobile phase containing CO_2 and isopropanol.

Pigments were determined in paprika, chili, and pepper by UHPS [25,26]. This is an important issue, as there is a public concern over the use of artificial dyes in foods and beverages, which has led to the switch to pigments sourced from natural products. Paprika oleoresins and saponified oleoresins were rapidly analyzed using a wide range of columns and modifiers [25]. The quality and value of paprika and paprika oleoresin depends on the intensity of the color, which is mainly due to two compounds (capsanthin and capsorubin), although a number of other colored components are present. In order to optimize the separation, different columns with the same dimensions and particle size (100×3 mm, 1.8 μm): SB-C_{18}, Eclipse Plus-C_{18}, Eclipse XDB-C_{18}, SB-CN, RX-Sil, and an Epic Diol HILIC and modifiers (isopropanol, methanol, acetonitrile, acetone, heptane, and ethyl acetate) were evaluated. The best separation of the paprika oleoresins was obtained by gradient elution on an SB-C_{18} column whereas bare silica provided the best separation for the saponified oleoresins. In both cases, the mobile phase was CO_2 containing isopropanol. Separations by SFC were six-times faster than HPLC and avoid the use of chlorinated solvents. Eleven illegal fat-soluble, red–orange dyes, most frequently found in chili-containing spices were analyzed by UHPSFC [26]. The separation was carried out on an Acquity UPC^2 CSH Fluoro-Phenyl (100×3 mm, 1.7 μm) column by gradient elution with different modifiers (acetonitrile, methanol, isopropanol, and ethanol). The most efficient separation with acceptable resolution was observed with a mixture of acetonitrile and methanol containing formic acid in CO_2. The UHPSFC method was faster than UHPLC. However, the UHPSFC method had poorer detection limits than UHPLC for the same type of detection.

17.4 OTHER ORIGINS

The use of SFC in the analysis of compounds in other food matrices, such as dietary supplements or beverages, for the period 2012–16 is summarized in Table 17.3. The consumption of dietary

TABLE 17.3 SFC Applications in the Analysis of Food of Other Origins

Matrix	Compounds of interest (number)	Stationary phase	Mobile phase conditions	Instrument	References
Dietary supplements	Carotenoids (9)	Acquity UPC² HSS C_{18} SB (150 × 3.0 mm, 1.8 μm)	A: CO_2, B: Methanol/ethanol (1:2, v/v) Gradient elution mode T^a: 35°C; F-R: 1.0 mL/min; BP: 2205 psi	UHPSFC-PDA	[4]
Dietary supplements	Terpene lactones (5) and ginkgolic acids (4)	Acquity UPC² BEH 2-EP (150 × 3.0 mm, 1.7 μm)	A: CO_2, B: Isopropanol/methanol (50:50) containing 10 mM ammonium acetate Gradient elution mode T^a: 30°C; F-R: 1.4 mL/min; BP: 1500 psi	UHPSFC-PDA/MS	[16]
Dietary supplements	Isoflavones (9)	Acquity UPC² BEH (100 × 3.0 mm, 1.7 μm)	A: CO_2, B: 0.05% phosphoric acid in methanol Gradient elution mode T^a: 50°C; F-R: 2.0 mL/min; BP:2175 psi	UHPSFC-PDA	[32]
PPC capsules	Phospholipids (11)	Acquity UPC² BEH column (100 × 3.0 mm, 1.7 μm)	A: CO_2, B: Methanol/acetonitrile/water (60:37.5:2.5 v/v), containing 10 mM ammonium acetate and 0.2% formic acid Gradient elution mode T^a: 50°C; F-R: 1.0 mL/min; BP: 1800 psi	UHPSFC-Ms/MS	[33]
Tocotrienol soft gelatin capsules	Tocopherols (7)	Acquity UPC² BEH column (100 × 3.0 mm, 1.7 μm)	A: CO_2, B: Methanol Isocratic elution mode T^a: 50°C; F-R: 2.5 mL/min; BP:1800 psi	UHPSFC-Ms/MS	[34]
Beverages	Sorbate, benzoate, and caffeine	Lichrospher Diol (250 × 4.6 mm, 5 μm)	A: CO_2, B: 0.1% Acetic acid in methanol Isocratic elution mode T^a: 50°C; F-R: 3.5 mL/min; BP: 2175 psi	SFC-DAD	[35]

A and B relate to the mobile phase components; BP, backpressure; DAD, diode array detector; F-R, flow-rate; MS, mass spectrometry; MS/MS, tandem mass spectrometry; PDA, photodiode array detector; SFC, supercritical fluid chromatography; T^a, temperature; UHPSFC, ultra-high performance supercritical fluid chromatography.

supplements has increased in the last few years as a way to improve general health. Although many of these product claim to be 100% natural, their adulteration has been reported many times. To determine the composition and to guarantee the quality and safety of dietary supplements, appropriate analytical methods are necessary. Traditionally high performance liquid chromatography (HPLC) or gas chromatography (GC) has been the first choice but SFC has emerged as a valuable alternative offering shorter separation times compared with HPLC. Carotenoids and isoflavones are found in many dietary supplements. Moreover, carotenoids have similar structures and exist in a number of structural isomer forms, which are difficult to separate. UHPSFC with sub-2-μm particle columns was used for the separation of nine carotenoids (including carotenes and xantophylles) [4]. Eight stationary phases were evaluated with a good separation by gradient elution obtained on an Acquity UPC^2 HSS C_{18} SB column (150 × 3.0 mm, 1.8 μm) with CO_2 and a mixture of methanol and ethanol as the mobile phase. Comparing with UHPLC, resolution was improved and the separation time reduced. In addition, α- and β-carotene were baseline separated and the elution order of the polar and low-polarity carotenoids reversed. UHPSFC also afforded a significant reduction in the separation time for isoflavones in dietary supplements as well as improved resolution [32]. A baseline separation of nine isoflavones of different polarity in 8 minutes was obtained on an Acquity UPC^2 BEH column (100 × 3.0 mm, 1.7 μm) with a mobile phase containing 0.05% (v/v) phosphoric acid in methanol in the gradient elution mode. The separation time was reduced threefold compared with UHPLC. The analysis of various types of phospholipids in polyene phosphatidyl choline extracts from soybean in capsules and bulk drug form were analyzed by UHPSFC using an Acquity UPC^2 BEH 2-EP (100 × 3.0 mm, 1.7 μm) column [33]. As might be expected the sensitivity and selectivity were better for the UHPSFC-MS/MS method compared with UHPLC-ELSD. Furthermore, the toxicity of the mobile phase was lower, the separation was faster, the peaks were sharper, and the pretreatment was simpler (dissolution without derivatization). UHPSFC-MS/MS was also evaluated for analysis of vitamin E in tocotrienol soft gelatin capsules [34]. The current official method for analyzing vitamin E in oils and fats employs normal-phase HPLC and requires the use of toxic solvents. In some cases, the poor resolution of vitamin E homologs is a problem for HPLC. An Acquity UPC^2 BEH column (100 × 3.0 mm, 1.7 μm) with an isocratic mobile phase consisting of CO_2 and methanol was used for the separation of the vitamin E homologs by UHPSFC. UHPSFC is five times faster than HPLC, greener and has lower costs. Moreover, no significant differences were found in vitamin E contents analyzed with both methodologies.

Food preservatives (benzoic and sorbic acids) plus caffeine were analyzed in several food matrices, including beverages, by SFC with diode array detection [35]. Although benzoic and sorbic acids inhibit the growth of molds and yeasts, and are also effective against a wide range of bacteria, some adverse reactions have been noted in humans, which have led to limitations on the concentrations that can be used in foods. Accordingly, benzoic and sorbic acids as well as caffeine were separated and quantified in a wide range of foods and beverages, with direct injection of the aqueous (though diluted) samples using a Lichrospher Diol column (250 × 4.6 mm, 5 μm) and a mobile phase of CO_2 and methanol containing 0.1% of acetic acid.

17.5 CONCLUSION

In this chapter we present an overview on the use of SFC in food analysis covering the period 2012–16. The number of applications in food has increased drastically in the last few years, mainly due to the coupling with MS detectors, and the commercialization of UHPSFC systems and columns. These new tools have made SFC an even more attractive alternative to GC or HPLC. SFC applications are still mainly focused on the analysis of lipid compounds in food from plant origin, although this trend is slowly changing thanks to new advances in SFC instrumentation. In addition, it has been observed and compared with earlier years (up to 2012), that the use of preparative SFC in the field of food analysis has not increased in the same way UPHSFC. Neither has applications of two-dimensional SFC nor SFC-HPLC to food matrices grown to a significant extent. To sum up, it can be concluded that the recent commercialization of UHPSFC system and the coupling to MS detectors have drastically contributed to the revival of SFC for food analysis. This trend combined with the on-going developments of new stationary phases, detectors, or instruments will most probably further increase the use of SFC.

References

[1] Bernal JL, Martín MT, Toribio L. Supercritical fluid chromatography in food analysis. J Chromatogr A 2013;1313:24–36.

[2] Armenta S, De la Guardia M. Green chromatography for the analysis of foods of animal origin. Trends Anal Chem 2016;80:517–30.

[3] Pano-Farias NS, Ceballos-Magaña SG, Gonzalez J, Jurado JM, Muñiz-Valencia R. Supercritical fluid chromatography with photodiode array detection for pesticide analysis in papaya and avocado samples. J Sep Sci 2015;38:1240–7.

[4] Li B, Zhao H, Liu J, Liu W, Fan S, Wu G, et al. Application of ultra-high performance supercritical fluid chromatography for the determination of carotenoids in dietary supplements. J Chromatogr A 2015;1425:287–92.

[5] Lísa M, Holčapek M. High-throughput and comprehensive lipidomic analysis using ultrahigh-performance supercritical fluid chromatography-mass spectrometry. Anal Chem 2015;87:7187–95.
[6] Ashraf-Khorassani M, Isaac G, Rainville P, Fountain K, Taylor LT. Study of ultrahigh performance supercritical fluid chromatography to measure free fatty acids without fatty acid ester preparation. J Chromatogr B 2015;997:45–55.
[7] Chen Y, Lehotay SJ, Moreau RA. Supercritical fluid chromatography-tandem mass spectrometry for the analysis of lipid A. Anal Methods 2013;5:6864–9.
[8] Chen Z, Dong F, Li S, Zheng Z, Xu Y, Xu J, et al. Response surface methodology for the enantioseparation of dinetofuran and its chiral metabolite in bee products and environmental samples by supercritical fluid chromatography/tandem mass spectrometry. J Chromatogr A 2015;1410:181–9.
[9] Bernal J, Ares AM, Pól J, Wiedmer SK. Hydrophilic interaction liquid chromatography in food analysis. J Chromatogr A 2011;1218:7438–52.
[10] Duval J, Colas C, Pecher V, Poujol M, Tranchant JF, Lesellier E. Contribution of supercritical fluid chromatography coupled to high resolution mass spectrometry and UV detections for the analysis of complex vegetable oil-application for characterization of *Kniphoria uvaria* extract. C R Chimie 2016;19:1113–23.
[11] Jumaah F, Sandahl M, Turner C. Supercritical fluid extraction and chromatography of lipids in bilberry. J Am Oil Chem Soc 2015;92:1103–11.
[12] Lee JW, Uchikata T, Matsubara A, Nakamura T, Fukusaki E, Bamba T. Application of supercritical fluid chromatography/mass spectrometry to lipid profiling in soybean. J Biosci Bioeng 2012;113:262–8.
[13] Tu A, Du Z, Qu S. Rapid profiling of triacylglycerols for identifying authenticity of edible oils using supercritical fluid chromatography-quadruple time-time-of-flight mass spectrometry combined with chemometric tools. Anal Methods 2016;8:4226–38.
[14] Qu S, Du Z, Zhang Y. Direct detection of free fatty acids in edible oils using supercritical fluid chromatography coupled with mass spectrometry. Food Chem 2015;170:463–9.
[15] Hori K, Matsubara A, Uchikata T, Tsumura K, Fukusaki E, Bamba T. High-throughput and sensitive analysis of 3-monochloropropane-1,2.diol fatty acid esters in edible oils by supercritical fluid chromatography/tandem mass spectrometry. J Chromatogr A 2012;1250:99–104.
[16] Wang M, Carrell EJ, Chittiboyina AG, Avula B, Wang YH, Zhao J, et al. Concurrent supercritical fluid chromatographic analysis of terpene lactones and ginkgolic acids in *Ginkgo biloba* extracts and dietary supplements. Anal Bioanal Chem 2016;408:4649–60.
[17] Song W, Qiao X, Liang WF, Ji S, Yang L, Wang Y, et al. Efficient separation of curcumin, demethoxycurcumin, and bisdemethoxycurcumin from turmeric using supercritical fluid chromatography: from analytical to preparative scale. J Sep Sci 2015;38:3450–3.
[18] Fernández-Ponce MT, Casas L, Mantell C, Martínez de la Ossa E. Fractionation of *Mangifera indica Linn* polyphenols by reverse phase supercritical fluid chromatography (RP-SFC) at a pilot plant scale. J Supercrit Fluid 2014;95:444–56.
[19] Gong X, Qi N, Wang X, Lin L, Li J. Ultra-performance convergence chromatography (UPC^2) method for the analysis of biogenic amines in fermented foods. Food Chem 2014;162:172–5.
[20] Bijttebier S, D'Hondt E, Noten B, Hermans N, Apers S, Exarchou V, et al. Automated analytical standard production with supercritical fluid chromatography for the quantification of bioactive C17-polyacetylenes: a case study on food processing waste. Food Chem 2014;165:371–8.
[21] Ishibashi M, Ando T, Sakai M, Matsubara A, Uchikata T, Fukusaki E, et al. High-throughput simultaneous analysis of pesticides by supercritical fluid chromatography/tandem mass spectrometry. J Chromatogr A 2012;1266:143–8.

[22] Ishibashi M, Izumi Y, Ando T, Fukusaki E, Bamba T. High-throughput simultaneous analysis of pesticides by supercritical fluid chromatography coupled with high-resolution mass spectrometry. J Agric Food Chem 2015;63:4457–63.
[23] Tao Y, Dong F, Xu J, Liu X, Cheng Y, Liu N, et al. Green and sensitive supercritical fluid chromatography-tandem mass spectrometric method for the separation and determination of flutriafol enantiomers in vegetables, fruits and soil. J Agric Food Chem 2014;62:11457–64.
[24] Chen X, Dong F, Xu J, Liu X, Chen Z, Liu N, et al. Enantioseparation and determination of isofenphos-methyl enantiomers in wheat, corn peanut and soil with supercritical fluid chromatography-tandem mass spectrometric method. J Chromatogr B 2016;1015-1016:13–21.
[25] Berger TA, Berger BK. Separation of natural food pigments in saponified and un-saponified paprika oleoresin by ultra high performance supercritical fluid chromatography. Chromatographia 2013;76:591–601.
[26] Khalikova M, Šatínský D, Solich P, Nováková L. Development and validation of ultra-high performance supercritical fluid chromatography method for determination of illegal dyes and comparison to ultra-high performance liquid chromatrography. Anal Chim Acta 2015;874:84–96.
[27] Ares AM, Nozal MJ, Bernal J. Extraction, chemical characterization and biological activity determination of broccoli health promoting compounds. J Chromatogr A 2013;1313:78–95.
[28] Laboureur L, Ollero M, Toubul D. Lipidomics by supercritical fluid chromatography. Int J Mol Sci 2015;16:13868–84.
[29] Hartmann A, Ganzera M. Supercritical fluid chromatography-theoretical background and applications on natural products. Planta Med 2015;81:1570–81.
[30] Ng MH, Choo YM. Packed supercritical fluid chromatography for the analyses and preparative separations of palm oil minor components. Am J Anal Chem 2015;6:645–50.
[31] El-Saeid MH, Khan HA. Determination of pyrethroid insecticides in crude and canned vegetable samples by supercritical fluid chromatography. Int J Food Prop 2015;18:1119–27.
[32] Ganzera M. Supercritical fluid chromatography for the separation of isoflavones. J Pharm Biomed Anal 2015;107:364–9.
[33] Jiang Q, Liu W, Li X, Zhang T, Wang Y, Liu X. Detection of related substances in polyene phosphatidyl choline extracted from soybeans and in its commercial capsule by comprehensive supercritical fluid chromatography with mass spectrometry compared with HPLC with evaporative light scattering detection. J Sep Sci 2016;39:350–7.
[34] Gee PT, Liew CY, Thong MC, Gay MCL. Vitamin E analysis by ultra-performance convergence chromatography and structural elucidation of novel α-tocodienol by high-resolution mass spectrometry. Food Chem 2016;196:367–73.
[35] Berger TA, Berger BK. Rapid, direct quantitation of the preservatives benzoic and sorbic acid (and salts) plus caffeine in foods and aqueous beverages using supercritical fluid chromatography. Chromatographia 2013;76:393–9.

CHAPTER

18

Physicochemical Property Measurements Using SFC Instrumentation

T.L. Chester

University of Cincinnati, Cincinnati, OH, United States

OUTLINE

Supercritical Fluid Chromatography.
DOI: http://dx.doi.org/10.1016/B978-0-12-809207-1.00018-5

18.1 INTRODUCTION

Workers developing new separation or processing methods involving condensed gases, whether supercritical or near-critical, pure or mixed, face the daunting task of optimizing the method. The search for optimal conditions necessarily requires knowing or estimating necessary physicochemical properties and establishing the limits of the parameter-value space that might be permissible. Some of these limits, such as maximum pressure, are system parameters that are primarily dependent on equipment capabilities. Some parameters are limited by a combination of equipment capabilities and fluid characteristics. For example, maximum flow rate may be limited either by the maximum delivery rate of a pump or by the pump pressure limit; this depends on the "resistance to flow" of the system and the viscosity of the fluid, which, in turn, depends on the fluid composition, temperature, and pressure. These interdependencies complicate decision-making and process optimization.

Literature can provide some guidance regarding bulk fluid properties and fluid interactions with solutes and surfaces [1]. However, the characteristics of any specific fluid system under study have often not been fully investigated. While it is possible to guess at the gross behavior of fluid systems from experience with similar systems, it is most enabling to have specific knowledge of the system under study and of the limits of operational parameters to allow the widest possible range of conditions to be explored in pursuit of the best method performance. Several important properties of solutes, fluids, and interactions can be estimated or, in some cases determined fairly accurately, using an analytical SFC system, and are particularly useful when specialized equipment for accurate determinations is not available.

18.2 A SIMPLE EXAMPLE: VISCOSITY ESTIMATION

Viscosity of a binary fluid can be estimated using a packed-column SFC with binary gradient pumping capability [2]. With the assumptions that the pressure drop due to extra-column components is negligible (or a correction applied) and the viscosity is nearly constant over a small range of pressure, there is a direct proportionality between viscosity, η, and pressure drop, ΔP, across a packed-column pumped at a flow rate, F, as described by Darcy's Law:

$$\eta = \frac{C\Delta P}{F} \tag{18.1}$$

where

$$C = \frac{Ad_p^2}{\phi L} \tag{18.2}$$

and A is the column cross-sectional area, d_p is the particle diameter, ϕ is a column resistance factor, and L is the column length. The proportionality constant, C, can be determined experimentally by pumping pure CO_2 at a temperature and outlet pressure where the viscosity is known [3]. Then, other fluid components can be added to the fluid using the SFC modifier pump while keeping the total flow rate constant, and the pressure drop will change in proportion with the viscosity of the fluid in the column. If the fluid of interest has two components and the viscosity of both pure components is known at the experimental temperature and pressure, then a two-point calibration will provide a better estimate. If the mixed fluid is much more viscous than pure CO_2, then the flow rate can be changed and accounted for in Eq. 18.1.

This example illustrates both the utility and potential shortcomings of using SFC equipment for such estimates. The quality of a viscosity estimate with this approach depends, among other things, on keeping the volumetric flow rate constant and in knowing the actual composition of a binary mobile phase. The pumps used in SFC instruments are precise, but the accuracy of flow rates of compressible fluids is fraught with uncertainty (see Section 18.7.2). So, this technique can provide nothing better than a fairly crude but nonetheless useful estimate.

This viscosity estimation technique, and most of the other techniques discussed later, will also fail if the mixed fluid separates into two or more phases in the course of the measurements. Similarly, phase separation while doing SFC can lead to unusable chromatographic results including distorted and unstable peak shapes, noise, and poor reproducibility. In contrast, purposely operating in the two-phase liquid–vapor (L–V) region under some circumstances can provide advantages not otherwise possible, such as solute focusing in open-tubular SFC (ot-SFC) [4–6]. Some understanding of phase behavior of mixed-fluid systems is necessary both to set practical limits on operational parameters and to provide the maximum flexibility during method development. The following section gives a short summary of Type I binary phase behavior, followed by ways to estimate the critical locus, phase boundaries, solubility, and binary diffusion coefficients using SFC instrumentation.

18.3 DETERMINATION OF PHASE BEHAVIOR

18.3.1 Phase Behavior of Type I Binary Systems

Liquid–liquid, liquid–vapor, liquid–liquid–vapor, and even vapor–vapor separation is possible in compressed gas systems depending on the fluid components, their concentrations, temperature, and pressure. Van Konynenburg and Scott described six types of binary mixtures of liquids [7]. The simplest of these, Type I, results when the two components are miscible as liquids. These simple systems are the most easily understood and predictable. They provide large single-phase ranges, and therefore are the most desirable for use as binary mobile phases in SFC and in many process applications. However, liquid–vapor phase separation remains possible in Type I systems and must be avoided during the chromatographic separation process by maintaining the pressure (*P*), temperature (*T*), and modifier concentration in the single-phase region of the phase diagram.

Fig. 18.1 begins to depict the phase diagram for CO_2–methanol using pressure (*P*), temperature (*T*), and composition (%*b*) coordinates. Note that in SFC the mobile phase composition is expressed as the volume

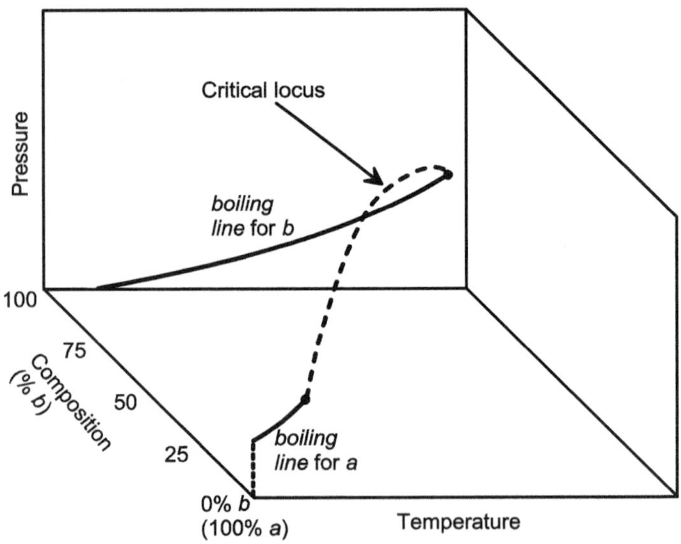

FIGURE 18.1 The locus of mixture critical points for a Type I binary system like CO_2–methanol spans the *P*–*T*–composition space connecting the two pure-component critical points. (The two-phase *L*–*V* region is added in Fig. 18.2). *Source: Adapted with permission from Chester TL. Determination of pressure–temperature coordinates of liquid–vapor critical loci by supercritical fluid flow injection analysis. J Chromatogr A 2004;1037:393–403 [6]. Copyright 2004 Elsevier.*

percent of modifier in the mobile phase measured at the pump outlets. Here, the symbols *a* and *b* identify the primary fluid (CO_2 usually) and modifier, respectively, %*b* is the volumetric percentage of modifier flow at the pump outlets, and *x* is the mole fraction of modifier in the mobile phase. The limiting planes at 0 and 100 %*b* are the familiar *P*–*T* diagrams for pure CO_2 and for pure methanol, and their boiling lines are the solid lines in their respective *P*–*T* plane at each end of the %*b* axis.

Each of these boiling lines ends at the critical point for the corresponding pure component. Every mixture of the two components spanning the composition dimension has its own critical point. The locus of mixture critical points joins the two pure-component critical points and spans the intervening space in the diagram. This *critical locus* is indicated by the dashed line in Fig. 18.1.

For the two pure fluids, the two-phase liquid–vapor equilibrium conditions exist only on the respective boiling lines, that is, a line in a two-dimensional (*P*–*T*) space. However, the region of liquid–vapor equilibrium for a binary mixture occupies a volume in a three-dimensional (*P*–*T*–composition) space as shown in Fig. 18.2. The two-phase region is the interior of the shaded surface. Several isotherms,

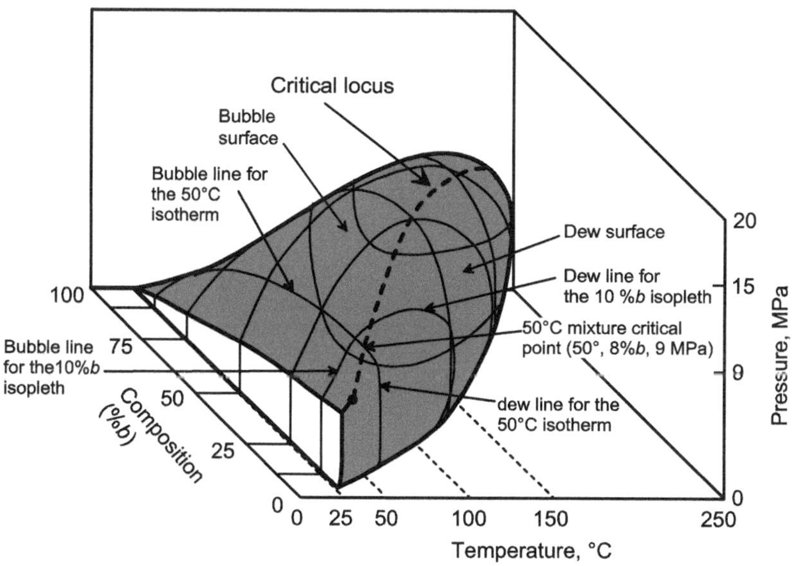

FIGURE 18.2 The *P*–*T*–%*b* depiction of phase behavior for CO_2–methanol. The two-phase *L*–*V* region is within the surface of the shaded figure. *Source: Adapted with permission from Chester TL. Determination of pressure–temperature coordinates of liquid–vapor critical loci by supercritical fluid flow injection analysis. J Chromatogr A 2004;1037:393–403 [6]. Copyright 2004 Elsevier.*

isopleths, and isobars are indicated, and the figure is truncated at 25°C to show the isotherm at that temperature.

Liquid–vapor equilibrium exists at all points within the shaded figure, and a homogeneous fluid cannot exist at *P–T–%b* coordinates inside the shaded figure. Instead, liquid and vapor phases separate. The compositions of the two phases for a given *P–T–%b* coordinate are given by the intersections of a tie line with the surface: the tie line is drawn through the coordinate point and is parallel with the *%b* axis. The vapor phase is the one richer in the more-volatile component, CO_2 in this case. The quantity ratio of the *L* and *V* phases depends on *%b*. The surface of the phase boundary on the CO_2 side of the critical locus is a dew point surface. Note that the dew point surface extends below the shaded region and terminates on the methanol boiling line. The upper surface on the methanol side of the critical locus is a bubble point surface. Most importantly, the volume outside the shaded figure is a continuous single-phase region through the *P–T–%b* space. This continuous space connects the liquid phase of the CO_2 with the vapor phase of the methanol. All of this continuous single-phase space can be used for SFC or for other processes that must avoid phase separation.

Projecting the critical locus into a *P–T* plot, as in Fig. 18.3, clearly indicates the *P–T* coordinates of the mixture critical points. This plot provides no specific information about the mixture composition other than for the boiling lines, which are for the pure components. However, the chart is simple to read and is still quite useful: The value of *%b* increases monotonically from zero at the CO_2 critical point to 100% at critical point of the

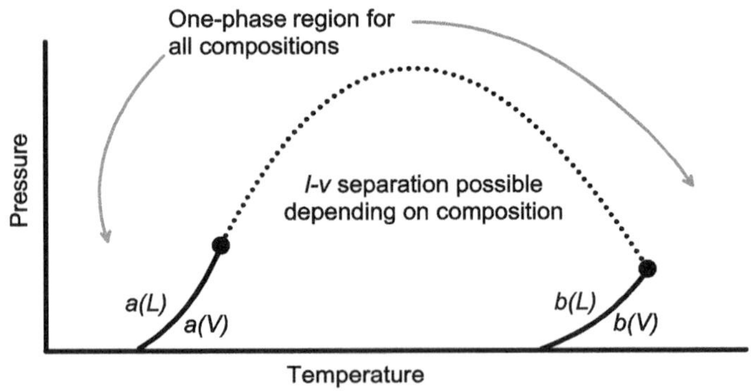

FIGURE 18.3 A projection of the critical locus into the *P–T* plane. The one-phase region above the critical locus connects the liquid phase of the more-volatile component with the vapor phase of the less-volatile component. Choosing pressure and temperature in this space avoids *L–V* phase separation regardless of *%b*. *L–V* phase separation is possible under the critical locus depending on *%b*. It is best to avoid working in this range unless more specific phase behavior data is available to establish the limits of *%b* for combinations of *P* and *T*. *Source: Adapted with permission from Chester TL. Determination of pressure–temperature coordinates of liquid–vapor critical loci by supercritical fluid flow injection analysis. J Chromatogr A 2004;1037:393–403 [6]. Copyright 2004 Elsevier.*

cosolvent, methanol in this case. Only one continuous fluid phase exists in the space above the critical locus extending from the liquid phase of pure CO_2 to the vapor phase of pure methanol. All of this $P-T$ space is potentially usable for chromatography regardless of the ratio of the two fluids.

In the $P-T$ space inside the critical locus, $L-V$ phase separation may occur depending on the composition of the fluid. Additional usable coordinate space may be available inside the critical locus, but care must be taken to avoid conditions resulting in $L-V$ separation. A $P-x$ isotherm, if available, will clearly show the composition and pressure limits of the single-phase region for a given temperature. Several isotherms for CO_2–methanol mixtures are shown in Fig. 18.4. Unfortunately, isotherms for systems of CO_2 and other solvents are relatively rare. With isotherms using mole fraction, remember that SFC instruments with binary pumping systems control the volumetric ratio of the fluids at the pump outlets.

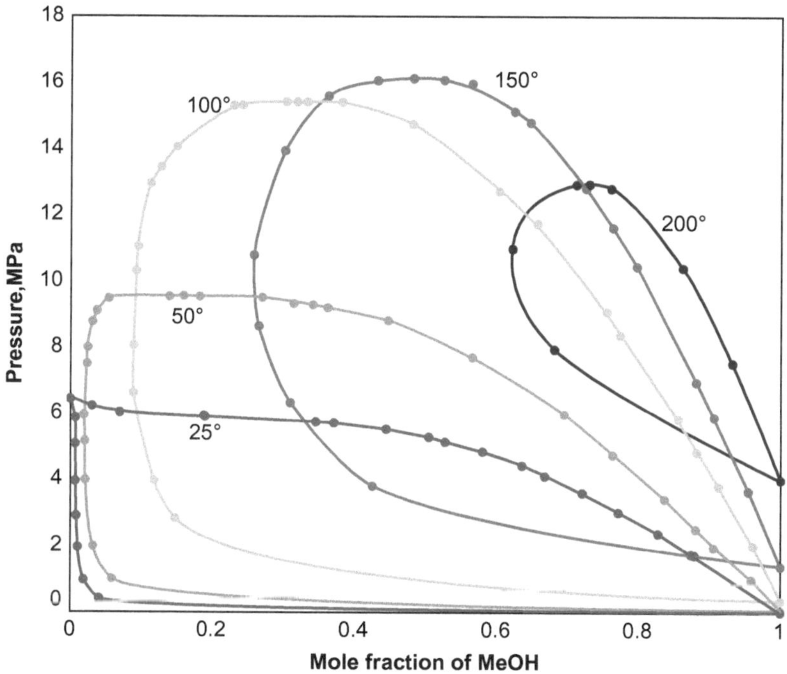

FIGURE 18.4 Isotherms for CO_2–methanol at 25, 50, 100, 150, and 200°C. The $L-V$ region is inside the loop for each temperature. Notice that 25°C is below the critical temperature of CO_2, and the isotherm intersects the left axis at the corresponding CO_2 vapor pressure. *Source: Adapted with permission from Brunner E, Hü ltenschmidt W, Schlichthärle G. Fluid mixtures at high pressures IV. Isothermal phase equilibria in binary mixtures consisting of (methanol + hydrogen or nitrogen or methane or carbon monoxide or carbon dioxide). J Chem Thermodyn 1987;19:213–91 [8]; Air Liquide. Gas Encyclopedia, <http://encyclopedia.airliquide.com/encyclopedia.asp>; 2013 [accessed 24. 08.16] [9].*

18.3.2 Estimating Critical Loci of Binary Mixtures with the Peak-Shape Method

Traditionally, critical points and other phase transitions of binary fluid mixtures are determined using cells that are filled with fluid mixtures of known composition, temperature, and pressure and are examined visually through a cell window. The procedure requires dedicated equipment, and it is often used to accurately map phase behavior in a narrow range of temperature and pressure. Finding mixture critical points is labor intensive because both pressure and temperature must be varied systematically to search for the pair of critical values (P_c and T_c) corresponding to the fluid composition.

Very little data existed for entire critical loci for CO_2−solvent mixtures prior to the development of a simple flow-injection technique utilizing peak shape observations to detect phase separation [4–6,10–13]. This peak-shape method can quickly find the critical pressure corresponding to a set temperature. This method has so far produced critical loci for 23 CO_2−solvent pairs. Mapping the $P-T$ critical locus for a new CO_2−solvent pair can be done in several hours using an ot-SFC instrument.

The experiment is relatively simple: the open-tubular column is replaced with an empty capillary tube connected directly to an ambient-temperature injector. The tube then runs through the column oven and is connected to a flow restrictor leading to the detector. In all the work so far a flame-ionization detector (FID) was used. Pure CO_2 is used as the mobile phase and is delivered with a pump operating in pressure-control mode. A plug of the b component is injected into the CO_2 stream.

Injection is done at ambient temperature. If the pressure is above the CO_2 vapor pressure, all Type I combinations using CO_2 will be a single liquid phase in the injector environment and the ambient-temperature section of the tube, and no phase boundaries will exist around the solvent plug at injection conditions. At a point early in the heated section of the capillary, the concentration of solvent begins at zero, increases to 100 $\%b$ as the plug passes, and then eventually returns to zero as the tail of the plug passes. In this manner the entire range of x is experienced as the plug passes the point. If a liquid phase separates when the solvent plug passes into the oven and is heated, then a liquid−vapor phase boundary will form on the leading edge of the plug, and another phase boundary will form again when the trailing edge of the plug reaches the heated zone. If the tube is wettable by the solvent, the action of the flow of CO_2 will produce a flooded zone, as depicted schematically in Fig. 18.5, with liquid dynamically coating the tube walls while vapor, driven by the flow of CO_2, passes through the center. In practice, the flooded zone is on the order of a meter in length for a 50-μm-i.d. capillary with an injection in the range of 0.2−0.5 μL.

The vapor exiting the flooded zone will be saturated with the solvent evaporating from the liquid phase, and the flooded zone will disappear from its upstream end as the liquid film is evaporated and transported by "fresh" CO_2 coming from farther upstream. CO_2 saturated with solvent vapor at the test temperature and pressure exits the flooded zone and is transported to the detector. (The detector is operated at a temperature higher than the oven to ensure that vapor will not condense before entering the FID flame.) The peak front begins to form as normal, but the signal becomes limited when the saturation level reaches the detector. The signal will remain essentially constant as long as saturation is maintained. When the last of the liquid film evaporates and the tail of the peak enters the detector, the signal decays as for a normal peak. Thus the peaks produced upon injection of a solvent plug will be flat-topped if liquid–vapor phase separation occurs.

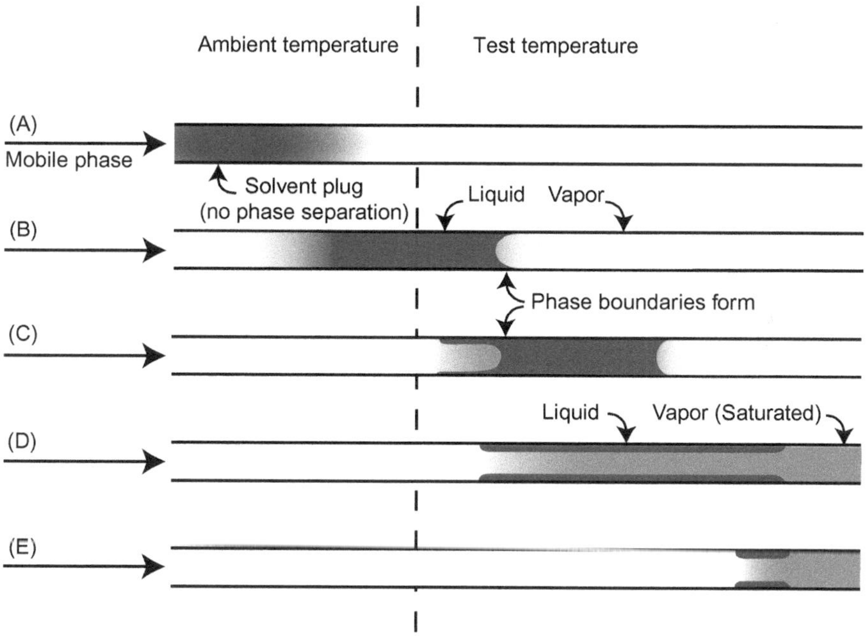

FIGURE 18.5 Schematic description of the creation of the flooded zone in the peak-shape method. (A) A solvent plug is injected into CO_2 at ambient temperature. (B, C) Upon reaching the heated zone, $L-V$ separation occurs and a phase boundary is formed on both ends of the solvent plug. (D, E) The CO_2 flow spreads the liquid as a film creating the flooded zone. Liquid evaporates, and saturated CO_2 is transported downstream to the detector. The solvent film evaporates from the upstream end. *Source: Adapted with permission from Chester TL. Determination of pressure–temperature coordinates of liquid–vapor critical loci by supercritical fluid flow injection analysis. J Chromatogr A 2004;1037:393–403 [6]. Copyright 2004 Elsevier.*

As the pressure is increased the peaks get taller but remain flat-topped as long as the experiment is in the two-phase region and a sufficient volume of liquid has been injected. When the test pressure is above the critical pressure, liquid–vapor separation and saturation of the vapor will not occur, and the peaks will be naturally rounded just as an ordinary overloaded solvent peak. Injections can be repeated at new pressures every few minutes, and the critical pressure corresponding to the test temperature can be quickly found by looking for the pressure where peak shapes change from flat-topped to rounded. Some care is necessary in selecting column dimensions, flow rates, and injection volume to get the flooded zone length appropriate with respect to the tube length, to make sure the quantity of solvent injected remains sufficient to form a flooded zone as successively higher pressures are used, and to avoid having the flooded zone exceed the length of the tube, in which case pure liquid solvent will reach the detector. It may also be necessary to deactivate the tube to ensure it is wettable by the liquid solvent in order to form a stable film.

Figs. 18.6 and 18.7 indicate points on the critical loci with CO_2 for nine popular solvents often used in mixtures with CO_2 and examined

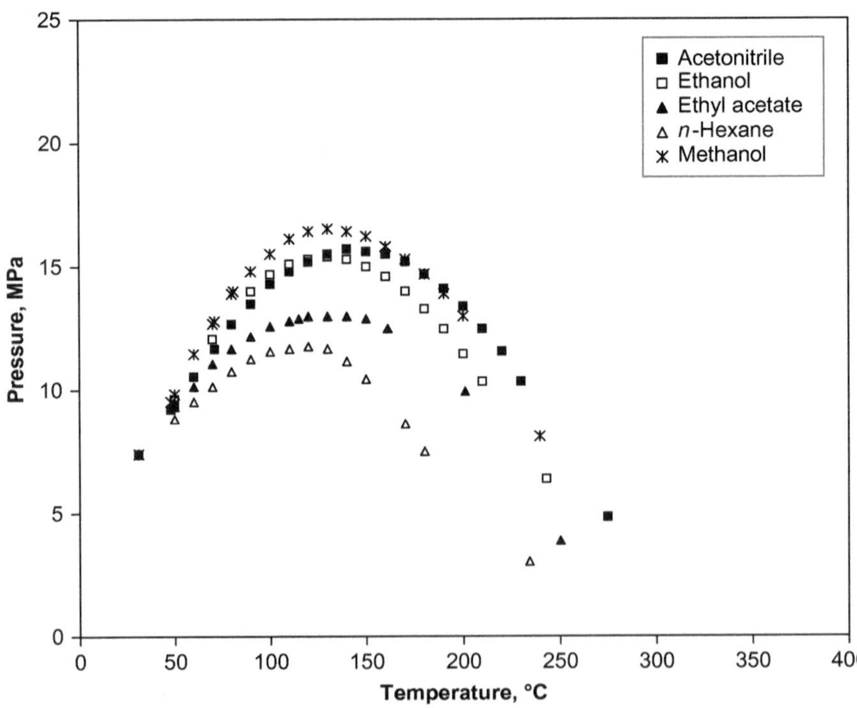

FIGURE 18.6 Points on the critical loci for mixtures of CO_2 with acetonitrile, ethanol, ethyl acetate, *n*-hexane, and methanol. *Source: Adapted with permission from Chester TL. Determination of pressure–temperature coordinates of liquid–vapor critical loci by supercritical fluid flow injection analysis. J Chromatogr A 2004;1037:393–403 [6]. Copyright 2004 Elsevier.*

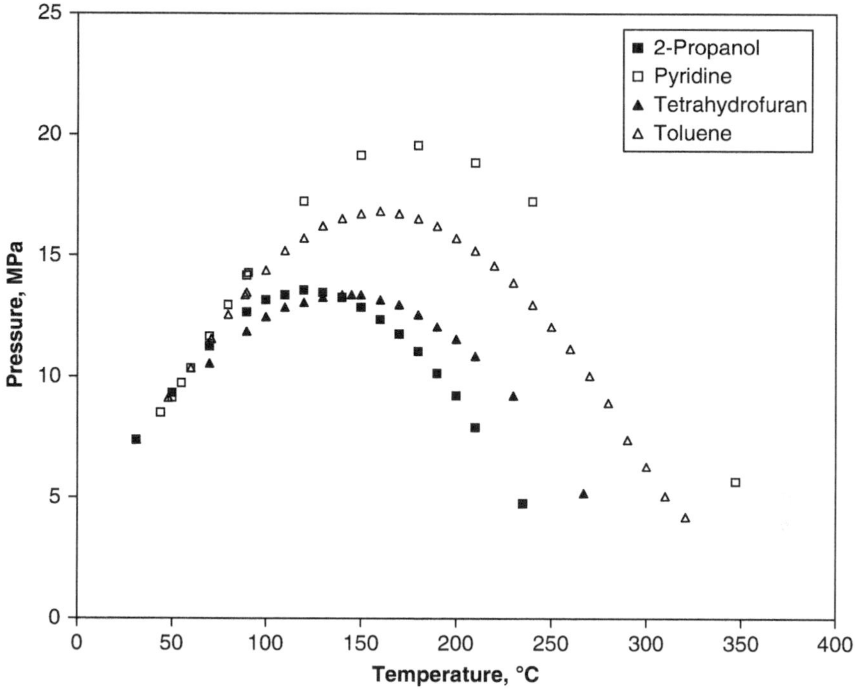

FIGURE 18.7 Points on the critical loci for mixtures of CO_2 with 2-propanol, pyridine, tetrahydrofuran, and toluene. *Source: Adapted with permission from Chester TL. Determination of pressure–temperature coordinates of liquid–vapor critical loci by supercritical fluid flow injection analysis. J Chromatogr A 2004;1037:393–403 [6]. Copyright 2004 Elsevier.*

by this method. Note that the boiling lines of the pure liquids are omitted for clarity, but the first and last points given for each solvent are the critical points for CO_2 and for the solvent reported. Experimental points for the remaining 14 solvents are given in reference [6].

If the only SFC available is a packed-column instrument, it could be used along with a gas chromatograph equipped with an FID with the addition of a high-pressure internal-loop injector and a restrictor, following the setup of the original work: [4,5,10] Pressure-regulated CO_2 can be obtained from the SFC by adding a tee just ahead of the back pressure regulator (BPR). Traces of modifier would have to be flushed out of the SFC. This may require disconnecting the modifier pump and capping its connection to the mixer.

18.3.3 Estimating Vapor Composition (Dew Points) Using the Peak-Shape Method Setup

The vapor-phase concentration of solvent can be estimated with a different interpretation of data from the peak-shape method. Here are the additional requirements: (1) the injection volume and density of the solvent being tested must be known; (2) the detector signal must remain proportional to the solvent mass rate of transport to the detector through the entire range of vapor-phase solvent concentration spanned in recording a peak; and (3) the mass flow rate of CO_2 must be the same everywhere throughout the system. This means that during the experiment the CO_2 and solvent are in equilibrium over most of the flooded zone. Dittmar et al. reported that diffusion of CO_2 into ethanol at the vapor–liquid interface appears to reach equilibrium within a few seconds [14]. This is likely to also be the case for other solvents with polarity and viscosity similar to ethanol.

To estimate dew points, the peak-shape experiment is conducted as described earlier with the injection of a volume, V, of solvent into a stream of CO_2. The ratio, R, of solvent mass to CO_2 mass associated with any detector signal level, s, during this experiment is:

$$R = \frac{s}{F\rho_{CO2}} \frac{V\rho_{solvent}}{A} \tag{18.3}$$

where m is the mass (g) of solvent injected, F and ρ_{CO2} are respectively the CO_2 flow rate (mL/s) and density (g/mL) at the pump outlet, $\rho_{solvent}$ is the density of the pure solvent, and A is the area (in units of signal-seconds) of the peak resulting from the injection. If the signal s is taken at the plateau of a flat-topped peak then R is the solvent-to-CO_2 mass ratio at the dew point for the temperature and pressure used.

The FID, by its nature, is a mass-sensitive detector that works accurately even for solvent peaks at the masses and delivery rates encountered in ot-SFC. If not overloaded, the FID produces the same peak area for a given injected mass regardless of flow rate. However, a concentration-sensitive detector like UV–Vis produces peak areas inversely proportional to the volumetric flow rate at the detector. Often, temperature is not well controlled in UV–Vis detectors, and fluid composition (and density, heat capacity, thermal conductivity, etc.) of course changes during this experiment. Thus, flow rate and UV–Vis detector response may vary whenever changes in temperature, pressure, or composition alter the fluid density, even for constant flow rate at the pump. So, there is additional uncertainty in the estimates whenever a concentration-sensitive detector is used, and an FID is preferred whenever possible.

18.3.4 Packed-Column SFC Techniques for Determining Phase Transitions

Ziegler noted that flooding and phase separation will occur in packed columns as well as empty tubes [12]. However, we are not aware of any work using the peak-shape method with packed columns. Other measurable changes in instrument parameters dependent on fluid properties can also be used to indicate phase transitions. For a packed-column SFC instrument, the easily measured fluid-related properties are the detector signal and the inlet pressure (or, equivalently, the column pressure drop) assuming pressure is controlled at the column outlet.

Akien et al. have used pressure drop to experimentally estimate the $P-T-x$ coordinates of dew points and bubble points in several Type I CO_2–solvent systems [15]. These measurements were done either statically or with gradients of outlet pressure, temperature, or modifier concentration using a custom-built instrument resembling a packed-column SFC. The static approach requires a progressive series of experiments that, individually, are isothermal, isobaric, and isocratic. The column is held at constant temperature and pressure, and the fluid of interest is continuously pumped through isocratically (specifically, at a constant mole ratio). The pressure drop is measured after the system reaches equilibrium. The parameter to be varied (P, T, or x) is then set at a new value, the instrument is re-equilibrated, and another measurement is made.

The gradient method involves ramping one parameter (either P, T, or x) while holding the other parameters and the overall flow rate constant and recording the inlet pressure or pressure drop. When there is vapor in the system, increasing the pressure compresses the vapor, reduces the volumetric flow rate and lowers the pressure drop required to maintain the set outlet pressure. Raising the temperature increases vapor viscosity and volume and thereby increases the pressure drop. Encountering a phase transition results in a discontinuous change in the slope of pressure drop versus the changed parameter (P, T, or x).

Akien et al. reported reasonable estimates for the upper surface of the CO_2–methanol envelope obtained using pressure ramps, but their results are several bars higher than view-cell values [8,16–19]. In contrast, they reported only one point on the lower dew point surface at 63 bar for CO_2–methanol with $x_{methanol} = 0.279$ at 100°C. This pressure differs significantly from the view-cell measurements of Brunner et al. [8], from whose work we can infer that the dew point actually occurs at a pressure below 28.3 bar.

Fig. 18.8 shows pressure-drops with temperature programming for CO_2–tetrahydrofuran mixtures at 100 bar. The signal-to-noise ratio and the clarity of the transitions here is far better than the pressure programming experiment, at least for the data reported thus far. Two discontinuous slope changes are clear in all but the last trace: the lower-temperature slope change corresponds to a bubble point upon crossing from a homogeneous liquid phase to a two-phase liquid–vapor region, and the higher-temperature slope change corresponds to entering the

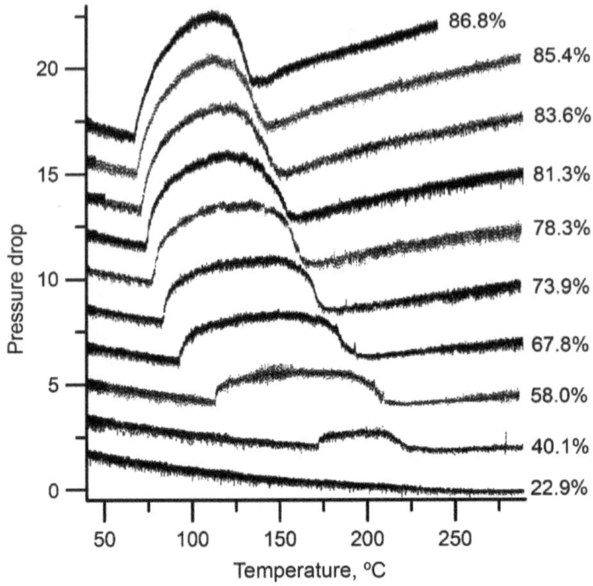

FIGURE 18.8 Traces of pressure drop versus temperature for the CO_2–tetrahydrofuran system. Temperature was programmed from 40°C at 1°C/min with the pressure held at 10 MPa (100 bar), and the mole percent of toluene as indicated. Changes in the slope indicate dew points or bubble points. *Source: Reprinted (adapted) with permission from Akien GR, Skilton RA, Poliakoff M. Pressure drop as a simple method for locating phase transitions in continuous flow high pressure reactors. Ind Eng Chem Res 2010;49:4974–80 [15]. Copyright 2010 American Chemical Society.*

one-phase vapor region. A transition is not visible using 22.9 mol% CO_2. The tetrahydrofuran data in Fig. 18.7 suggests that 100 bar is very near or above the critical pressure for this mixture, and if the experimental conditions are above the critical locus then no transition will occur.

In both the pressure-changing and temperature-changing techniques, when phase transitions are encountered they do not occur at the same time everywhere on the column because of the pressure gradient. Thus there is some uncertainty in the indicated $P–T$ coordinates of a transition. Programming the fluid composition is also possible but introduces more uncertainties due to gradient delay, and spatial variation in x along the length of the column due to the gradient. If temperature programming is not possible due to equipment limitations (many SFC instruments do not heat fast enough to program the temperature), then a series of equilibrium measurements stepping through a range of fluid compositions may be the most useful approach. Otherwise, it may be more fruitful to move the column into an adjacent gas chromatograph oven to perform temperature programming.

18.4 DIRECT SOLUBILITY ESTIMATION

Solubility of a solute in a fluid can be measured directly in an SFC instrument by replacing the SFC column with a saturation cell. This can be a small column containing a sufficient quantity of solute dispersed in a nonreactive adsorbent. The cell is simply pumped with the fluid of interest, and the saturated fluid is transported to the SFC detector. For example, Miller and Hawthorne measured the solubility of organic solutes in CO_2 by directly introducing solute-saturated CO_2 into an FID [20]. They replaced the usual column in what amounts to an ot-SFC with a commercial saturation cell that they loaded with the solute. They also added an auxiliary heating cell just ahead of the detector to ensure no solute condensation occurs before detection. The FID was separately calibrated by integrating the signals of known masses of solute. During a measurement, the saturated mobile phase produces a steady FID signal that is proportional to the mass rate of solute. The mass rate of CO_2 was separately determined, and calculation of solubility was straightforward.

Zhou et al. used a micro-SFE cell within a packed-column SFC instrument [21]. The extraction cell was arranged parallel with the SFC column, and both fed directly to a single flow restrictor interfaced to an FID. One stream or the other could be selected. Pure sample was loaded with silica or alumina into the extraction cell, and a flow of CO_2 delivered a saturated solution to the FID. Calibration was done using the SFC column to deliver known quantities of a standard to the detector using the same outlet pressure and flow restrictor so that the flow rate into the detector was constant and the calibration could be applied without any corrections.

Sampling an extraction cell is also possible [22,23]. For example, Gahm et al. packed a 30-mm × 21-mm-i.d. steel column with silica gel and up to 100 mg of the solute under test [22]. After equilibrating in the desired mobile phase using an SFC instrument, a 5- to 10-μL sample was isolated using valves, and this was then pumped, using the same mobile phase, through a UV detector for quantitation.

It is conceivable that solubility could be estimated, especially for poorly soluble solutes, using a packed analytical column instead of a saturation cell, and by injecting a concentrated solution of solute in the usual manner. The sample would have to be more soluble in the sample solvent than in the test fluid. The column packing should be chosen to have low retention for the solute in order to promote saturation in the mobile phase. What is required to happen on the column is the formation of a solute-overloaded zone near the column inlet, separation of the injection solvent from the solute, and development of a solute saturation zone in the mobile phase downstream from the overloaded zone.

The flow rate will have to be respectful of the evaporation rate of solute from the overloaded zone in order to produce saturated mobile phase. Then mobile phase flow through the overloaded zone will saturate the remainder of the column, and a steady-state signal will result when the saturation zone breaks through the column outlet. The signal will last until the solute is depleted. Considerable attention will be required to sort out dimensions and flow rates to ensure the development of the required saturated zone.

18.5 ADSORPTION ISOTHERM DETERMINATIONS

Knowledge of an adsorption isotherm is essential in understanding mass transfer in preparative separations. There are numerous static and dynamic procedures for determining isotherms in systems with supercritical fluids [24], but with SFC equipment the most straightforward method is frontal analysis [24–30].

Kamarei et al. removed excess extra-column volume from their SFC instrument, particularly from the mixer, in order to sufficiently sharpen the solute fronts that must be transported to the detector in frontal analysis [29]. The authors also added a mass flow meter between the CO_2 supply and its pump inlet. This was necessary so that the CO_2 mass flow rate could be monitored throughout the experiments. The adsorbent was packed in a column. The CO_2 pump and the (primary) modifier pump were used in normal fashion. The authors added a second modifier pump (we will call it the solute pump), which was used to deliver a solute dissolved in modifier. Typically the system is equilibrated with zero flow from the solute pump. At time $t = 0$ the solute pump flow is stepped to the desired value with a concomitant reduction in the modifier pump flow so that the combined flow of the modifier and solute pumps is held constant.

The equilibrium concentration, q, of solute adsorbed on the adsorbent phase can easily be derived from the time required for breakthrough of the front from the column outlet following the step in the solute concentration in the mobile phase:

$$q = \frac{C_0}{\left(\frac{1-\varepsilon_t}{\varepsilon_t}\right)} \frac{(t_R - t_0)}{t_0} \tag{18.4}$$

where C_0 is the equilibrium concentration of solute in the mobile phase at the column inlet following the step change in the solute concentration, t_R is the retention time of the front (usually measured at the inflection point of the step), t_o is the column void time, and ε_t is the total porosity of the column. Thus, points (C_0, q) on the isotherm can be determined from retention time measurements. Fig. 18.9 shows a series of breakthrough peaks following 5-minute-long injections of various solute concentrations.

For accurate results, corrections must be made to account for the changes in the volumetric flow rate and volumetric ratios of the mobile phase components occurring between the pump and the column. This is necessary, e.g., to accurately calculate C_0 (at the column inlet) from the solute concentration in the methanol solution. Details are provided in reference [29].

The advantage of adding the second modifier pump is that a stock solution of solute in modifier can be prepared, and then the various injections can be made by programming the blending ratio of the two modifier pumps. The modifier pump flow ratios are chosen to produce a series of injections with the desired concentration steps. This could then be done unattended under program control.

When a second modifier pump is not available, it seems conceivable that the experiment could be carried out by using a loop injector inserted

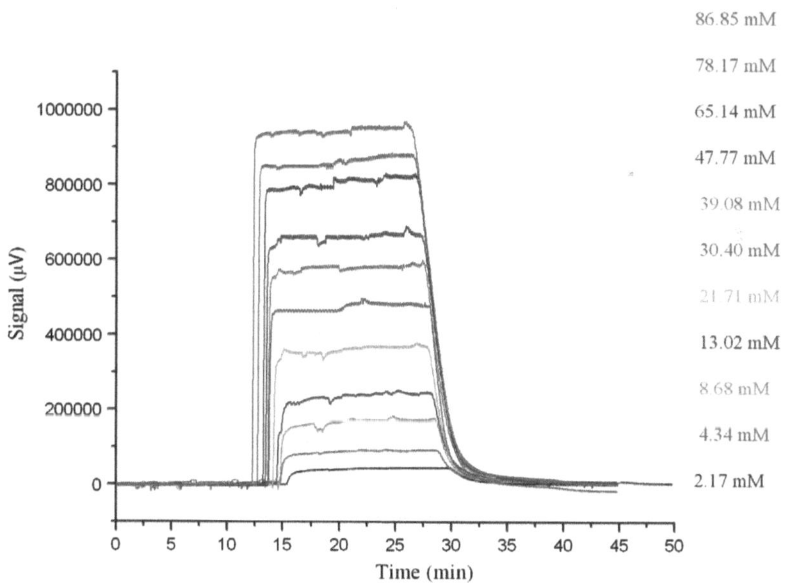

FIGURE 18.9 Breakthrough peaks for S-naproxen at various concentrations at 310.15 K, 210 bar, and 1 mL/min with 80/20 (%, v/v) CO_2/methanol using a 250 × 4.6-mm (R, R)-Whelk-O1 column [29]. *Source: Adapted with permission from Kamarei F, Gritti F, Guiochon G, Burchell J. Accurate measurements of frontal analysis for the determination of adsorption isotherms in supercritical fluid chromatography. J Chromatogr A 2014;1329:71–7 [29]. Copyright 2014 Elsevier.*

between the modifier pump and the mixer to substitute a volume of solute solution for pure modifier. A series of test solutions of various solute concentrations would have to be prepared in otherwise pure modifier, and one of these solutions would be loaded into the loop for each measurement. The loop volume would have to be sufficient to produce equilibrium conditions on the resulting solute plateau that is detected. For example, a 1-mL loop in the modifier stream with an overall flow rate of 1 mL/min but with 20% modifier in the mobile phase would produce a 5-minute-long injection, thereby matching the conditions in Fig. 18.9.

Kamarei et al. used frontal analysis and also what they called the *inverse method* to determine adsorption isotherms [30]. This technique is applicable to systems with more than one adsorption site and with competitive binding from two solutes. The authors used a bi-Langmuir model to account for both nonspecific and specific adsorption sites on a chiral stationary phase, and they also examined competition between the (R) and (S) enantiomers of the solute for the nonspecific sites. This method is called *inverse* because a model is proposed and then fit to the data with adjustment of the model parameters to minimize the differences between the model and the data.

The main advantage of the frontal analysis method is the direct production of points in the solute adsorption isotherm. It also works for both pure and mixed mobile phases, and it works with any stationary phase or packing. The disadvantages are the need for an additional modifier pump for programmed measurements, and the need to carefully measure the mass flow rates of the mobile phase components.

18.6 DIFFUSION

Medina recently reviewed experimental methods for determining diffusion coefficients [31]. Some of these methods can be implemented on SFC instruments, and all of these derive diffusion coefficients from measurements of peak dispersion. The Taylor-Aris method involves measuring the broadening of a small injection of solute as it is transported through an uncoated open tube [32–38]. The Chromatographic Impulse Response (CIR) method, when used for measuring diffusion coefficients, is somewhat similar to the Taylor method but substitutes a wall-coated open tube in part or all of the sample path [39–41]. This method setup resembles ot-SFC but typically uses relatively large-diameter tubes, e.g., a 15.28-m × 0.502-mm-inside-diameter tube with a 1-μm-thick chemically bonded polyethylene coating [40]. (Note that this column diameter is much too large for analytical ot-SFC.) Variations of the CIR method can also be used to measure solubility and partial molar volumes [41].

The pulse-response moment analysis method [42–46] is similar but sufficiently detailed in its treatment to make it applicable to packed columns. In all of these techniques, solute diffusion rates can be calculated from the peak dispersion.

However, with packed-column SFC instrumentation the simplest and potentially most accurate method is the peak parking method [24,47–49]. This involves injecting a solute and transporting it onto the column in the usual manner, but stopping the mobile phase flow, thus parking the peak, while the solute is on the column. The conditions are held for a parking period during which the solute peak continues to broaden axially by diffusion. After the parking period the flow is restored and the broadened peak is recorded.

When the mobile phase is flowing, pressure and density gradients will naturally form on the column. When the pumps are stopped, these gradients will relax to their average values over the entire column. The gradients will be reestablished when the flow resumes. While the pumps are stopped it is essential that the flow is completely stopped at both ends of the column and that, once the gradients are relaxed, the pressure remains constant on both ends of the column during the remainder of the parking period. If the BPR and the pump check valves are not capable of completely stopping the flow then it will be necessary to add hard (low-dispersion) shutoff valves on each end of the column. Some care must be taken in selecting flow rates and the parking location on the column to minimize the effects of relaxation of the pressure gradient on the column when the flow is stopped [24].

What is measured is the observed peak variance versus parking time. Artifacts due to stopping and restarting the flow should not affect the slope of this plot because only the parking time is varied in a series of measurements. The apparent diffusion rate is then given by

$$D_{\mathrm{app}} = \frac{u^2 \Delta\sigma^2}{2(1+k)^2 \Delta t} \tag{18.5}$$

where u is the mobile phase velocity on the column when the flow is on, k is the retention factor for the solute under the test conditions, and $\Delta(\sigma^2)/\Delta t$ is the slope of a plot of the peak temporal variance versus parking time [47].

18.7 LIMITATIONS OF ESTIMATES

18.7.1 Uncertainties of the Stationary Phase

It is possible to determine several thermodynamic properties from measurements in chromatographic systems. For example, partition coefficients can be derived from solute retention if the phase ratio is known with sufficient accuracy. Likewise, solubility, partial molar volumes, and free energy changes are all indicated by retention, but all require additional measures of phase volumes and knowledge of any changes occurring in the chemical properties of the stationary phase [1].

Thermodynamic measurements based on solute retention in gas chromatography are reliable. This stems to some extent from the fact that the stationary phase is well defined, it is not subject to interactions with the mobile phase, and the phase ratio is robust. However in HPLC, and especially in SFC, the stationary phase may undergo changes when other conditions are changed.

There are at least two possible mechanisms leading to stationary phase changes: (1) in the case of polymeric stationary phases the polymer may swell by absorption of mobile phase components thereby changing the stationary phase volume, the phase ratio, and the chemical properties [1]. (2) Mobile phase components can adsorb onto stationary phase surfaces from liquid mobile phases under HPLC conditions [50–53], and likewise from CO_2 and methanol under supercritical conditions [1,54,55]. In HPLC these layers of excess mobile phase components function as part of the stationary phase. For example, Hydrophilic Interaction Chromatography (HILIC) requires a surface excess of water adsorbed on the stationary phase surface and functioning as a dynamic part of the stationary phase to retain polar solutes. The amount and concentration of a surface excess changes with mobile phase composition and with temperature. Because the mobile phase is compressible in SFC, the thickness and composition of a surface excess of adsorbed mobile phase components may also change with pressure. Whenever the stationary phase changes in volume or composition, especially its composition at the interface with the mobile phase, property determinations requiring an unchanging stationary phase or a stationary phase assumed not to interact with the mobile phase are subject to error unless corrections can be applied.

18.7.2 Uncertainties of the Mobile Phase

A perfect reciprocating pump will deliver an incompressible fluid at the displacement rate of the pump. Most HPLC mobile phase components, like most normal liquids, have relatively small compressibility. Most modern HPLC pumps have some form of compressibility compensation in order to lower the inaccuracies in the actual pump delivery rate compared to the flow rate set point. However, with compressible fluids the pump must spend a large and variable (depending on the outlet pressure) fraction of its displacement to compress the fluid and raise its pressure to match the downstream pressure before the outlet check valve will open to commence the flow [56]. As a consequence, the delivery rate of pressurized fluid can be much lower than the volumetric displacement rate or set point of the pump if compensation is inadequate or ignored.

Many CO_2 metering pumps have some means to at least partially correct for compressibility. In these the pump outlet pressure is measured or estimated from the loading of the pump motor, and the displacement rate is increased as necessary to maintain a constant volumetric delivery rate over the specified outlet pressure range of the pump. However, the effectiveness of this approach over the full range of operating conditions is in doubt [56]. Cooling the CO_2 stream ahead of the pump will lower the compressibility and reduce the errors, but the accuracy of various pumps used in this configuration is not well known. Prepressurizing the CO_2 feed stream with a booster pump to (or near) the pressure required downstream eliminates the need for compression compensation in the metering pump, so the volumetric delivery rate will match the pump displacement rate [56].

Another problem remains even if the pumps accurately deliver the desired flow rates at their outlets: the volumetric ratio of a mixture of CO_2 and a modifier solvent measured at the temperature of the pump outlets will not match the volumetric ratio at the column if the temperature there is different. For physicochemical property measurements, and especially for thermodynamic property measurements, regulating mass flow in the pumps would be superior to volumetric flow control since mass flow is invariant regardless of the downstream conditions. This would allow modifier concentration to be specified as mole fraction, which would remain constant all the way to the system outlet. Regardless, if the mass flow rates are known, then when knowledge of volumetric flow rate on the column is required, the application of an equation of state that is validated for the mixture and the specific conditions would be the best bet.

The CO_2 pumping setup of Akien et al. [15] was unusual, at least with respect to commercial SFC equipment, because a separate BPR was installed between the CO_2 pump outlet and the mixer. (Another BPR was used in the normal fashion to control the column outlet pressure.) The stated purpose for using the additional BPR was to allow the system downstream to be operated at low pressures, i.e., even at pressures below that of the stream feeding the CO_2 pump. However, if the upstream BPR is set to a pressure higher than the maximum required column inlet pressure, the CO_2 pump will always work into a constant back pressure at all flow rates. Therefore operating the CO_2 pump at a constant flow rate in this setup would also result in constant mass flow of CO_2. No form of mass flow control is currently offered on the CO_2 pumps in commercial analytical SFC systems. Such a capability would greatly improve the prospects of making thermodynamic measurements on SFC equipment. Until then, much care will be necessary to adequately understand the composition and conditions of the mobile phase on the column.

18.8 CONCLUSION

An analytical SFC instrument can be used to provide estimates or even accurate determinations of many physicochemical properties. Phase transitions, critical points, solubilities, adsorption isotherms, and diffusion rates can be determined with good accuracy in many cases. Viscosity can be easily estimated. Even estimates, trends, and comparisons of conditions can be useful in guiding process development and especially in problem solving when more specialized equipment is not available.

Some measurements must account for the compressibility of fluids and the volume and flow rate changes that occur in SFC systems with temperature and pressure changes along the flow path. The use of retention times for estimating thermodynamic properties must not only correct for these changes in the fluid, but also changes in the effective phase ratio and the surface excess of adsorbed mobile phase components.

SFC is more complex than HPLC, but this complexity adds flexibility and capability for making contributions not otherwise possible. With knowledge of the underlying chemistry, SFC can do much more than producing routine chromatograms.

References

[1] Roth M. Determination of thermodynamic properties by supercritical fluid chromatography. J Chromatogr A 2004;1037:369–91.

[2] Fields PR, Chester TL, Stalcup AM. Viscosity estimation in binary and ternary supercritical fluid mixtures containing carbon dioxide using a supercritical fluid chromatograph. J Liq Chromatogr 2011;34:995–1003.

[3] Stephan K, Lucas K. Viscosity of dense fluids. New York: Plenum Press; 1979.

[4] Chester TL, Innis DP. Dynamic film formation and the use of retention gaps with direct injection in open-tubular supercritical fluid chromatography. J Microcol Sep 1993;5:261–73.

[5] Chester TL, Innis DP. Quantitative open-tubular supercritical fluid chromatography using direct injection onto a retention gap. Anal Chem 1995;67:3057–63.

[6] Chester TL. Determination of pressure–temperature coordinates of liquid–vapor critical loci by supercritical fluid flow injection analysis. J Chromatogr A 2004;1037:393–403.

[7] Van Konynenburg PH, Scott RL. Critical lines and phase equilibria in binary Van der Waals mixtures. Phil Trans R Soc Lond A 1980;298:495–540.

[8] Brunner E, Hültenschmidt W, Schlichthärle G. Fluid mixtures at high pressures IV. Isothermal phase equilibria in binary mixtures consisting of (methanol + hydrogen or nitrogen or methane or carbon monoxide or carbon dioxide). J Chem Thermodyn 1987;19:213–91.

[9] Air Liquide. Gas Encyclopedia, <http://encyclopedia.airliquide.com/encyclopedia.asp>; 2013 [accessed 24. 08.16].

[10] Ziegler JW, Dorsey JG, Chester TL, Innis DP. Estimation of liquid-vapor critical loci for CO_2-solvent mixtures using a peak-shape method. Anal Chem 1995;67:456–61.

[11] Ziegler JW, Chester TL, Innis DP, Page SH, Dorsey JG. Supercritical fluid flow injection method for mapping liquid-vapor critical loci of binary mixtures containing CO_2. In: Hutchinson KW, Foster NR, editors. Innovations in supercritical fluids science and technology. ACS Symposium Series, vol. 608. Washington: American Chemical Society; 1995. p. 93–110.

[12] Ziegler JW. "Ain 'T Misbehavin'"—Fundamental investigations into supercritical fluid binary mixtures containing CO_2 and supercritical fluid chromatography. Ph.D. University of Cincinnati, 1996.

[13] Chester TL, Haynes BS. Estimation of pressure-temperature critical loci of CO_2 binary mixtures with methyl-*tert*-butyl ether, ethyl acetate, methyl-ethyl ketone, dioxane, and decane. J Supercrit Fluids 1997;11:15–20.

[14] Dittmar D, Oei SB, Eggers R. Interfacial tension and density of ethanol in contact with carbon dioxide. Chem Eng Tech 2002;25:23–7.

[15] Akien GR, Skilton RA, Poliakoff M. Pressure drop as a simple method for locating phase transitions in continuous flow high pressure reactors. Ind Eng Chem Res 2010;49:4974–80.

[16] Ohgaki K, Katayama T. Isothermal vapor-liquid equilibrium data for binary systems containing carbon dioxide at high pressures: methanol-carbon dioxide, *n*-hexane-carbon dioxide, and benzene-carbon dioxide systems. J Chem Eng Data 1976;21:53–5.

[17] Hong JH, Kobayashi R. Vapor-liquid equilibrium studies for the carbon dioxide-methanol system. Fluid Phase Equil 1988;41:269–76.

[18] Yeo S-D, Park S-J, Kim J-W, Kim J-C. Critical properties of carbon dioxide + methanol, + ethanol, + 1-propanol, and + 1-butanol. J Chem Eng Data 2000;45:932–5.

[19] Joung SN, Yoo CW, Shin HY, Kim SY, Yoo K-P, Lee CS, et al. Measurements and correlation of high-pressure VLE of binary CO_2–alcohol systems (methanol, ethanol, 2-methoxyethanol and 2-ethoxyethanol). Fluid Phase Equil 2001;185:219–30.

[20] Miller DJ, Hawthorne SB. Determination of solubilities of organic solutes in supercritical CO_2 by on-line flame ionization detection. Anal Chem 1995;67:273–9.

[21] Zhao S, Wang R, Yang G. A method for measurement of solid solubility in supercritical carbon dioxide. J Supercrit Fluids 1995;8:15–19.

[22] Gahm KH, Tart W, Eschelbach J, Notari S, Thomas S, Semin D, et al. Purification method development for chiral separation in supercritical fluid chromatography with the solubilities in supercritical fluid chromatographic mobile phases. J Pharm Biomed Anal 2008;46:831–8.

[23] Li B, Guo W, Song W, Ramsey ED. Determining the solubility of organic compounds in supercritical carbon dioxide using supercritical fluid chromatography directly interfaced to supercritical fluid solubility apparatus. J Chem Eng Data 2016;61:2128–34.

[24] Guiochon G, Tarafder A. Fundamental challenges and opportunities for preparative supercritical fluid chromatography. J Chromatogr A 2011;1218:1037–114.

[25] James DH, Phillips CSG. The chromatography of gases and vapors. III. Determination of adsorption isotherms. J Chem Soc 1954;1066–70.

[26] Schay G, Székely G. Gas-adsorption measurements in flow systems. Acta Chim Hung 1954;5:167–82.

[27] Jacobson J, Frenz J, Horváth C. Measurement of competitive adsorption isotherms by frontal chromatography. Ind Eng Chem Res 1987;26:43–50.

[28] Gritti F, Guiochon G. Critical contribution of nonlinear chromatography to the understanding of retention mechanism in reversed-phase liquid chromatography. J Chromatogr A 2005;1099:1–42.
[29] Kamarei F, Gritti F, Guiochon G, Burchell J. Accurate measurements of frontal analysis for the determination of adsorption isotherms in supercritical fluid chromatography. J Chromatogr A 2014;1329:71–7.
[30] Kamarei F, Vajda P, Gritti F, Guiochon G. The adsorption of naproxen enantiomers on the chiral stationary phase (R,R)-whelk-O1 under supercritical fluid conditions. J Chromatogr A 2014;1345:200–6.
[31] Medina I. Determination of diffusion coefficients for supercritical fluids. J Chromatogr A 2012;1250:124–40.
[32] Taylor G. Dispersion of soluble matter in solvent flowing slowly through a tube. Proc Roy Soc A 1953;219:186–203.
[33] Taylor G. Conditions under which dispersion of a solute in a stream of solvent can be used to measure molecular diffusion. Proc R Soc A 1954;225:473–7.
[34] Aris R. On the dispersion of a solute in a fluid flowing through a tube. Proc R Soc A 1956;235:67–77.
[35] Ouano AC. Diffusion in liquid systems. I. A simple and fast method of measuring diffusion constants. Ind Eng Chem Fundam 1972;11:268–71.
[36] Liong KK, Wells PA, Foster NR. Diffusion in supercritical fluids. J Supercrit Fluids 1991;4:91–108.
[37] Levelt Sengers JMH, Deiters UK, Klask U, Swidersky P, Schneider GM. Application of the Taylor dispersion method in supercritical fluids. Int J Thermophys 1993;14:893–922.
[38] Akgerman A, Erkey C, Orejuela M. Limiting diffusion coefficients of heavy molecular weight organic contaminants in supercritical carbon dioxide. Ind Eng Chem Res 1996;35:911–17.
[39] Funazukuri T, Kong CY, Kagei S. Impulse response techniques to measure binary diffusion coefficients under supercritical conditions. J Chromatogr A 2004;1037:411–29.
[40] Kong CY, Funazukuri T, Kagei S, Wang G, Lu F, Sako T. Application of the chromatographic impulse response method in supercritical fluid chromatography. J Chromatogr A 2012;1250:141–56.
[41] Kong CY, Siratori T, Funazukuri T, Wang G. Infinite dilution partial molar volumes of platinum(II)2,4-pentanedionate in supercritical carbon dioxide. J Chromatogr A 2014;1362:294–300.
[42] Miyabe K, Guiochon G. Fundamental interpretation of the peak profiles in linear reversed-phase liquid chromatography. Adv Chromatogr 2000;40:1–113.
[43] Miyabe K, Guiochon G. Measurement of the parameters of the mass transfer kinetics in high performance liquid chromatography. J Sep Sci 2003;26:155–73.
[44] Ruthven DM. Principles of adsorption and adsorption processes. New York: John Wiley and Sons; 1984.
[45] Suzuki M. Adsorption engineering. Tokyo: Kodansha/Elsevier; 1990.
[46] Guiochon G, Golshan-Shirazi S, Katti AM. Fundamentals of preparative and nonlinear chromatography. Boston: Academic Press; 1994.
[47] Gritti F, Guiochon G. Effect of the surface coverage of C18-bonded silica particles on the obstructive factor and intraparticle diffusion mechanism. Chem Eng Sci 2006;61:7636–50.
[48] Miyabe K, Matsumoto Y, Guiochon G. Peak parking-moment analysis. A strategy for the study of the mass-transfer kinetics in the stationary phase. Anal Chem 2007;79:1970–82.

[49] Miyabe K, Ando N, Guiochon G. Peak parking method for measurement of molecular diffusivity in liquid phase systems. J Chromatogr A 2009;1216:4377–82.
[50] Kazakevich YV, McNair HM. Study of the excess adsorption of the eluent components on different reversed-phase adsorbents. J Chromatogr Sci 1995;33:321–7.
[51] Kazakevich YV, LoBrutto R, Chan F, Patel T. Interpretation of the excess adsorption isotherms of organic eluent components on the surface of reversed-phase adsorbents. Effect on the analyte retention. J Chromatogr A 2001;913:75–87.
[52] Rustamov I, Farcas T, Ahmed F, Chan F, LoBrutto R, McNair HM, et al. Geometry of chemically modified silica. J Chromatogr A 2001;913:49–63.
[53] Gritti F, Guichon G. Adsorption mechanism in RPLC. Effect of the nature of the organic modifier. Anal Chem 2005;77:4257–72.
[54] Strubinger JR, Song H, Parcher JF. High-pressure phase distribution isotherms for supercritical fluid chromatographic systems. 1. Pure carbon dioxide. Anal Chem 1991;63:98–103.
[55] Strubinger JR, Song H, Parcher JF. High-pressure phase distribution isotherms for supercritical fluid chromatographic systems. 2. Binary isotherms of carbon dioxide and methanol. Anal Chem 1991;63:104–8.
[56] Berger TA. Instrumentation for analytical scale supercritical fluid chromatography. J Chromatogr A 2015;1421:171–83.

Index

Note: Page numbers followed by "*f*" and "*t*" refer to figures and tables, respectively.

G

H

N

T

FSC
www.fsc.org
MIX
Papier | Fördert gute Waldnutzung
FSC® C083411

Kolibri 360 GmbH
Ferdinand-Jühlke-Straße 7
99095 Erfurt, Deutschland
produktsicherheit@kolibri360.de